94 Notes on Numerical Fluid Mechanics and Multidisciplinary Design (NNFM)

FLOMANIA - A European Initiative on Flow Physics Modelling

Results of the European-Union funded project, 2002 - 2004

Werner Haase
Bertrand Aupoix
Ulf Bunge
Dieter Schwamborn
(Editors)

Springer

Dr. Werner Haase
EADS Military Air Systems
Dept. OPEA3 / Build. 70.N
D-81663 München
Germany

Dr. Bertrand Aupoix
ONERA
BP 4025
F-31055 Toulouse Cedex 04
France

Dr. Ulf Bunge
IVM Automotive Wolfsburg GmbH
Wolfsburger Landstraße 22
D-38442 Wolfsburg
Germany

Dr. Dieter Schwamborn
DLR
Institut für Aerodynamik und Strömungstechnik
Bunsenstraße 10
D-37073 Götingen
Germany

Library of Congress Control Number: 2006924581

ISBN-10 3-540-28786-8 Springer Berlin Heidelberg New York
ISBN-13 978-3-540-28786-5 Springer Berlin Heidelberg New York

Springer is a part of Springer Science+Business Media
springer.com

Typesetting: Digital data supplied by editors
Final processing by PTP-Berlin Protago-TEX-Production GmbH, Germany (www.ptp-berlin.com)
Cover-Design: deblik, Berlin
Printed on acid-free paper 89/3141/Yu – 5 4 3 2 1 0

NNFM Editor Addresses

Preface

In aircraft design, efficiency is determined by the ability to accurately and reliably predict the occurrence and to model the development of turbulent flows. Hence, the main objective in industrial computational fluid dynamics (CFD) is to increase the capabilities for an improved predictive accuracy for both complex flows and complex geometries. At the same time it proves necessary to take notice of a current change in paradigm in CFD, meaning that the prediction of delta-value between two consecutive solutions is going to be replaced by the need to provide absolute values, e.g. for force coefficients. As a direct consequence, highly resolved meshes with 20 million and more nodes are needed and used, leading to an approach giving rise to new challenges in CFD with respect to mesh generation, convergence of the chosen numerical methods chosen, and – the most crucial topic - the selection of robust and reliable turbulence models.

When applying CFD methods to improved design, better flight performance and enlarged flight envelopes the key aspect is always enhancement of accurate flow-physics modelling. Keeping the latter in mind, three main topics have been identified in the FLOMANIA (FLOw-physics Modelling – AN Integrated Approach) EU project to overcoming these still existing problems in CFD and in particular in turbulence modelling:

1. Improvement of current industrially used turbulence models striving for more robust and reliable models,
2. Employment of sophisticated (differential/full) Reynolds stress models, which are seen to represent flow physics in the whole flow domain much better,
3. Detection of application limits in (unsteady) Reynolds averaged Navier-Stokes ((U)RANS) computations by applying the new Detached-Eddy Simulation (DES) method when massively separated flows are concerned.

The present book describes the complete outcome of the FLOMANIA project with its general aim of increasing the capabilities of the European aeronautics industry in the area of aircraft design by enhancing the ability to more accurately predict aerodynamic flows - and the editors do hope that this book clearly and effectively contributes, with respect to new and innovative results and methods, to that area of interest.

Moreover, we hope that the chapter on turbulence models and the assessment of models including the ‘lessons learnt’ will help for more advanced future investigations and validation work.

The close collaboration between the aeronautical industries, research institutes and universities across Europe - stimulated by the financial support from the European Union - is claimed to have promoted and accelerated the enhancement of CFD approaches for aeronautical applications, and to a far greater extent than would have been possible with the partners progressing their own capabilities in isolation.

An important aspect of the very successful and highly prolific collaboration was the workshop held in St. Petersburg in July 2003 and the final meeting in

Brussels in September 2004 – both meetings attended by industrial observers as well as academics from outside FLOMANIA. Moreover, Airbus (France) was an official industrial observer participating in all technical meeting

Thanks are due to all FLOMANIA partners who have contributed in an open and collaborative manner to ensure success and by this making it a pleasure for the editors to summarize the programme's technical achievements in the present book. Special thanks are going in particular to K. Weinman from DLR and C. Mockett from TU Berlin for their contributions to modelling aspects and 'proof-reading'.

Finally, thanks are also due to R. Simonini, the European Commission's Scientific Officer of the FLOMANIA project, who was providing every help at any time for making this programme a success. Moreover, all FLOMANIA partners appreciated the financial support received from the European Union and are, furthermore, grateful to the KATnet (a Commission's Thematic Network) managing board for supporting the present publication financially.

Last but not least, the editors of this book would like to express their gratitude to E.H. Hirschel, the General Editor of the Springer series 'Notes on Numerical Fluid Mechanics and Multidisciplinary Design', for his kind agreement to publish the FLOMANIA technical results but even more for his editorial advice.

January 2006

Werner Haase München
Bertrand Aupoix Toulouse
Ulf Bunge Wolfsburg
Dieter Schwamborn Göttingen

Table of Contents

Page

I The FLOMANIA project

1 Summary

The primary purpose of the FLOMANIA proposal was to support the European aeronautics community (industry, research labs and universities) in the area of flow-physics modelling by bridging the gap between the most advanced research in this area, in particular turbulence modelling, and its applications in daily industrial practice. To properly cope with the main aspects involved, the following issues have been the basis for the FLOMANIA project:

1. Make the European industry benefit from the latest advances in turbulence modelling, considering reliability, robustness and the knowledge of limitations of currently used eddy viscosity models and methods.
2. Generate a validated integration of new non-linear and EARSM models in industrial aerodynamic CFD codes and establish limits of validity of the various models dedicated for future industrial applications.
3. Perform a validated integration of full differential Reynolds stress models (RSM) in industrial aerodynamic CFD codes.
4. Extend these implementations to the new generation of unstructured solvers.
5. Create an industrially feasible transition process towards Detached-Eddy Simulation (DES), suggested to close the gap between RANS and LES methods.
6. Ensure a high level of focused technology transfer from model development via model implementation to validation strategies in industrial codes and for all relevant industrial challenging applications.

In order to give rise to accurate and reliable validation of the new models in current industrial methods, complex industrial test cases (together with more generic ones) have been treated, even when intellectual property rights did not allow to distribute all desired geometries and full results. Nevertheless, "informal" results have been provided (in somewhat restricted form). The reason for including such cases in the test-case list was because of their utmost importance on demonstrating the improvements obtained in turbulence modelling in current industrial environments.

As pointed out in the list given above, all FLOMANIA partners implementing new models have received technical support by all those partners involved in the development of turbulence models.

The both challenging and innovative FLOMANIA project was aiming, in accordance with the obtainable improved predictive accuracy in the area of aircraft aerodynamics, at robustness and reliability enhancements as well as an enlarged domain of applicability, which now results in reduced simulation effort and design time and costs, and is allowing for faster-time-to-market strategies.

During the context of the FLOMANIA project, it became clear that - despite a 40-50% increase in computation (CPU) time when using a full RSM turbulence model – improved robustness, reliability and accuracy of the new approaches is really decreasing turn-around times because often needed "re-runs" with different turbulence models in order to reach final (converged) results can be omitted by using new and/or advanced models.

To conclude, the FLOMANIA project was considered to be a unique attempt to group the best experts in research and CFD developers in Europe, in order to transfer their advanced knowledge to the aeronautical industry in a controlled, objective-oriented way. In addition, referring the reader to the following chapters, it has generated a unique evaluation and synthesis effort towards a European "consensus" on the range of validity of advanced turbulence models, the best rules for their implementation, that was – and still is - shared and transmitted in an organized way to industry.

2 Introduction

2.1 The Basis for the FLOMANIA project

Future air transport systems have to support European growth and to attract and foster business, whilst meeting the needs of European citizens for travel and transport, with a major aim of preserving both environment and quality of life.

Today, airlines are operating close to 10,000 medium to large jet aircraft. With an estimated average growth of 5% per year air traffic volume will triple by 2020. In close correlation to the air traffic growth rate which is well beyond general economic growth rates there are important and demanding aims directly influencing the economic perspectives:

- Five fold reduction in accidents
- Halving perceptive aircraft noise
- 50% cut in CO_2 emission per passenger kilometre (halving fuel consumption)
- 80% cut in NO emission
- Air traffic system capable of handling 16 million flights a year with 24h operation at airports and more comfort for passengers

Of course, accurate CFD tools are playing an important role in aiming at a reliable and general simulation of manoeuvring aircraft. With the need to extend flight regimes and envelopes to even more reduce e.g. drag, more accurate and reliable CFD approaches form the basis for predicting flight performance with impact on reducing pollution and aiming at an improved environment.

An often used estimate of the global investment in new aircraft development might yield a figure of the order of 500 billion EUR. Assuming that the improvement of capabilities in CFD - by improving flow-physics models as it was the major aim of the FLOMANIA project - would lead to an increased competitiveness of European aircraft by only 0.5%, this would imply a business increase of 2,500 million EURO for the European aircraft industry. In that respect the FLOMANIA project has well contributed to the economic growth in Europe.

In addition, the new turbulence modelling technology obtained in FLOMANIA can be easily transferred to other industries - besides those already incorporated in the project, i.e. turbo-machinery and train industry - dealing with aerodynamic simulations, as there are helicopters, aero-engines and automobiles, and it supports smaller enterprises by improving their capabilities in aeronautics.

Moreover, the education of young scientists being involved in the FLOMANIA consortium has been directly supported according to the knowledge gained from the FLOMANIA trans-national project.

2.2 State of the art

At the beginning of the new century it was well recognised that turbulence modelling is still a crucial topic in CFD. At a NASA-ICASE workshop (Rubinstein et al, 2001) a discussion took place with US colleagues from universities, research labs and industry on past and future activities in CFD. It became clear that problems identified were (and they still are) identical for all participants, lasting from mesh generation via numerical issues to turbulence modelling aspects. Thus, the conclusion was a concise statement: "**Think flow physics**"! It was the meeting itself and the "lessons learnt" that resulted in the crystallisation point for the FLOMANIA project with the aim to foster highly sophisticated Reynolds stress models (RSM) for use in aeronautics.

The item "flow-physics modelling", however, was put up earlier by a group working in the ERCOFTAC Pilot Centre Germany South and was presented already in 1999 in a lecture on, "Present and Future Aerodynamic Process Technologies", by E.H. Hirschel (unpublished). Many of the items that are going to be discussed in the present book have been pointed out at that time already.

Obviously, a technology gap existed/exists between the upstream development of turbulence models being in use by the aeronautics industry. For instance, highly sophisticated Reynolds stress models have been developed and used in the research world with good success, but the advances were not transferred to the aeronautics industry.

To support computational fluid dynamics aspects in the aeronautics industry, efforts on improvements of numerical methods, grid generation and adaptation have been carried out intensively over the past decades. And CFD is becoming a reliable tool for interdisciplinary work related to design optimisation, investigation of aero-elasticity or taking flight mechanics and loads into account. However, the weakest link, and this is first of all flow-physics (turbulence) modelling, is the most crucial part in the context of CFD for the aeronautics industry.

Due to bigger and faster computers, finer meshes and more complex applications are treated, where turbulence modelling errors are more visible and have more consequences today than they had 10 years ago. Together with a change in CFD-paradigm taking place in the aeronautics industry, CFD simulations are now often used in design, in verification, and in support of flight tests. Thus, the capability of CFD as a design tool is requesting needs for high accuracy, reliability and robustness.

Therefore, the major challenge for CFD is no longer to merely deliver results on a qualitative basis but to provide accurate quantitative values. Moreover, as these quantitative answers are requested on complete/complex aircraft and aircraft components, highly resolved meshes with often more than 20-30 million mesh points are becoming standards in particular for the current technology drivers, i.e. European aircraft like the Airbus A380 and high-speed (military) aircraft.

The technological effort versus complexity of application is sketched in Fig. 1. The application range of FLOMANIA has concentrated on 3-D complex flows, but has also incorporated - for the ease of testing of sophisticated turbulence models - some generic and "simpler" cases. Again, accurate CFD results offer a sound basis for any disciplinary work including all aspects of optimisation.

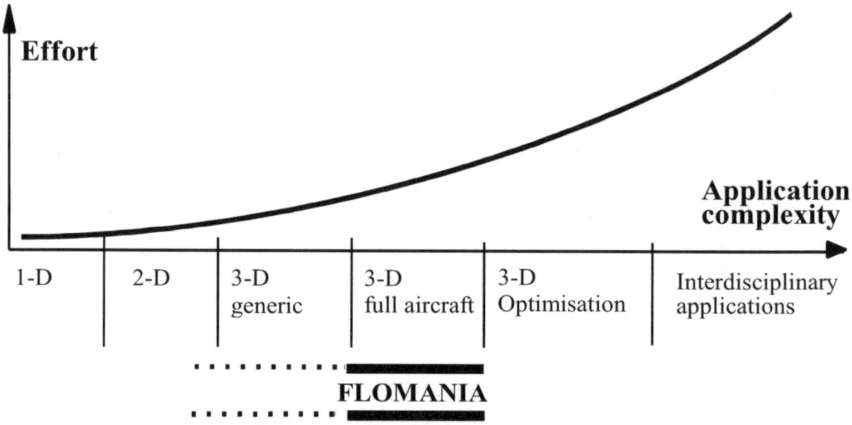

Figure 1 Effort vs. application complexity in CFD applications and range of FLOMANIA applicability

Additionally, Fig. 2 provides a sketch of effort versus the physical completeness of methods. It also shows the current edge of technology generally used by aeronautics industries and, furthermore, demonstrates the working area of FLOMANIA, being clearly beyond that current edge.

As almost all turbulence models - and in particular one- and two-equation models - are linked to limitations and restrictions, it is evident that these limitations are a source of weaknesses in any prediction. To achieve converged solutions, say in cases where flows are considered as steady, these weaknesses often lead to an absolutely unacceptable situation, i.e. failure of convergence and/or failure of the model to predict certain types of flow. Obviously, this causes extra and time-consuming efforts, i.e. re-computation of the problem of interest mostly accompanied by a change to a more robust, and often less accurate, model.

Hence, in accordance to the exponentially rising needs for predictive accuracy together with reliability and computational robustness in industrial applications, there was this clear demand for FLOMANIA on stepping forward with turbulence modelling by investigating and applying more sophisticated models ranging from non-linear two equation models (validated for the aeronautical range of flows) to the RSM family of models. Of course, RSM is not the final answer to turbulence modelling, as it can be well seen in Fig. 2, but it is suggested to be used in complex applications where two- or one-equation models fail because of specific weaknesses in predicting all relevant flow features. Thus, RSM models will definitely improve the predictive capabilities of RANS applications - until LES after about four to five (?) decades might take over.

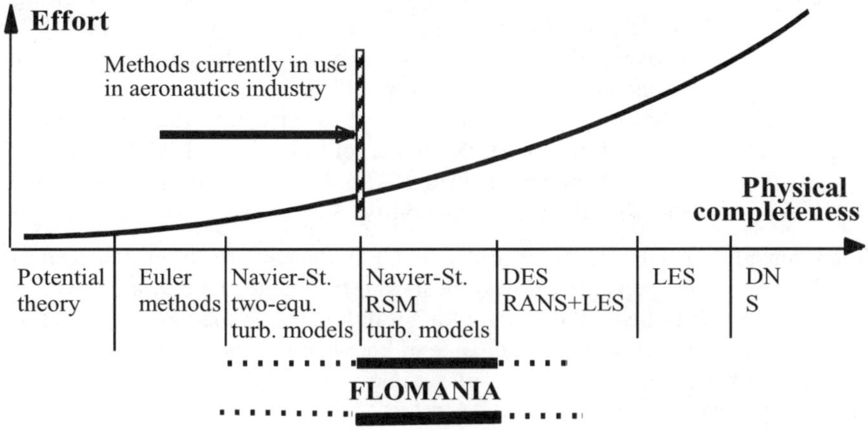

Figure 2 Physical completeness of CFD methods and range of FLOMANIA work items

The most advanced edge of the FLOMANIA work items was definitely a first attempt on using and validating Detached-Eddy Simulation methods as well as other hybrid RANS-LES methods and to provide some means for receiving answers on limitations for (U)RANS and DES. The DES method is seen to be able to close the technology gap between (U)RANS and LES in current industrial applications. Thus, it is of utmost interest for the aeronautics industry to take care of the ultimate goal: having advanced computational tools at hand in order to be able to predict absolute flight performance over the entire flight envelope.

3 Technical project description

3.1 Research approach and technical achievements

The innovative elements of the FLOMANIA project are concerned with the improvement and enhancement of flow-physics modelling by taking care of current (and future) industrial needs and requirements which will in turn help to design new transportation means better preserving the environment as well as quality of life. This was achieved by an integrated approach:

1. Provision of fast, reliable and accurate turbulence models going to be used by industry => key item: *Robustness of today's models*.
2. Closing the gap between the level of sophistication, availability of models, and the old(er) status of use of turbulence models in industry => key item: *RSM integration*.
3. Taking upstream research into consideration to avoid the generation of a new technology gap => key item: *DES approach*.

Thus, three main goals had been identified:

Short-term: Enhancement of a pre-specified set of models used by industry (RANS and URANS).

Medium-term: Reduction of identified weaknesses in turbulence modelling by using (full) Reynolds stress models and by this closing the gap between the level of sophistication on the model development strand and the industrial capabilities.

Long-term: Formation of a "smooth" and rational transition - at industrial level - from the (U)RANS level towards the LES/DES level, i.e. taking available hybrid RANS-LES methods into account and identify limits of applicability.

When considering DES to become a new CFD standard in the next decade (for unsteady, massively separated flows), near-wall turbulence modelling - in the area where RANS has to be applied – provided a link between medium- and long-term goals, thus there was a constant exchange of knowledge between these strands.

The following list provides a brief summary of achievements obtained during the context of work in the FLOMANIA project:

1. More accurate and reliable prediction of steady and unsteady flow due to the use of more sophisticated nonlinear two-equation models and RSM. The latter offering a realistic hope to using them for various and complex flow predictions by circumventing the often applied approach to run different models on different types of flow.
2. Recommendations of best practice for incorporating turbulence models in the context of structured and, moreover, unstructured approaches.
3. Incorporation of results, knowledge and expertise from both the DNS and DES side in order to feed knowledge back for the sake of RANS and URANS methods and applications.
4. Supporting interdisciplinary investigations where CFD is involved. Thus, the assessment of the FLOMANIA flow-physics modelling activity was a step forward towards the ability of the European industry to properly cope with the next-generation aero-elastic problems on highly optimised designs.
5. FLOMANIA has enabled the European aircraft industry to keep a leading role in international competition by standardisation and knowledge dissemination. Thus, the European industry benefited from the latest advances in turbulence modelling concerning reliability, robustness and accurate knowledge of the limitations of currently used turbulence models.
6. FLOMANIA has generated a validated integration of new nonlinear and EARSM models in industrial aerodynamic CFD codes and, moreover, has given rise to a validated integration of full RSM models in industrial aerodynamic CFD codes
7. Last but not least, FLOMANIA has initiated a transition process towards hybrid RANS-LES approaches as being a next/future generation of CFD methods with advanced turbulence modelling and has established limits of validity and applicability of hybrid models versus "conventional" (U)RANS models for industrial applications.

3.2 Dissemination and exploitation

In the FLOMANIA project a new/improved and to a certain extent standardised innovative European turbulence modelling technology for aerodynamic applications was provided and is open for being exploited for future use of all simulation environments where CFD is playing a major role. The new turbulence modelling approaches developed and implemented serve as the main item for being exploited by the FLOMANIA partners, the aeronautics industry, the non-aeronautics industry partners, research labs and by the involved universities – in support of the technology drivers, e.g. the new Airbus 380.

Consequently, as an accurate investigation of the whole range of aerodynamic problems is aiming at future aircraft with less fuel consumption and low (but accurately predicted) drag, at reduced maintenance, increased lifetime, enlarged flight envelopes and a positive impact on environmental issues, a successful exploitation of the FLOMANIA turbulence approaches (including RSM models) is still a major aim. Thus, the use of highly sophisticated turbulence models will have a major impact on the development of the industrial and scientific projects and applications in the area of CFD, and the dissemination of highly advanced computational tools will now close the mentioned technology gap.

An important task related to knowledge dissemination and exploitation was to establish rules for the intended validation work and to initiate and steer inter-partner cooperation in order to provide a "service" for those partners implementing new (RSM) turbulence models. This was perfectly achieved by including in addition:

1. Specification and documentation of mathematical models to be implemented by partners,
2. Provision of advice on the implementation of models and methodologies,
3. Exchange of algorithms among partners to implement new capabilities,
4. Exchange of description of algorithmic methods to test certain physical concepts.

It is in the nature of such work that many of the interactions were bilateral or involved only a small subset of partners. Thus, the transfer of algorithms, models and implementation details was highly partner-specific and had often enabled very targeted and specific exchanges. On the other hand, some aspects of technology and software transfer were more wide-spread and were distributed amongst the entire partnership. Examples are:

1. Specification of common models, in particular the RSM,
2. Definition of data-exchange standards and routines, facilitating the exchange of software and numerical data,
3. Definition of a framework for collecting data for computational test cases and associated collective graphical representation of the results,
4. A workshop on advanced turbulence modelling in co-operation with the ERCOFTAC Association,
5. An established link between the AIAA Drag-Prediction Workshop (DPW) and FLOMANIA,
6. A workshop on DES methods and results at STPU in St. Petersburg.

More specific when considering the latter three items, a six-day Summer School and Workshop on the implementation of turbulence models and its numerical aspects was organized by the FLOMANIA partner UMIST. The workshop was attended by all partners, and several of them contributed by giving a number of lectures. The programme covered a broad range of topics, including the implementation of two-equation, eddy-viscosity, elliptic relaxation and Reynolds-stress models. Alongside RANS methods, LES and DNS approaches were treated too. New results on the implementation of advanced wall functions and low Reynolds number models were reported, and numerical aspects of the implementation of the models mentioned above were considered in detail. A general strategy for developing and using industrial and commercial software, such as CFX, FLUENT and SATURN, was reported by the attending developers.

The AIAA Drag-Prediction Workshop - linked to FLOMANIA - provided the possibility to work on the DLR-F6 test case (see chapter IV.20), an Airbus-like wind-body-pylon-nacelle configuration and to compare with workshop results. Obviously, according to the many workshop results, this test case provided a comprehensive data base for validation.

The workshop on DES held in conjunction with the midterm meeting of the project showed the great interest of all partners in this strand and with a considerable number of presentations indicated that DES was not such a long-term goal as had been previously anticipated.

Last but not least, links to other EU projects were established, e.g. to KATnet and QNET-CFD. Via the first project is was possible to disseminate knowledge gained in FLOMANIA to a larger scientific and industrial community in Europe, while FLOMANIA supported QNET-CFD partners on setting up and describing so-called application challenges. It should be acknowledged at this point again that FLOMANIA was supported by KATnet in issuing the present book. Moreover, a link to the M-DAW project was established leading to an exchange of numerical grids for test case 1 and a lively discussion on the results obtained.

3.3 Description of tasks

A graphical overview describing the FLOMANIA project is provided in Fig. 3.

3.3.1 Task 1.1: Project management

The overall objective of this task was to manage the ambitious and innovative FLOMANIA project, i.e. to monitor success as well as risk. A major means for information dissemination was the FLOMANIA Web site. All relevant information, including reporting, data base set-up and administrative issues (like meetings) were handled via the Web site. A Project Management Group was established in order to steer and monitor the project.

3.3.2 Task 1.2: Web server

The Web site, see also chapter I-5 below, had been constructed to contain two areas, a public and a consortium restricted one.

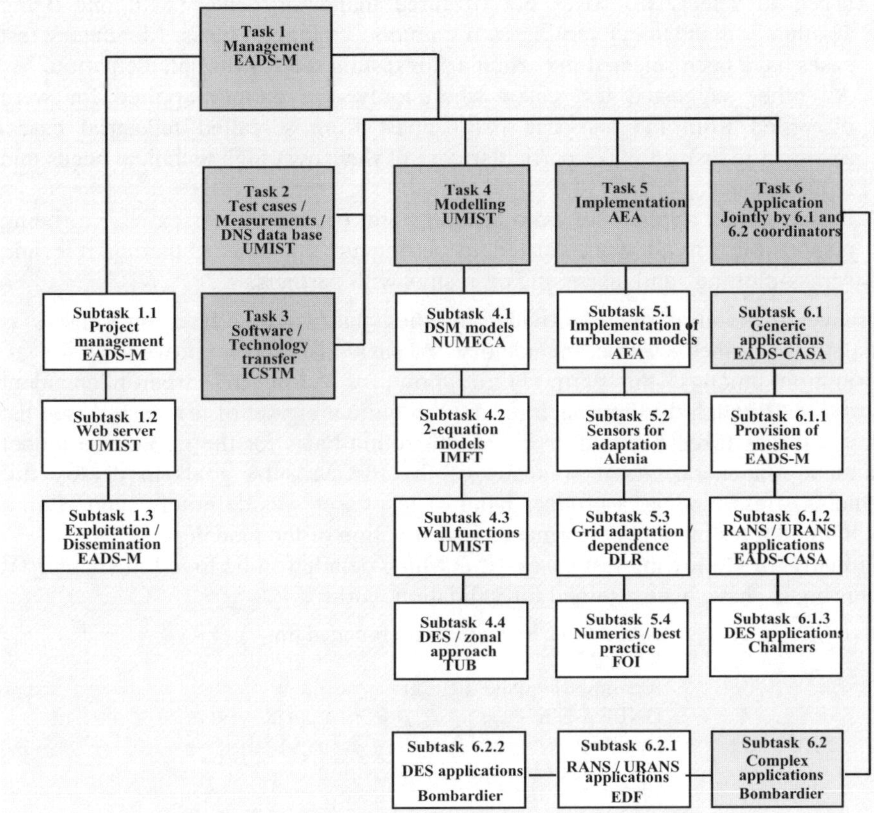

Figure 3 Task layout of the FLOMANIA project

The first (public) area was – and still is - accessible to any web user and, after the welcome page, consists mainly of pages which outline the aims, purpose and work content of the FLOMANIA project. This part also provides contact details for each partner.

To access the second, private, area a project username and password was required. This part of the Web was dedicated to communications and connections between partners, test-case definition and data storage for use throughout the FLOMANIA consortium.

3.3.3 Task 2: Test case data base

At FLOMANIA kick-off it was decided to gather information on possible test cases and – when having all relevant information assembled – to rate these cases in order to achieve a set of test cases worthwhile to be taken for the validation/calibration work intended. The following rules had been applied:

1. In order to provide a sound basis for validation work, and at the same time to keep the full amount of validation work as low as possible, partners were

urged to select two cases out of three mandatory ones (with one being absolutely mandatory) serving as a common validation basis. Mandatory test cases have been selected in particular for testing the RSM implementation.
2. All other suggested test cases where only one partner applied for were discarded from the test-case list - apart from so-called industrial cases, allowing industry to take particular care of their own high technical needs and requirements.
3. Partners then applied for co-ordinating one (or more) test cases, i.e. being responsible for all work related to setting up the case, gathering relevant, "cross-plotting" and discussing on results with partners.

Moreover, several partners provided meshes that were – later on - used as mandatory meshes. These mandatory meshes had been proven (prior to distribution amongst the partners) to allow for "close to" mesh-independent solutions. Although this approach resulted in quite a big set of test cases – see list below - it was taken because it provided a sound basis for the turbulence model treatment in general. As it was one of the FLOMANIA goals to deploy full Reynolds stress models, a broader band of test cases was definitely supporting a much broader testing and performance investigation of the model(s).

Finally, the following test cases – for which details can be found in chapter IV of this book - have been retained for validation work:

1. NACA0012 half wing with rounded tip
2. Rotor 37
3. Asymmetric plane diffuser
4. ONERA M6
5. OAT15A
6. NACA0012 beyond stall
7. MRTT
8. Generic train
9. Generic car mirror
10. TUM forward swept wing aircraft
11. TUM delta wing
12. 3D circular cylinder
13. AS28 w/b
14. Generic air-intake
15. Ahmed car body
16. 3D low hill with complex separation
17. Separation behind 2D hills
18. Aerospatiale A-airfoil
19. NACA0012 – DNS approach
20. DLR-F6 aircraft

It must be noticed at that point that one of the FLOMANIA partners, IMFT (see address list), carried out the measurements for test case number 12, the 3D circular cylinder, and uploaded the data to the data base. When this test case was later simulated employing DES, the experimental data had to be reconsidered due to the findings in these simulations.

Another test case not used from the beginning on and therefore not in the above given list was the computation of the decay of isotropic turbulence that was used to calibrate the different partners' DES implementations and models. Due to the interesting results and generally accepted importance of the calibration of a DES and in particular its LES part, it is now taken over by the EU DESider project as a mandatory test case.

3.3.4 *Task 3: Software and technology transfer*

As three different goals were considered in the FLOMANIA project - described above as the so-called short-, medium- and long-term goals - partners working on the development/improvement of models were encouraged to implement turbulence models of interest on a common (!) basis. Thus, by help of the technology support group a common RSM was implemented, enabling a comprehensive and comparable validation work. As a matter of fact, the Speziale-Sarkar-Gatski (SSG) RSM was chosen for implementation, because this model was identified as a very robust and reliable model.

This software/technology transfer was initiated by a requirements catalogue and did circumvent redundant work as it was the case in past projects. Hence, all industrial partners benefited from the same/common implementation. Obviously, this technology transfer included on-site support – which was provided by partners participating in task 4 (see below). Hence, "supplier" partners met with industry or research labs for (common) implementation of models into corresponding codes as an "on-site" help and support.

3.3.5 *Task 4.1: Modelling of full Reynolds stress models (RSM)*

The main objective of the FLOMANIA project was definitely related to turbulence modelling issues. This included work on "some" improvements of models currently in use by industry with the aim to improve robustness – and was mentioned as the short-term goal - for speeding up convergence of current complex industrial applications.

Unfortunately, RSM applications for industrial external aerodynamics are quite limited and most of the RSM applications have been carried out for internal aerodynamics. Thus the modelling of full Reynolds stress models for usage in an industrial aeronautics environment was the most important aim of the FLOMANIA project.

It was not the aim to develop a new Reynolds stress model but to use an appropriate and well understood RSM. Hence, the mentioned SSG model was selected for implementation. Of course, the use of the SSG model in a prescribed form did not exclude a necessary "adaptation" of the model to the corresponding code and its numerical scheme. The initial specification was to take the k-ε version, it turned out, however, that for "numerical convenience" also a k-ω version was defined and implemented – but the latter version was not "common".

The SSG model was implemented with a so-called 'automatic' wall function which automatically allowed for both low-Re behaviour in case of fine meshes and wall-function treatment in case of coarser meshes.

3.3.6 Task 4.2: Two-equation modelling

In order to fulfil convergence and non-failure requirements on currently used two-equation turbulence models, improvements of models were and are still necessary at the current industrial "state-of-the-art" environment. Thus, it seems to be possible now to circumvent the costly procedure to start with one model and in case of failure to re-run cases with other model(s). Of course, the knowledge on model performance included the knowledge about model weaknesses, a crucial point when absolute flow values, e.g. force coefficients, are requested.

This task was not aiming at improving two-equation turbulence models to better cope with particular applications - and as a result of the initial assumptions to fail in other areas - but to improve robustness and reliability of these models to directly support today's industrial needs. The latter included the provision of low-Re models that are less sensitive to location of the first mesh node.

3.3.7 Task 4.3: Sophisticated wall functions

As pointed out in the previous section, the treatment of the near-wall regions is of utmost importance for improving convergence and processing times of CFD applications in general and the accuracy of the simulation results in particular. Thus, improvements obtained in the wall-function development had to be correlated with accuracy control.

Hence, further improvements of existing, highly sophisticated, wall functions were considered in this task. Implementation of these functions, and in particular the automatic wall function used in conjunction with the SSG RSM model, allowed for running realistic cases in acceptable times. In turn, a universal high/low-Re boundary condition is particularly important for the simulation of complex industrial flows, where a fair and continuous resolution of the semi-viscous sub-layer is neither desirable nor feasible.

3.3.8 Task 4.4: DES and zonal approach

The so-called long-term goal of FLOMANIA was concerned with DES and hybrid RANS-LES approaches, very likely being the next-generation simulation tool in industry. In contrast to RANS-based closures, Large-Eddy Simulation offers much more predictive prospects and hybrid RANS-LES methodologies, i.e. OES, DES, LNS, VLES, have shown to be a feasible approach to predict unsteady turbulent flows of industrial significance. And it should be noted that DES is/will be used by industry not only when flow unsteadiness is dominant due to (large) separated regions, but also for generating in that cases more accurate – compared to URANS – mean values, e.g. on force coefficients because this will allow in turn to provide an improved predictive accuracy for aerodynamic loads in fluid-structure interaction. In addition it is well-known that RANS results for aero-acoustics are by far too dissipative, thus DES results are seen to be much more accurate, but aero-acoustics cases have not been treated in FLOMANIA.

FLOMANIA work concentrated on an accurate prediction of complex, three-dimensional flows which were afflicted by severe unsteadiness, e.g. manoeuvring aircraft, side wind on trains, car side mirror. In such circumstances, the flow's gross characteristics are governed by seemingly subtle details, pertaining to the representation of mean transient and residual turbulence interaction, in particular the predictive capabilities of the employed turbulence closure models.

The DES task was carried out taking into account the improved turbulence modelling capabilities provided in the previous tasks 4.2 and 4.3.

3.3.9 Task 5.1: Implementation of turbulence models

Improved turbulence models including the newly employed RSM model, wall functions and sensors for adaptation, all these items have been implemented into industrial codes to form the basis for an advanced simulation platform which can be used in all future investigations whatever complexity of flow and/or geometry is needed. It was mentioned already that task 3 was strongly related to implementation work by providing support for a "comparable" implementation of models and tools. Of course, this provided a proper basis for a comprehensive validation and testing of the incorporated models compared to projects in the past, "purely" dealing with validation of codes and models. Partners working in task 3, "software/technology transfer", provided on-site support, to ensure that all partners received the same (or at least according to the different 'numerics', a similar) implementation of turbulence models.

Obviously, as DES approaches rely on RANS turbulence models in the near wall and mainly attached-flow regime, one of the task objectives was concerned with implementation/improvement of turbulence models. Work in this task did not start from "scratch" but was making use of the currently available CFD approaches, featuring structured as well as unstructured meshes.

3.3.10 Task 5.2: Sensors for adaptation

An important effort was placed on sensors, which are needed for the following items holding for both structured and unstructured meshes:

1. Investigation of grid independent solutions based on flow-adjusted meshes taking care of a good resolution of all relevant flow features, including particular refinement and de-refinement strategies.
2. Investigation of meshes when wall functions are in use to allow for larger wall-adjacent step sizes.
3. Investigation of blending between RANS and LES parts for a "hybrid" approach, i.e. DES in conjunction with mesh dependence investigations.

One of the objectives of this task was to find an appropriate sensor for mesh enrichment on hybrid grids, to obtain better mesh resolutions in shear layer and wakes. Moreover, anisotropic mesh refinement strategies have been applied, which are based on the Hessian for adaptation sensors. Such techniques now allow individuating the appropriate directions along which new grid points should be added and new error estimator formulations have been found for reducing the sensitivity of adaptation criteria to user-defined parameters.

In DES methods rather different demands for RANS and LES "numerics" have to be considered. Apart from the different schemes for RANS and LES, a correct choice of blending functions between upwind and central discretisation is essential. Thus, though a statement such as "upwind schemes are unacceptable for LES" cannot be correct (any numerical method, if consistent, results in an acceptable accuracy with a grid fine enough), the issue of numerical dissipation in DES does exist, and requires attention: although the excessive dissipation does not result in an unstable or meaningless solution, it prevents the solution from taking full advantage of the grid provided and so makes DES more "grid-demanding".

3.3.11 Task 5.3: Grid adaptation and grid dependence

It should be noted first, that all test case computations required extensive study of grid dependence in order to be sure that the results obtained are well comparable and to obtain what was called a mandatory grid.

Grid adaptation and grid dependence issues were carried out hand in hand. Although work, knowledge, and expertise necessary for structured and unstructured meshes are different (but fertilising each other), the main goal was to arrive at an as high as possible predictive accuracy for any particular application. When it was mentioned earlier that a change in CFD-paradigm has taken place because of the need to provide accurate, absolute and quantitative results for any flow, instead of just providing so-called delta-values or qualitatively accurate results in the past, work carried out in this task received utmost importance.

Of course, it is possible in many cases to use very fine meshes of about 20-30 million nodal points, however, today's requirements to provide simulations at least over night, aim at a reduction of mesh sizes – by "hopefully" keeping the predictive accuracy. The time constraints hold in particular for current and future designs of aircraft and transportation means due to the fact that sophisticated CFD methods are more and more becoming a tool even in pre-design phases where quick answers are needed. When it is recalled that a design freeze is normally taking place when only roughly 15% of the overall knowledge is available, predictive accuracy in pre-design phases is doubtless of utmost importance.

It was an additional effort in this task to investigate on limitations of RANS/URANS approaches, as it is well known that finer and finer meshes often produce unsteady flows in contrast to coarse meshes where the flow is still steady.

3.3.12 Task 5.4: Numerics and "best-practice recommendations"

The major objective of this task was to support all partners (in most cases industrial partners) on numerical implementation of turbulence models in a common way. Moreover, this task was dedicated to a production of so-called best-practice guidelines for an ease of use of the highly-sophisticated model(s).

It was not proposed to provide RSM with different coefficients for different applications and/or partner codes, but the complexity of the new models made it more than worthwhile to provide a set of best-practice recommendations for use, see also chapter V, and to a certain extent presentation of the different applications in chapter IV . Reporting in this task emphasised on the implementation of models and tools and resulted in an as transparent as possible description of methods.

Additionally, FLOMANIA provided help for the QNET-CFD project by offering support for working out best-practice guidelines for the NACA0012-wing-tip application challenge, see chapter IV.1.

3.3.13 Task 6.1: Applications – generic cases – provision of meshes

Work in this task settled around computations for generic test cases, comparison and validation. Partners had teamed up to run computations on the finally selected generic test cases by using mandatory meshes and comparing results on the basis of the common RSM. In view of the objectives of the project, aiming at industrial applications using advanced turbulence models (NLEVM, EARSM, RSM in (U)RANS approaches), main emphasis was placed on accuracy and robustness for a comprehensive series of flow configurations and flow phenomena, featuring e.g. separation, shock-BL interactions, shock-shock interaction, shear flows, wall-bounded flows, and, besides others, effects of incidence.

3.3.14 Task 6.1.1: Provision of meshes

To obtain the main goals conveyed above, a major task was dedicated to the provision of a mesh catalogue for the generic applications. Meshes were provided taking care of the different approaches, they were adapted to the models used, e.g. solutions with or without wall functions, in order to accurately compare performance and accuracy of results. In order to cope with different wall functions or running a low-Re model, meshes have been constructed that differ only in the near-wall region but that were not changed outside of the viscous layer(s).

As mentioned before, meshes have been provided – structured or unstructured – that had been pre-adapted to achieve mesh independent solutions. However, even for an experienced group of CFD-people generation of an "optimal" grid is not an easy task, as can be seen in the discussion of test case 1, see chapter IV.1.

3.3.15 Task 6.1.2: RANS and URANS applications

As the major emphasis of the FLOMANIA project was related to turbulence modelling for RANS application as today's workhorse in industrial aerodynamics, the present task was a major source of knowledge gathering. Even more when recalling that (U)RANS computations on very fine meshes - in order to accurately predict absolute flow values and allow for flow simulation around complete aircraft - are the current procedure of the aeronautics industry.

Considering aerodynamics results as the potential basis for interdisciplinary approaches, e.g. (static or) dynamic aero-elasticity, it is of utmost importance to validate performance of the RSM for steady and unsteady flows. Once again, the use of RSM model(s) for external aerodynamics is based on the hope to obtain a model which features all flow regimes. And from an industrial point of view it is also necessary to invoke unsteady approaches (URANS) because reliability of turbulence models for unsteady flows is still a real challenge.

It is self-evident that most of the partners have been working in this task as it was considered to be a key topic for performance tests of the new/improved turbulence models and gaining insight in the complex physics by using the new models. Thus, code developers and end-users co-operated and communicated on the basis of results obtained on model performance and physical issues.

3.3.16 Task 6.1.3: DES applications

Besides the further improvement of RANS simulations described in the previous tasks, the objective of this task was to begin with DES computations in general. However, it was not intended to run a validation exercise on DES - because this is a laborious part of work and was intended to be carried out in a separate project, particularly emphasising on upstream research covering hybrid RANS-LES topics. It should be added at this point, that such a project is currently running as a new EU project that was kicked-off in January 2004 under the acronym DESider.

For the FLOMANIA project it was decided to gain “initial” knowledge about DES and hybrid RANS-LES approaches in order to achieve “first guesses” on when and how DES can be applied and where possible limits of applicability arise.

3.3.17 Task 6.2: Complex applications

The major concern of this task was to compute complex and maybe proprietary cases being real-world application of the industrial partners. For various reasons this task was kept separate from the generic-application task 6.1 because it was even allowed to test the new turbulence modelling capabilities on confidential industrial in-house applications.

Whether or not these cases were confidential, a report - omitting the scales in case of confidentiality of course - was provided by all industrial partners to disseminate knowledge gained by these exercises. In other words: The procedures taken were transparent, the quantitative results and the geometry not necessarily.

3.3.18 Task 6.2.1: RANS and URANS applications

As mentioned above, for complex industrial test cases RANS approaches are today’s workhorse. Thus, the most challenging validation part of the FLOMANIA project was concerned with the validation of such real-world, i.e. complex and challenging test cases. This included the use of meshes with high resolution.

3.3.19 Task 6.2.2: DES applications

Similar to task 6.1.3, the overall objective of this task was to give rise to one or another complex DES computation to gain some “preliminary” knowledge on the use of DES and the performance of this new numerical tool.

As already mentioned, DES applications were so-called long-term research items in FLOMANIA and used for feeding back information from the DES side to the RANS application strand. As future use of DES in an industrial environment will possibly be envisaged in a shorter period, first knowledge on DES when running industrial/complex test cases was of great interest.

3.4 Task-per-partner matrix

Partner \ Task	1	1.1	1.2	1.3	2	3	4	4.1	4.2	4.3	4.4
ANSYS/CFX		x		x	x	x				x	
Alenia		x		x	x						
Bombardier		x		x	x					x	
Dassault		x		x	x			x	x	x	
EADS-CASA		x		x	x						
EADS-M	X	X	x	X	x						
EDF		x		x	x	x		x	x	x	
NUMECA		x		x	x	x		X	x	x	
DLR		x		x	x				x		x
FOI		x		x	x			x		x	
IMFT		x		x	x			x	X		x
ONERA		x		x	x			x		x	
Chalmers		x		x	x	x			x	x	x
ICSTM		x		x	x	X		x	x	x	x
SPTU						x					x
TUB		x		x	x	x					X
UMIST		x	X	x	X	x	X	x	x	X	

Partner \ Task	5	5.1	5.2	5.3	5.4	6.1	6.1.1	6.1.2	6.1.3	6.2	6.2.1	6.2.2
ANSYS/CFX	X	X	x		x		x	x			x	
Alenia		x	X	x	x			x				
Bombardier		x	x					x	x	X	x	X
Dassault											x	
EADS-CASA		x		x		X	x	X			x	
EADS-M				x	x		X	x			x	
EDF		x			x			x			X	
NUMECA		x	x	x	x			x			x	
DLR		x	x	X				x	x		x	
FOI					X			x				
IMFT								x	x			
ONERA		x			x			x			x	
Chalmers								x	X			
ICSTM		x						x				
SPTU		x	x	x	x		x		x			
TUB				x	x				x			x
UMIST		x			x			x				

The table above provides a cross reference of work load per task and per partner. The matrix summarises partner contributions. Please note that crosses in bold indicate who was responsible for co-ordinating the corresponding task.

3.5 List of partners and addresses

In the following please find the list of partners (industry, research labs, and universities). Participant names in bold denote the official point-of-contact.

Organisation	Participants	Address
EADS Military Aircraft	W. Fritz **W. Haase** S. Leicher	Dept. MT63 / Build. 70.N D-81663 München Germany
Bombardier Transportation	A. Orellano **T. Rung**	MLN/ESA Am Rathenaupark D-16761 Berlin Germany
ANSYS/CFX	M. Kuntz **F. Menter**	Staudenfeldweg 12 D-83624 Otterfing Germany
Alenia Aeronautica	**N. Ceresola**	C.P. 432 Corso Marche, 41 I- Turino Italy
Dassault	A. Davroux **M. Mallet**	DTIAE/AERAV 78, quai Marcel Dassault F-92214 Saint Cloud France
EADS-CASA	**J.J. Guerra** V. Ibanez L. Ruiz-Calavera	Dept. Aerodinamica Avda. John Lennon s/n 28906 GETAFE, MADRID Spain
EDF	**D. Laurence** T. Pasutto	EDF R&D 6 quai Watier F-78400 Chatou France
NUMECA	**C. Hirsch** A. Patel B. Leonard	5 Av. Franklin Roosevelt B-1050 Brussels Belgium
DLR	B. Eisfeld **D. Schwamborn** K.Weinman	Institut für Aerodynamik und Strömungstechnik Bunsenstr. 10 D-37073 Göttingen Germany
FOI/FFA	P. Eliasson S-H Peng **S. Wallin**	Aerodynamics Division FFA Computational Physics SE-17290 Stockholm Sweden

IMFT	**M. Braza** Y. Hoarau	EMT2 Group Avenue du Professeur Camille Soula F-31400 TOULOUSE France
ONERA	**B. Aupoix** P. Ferrey C. Gleyzes R. Houdeville C. Marmignon	Centre de Toulouse 2, Avenue E. Belin BP 4025 F-31055 Toulouse Cedex 04 France
Chalmers University	**L. Davidson** S. Krajnovic	Department of Thermo- and Fluiddynamics Horsalsvagen 7 SE-41296 Gothenburg Sweden
Imperial College	**M. Leschziner** C. Wang	Aeronautics Department Prince Consort Road London SW7, 2AZ England
TU Berlin	**U. Bunge** A. Martin C. Mockett F. Thiele	Hermann-Foettinger-Institut fuer Stroemungsmechanik, Sekr. HF1 Mueller-Breslau-Str. 8 D-10623 Berlin Germany
UMIST	T. Craft **D. Laurence** A. Revell J. Uribe S. Utyuzhnikov	Department of Mechanical, Aerospace & Manufactoring Engineering PO Box 88 Manchester, M60 1QD UK
St. Petersburg Technical University	A. Garbaruk D. Magidov **M. Strelets** M. Shur A. Travin	Faculty of Physics and Mechanics Hydro-Aerodynamics Department 29 Polytechnicheskaya Str. 195251 St.-Petersburg Russia

4 Conclusion

The European Commission funded FLOMANIA project, with 17 partner organisations, running for 33 months, has proven to be very successful as it has enabled the European partners to improve and enhance their knowledge on turbulence modelling. Most of all, the first use of RSM in an aeronautical industrial environment was and is seen as being a big step forward to improved and more reliable flow simulations.

Secondly, the first industrial attempt using DES (hybrid RANS-LES) method(s) for unsteady flow simulation was taking place. Results achieved were very promising and paved the way for the currently running EU project DESider, which is based on improving hybrid RANS-LES approaches and making them usable for industry.

Technical achievements apart, the FLOMANIA project offered a sound organisational framework promoting understanding and co-operation between partners as well as with third parties that have been supported by expertise that arose from FLOMANIA, enabling dissemination and exploitation of results at the same time. Collaboration with FLOMANIA was based on tightly controlled information-exchange protocols, clear descriptions of test cases, provision of mandatory meshes, model implementation support, centralised collection of results, a comprehensive collection of information via deliverables, and a fruitful co-operation between all partners. Last but not least it is also documented by the present book.

All test case results are provided in chapter IV of this book, together with brief method descriptions of approaches and methods used by partners in chapter II, the presentation of turbulence models that have been used in chapter III, and – summarising the findings - an attempt to give some advice and best-practice guidelines on the usage of these models in chapter V.

To conclude, it can be stated (again) that the collective outcome is seen to be by far greater than what could have been expected to arise from partners working on their own resources only. Evidently, the FLOMANIA project – as a European "network" - tended to be far richer and more diverse in approach, in new and challenging ideas, as well as in the different partner-related ways looking at the same problems.

5 The FLOMANIA Web site and the access to it

As pointed out above, the FLOMANIA Web site provided two areas, a public area that is accessible to any web user and, after the welcome page, consists mainly of pages which outline the aims, purpose and work content of the FLOMANIA project and, secondly, a private area that requires a project username and password.

For dissemination purposes, the FLOMANIA consortium has decided to release appropriate information, i.e. to shift documents from the private to the public part of the Web site. All FLOMANIA partners hope that this additional source of information which is issued in parallel with the present book is helpful in further attempts to run own validation exercises using the test cases described in chapter IV.

For reasons of efficiency, and to maximise the usage of existing facilities, the project web site will be maintained on the web server of the CFD/Turbulence Modelling research group at UMIST.

The web site is available at

http://cfd.me.umist.ac.uk/flomania

and all the FLOMANIA participants hope that the reader will find visiting it as being valuable for her or his own validation and turbulence model assessment work.

II Technical, partner-related reports – methods, models and applications performed

The following chapter contains brief descriptions of method and tools used by the different partners as well as highlights of work performed in and for the FLOMANIA project.

All these presentations are given in alphabetical order.

1 Contribution of ANSYS: Main achievements in FLOMANIA

M. Kuntz, F. R. Menter, ANSYS Germany GmbH

Abstract

The main targets for ANSYS in FLOMANIA are turbulence model improvements and the validation of established turbulence models. These issues are:

- Development of an improved DES model based on the SST turbulence model
- Evaluation of the DES model for separated flows
- Support the definition of common RSM formulation
- Provide know-how on advanced wall treatments to consortium.
- Testing common Reynolds stress model for generic applications
- Testing and validation of eddy viscosity models for complex aerodynamic applications

All models developments and applications are carried out with the fluid mechanics software CFX-5 of ANSYS Inc. (CFX-5, 2004).

1.1 DES model development

During the FLOMANIA project an SST based DES model is implemented in CFX-5 (Menter et al., 2002). The model is a hybrid RANS-LES model based on the SST model of Menter (1994). The blending technique between RANS and LES mode works in general like the DES model of Strelets (2001), but adds a modification to reduce the grid sensitivity of the model. The dependency on the grid is caused by the grid scale included in the blending switch. This switch gives the turbulence model an estimate, whether it is possible to resolve turbulent structures on the grid. But in case of a local grid refinement, it can also cause the onset of the LES mode in boundary layers, and therefore a dependency of the flow separation on the local grid density. An example for grid induced separation is shown for a NACA0012 airfoil in Figure 1. The improved zonal SST-DES model of CFX includes an additional shielding function based on the SST blending functions to force the RANS mode in the boundary layer. The zonal DES formulation leads to a substantial change in the characteristics in the DES formulation in general, as it enables the DES approach to be applied in complex industrial geometries. The zonal concept has now been adopted also by the originators of the DES approach (Spalart, 2005) albeit with a different name (DDES). A description of the zonal method is given in chapter III about hybrid turbulence models.

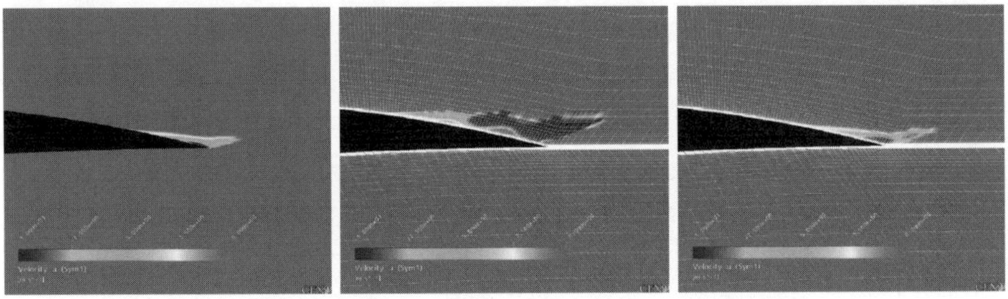

Figure 1 Grid induced separation (left: RANS model, middle: standard DES model, right: zonal DES model)

1.2 DES application

The computations for the Ahmed car in the frame of the FLOMANIA project are described in chapter IV/15. This geometry is applied to demonstrate the improved capabilities of the DES model for the prediction of separated flows. The experimental flow field around the geometry with 25° slant angle shows a separation with a reattachment in the middle of the slant. No RANS model produces the correct flow field results, the flow topology is either fully attached or fully separated. Therefore, this case is a good candidate for the evaluation of the improvement achieved by the DES formulation in the separated flow region. The goal is the resolution of the unsteady features of the shear layer separating from the edge of the slant. The vortex-shedding and break-up is expected to be the main mechanism to increase the turbulent energy in the separated shear layer, thereby forcing a reattachment of the flow. However, as the high levels of turbulence in the measurements cannot be explained by classical shear layer physics, the next question is, if the large-scale unsteadiness observed in the experiments can be triggered by the turbulent structures emanating from the slant.

The DES simulations require a significantly refined grid in the slant region, particularly in the spanwise direction, to activate the DES limiter. In order to keep the total number of nodes at an acceptable level, two steps of local grid refinements have been introduced in the slant region. The total number of nodes is 1.783.000, and the number of spanwise cells is 70 on the slant. The time step for the simulation is $\Delta t=10^{-4}$[s], which corresponds to 250 time steps for one passage of the car at free stream velocity.

The main change in the DES simulation compared to RANS can be seen in Figure 2, showing the local flow topology on the slant of the body. While the SST-RANS solution gives a fully stalled topology, the DES method is in agreement with the experiment. The main difference between DES and experiment is that the DES separation zone is overpredicted in the center of the slant (see velocity profiles below). Nevertheless, the DES solution shows the correct overall flow features, including the strong side vortices. The current DES simulation was carried out with limited computing power and is it's not clear, if longer running times, further mesh and time step reduction, would have resulted in a smaller separation zone.

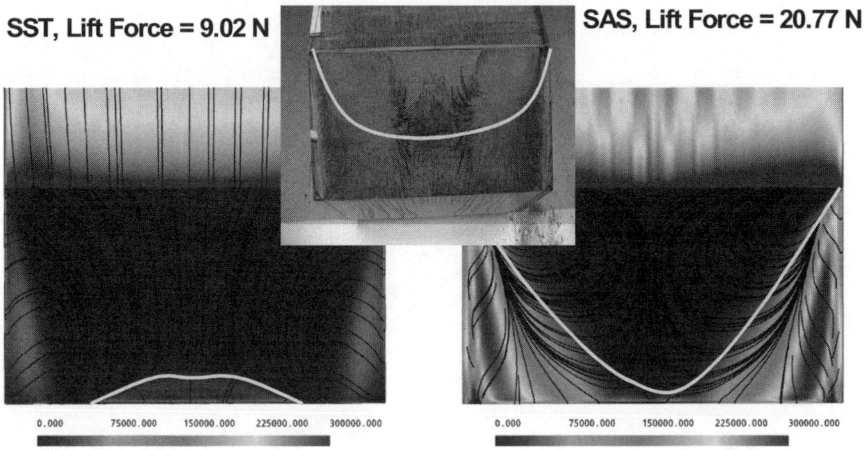

Figure 2 Flow Topology on the Slant of Ahmed Car (left: SST solution, right SAS solution, middle: experiments)

The velocity profiles in the symmetry plane on the slant and in the wake are presented in Figure 3 and Figure 4. In the symmetry plane, the advantage of the DES simulation is not very pronounced, but in the off-symmetry plane and in the wake, the change in flow topology and the improvement in the predicted results is apparent. The unsteady simulations give an indication of the underlying mechanism of the large-scale unsteadiness of the flow field. Figure 5 shows an instantaneous picture of the flow topology in the slant region. At the given instant in time, the separation bubble is interacting with the left side vortex, leading to a vortex break-down. This might be one of the mechanisms for the large fluctuation levels observed in the experiments in the slant region.

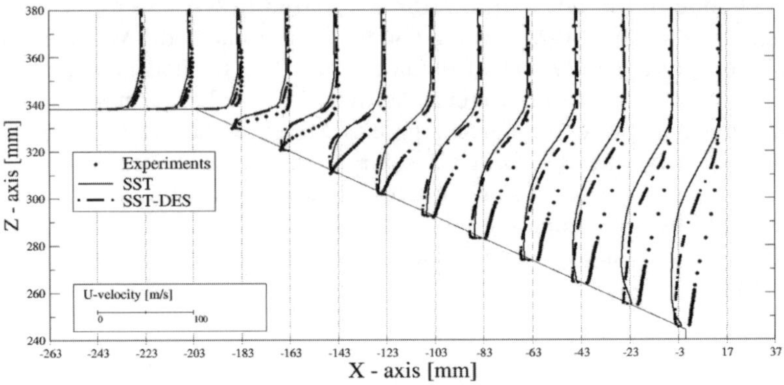

Figure 3 Velocity profiles on the slant in the symmetry plane

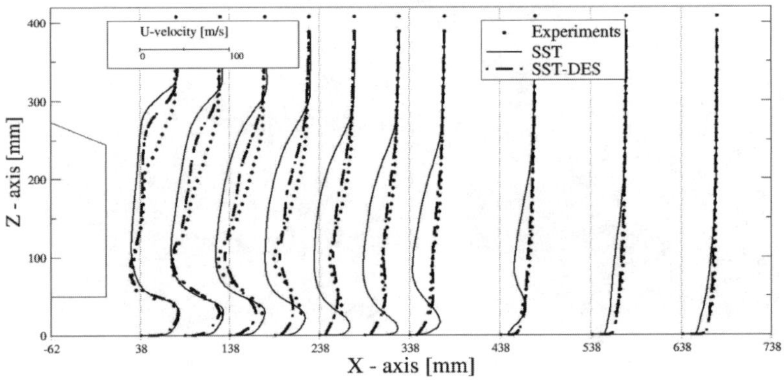

Figure 4 Velocity profiles of Ahmed car body (25° slant angle) in the wake in the symmetry plane

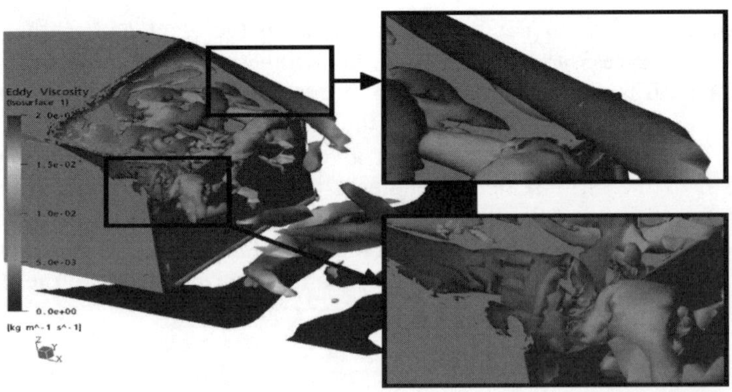

Figure 5 Vortex structures on slant of Ahmed car body (25° slant angle)

1.3 RANS Applications

1.3.1 DLR-F6 Configuration

The DLR-F6 configuration is one of the complex applications in FLOMANIA, described in detail in chapter IV/20. This test case was also chosen for the 2nd drag prediction workshop organized in 2003 by AIAA. The experimental test campaigns were performed between 1993 and 1996 in the ONERA S2MA pressurized wind tunnel. The model was sting mounted in the transonic test section and the Mach number was varied between $Ma_\infty = 0.6$ and 0.8 while the Reynolds number (based on aerodynamic mean chord, c) was held constant at $Re = 3x10^6$. Two different configurations are computed: wing and body (WB) and wing, body, nacelle, pylon (WBNP), see Figure 6.

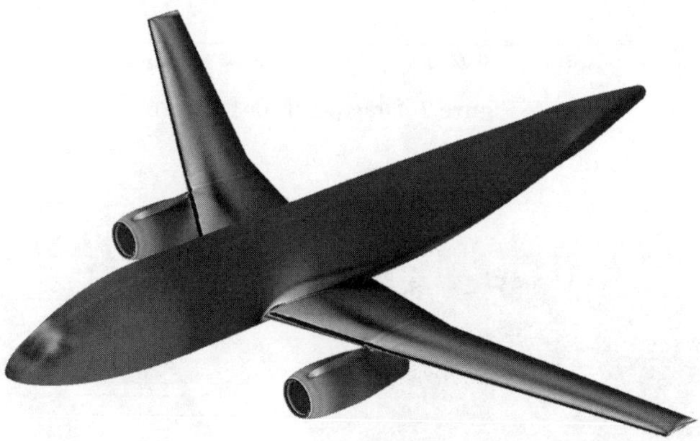

Figure 6 DLR-F6 wing/body/nacelle/pylon configuration

Computations are carried out on the three standard grids provided by the workshop organizers (created with ICEM-CFD HEXA). Different angles of attack were computed on the medium grid to compute the drag polar. The drag polar for both configurations obtained on a grid of about 6 million nodes is compared to experimental data in Figure 7. A good agreement is achieved. The comparison between the experimental oil-flow visualization and the CFD predictions for separated regions is shown in Figure 8. The size of the separation at the wing-engine pylon junction appears to be somewhat larger in the simulation than the experiment. This could potentially explain why in Figure 7 the CFX-5 drag prediction was on average 5 percent higher than the experimental value. More details of the simulation can be found in Menter and Langtry (2004).

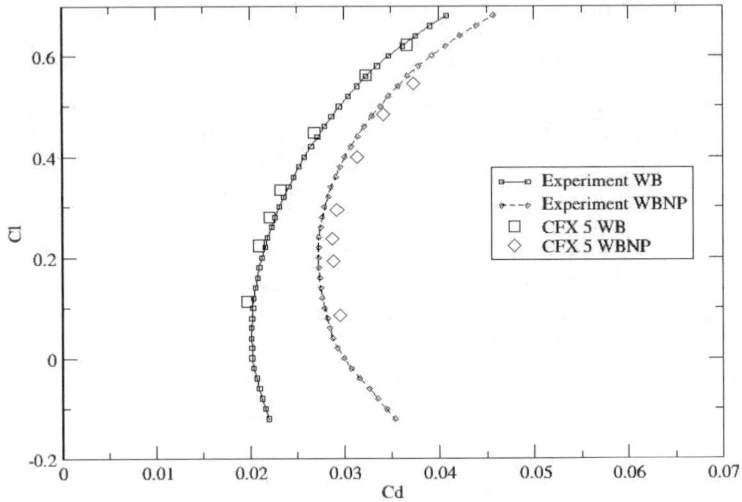

Figure 7 Drag polar for DLR-F6

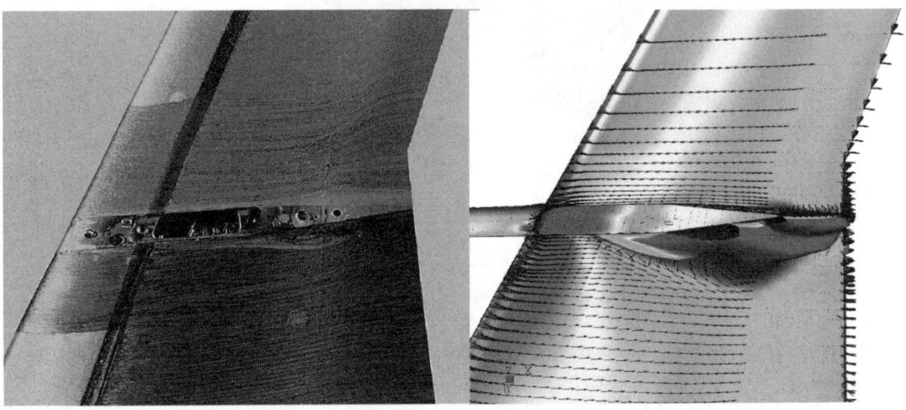

Figure 8 Comparison between experimental oil-flow visualization and CFD prediction of the wing-engine pylon separation (DLR-F6)

1.3.2 Forward Swept Wing Configuration

The forward swept wing configuration was proposed by the partner EADS-M for the FLOMANIA project. Experimental data were provided by Breitsamter (Breitsamter and Laschka, 1998). Hexahedral grids generated with ICEM-CFD HEXA by EADS-M were used for the computations with CFX-5 by ANSYS. Three different grid sizes were available (fine grid: 9.907.520 elements, medium grid: 1.238.440 elements and coarse grid: 154.805 elements). Convergence was obtained after about 60 iterations. A grid convergence study and a study on the influence of the turbulence model was carried out (k-ε, k-ω and SST models). An example for of the flow field results computed on the fine grid using the SST turbulence model is shown in Figure 9.

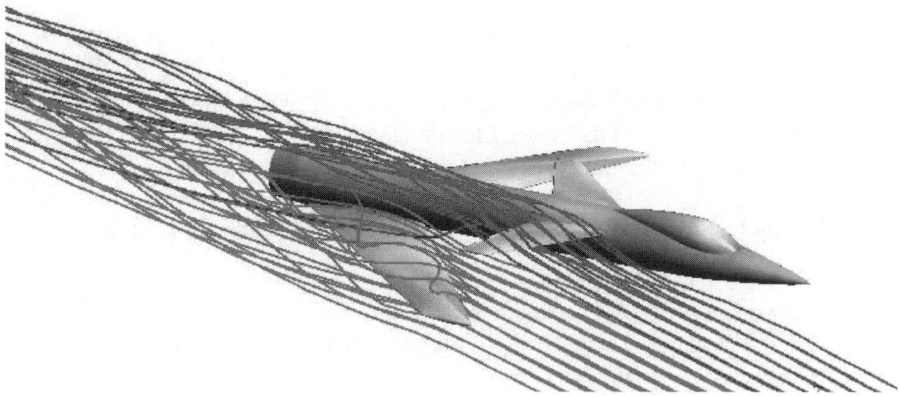

Figure 9 Flow solution on fine grid

Figure 10 (left) shows the drag polar for an angle of attack varied between 10 to 45 degrees for three different grids using the SST turbulence model. Only small differences are obtained for the high angle of attack, a better agreement with experimental data is given for the fine grid. The right part of this figure shows the drag polar computed on the medium grid for three different turbulence models. The k-ε and the k-ω model show some disagreement also for lower angle of attack, whereas the SST model agrees well in the whole range of angles.

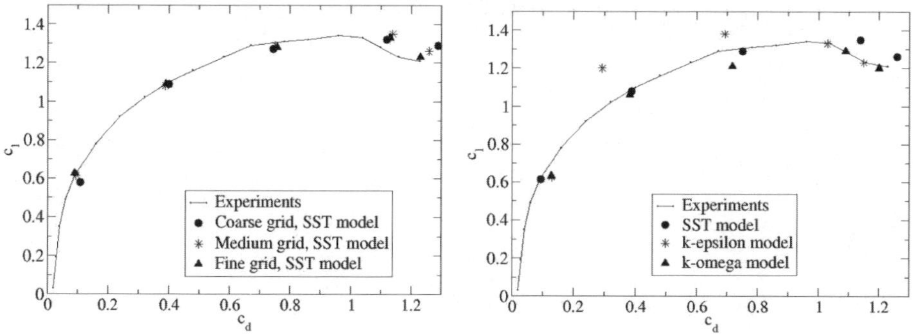

Figure 10 Drag polar for forward swept wing

1.4 Conclusions

ANSYS developed and provided the following technologies during the Flomania project:

- Provided advanced wall treatment
    - Scalable wall functions for ε-equation (Grotjans and Menter, 1998)
    - Automatic wall treatment for ω-equation (Esch et al., 2003)
- Contributions to formulation of common RSM model
- Support for RSM model implementation of RSM model based on ω-equation into the DLR code Flower.
- Formulation of zonal DES approach.

The following test cases have been computed with CFX during the project:

- OBI diffuser
- Ahmed car body (25° and 35° case)
- DLR F6 wing-body and wing-body-pylon-nacelle configurations
- Forward-swept wing test case

2 Contribution of Alenia: Main results obtained within the project

N. Ceresola, Alenia Aeronautica

Abstract

In this chapter the main outcomes of the activity made by Alenia in the framework of Flomania project are outlined.

The main technical achievements that have been obtained are: 1. Validation of CFD methods and assessment of their application envelope in cases which present critical features with respect of turbulence modelling. 2. Improvement of the quality of CFD simulation codes including advanced RANS turbulence modelling, under the aspects of reliability and robustness. 3. Assessment of the influence of computational grid and of the benefits induced by the use of mesh adaptation techniques.

In addition, the participation to the project work has contributed to individuate directions for the development of company's CFD codes in the near future, accordingly with the foreseen needs of aerodynamic design.

2.1 Description of numerical techniques and turbulence models implemented in code UNS3D

2.1.1 The UNS3D code

The computations have been performed using the code UNS3D. The solution algorithm is based on a finite volume, node centred approach operating on an unstructured grid. The integration of the fluxes on the cell surfaces is made on a segment basis; the solver can deal therefore without distinction with tetrahedra, hexahedra and prisms, even mixed to form hybrid grids.

The artificial dissipation model is derived from the nonlinear scheme of Jameson. The equations are integrated in time with a second order backward difference and the dual time stepping technique. A five stage Runge-Kutta scheme is used to drive toward zero the residual at each time step. With the use of implicit residual averaging, a local CFL number of 4.9 could be employed in the five stage multistage subiteration process.

2.1.2 Low Mach number preconditioning

The Weiss and Smith version of low Mach number preconditioning is implemented in the code. For the computation of Flomania test cases with a low free stream Mach number, its application is necessary in order to find a converged solution in a reasonable number of iterations. It is also necessary to introduce a sensor depending on cell Reynolds number to avoid applying preconditioning inside viscosity-dominated flow regions

2.1.3 The k-Rt –EARSM turbulence model

A two-equation model is currently implemented in UNS3D, that is based on the solution of the transport equations for turbulence kinetic energy k and turbulence Reynolds number Rt (Arina, Ceresola, Piantà 1996). The rationale for having developed and chosen this model was that it is wall-distance free, with simple boundary conditions to be easily applied to unstructured grids on complex 3D geometries. The Menter's Shear Stress Transport (SST) technique is applied to limit turbulence production in adverse pressure gradients. The cross diffusion term arising from the transformation from the k-ε equations is retained in the model.

The Wallin-Johansson Explicit Algebraic Stress Model (WJ-EARSM) is implemented using k.Rt as the basis RANS model.

In Flomania, the full anisotropic version of the model was used.

2.1.4 Streamline curvature correction

To neglect the Reynolds stress transport terms implies not taking in account physical phenomena such as the effect of flow rotation on the turbulent field. To try to account for such an effect, the streamline curvature correction devised by Wallin and Johansson has been implemented in the model. The streamline curvature correction can be optionally switched on in case the effect of streamline curvature is important, e.g. for vortical flows. Such correction has been applied in the computation of TUM delta wing test case.

2.1.5 2. The k-ω–EARSM turbulence model

A k-ω turbulence model was implemented, that have been recently developed by Hellsten. The model was calibrated requiring consistent behaviour near the edges between turbulent and laminar flow, inside shear flows and zero pressure gradient wall flows. In particular, the calibration have been considered taking in account a variable c_μ , as it is the case if an algebraic stress model (EARSM) is included.

2.1.6 An adaptation sensor for shock waves, shear layers and wakes

The purpose of this work, made in the framework of Flomania, was to find an appropriate sensor for mesh enrichment on hybrid grids, to obtain a better resolution of shear layer and wakes.. After some attempts, a sensor derived from a function representing the local entropy dissipation of energy was adopted, see Tognaccini and Paparone.

Once obtained the value of the sensor function on the flow field, an anisotropic mesh refinement strategy is applied, which is based on the computation of the Hessian of the adaptation sensor. Such technique allows individuating the most appropriate direction along which new grid points should be added.

The technique is described in more detail in Flomania deliverable D5.2-27.

2.2 Grid adaptation on NACA64010 transonic airfoil

The test case considered for grid adaptation was the flow past NACA64010 airfoil, at Mach=0.789, $Re=11\cdot10^6$ and an angle of incidence $\alpha=4°$. For this case, a strong shock-boundary layer interaction is present, with shock-induced separation and a thick wake past the airfoil. The computations were made using Hellsten's k-ω turbulence model, with the Wallin-Johansson algebraic Reynolds stress model.

The basic grid is made of 16720 nodes. Two refinement cycles have been performed starting from the same computed flow field. After that, a refined grid resulted which was made of 20010 nodes. The original and adapted grids are shown in Figure 1. Anisotropic refinement approximately in the direction perpendicular to the flow has been made both in the separated region and in the wake. Across the shock, the refinement was performed preferably in the shock normal direction. As a result of the adaptation, the resolution of flow gradients in the shock wave, in the external part of boundary layer and into wake appeared to be greatly increased. Wiggles in the pressure field inside the separated region, present in the original result, disappeared in the computation on refined grid.

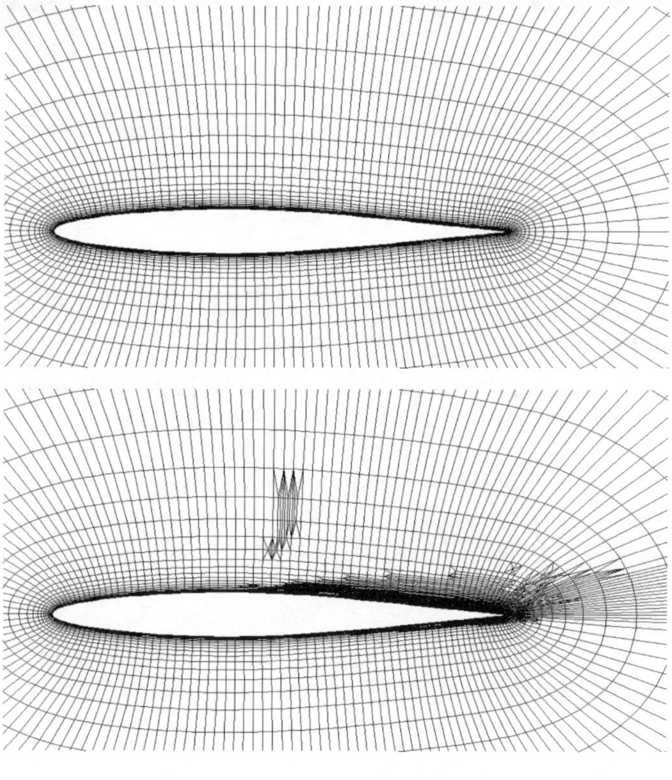

Figure 1 Basic(top) and adapted (bottom) grids on NACA64A010 airfoil

2.3 Computations on NACA0012 with rounded tip.

The 700k points grid was used for the computations, even it was demonstrated to be too coarse for a proper resolution of the tip vortex.
As the inflow Mach number have been set to 0.05, the Weiss-Smith low Mach preconditioner have been used, in order to enhance convergence and robustness of the computations.

Two turbulence models have been used for the computations, to assess the effect of Wallin-Johansson streamline curvature correction against the standard model (see D5.1-30). In Figure 2 the total pressure coefficient in the two cases is shown.

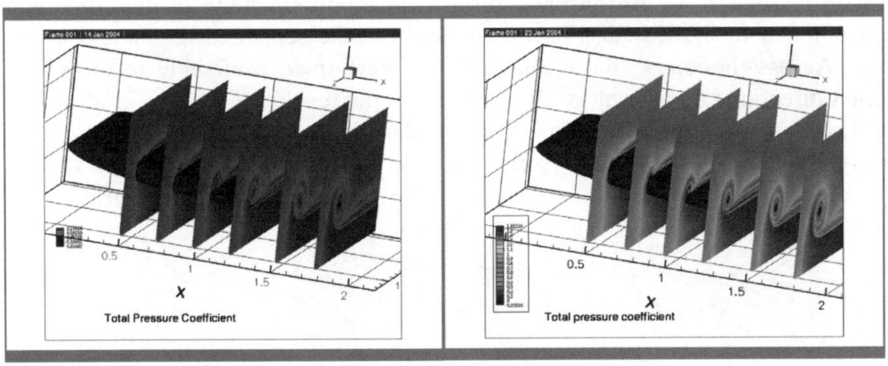

Figure 2 Total pressure coefficient, without (left) and with (right) streamline curvature correction

2.4 Computations on TUM Delta Wing

The turbulence models used for the computations were the 2-equations k-Rt model with the explicit algebraic stress model of Wallin and Johansson (WJ EARSM), including streamline curvature correction (see D5.1-30). The main purpose was to validate the curvature correction against a schematic vortical flow problem, which arises frequently and in many situations in the aerodynamic design. Two angles of incidence have been computed, namely α=25° and α=35°.

At 25° angle of incidence, the experimental flow presents a strong leading edge vortex with no vortex breakdown. A comparison between the experimental and computed wall pressures on several spanwise sections are presented in Figure 3.

The agreement with the computations seems to be good.

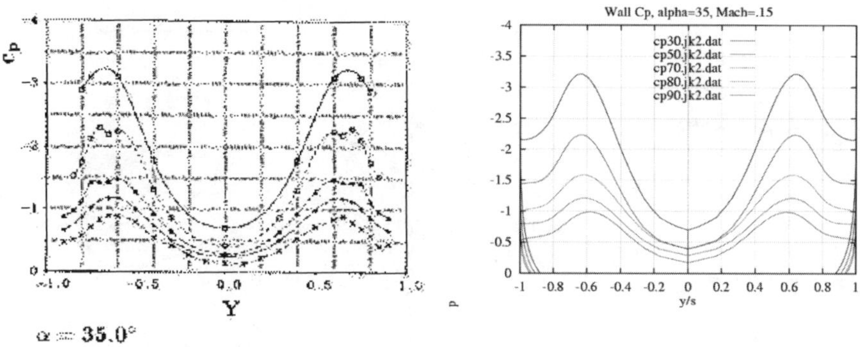

Figure 3 TUM delta wing, α=25°, experimental (left) and computed (right) Cp contours

At α=35° a vortex breakdown is present in the experiment. From the wall pressure cuts in Figure 4 , the breakdown is represented in the numerical simulation as well, and the agreement seems to be particularly good.

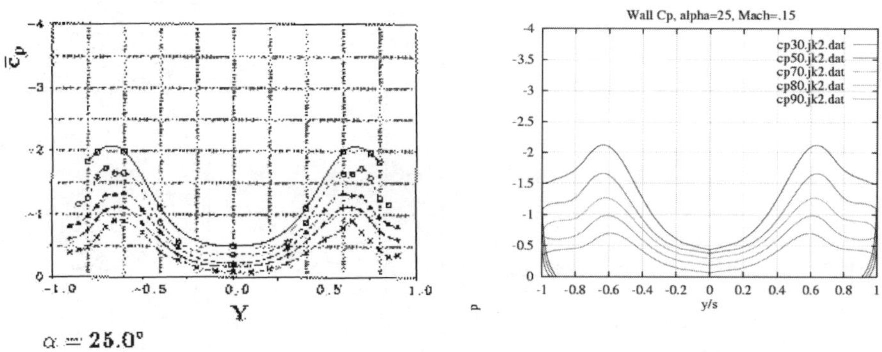

Figure 4 TUM delta wing, 25°, experimental (left) and computed (right) Cp contours

2.5 Computations on OAT15A airfoil in wind tunnel

The Helssten k-ω-EARSM turbulence model was used in the computations. The Cp distributions along the profile at symmetry plane and at wind tunnel wall intersection are shown in Figure 5, in comparison with experimental data. The shock wave is positioned slightly upstream with respect to the experiment, but the overall agreement is good.

The Cp values on the upper and lower walls, respectively in Figure 6, are also well predicted, showing that the effect of the presence of wind tunnel walls were properly taken in account by the present model.

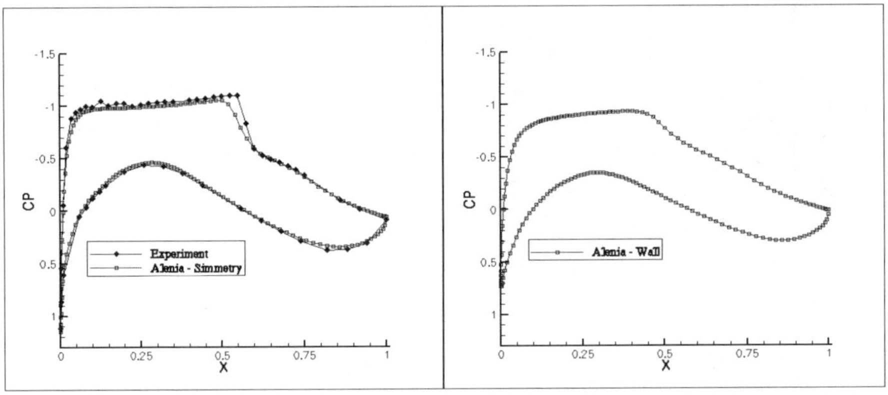

Figure 5 Cp on profile section at symmetry (left) and section at intersection with wind tunnel wall (right)

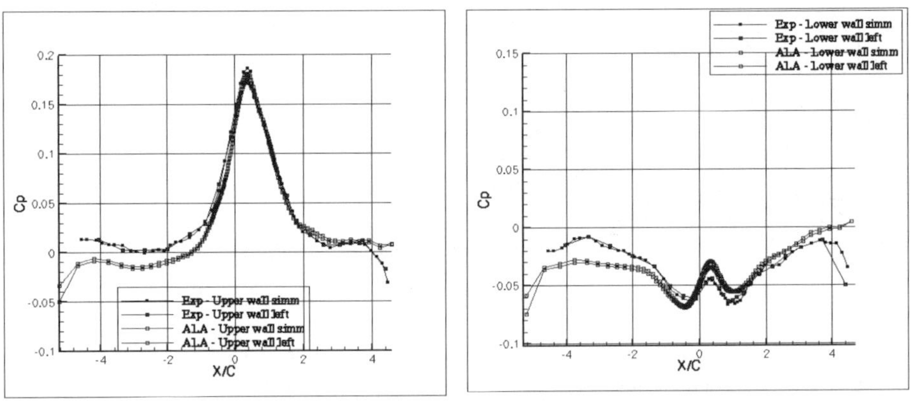

Figure 6 Cp on upper (left) and lower (right) walls, at symmetry plane and at wall intersection

3 Predictive Prospects of DES in Industrial External Aerodynamic Flow Simulations

T. Rung and D. Wu, Bombardier Transportation (BT)

Abstract

The contribution of Bombardier Transportation (BT) seeks to scrutinize the prospects of *Detached-Eddy Simulations* (DES) for the investigation of external aerodynamic flows in an industrial context. The focal point is on inherently transient, massively separated flows at $Re = 10^5 - 10^7$, which feature a considerable amount of turbulent-transient interaction. Examples investigated refer to calibration and verification exercises, baseline validation and exemplary applications using a commercial software package. Results are compared against companion (unsteady) RANS simulations next to experimental and numerical data taken from literature. Findings convey that DES is able to accurately predict transient phenomena in massively separated turbulent flows. Moreover, the approach clearly outperforms unsteady RANS simulations at comparable computational expenses. Although the DES method is significantly more efficient than LES and yields noticeable potential for industrial CFD, the computational effort of the inherently 3D and unsteady approach is still considerable.

3.1 Objectives and Motivation

The effort is motivated by the inadequacy of conventional Reynolds-averaged (RANS) approaches for the simulation of flows governed by transient, large-scale turbulent eddies and the prohibitive computational expenses still associated with physically more appropriate Large Eddy Simulation (LES) techniques for wall-bounded flows. DES aims to combine the benefits of the two methodologies at moderate computational efforts and thus promises to offer a viable, application-orientated compromise. Investigations reported in this chapter refer to fundamental calibration and verification and baseline validations. Moreover, two generic industrial applications, featuring a generic ICE2 train exposed to 30o cross-wind and the flow over a generic car mirror are computed. Results of these applications are reported in chapters IV.8 and IV.9, respectively.

3.2 Computational Framework

The numerical framework of the study is based on the commercial flow-simulation package STAR-CD V3.15. The algorithm is based on the strong conservation form of the finite-volume methodology within general coordinates and employs a fully co-located storage arrangement for all transport properties. The accuracy of the employed procedure is first-order implicit in time. Diffusion terms are approximated by means of 2nd order accurate central differences, whereas advective fluxes are approximated by either conventional 2nd order central differences blended with 5% first order upwind differences (DES) or 3rd order bounded upwind-biased schemes (U/RANS). The odd-even decoupling problem of the cell centred formulation is suppressed by a fourth-order artificial dissipation pressure term. The solution is iterated to convergence using a pressure-correction scheme. The procedure is parallelised by means of a domain decomposition technique.

All investigations are confined to two linear background eddy-viscosity models, i.e. the Spalart-Allmaras one-equation model and the k-ε two-equation model. Both models are operated in their low-*Re* mode and DES variants refer to the standard formulations as described in chapter III.2. Due to the prohibitive computational expenses associated with pure low-*Re* practices, hybrid law-of-the wall boundary conditions which are compatible to both low-*Re* and high-*Re* situations and allow for a local grid refinement at selected faces have been utilized in line with the descriptions found in chapter III.3.

3.3 Verification and Validation

Figure 1 illustrates the considered validation cases computed by BT in the framework of FLOMANIA. The focal point is the predicted decay of homogeneous isotropic turbulence (DHIT). Supplementary, an assessment of a low-*Re* step flow, the computed mean forces for a NACA0012 low-aspect ratio wing beyond stall were used to validate the computational model.

Figure 1 Verification and validation cases computed under the aegis of the FLOMANIA project (DHIT – left; backward facing step at $Re = 5000$ – middle; and NACA0012 wing beyond stall at $Re = 10^5$ – right)

3.3.1 Decay of Homogeneous Isotropic Turbulence

Similar to the LES approach, the DES methodology requires the calibration of a subgrid scale 'Smagorinsky' constant, viz. $C_{DES.}$ The assignment of the latter has to be performed in accord with the numerical properties of the utilised flow solver. In an analogy to the LES approach, it is calibrated by the decay of homogeneous, isotropic box turbulence in comparison to experimental data reported by Comte-Bellot and Corrsin (1971) and literature reported results (Shur, 1999) for the same case. Emphasis is given to the comparison of the turbulent kinetic energy spectra after a short period of decay. In this context, the DES approach is exercised in pure LES mode, targeting to show that the modified turbulence closure provides a fair predictive accuracy in the inertial subrange. The DHIT has been simulated with three uniform grids (16^3, 32^3 and 64^3 CV's) representing a cube with the size

$2\pi \times 2\pi \times 2\pi$. Periodic boundary condition is imposed in all three directions. The time integration is first-order implicit with $CFL = 2 \cdot 10^{-2}$.

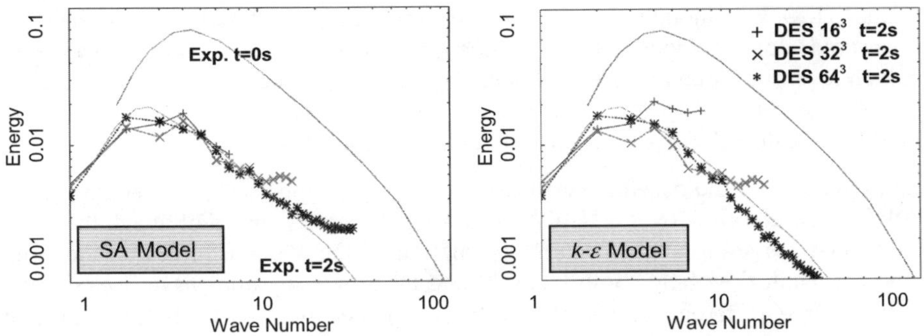

Figure 2 Energy spectra of the DHIT obtained with the SA turbulence model (left, C_{DES} = 0.65) and the k-ε turbulence model (C_{DES} = 0.73).

Figure 2 displays the resolved turbulent kinetic energy spectra as a function of the wave number returned in conjunction with the SA turbulence model (left) and the k-ε (right) model. An improvement of the resolved range of frequencies due to grid refinement is clearly achieved by the series of consecutively refined meshes. Note that the results are obtained with a CFL-number that is roughly one order of magnitude lower than the one mentioned by Shur et al. (1999) due to the employed low-order temporal discretisation scheme. Nonetheless, the assignment of the C_{DES} constant to $C_{DES} = 0.65$ corresponds to the value suggested by reference to higher-order schemes in the literature (Shur et al. 1999).

The spectral slope in the inertial subrange is favourably captured by the SA-SGS models in the low-wave number region. However, the numerical solution fails to mimic the energy distribution correctly at the respective upper ends of the resolved frequencies. This is in accord with the findings of other partners of the FLOMANIA project. The reason for this unfavourable discrepancy is identified to be the local low-Re functions of the SA model. When the grid refinement exceeds a threshold level, the SA model interprets it as a near-wall region and accordingly initialises a fast non-linear drop of the turbulent viscosity. Strelets et al. (2004) offer a modification for the subgrid SA model that is consistent with both RANS and DES methodology. A small modification of the employed DES constant CDES has been imposed on the k-ε model for a similar reason. The employed constant CDES = 0.73 is slightly increased when compared to the literature value of CDES = 0.61 (Travin et al. 2002). The rationale hereof is provided by the fact that the employed k-ε closure model inheres a non-local low-Re modification ($f_{\varepsilon 2}$) which effectively gives rise to a smaller destruction coefficient ($c_{\varepsilon 2}\, f_{\varepsilon 2} \approx 1.55$) close to the cut-off. Accordingly, the relation

$$C_{DES} = C_{smagorinsky} \sqrt{\left(\frac{c_{\varepsilon 1}}{f_{\varepsilon 2} c_{\varepsilon 2} c_{\mu}}\right)^{1.5}} \quad \text{with } C_{smagorinsky} \approx 0.125$$

returns a roughly 20% higher value for C_{DES}. The predicted energy distribution obtained with the finest grid presents a tendency much more alike to the experiments at higher wave numbers, justifying the augmentation of the C_{DES} value. However, though the predicted energy decay is slightly higher than the measured slope, it does denote a slope of -5/3, unlike the measured spectrum which reveals a slope in the inertial subrange of -5/4 for the considered time step.

3.3.2 NACA0012 1:1 Aspect Ratio Wing Beyond Stall

Subsequently to the baseline verification of the DES approach by reference to the DHIT case, the flow over a NACA0012 wing at various angles of attack has been simulated, targeting at the accurate prediction of the average forces on the wing and the mean flow field. Both URANS and DES investigations have been carried out at three different angles of attack (AoA), i.e. at $\alpha = 12°$ (before stall), at $\alpha = 45°$ ($C_D \approx C_L$) and at $\alpha = 60°$ (beyond stall). A considerable amount of force data is available from the literature at higher Reynolds-numbers $Re > 10^6$ (Hoerner, 1958). In the present study, the investigated *Re*-number is $Re_c = 10^5$. Due to the small *Re*-number sensitivity, the discrepancy of the *Re*-number is of minor importance when attention is restricted to the post-stall regime. Results were assessed by means of the predicted time-mean values of lift and drag forces in comparison to literature reported values published by Shur et al (1999) for the same mesh.

The computational domain spans roughly 30 chord lengths in radial direction. The spanwise extension is equal to the chord length *c,* the non-dimensional distance to the wall of the first cell centre is between $0.1 < Y^+ < 1$. The airfoil is being surrounded by 140 cells in O-grid arrangement while the spanwise and normal direction are resolved by 24 cells respectively. The mesh features about 200 000 grid points in total and is displayed in Figure 1. Periodic boundary conditions are applied in the homogeneous (spanwise) direction. The far-field boundary of the domain is split into two almost identical sections under a certain angle depending on the angle of attack. The upstream section is allocated as the inflow of the domain. The downstream half of the ring is allocated as constant pressure boundary condition in conjunction with zero gradients for the turbulence scalars. The flow was considered to be incompressible.

The non-dimensional time step is prescribed as $\Delta t = 0.0025 \times c/U$ ($CFL_{max} \sim 0.125$). DES investigations have been carried out with both the hybrid (3rd-order upwind biased TVD vs. 2nd-order CDS) differencing scheme, a central differencing scheme blended with 5% 1st order UDS and a pure central differencing scheme (CDS) for the approximation of advective fluxes respectively, whereas the CDS was imposed for the URANS simulation. Reported results are restricted to the SA background closure in favour of a compact presentation. Analogues findings are returned by the k-ε closure. Figure 3 depicts the predicted time-mean values for the lift and drag coefficients in comparison to experimental investigations for the three investigated AoA. Results confirm the findings of Shur et al. (1999) which clearly reflect the superiority of DES for flows featuring massive open vortical separation.

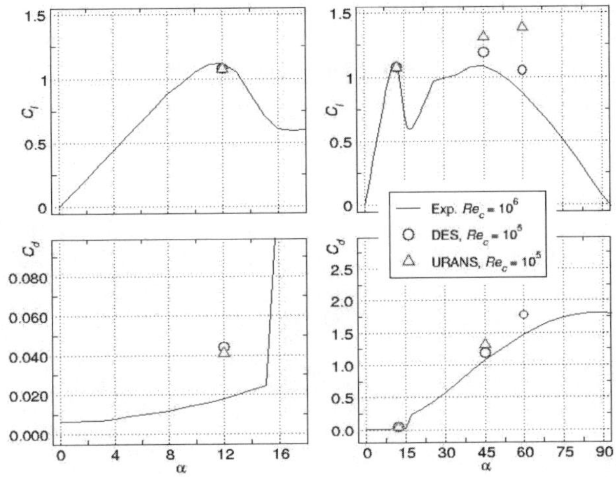

Figure 3 Results for predicted lift and drag coefficients of the NACA0012 1:1 aspect ratio wing for various angles of attack (SA model).

It should be noted that URANS and DES are in almost perfect agreement for the non-stalling case. As indicated, the predicted values returned by the DES model are in close proximity to the results of the experiments. Figure 4 reveals the origin of the predictive disparity outlined in table 1.

Table 1: Comparison of measured and predicted mean-force coefficients (Literature values taken from Shur et al., 1999)

AOA α	Parameter	URANS		DES		Exp.
		BT	Literature	BT	Literature	
α = 12°	C_{lift}	1.08	-	1.08	1.06	1.20
	C_{drag}	0.04	-	0.04	0.04	0.02
α = 45°	C_{lift}	1.31	1.30	1.1	1.20	1.15
	C_{drag}	1.31	1.30	1.1	1.20	1.15
α = 60°	C_{lift}	1.38	1.30	0.99	1.00	0.91
	C_{drag}	2.34	2.25	1.65	1.70	1.65

Distinct from the DES results – which predict a completely stalled airfoil – URANS simulations display a significant suction peak developing downstream of mid-chord. The present pressure distribution is in fair agreement with result reported by Shur et al. (1999) employing the same methodology and mesh. The right part of Figure 4 presents a snapshot of the predicted streamwise vorticity contours, which indicates that the spanwise extension (assigned periodicity) is sufficient due to the completeness of at least a few visible structures for the DES.

Although the URANS simulations have been performed with the same numerical framework and mesh, Figure 4 conveys no indication for any 3D-motions. Figure 5 depicts the contours of the instantaneous ω_z-vorticity contours for the DES solution, both using strictly 2nd-order CDS for the advective terms and flux blending (95% CDS + 5% 1st order UDS) in the mid-span plane at $\alpha = 45°$ flow incidence.

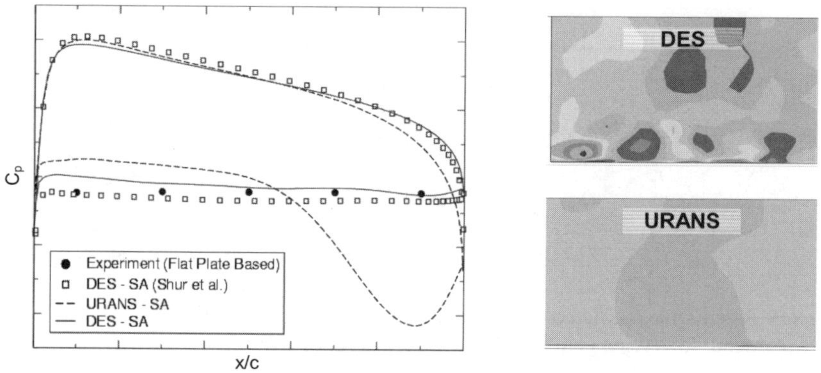

Figure 4 Distribution of spatially averaged mean pressure coefficient (left) and predicted instantaneous vorticity contours ω_x at mid chord ($x/c = 0.5$) (right). Results refer to the SA model.

Evidently, the visible structures on the pressure side of the airfoil for the DES are a clear indication of numerical wiggles. The latter can be attributed to the impact of the employed CDS scheme in the RANS regime. The unfavourable structures arisen in the DES solution with pure CDS are eliminated when introducing a small amount of upwind differencing.

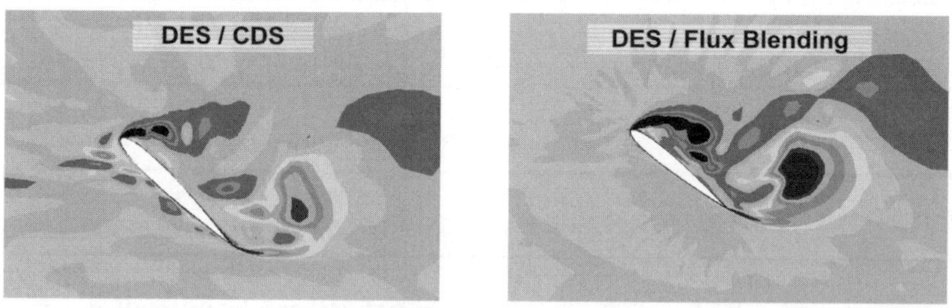

Figure 5 Distribution of the predicted instantaneous vorticity contours ω_z at mid span ($z/c = 0.5$) in response to the applied approximation of advective fluxes. Results refer to the SA model.

4 Contribution of Chalmers: LES and Hybrid LES-RANS

Siniša Krajnović and Lars Davidson,
Chalmers University of Technology

4.1 Introduction

Chalmers has worked on two main subjects: LES to simulate the flow around an Ahmed body with a 25 degrees angle of the rear slanted surface and development of a new hybrid LES-RANS method.

Although a large number of RANS simulations (using different turbulence models) and one LES simulation have been presented in the literature on the Ahmed body (25 degrees), they were all unsuccessful. The aim of Chalmers LES was to investigate if the LES can predict this flow and to explain the reasons for failure of previous simulations. If successful, the results of our simulations could be used for improvement of RANS turbulence models.

In hybrid LES-RANS, unsteady RANS (URANS) is used near walls and LES is used away from walls. Chalmers has developed a method for improving standard LES-RANS. The improvement consists of adding instantaneous turbulent fluctuations (forcing conditions) at the matching plane between the LES and URANS regions in order to trigger the equations to resolve turbulence.

4.2. Large eddy simulations of the flow around the Ahmed body

As the separation of the flow at the rear is determined by the sharp edges, the Reynolds number effects are minimized. Thus the Reynolds number of 768 000 (based on the height of the body and the velocity at the inlet) used in the experiments could be reduced to 200 000 in our LES. The combination of this lower Reynolds number and a clever grid arrangement enabled resolution of the important near wall coherent structures. Numerical accuracy was established by making three LES on different computational grids containing 3.5, 9.6 and 16.5 nodes. The influence of the small unresolved scales to the resolved ones was modelled with the standard Smagorinsky model and the value of the Smagorinsky constant of 0.1. The experimental set-up included a turbulent trip wire at the front of the model. The exact position and shape of the trip wire was not known from the experiment and it was not included or modelled in our numerical model.

4.2 Results

4.3.2 Comparison with the experimental data

An extensive comparison of our LES results and the experimental data was made and very good agreement was found for both velocities and Reynolds stresses. Only small discrepancies were found between our LES results and experimental data, and some of these are believed to be problems of the experiments. From the agreement of our LES and the experimental results we conclude that the flow around an Ahmed body (end especially the flow after its sharp edge separations) can be simulated using LES at lower Reynolds number. The main reason for the relative insensitivity of the flow around and behind the rear of the body to the Reynolds number, are the sharp edge separations in combination with the tripping of the turbulent boundary layer in the experiments.

4.3.3 Understanding the flow

Now when it was shown that our LES predicted flow in agreement with experimental observations, the results from LES were used to study both the instantaneous and time-averaged flows. It was found that the instantaneous flow contains large variety of the time and the length scales. This is the chief reason for the failure of the steady RANS simulations in which the turbulence model is unable to deal with such a variety of scales. Among the instantaneous coherent structures found in the flow two structures distinct themselves, cone-like vortices along the rear slanted edges and the hairpin-like vortices in the region of separated flows on the front and the rear slanted surface of the body. While the cone-like vortices are of the size of the body and are relatively steady, the hairpin-like vortices are small structures which change their shape during short period of time.

One problem with previous RANS simulations was their inability to mimic the unsteadiness on the rear slanted surface resulting in high Reynolds stresses in this flow region. Spohn and Gillieron (2002) suggested that the unsteadiness in this flow region are due to oscillations of the trailing cone-like vortices and their interactions with vortices shed from the front. We found in our LES the trailing vortices to be relatively steady with their cores at approximately same position during a long time. It is more likely that the interaction between hairpin-like vortices (found in our LES) is the reason for the high level unsteadiness at the lower part of the slanted surface. As the hairpin-like vortices traveling in the streamwise direction approach the region of the re-attachment, they break up and meet the hairpin-like vortices coming from the upstream corners. These meetings result in large number of collisions which together with the breakup of the hairpin-like vortices upstream of the region of the collisions make this region of the flow very unsteady. Thus the unsteady character of the flow in the lower part of the slant is an indirect consequence of the trailing vortices on the hairpin-like vortices, making them change their path and interact with hairpin-like vortices close to the symmetry plane of the body.

Flow around the front of the body was for the first time presented showing coherent structures on all four sides in both instantaneous and time-averaged flows. Another finding from our LES is that the flow on the rear slanted surface contains three pairs of cone-like vortices compared to only two pairs known from previous experimental studies.

The picture of the time-averaged flow was also corrected. The inaccurate position of two horseshoe vortices assumed in the experimental studies to be located one above another in the wake with their legs stretching in the streamwise direction was corrected here. Our LES results showed that the legs of the lower horseshoe vortex extend from the floor to the ceilings.

This is only a brief selection of the results from our LES. More extensive presentation of the results can be found in (Krajnović, Davidson 2005a, b, c).

4.4 Hybrid LES-RANS

The object of hybrid LES-RANS is to eliminate the requirement of high near-wall resolution in wall-parallel planes. In the near-wall region (the URANS region), a low-Re number RANS turbulence model (usually an eddy-viscosity model) is used. In the outer region (the LES region), the usual LES is used. The idea is that the effect of the near-wall turbulent structures should be modelled by the RANS turbulence model rather than being resolved. In the LES region, coarser grid spacing in wall-parallel planes can be used. The unsteady momentum equations are solved throughout the computational domain. The turbulent RANS viscosity is used in the URANS region, and the turbulent SGS viscosity is used in the LES region.

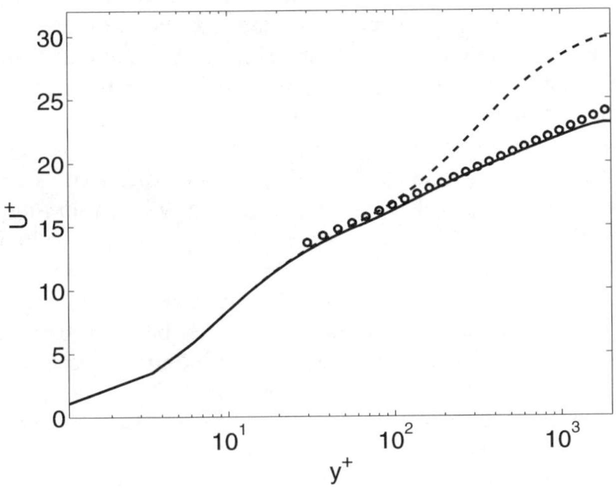

Figure 1 Streamwise velocities. Solid lines: forcing; dashed lines: no forcing; o: log-law

In the new method we add fluctuations to the momentum equations at the interface. The turbulent fluctuations are taken from synthesized homogeneous turbulence assuming a modified von Karman spectrum. The object is to trigger the equations to resolve turbulence. The momentum equations are solved in the entire domain and the turbulent viscosity is in both regions obtained from a one-equations k_{sgs} equation and an algebraic length scale. Interesting, and rather similar, approach was recently presented by Piomelli (2003) and Batten et al (2004). In the former work forcing was introduced in the interface region using white-noise fluctuations. In the latter work synthetic turbulent fluctuations were used to trigger the resolved turbulence when going from an URANS region to an LES region.

Adding fluctuations in order to trigger the equations to resolve turbulence is actually very similar to prescribing fluctuating turbulent inlet boundary conditions for DNS or LES (or hybrid LES-RANS). If no triggering inlet boundary conditions are prescribed in DNS or LES, the resolved turbulence near the inlet will be too small and a large streamwise distance is required before the equations trigger themselves into describing turbulent flow. This is also the case in hybrid LES-RANS: if no triggering (forcing) is applied at the interface between the LES region and the URANS region, the resolved turbulence in the LES region near the URANS region will be too small.

4.5 Results

A coarse mesh was purposely chosen. The mesh has 32 cells in both the streamwise (x) and the spanwise (z) direction. The size of the computational domain is $x_{max}=4\pi$, $y_{max}=2$ (geometric stretching of 17%) and $z_{max}=2\pi$. This gives a Δx^+ and Δz^+ of approximately 785 and 393, respectively. The location of the matching plane is at y=0.031 (lower wall), which corresponds to $y^+=62$ and 11 cells in the URANS region at each wall. Care has been taken to ensure that all results presented are independent of the size of the computational domain. It should be mentioned that on this mesh a pure LES is not able to sustain any resolved turbulence at all, but a steady solution is obtained.

Fully developed channel flow at $Re_\tau=u_\tau\, \delta/\nu=2000$ (δ denotes the channel half width) is used as a test case to evaluate the effect of different forcing conditions. The streamwise velocity profiles obtained with and without forcing are compared in Fig. 1 with the log-law. It can be seen that the centerline velocity is strongly over-predicted when no forcing is used, whereas forcing gives excellent agreement with the log-law. The reason for the overly large velocities without forcing is that the resolved shear is too small. It can be seen in Fig 2 that it is the resolved shear stress that increases when forcing is introduced, indicating that the resolved shear stress without forcing is too small. This was also observed by Piomelli et al (2003): when forcing is introduced, the resolved shear stress increases, which reduces the bulk and centerline velocity.

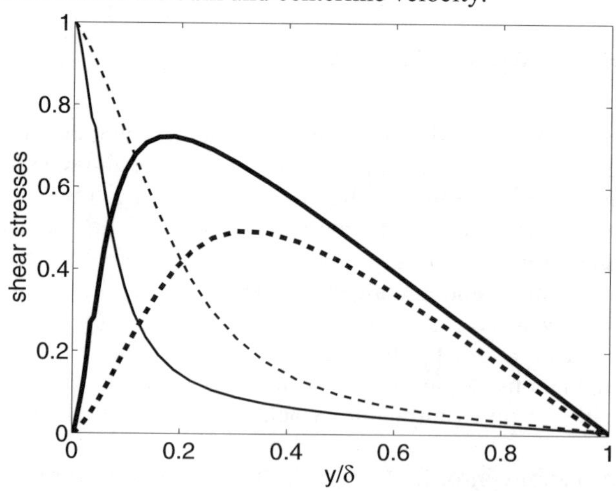

Figure 2 Shear stresses. Solid lines: forcing; dashed lines: no forcing. Thick lines: resolved; thin lines: modelled.

5 Dassault Aviation's main achievements within Flomania

F. Chalot, A. Davroux, M. Mallet, M. Ravachol

Abstract

Since CFD is crucial for an accurate aerodynamic design of aircraft, Dassault Aviation was involved in many of the major European projects dedicated to the improvements of CFD technologies, such as EUROVAL, ECARP, AVTAC, LESFOIL, UNSI, IDeMAS. The ambitious goals of Flomania, together with the expertise of the partners involved, naturally motivated us to take an active part into the project.

Developments performed within this project aimed at the enhancement of the turbulence models currently used They concern the improvement of wall functions, and the evaluation of two turbulence models: the v^2-f and an improved version of the k-l. Computations were also performed in order to evaluate the new developed models on an industrial configuration (3D transonic wing and walls of the wind tunnel), and to provide a basis for detailed comparisons between the most advanced two equations models (EARSM) and RSM. In this respect, two test-cases were computed: a 3D transonic wing (M6 wing) and a delta wing at two different angles of attack.

5.1 Dassault Aviation's in-house Navier-Stokes solver Aether

The CFD code used at Dassault Aviation is an in-house code called Aether, which solves the compressible Navier-Stokes equations. The equations are symmetrized using the entropy variables introduced by Hughes et al.. The numerical method is based on a space time Galerkin/Least Squares finite element formulation. As a result, the code can handle a mixture of several types of elements (triangles, and quadrilaterals in 2D, tetrahedra, bricks and prisms in 3D), though most of our meshes are based on triangles or tetrahedra. The mean flow solver and the turbulence solver are segregated, resulting in an easier and more practical management of the turbulence models implemented, and allowing a specific numerical treatment of the turbulence equations. Lastly, the code is fully vectorized and parallelized, and has been successfully ported on many different computer architectures. For more details, one can refer to (Chalot et al., 1994), (Chalot et al., 1997) or to (Stein et al., 2004).

5.2 Towards enhanced wall functions

Though computations of full aircraft configurations with low-Reynolds number models now run everyday, there is still a need for industry to have the capability to rely on wall functions, mainly in order to save computational time, e.g. in order to be able to rely even more on Navier-Stokes computations during the conception process, or to fasten computationally expensive unsteady simulations.

Two main flaws of the wall functions were addressed: the first one examines their dependence on the size of the first mesh cell, and the second one attempts to widen their field of application beyond the attached boundary layer hypothesis which no longer holds for real life industrial applications. Two different kinds of wall functions were considered, which are described in details in Section III. The two approaches mainly differ in the location of the wall: in one approach, the wall lies at the frontier of the mesh (Rubesin et al.) whereas a small distance to the wall is introduced in the other approach (Grotjans et al). Both approaches have been implemented and thoroughly tested, and the first one was finally selected (i.e. the wall lies at the frontier of the mesh). In addition, the numerical treatment of the wall functions was also improved, by enforcing the fluxes instead of the value at the wall in the finite element code.

Application of these wall functions on test cases of increasing complexity shows that the new developed wall functions are less sensitive than those previously used, and are able to satisfactory compute complex 3D boundary layer interaction.

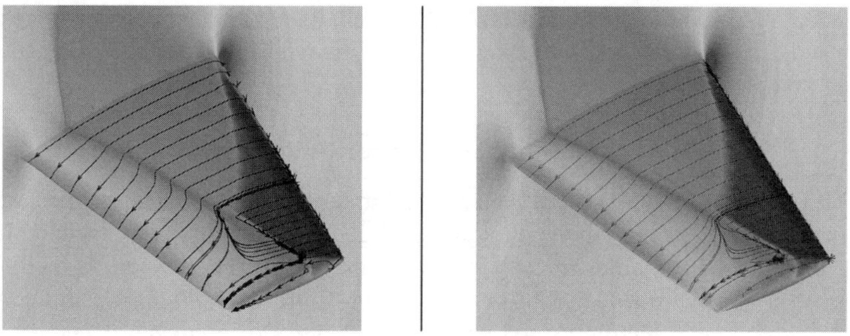

Figure 1 M6 wing skin friction lines - left: low Reynolds k-ε computation right: Wall functions computations - Results are almost identical

The new numerical treatment of the wall functions resulted in an unexpected improvement of the results on the TU Munich delta wing test case. The explanation lies in the way the leading edge singularity is computed. As a result, better predictions of the vortex location and of the pressure level on the suction side were obtained.

5.2.1 Driving two-equation models to their limit

During the Flomania project, two different turbulence models were developed: the v^2-f model, and a non-equilibrium version of the k-l proposed by Smith.

The v2f model

The v^2-f model is described in Section III. It has been shown to have an added value compared to more conventional k-ε models in many applications, and in particular in aerospace configurations. Among the different versions of this model, the ones proposed by Lien and Durbin, and by Laurence et al were implemented, the latter resulting in a friendlier near-wall behavior in $O(y^2)$ instead of $O(y^4)$. Though it was successfully applied to simple test cases, an issue was found that concern the behavior of the Phi variable at the edge of the boundary layer (namely a discontinuity of ϕ). This issue was checked with the active collaboration of others partners involved in this development, and will need further in-depth research

The non equilibrium k-l

The non-equilibrium correction proposed by Smith for his k-l model aims at improving the behavior of the model in regions where separation occurs. The principle is very similar to Menter's SST correction, which limits the turbulent viscosity when production is greater than dissipation. This kind of correction is already used at Dassault Aviation in the k-ε model, and was shown to be efficient, especially for transonic flows. Within Flomania, this non-equilibrium correction of the k-l model was implemented in our code and validated. Improvements of the results obtained on an industrial configuration proved the significant added value of this development, and motivated its integration in our design tools.

5.4 Performing EARSM computations for comparison with RSM

In order to support partners who implemented a full Reynolds Stress Model and to establish a base of computations, with which RSM results can be compared, Dassault Aviation performed computations with its two layers k-ε and with its EARSM WJ. Two test-cases were under consideration: the delta wing (chapter IV-11) and the M6 wing (chapter IV-04). For the latter, it was shown that using RSM modeling results in a better resolution of the wing-tip vortices. The former test case was chosen on purpose, since the two vortices emanating from the sharp leading edges are turbulent structures where the classical Boussinesq's assumption is expected to fail. Consequently, it was shown that the EARSM results are better than the k-ε's ones, especially on the more challenging 35° case.

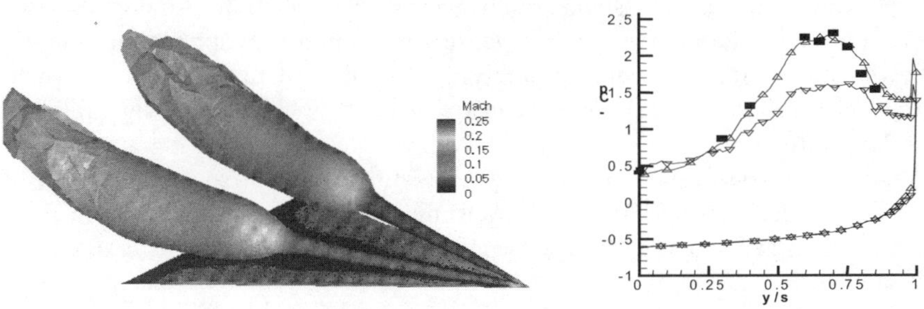

Figure 2 TU Munich delta wing at 35° aoa (Test Case n°20); improvement of EARSM (Δ)over to the k-ε (∇)

5.5 Evaluations on real-life application

In order to apply and to evaluate the developments performed within the Flomania project, a real-life, industrial application was chosen: the OAT15A (chapter IV-05). The purpose is to compute the wind tunnel and the transonic wing as a whole. This configuration was specifically chosen to highlight the benefit of the non-equilibrium correction. In fact, the shock-induced boundary layer separation downstream of the shock is exactly the kind of configuration in which the SST modification is expected to enhance the predictions. As a result, the new version of the k-l improves the prediction of the position of the shock, thick line in Fig. 3 right, with a very good agreement with the experiments whereas the shock was 3% too far downstream with the previous version of the k-l.

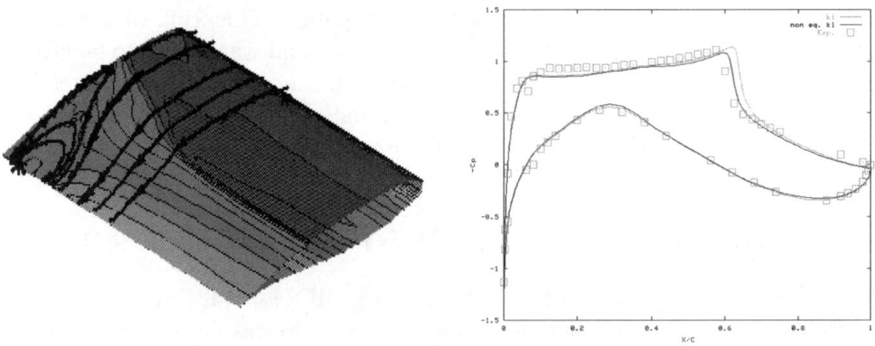

Figure 3 Improvement of the prediction of the shock location (3% of the chord) with the non equilibrium k-l

5.6 Conclusions and perspectives

The developments performed within the Flomania project are very valuable for Dassault Aviation. The new wall function behavior was shown to be satisfactory, less dependent on the size of the first mesh cell and thus more applicable to complex flows such as 3D shock boundary layer interaction. Among the two models under consideration, one is not yet ready for industrial application (the v2-f); the other one (the non equilibrium version of the k-l model) was shown to perform very well on several complex test cases and is about to be integrated in our design tools.

Computations performed provided a basis for a comparison with the RSM results from others partners on a 3D transonic wing (M6 wing) and on a delta wing. An additional outcome was the enhancement provided by the EARSM on the delta wing.

Last but not least, preliminary work on DES performed by a few "pioneer" partners provides very valuable guidelines as well as the dedicated workshop in Saint Petersburg. The DESider project, in which Dassault Aviation is also involved, is a continuation of the preliminary work on this promising technique.

6 DLR's Contribution to FLOMANIA – Methods, Models, Applications

B. Eisfeld, K. Weinman, and D. Schwamborn, DLR-G

Abstract

Here we briefly review the two numerical methods used by DLR during the FLOMANIA project as well as some remarks on the turbulence models available in these codes including some remarks on the implementation of DES and RSM. Furthermore some additional results obtained with the latter models are presented which are not mentioned elsewhere in this book.

6.1 Numerical methods

The DLR has used two different flow solvers for work performed within the scope of FLOMANIA. The block-structured FLOWer code has been used as the basis for Reynolds Stress Model implementation whilst the hybrid unstructured TAU code has been used as a vehicle for the introduction of Detached Eddy Simulation methods into the DLR. Both codes are briefly described in the following sections.

6.1.1 The DLR TAU code.

The DLR TAU code is an unstructured finite volume code which solves the compressible Navier-Stokes equations (Gerhold, 1999; Schwamborn, 1999). The code is capable of solving both two and three-dimensional problems. Code development has concentrated on industrially relevant problems, principally in the field of external aerodynamics. The code consists of several modules including a pre-processor module, a solver module, an adaptation module, and various additional post-processing utility tools.

The TAU code uses an edge-based dual-cell approach, i.e. a vertex-centred scheme, where inviscid terms are computed using a second-order central scheme, or a variety of upwind schemes which may be extended to second-order (in space). Viscous terms are computed with a second-order central scheme. Scalar or matrix artificial dissipation may be chosen by the user. Low Mach number preconditioning has also been implemented so that the code can be extended into the incompressible flow range.

Time integration is performed either explicitly using various Runge-Kutta schemes, or by using an implicit Krylov subspace method or an implicit LU-SGS method which is only marginally more expensive in memory than the explicit Runge-Kutta schemes. Convergence acceleration is achieved by a multi-grid algorithm. For time accurate computations the dual time stepping approach of the Jameson is employed. Additional options within the pre-processor module allow treatment of meshes with hanging nodes, and deforming or overlapping grids (Chimera technique). The code has been optimised for both vector and parallel environments on distributed memory architectures.

6.1.2 The DLR FLOWer code

The DLR FLOWer code is designed for solving the unsteady three-dimensional compressible Reynolds-averaged Navier-Stokes (URANS) equations on block-structured grids. It has been particularly developed for application to industrially relevant problems of external aerodynamics, mainly in the field of aeronautics. Its numerical method is based on a finite volume formulation, using either a cell centered or a cell vertex approach. The convective terms of the RANS equations are discretised either by a central scheme of second order accuracy with scalar or matrix artificial dissipation or by a variety of upwind schemes. The diffusive terms are discretised by a central scheme of second order accuracy. Integration in time is carried out by a hybrid explicit Runge-Kutta scheme, where convergence is accelerated by local time-stepping, implicit residual smoothing and a multigrid algorithm. For low Mach number flows a preconditioning technique is available.

The convective terms of the turbulence transport equations are discretised by a first order upwind scheme, while the diffusive terms are discretised by a second order central scheme. Integration in time is carried out by a special implicit scheme on the finest grid level only, where the time step is adapted to the multigrid algorithm for the RANS equations.

For time accurate computations a dual time stepping approach is available, where all the above techniques are applied in pseudo time, ensuring a high performance. Special options allow the treatment of meshes with hanging nodes at block boundaries, deforming or overlapping grids (Chimera technique). For rotor or propeller flows a rotating frame of reference is available, allowing efficient simulations. The code is particularly optimised for vector machines; parallel computations are based on MPI, exploiting the block structure of the grid.

6.2 Turbulence modelling in TAU and FLOWer

6.2.1 Turbulence models in TAU

The turbulence models implemented within the TAU code include both linear and non-linear eddy viscosity models spanning both one- and two-equation model families. The Spalart-Allmaras turbulence model forms the basis for the one-equation model implementations whilst the Wilcox k-ω model provides the basis for TAU two equation model development.

Within the FLOMANIA project, the DES method has been used to extend the Spalart-Allmaras turbulence model, as well as the Menter-SST and TNT k-ω turbulence models. Since the DES method is a development strategy that can be applied, in principal, to any eddy viscosity turbulence model, the original implementation within the TAU code required only the calculation of additional terms and a suitable switch to activate the DES model within the turbulence model source term calculation. Compared to the higher-order (in space) structured schemes utilized by some partners, the level of numerical dissipation is somewhat higher in a nominally second order unstructured solver which can be critical for DES. In computational terms, the overhead for the solution of a time step using

Detached Eddy Simulation is negligible. However, the true cost arises from the need to sufficiently resolve the temporal scales such that unsteadiness in a solution can grow in a physical way.
The computational cost of Detached Eddy Simulation is bounded by URANS from below, and by LES from above, and this statement is somewhat descriptive of the DES method itself. While the method may be trivial to implement, there are significant concerns that arise in the nature and implementation of other aspects of the numerical method which directly influence the performance of the DES. If the algorithms implemented in the base numerical methods are not matched to the requirements for the LES branch of the DES, then the DES implementation will not offer any significant more functionality than a URANS implementation. It is in this work, extending the performance of existing algorithms so that LES performance is optimised, where the majority of effort in a DES implementation must be expended.

6.2.2 Turbulence models in FLOWer

The DLR FLOWer code provides a variety of turbulence models, where the majority of models is based on the eddy viscosity concept.

Besides the algebraic Baldwin-Lomax model for providing rapid solutions, a couple of Spalart-Allmaras type one-equation turbulence models are available in FLOWer. Nevertheless in the past the applications with FLOWer have been dominated by two-equation models based on k and ω, where the Wilcox k-ω model serves as numerically robust baseline model. For transonic flows improvements have been observed with respect to the prediction of shock positions, when using the linear truncation of an EARSM, called the LEA k-ω model. The prediction of separated flows is improved by the Menter SST k-ω model. Furthermore the Wallin-Johansson EARSM based on the Wilcox ω-equation is available in FLOWer.

Within FLOMANIA two different Reynolds stress models have been implemented into FLOWer. The Wilcox stress-ω model has been selected for a baseline implementation, since it is based on ω as length scale variable, which has been proven to be favourable for external aerodynamics in aeronautics. Note that the Wilcox stress-ω model corresponds to the LRR model without wall reflection terms. Various ω-equations have been implemented, which can be freely combined with the Reynolds stress equations of the Wilcox model.

In contrast the FLOMANIA consortium selected the SSG model as common Reynolds stress model to be implemented by all partners. However the SSG model is based on ε as length scale, which is not favoured in the context of external aerodynamics. Therefore the hybrid SSG/LRR-ω model has been developed, combining the SSG model in the far field with the LRR model without wall reflexion terms close to solid walls. The model is attached to Menter's BSL ω-equation which provides the transition between length scales ω at the wall and ε further apart. Simple as well as generalised gradient diffusion models have been implemented that can be freely combined with the Wilcox stress-ω and the SSG/LRR-ω model.

After a lengthy debugging period and after extending the implicit scheme to the Reynolds stress model equations no particular numerical problems have been encountered for any application so far. The convergence depends on the flow problem and is not necessarily degraded with respect to eddy viscosity models.

The implementation appears to be rather efficient, showing an increase in CPU time per iteration by a factor of 1.5-2 with respect to two-equation models, depending on the test case, the computer platform and the particular models compared. The memory requirement is approximately doubled with respect to two-equation models.

6.3 Application highlights for the TAU Code

The following section illustrates typical calculations that have been computed using DES implementations in TAU. In order to calibrate the DES model, a decaying isotropic turbulence problem is usually computed. The computed energy spectrum, obtained on a 32^3 mesh, is shown in Figure 1 at two instants of time.

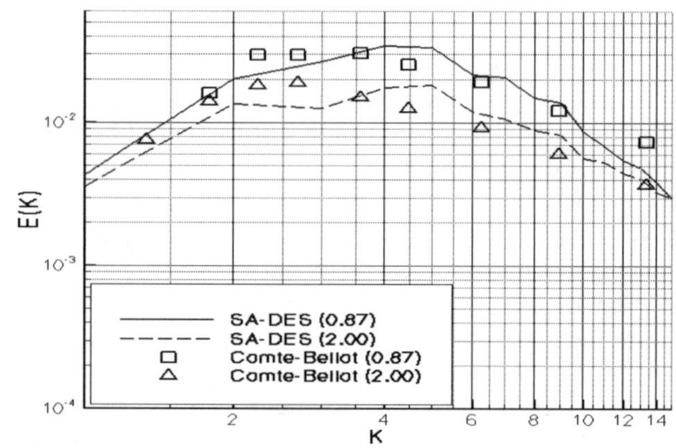

Figure 1 Comparison of computed the 3D Energy Spectrum against experiment of (Comte-Bellot, 1971) using a 32^3 mesh.

For the calculation shown, the SA-DES model has been used, with a C_{DES} constant of 0.45. Note that the calculation can only properly resolve the first 15 wave numbers for the given mesh (N=32) and agreement with the Comte-Bellot data is acceptable for wave numbers less than about twelve. However at higher wave numbers the spectra decay quickly in comparison with experiment. There is a rapid fall off in the spectral energy at the intermediate time, so that the effective viscous cut-off for the LES branch occurs at much lower wave numbers in comparison to experiment. The small scales which occur in nature are numerically damped in the calculation. In order to shift the effective cut-off wave-number as far to the right as possible, it was necessary to used matrix dissipation instead of a standard scalar dissipation algorithm for the central scheme. Thus, for a scheme of nominally second order both in space and time it appears that matrix dissipation and other methods of reducing the effective numerical diffusion are very useful.

The mandatory test case examined in the FLOMANIA project is that of flow about the NACA0012 airfoil at 60 degrees angle of attack. This test case is described in depth in Chapter V, Section 3, however some aspects of the TAU result for this case will be discussed in this section.

Figure 2a shows the estimated Power Spectral Density based on the lift coefficient from the calculations. The correct range of shedding frequencies seems to be returned, as observations of (Schewe, 1983) suggest a shedding frequency of about 16-18 Hz, however, a significant portion of energy exists at very low frequencies. The computed surface pressure distribution, shown in figure 2b, matches the data from (Abernathy, 1962) well. This suggests that the larger scale flow physics is being correctly modelled. In particular, the influence of the near field vortices on the airfoil pressure distribution is better resolved with the DES method than with URANS. The slight disturbances in the computed surface pressure distribution are an indicator of insufficient sampling time in generating the mean surface pressure data. The integral value of lift returned from this calculation is 0.92, which agrees well with published values of 0.96. However, the computed drag coefficient is low, returning a value of 1.55 in comparison to the experimental value of 1.65. Nevertheless, these estimates are significant improvements over the URANS estimations as noted in Chapter IV.

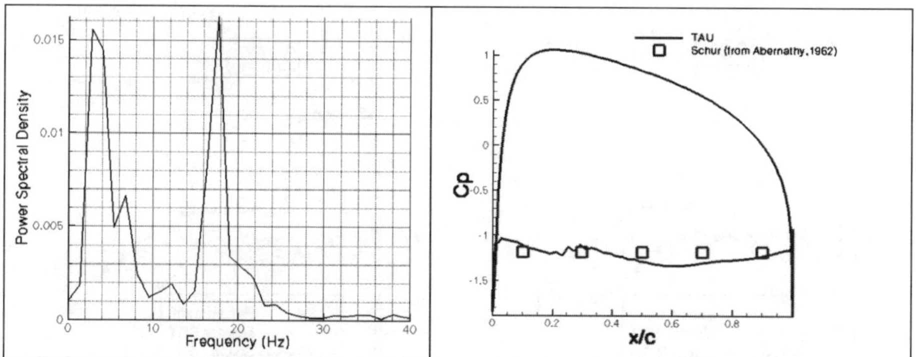

(a) Power Spectral Density based on lift

(b) Surface pressure distribution compared with experiment of (Abernathy, 1962).

Figure 2 Results obtained using the SA-DES implmentation of the TAU code

6.4 Applications highlights within FLOWer

During the FLOMANIA project the range of application of the Reynolds stress models implemented into FLOWer has been systematically investigated, starting with simple two-dimensional flows and ending with industrially relevant configurations in aeronautics. In the following some highlight applications are presented.

Generic transport aircraft AS-28

By courtesy of Airbus Germany the generic transport aircraft AS-28 has been tested, consisting of a wing-fuselage configuration with a pylon-mounted through-flow nacelle. The grid provided by Airbus contains $4.3 \cdot 10^6$ cells, distributed to 56 blocks. Flow conditions have been defined by a Mach number of $Ma = 0.80$, a Reynolds number of $\mathrm{Re} = 1.413 \cdot 10^6$ based on the mean chord and an incidence of $\alpha = 2.2^0$, representing cruise conditions. Figure 3 shows the geometry of the AS-28 configuration.

Figure 3 AS-28 configuration.

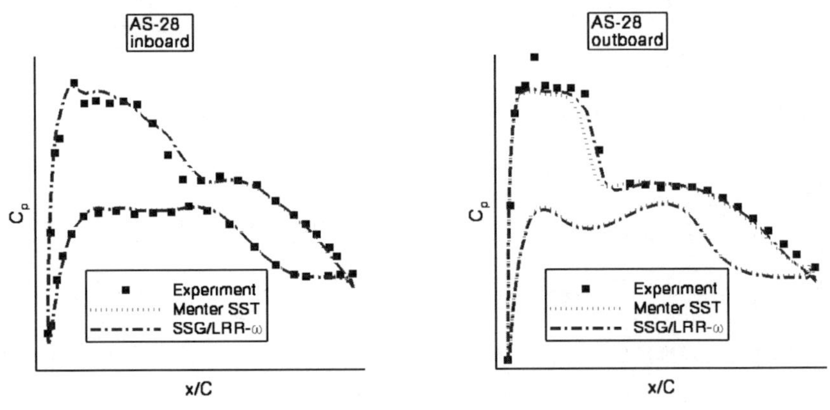

Figure 4 Inboard and outboard pressure distribution of AS-28

Figure 4 shows the inboard and outboard pressure distributions obtained with the Menter SST two-equation model and the SSG/LRR-ω RSM in comparison to the experiments provided by Airbus. As one can see close to the wing root, there is not much difference between the predictions, both models yielding the shock a bit downstream of the experimental position. In contrast towards the wing tip the SSG/LRR-ω model yields the shock position in good agreement with the measurement, while the SST model predicts the shock position upstream of the experiments. Note that this effect is observed from almost midspan to the wing tip. Thus the SSG/LRR-ω model demonstrates the improvement through Reynolds stress modelling for an industrially relevant configuration in transonic flow. Note that the reliability of the SSG/LRR-ω model for such type of configuration is confirmed by application to the DLR-F6 configuration (test case 30).

Dauphin helicopter fuselage

The Dauphin helicopter fuselage has been a complex test case in the French/ German co-operation CHANCE (ONERA, DLR, EC SAS, ECD, University of Stuttgart). Within FLOMANIA the applicability to and the potential of Reynolds stress models for such type of configurations have been investigated. The grid provided by ONERA consists of $6.8 \cdot 10^6$ cells distributed to 97 blocks. The flow conditions are defined by a Mach number $Ma = 0.0441$, a Reynolds number $\mathrm{Re} = 1.074 \cdot 10^6$ and an incidence $\alpha = -3.0^0$, I.e. subsonic flow. Thus low Mach number preconditioning has been applied in the computations. Figure 5 shows the configuration, which should be noted to be asymmetric.

Computations have been carried out with the Kok TNT k-ω model and the SSG/LRR-ω model. Figure 6 shows the horseshoe vortex developing in the region, where the engine fairing intersects with the roof of the fuselage. While with the Kok TNT model the streamlines show only a slight rotation around the bent vortex axis, the SSG/LRR-ω model predicts a much more complex vortex structure. In fact there are two separate vortices, merging rapidly, when bending around the fairing.

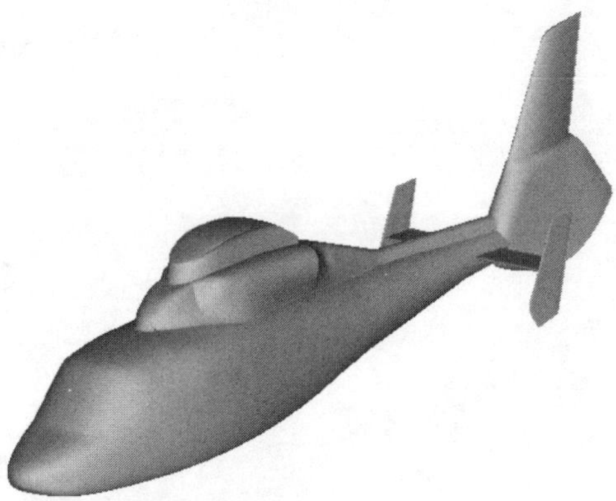

Figure 5 Dauphin helicopter fuselage.

Figure 7 shows the vortical structures behind the (closed) fenestron at the tail of the Dauphin fuselage. Although there is no comparison with experiments, it is evident, that the SSG/LRR-ω model resolves much more details than the Kok TNT k-ω model. In particular the Reynolds stress model is much less diffusive with respect to the trailing vortices.

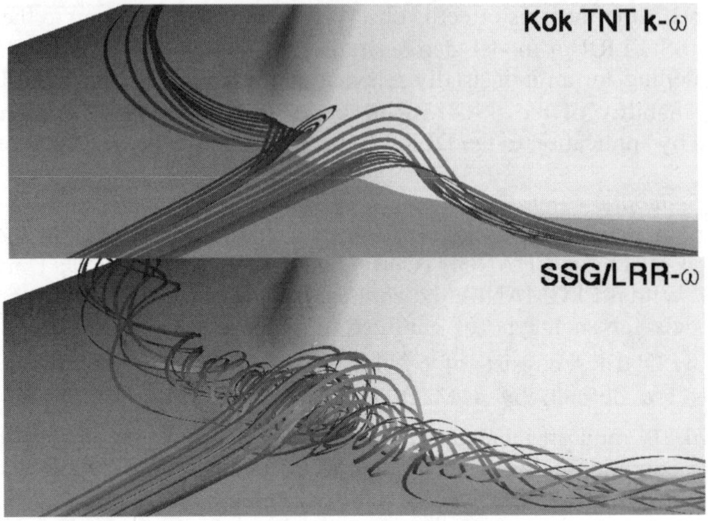

Figure 6 Horseshoe vortex forming at the intersection between the engine fairing and the fuselage roof of a Dauphin helicopter.

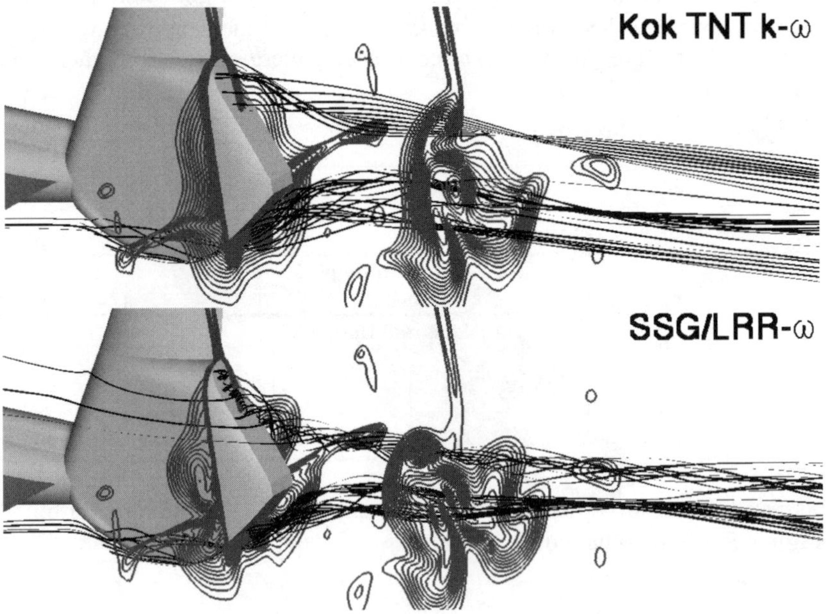

Figure 7 Vortical structures behind the (closed) fenestron of a Dauphin helicopter fuselage.

7 Contribution by EADS-CASA

J. J. Guerra, V. Ibáñez, EADS-CASA,
J. Ponsin, C. Lozano, INTA

Abstract

This chapter summarizes the activities of EADS-CASA in the FLOMANIA Project, providing some highlights about the implementations and results carried out using the in-house EMENS code. Among the computations performed, the MRTT (which is reported separately), the NACA0012 with round tip, the Obi diffuser and the RAE2822 airfoil are reported. The implementations include the SALSA TM and two types of wall functions: the Scalable WF and the Automatic WF. We are grateful to the TUB for their support with the SALSA; and we are also grateful to the INTA, which shares with EADS-CASA the EMENS code, for their collaboration in many aspects of the implementations.

7.1 Introduction

In industrial contexts it may happen, in some cases, that the use of more complex turbulence models is not worthwhile, if we take into account the added computational cost in comparison with the improvement of the results.

Generally speaking, the requirements from the aeronautical industry target in different aspects, accuracy definitely being one of them, but not the unique. For the aeronautical industry, complex flows are obviously a challenge, but most of times they occur at off-design points. This does not mean that these situations are not relevant, but it means that the analysis of the design inner loop cases demands, in addition to the accuracy, very high efficiency. And at this point another issue emerges: everything related to the meshes is, still nowadays, a bottleneck for many applications.

In this context, EADS-CASA focused, on one hand, on the implementation of the SALSA TM into the EMENS code, that has provided improved results concerning detached flow (compared with k-g), as will be shown below. And, on the other hand, on approaches to advanced wall function treatment, carrying out a couple of implementations: the Scalable Wall Functions and the Automatic Wall Functions. The advanced WF fit very well the aeronautical needs because of the wide range of Reynolds numbers that have to be computed in most configurations, allowing the same grid to be reused at different Reynolds numbers.

7.2 Implementations into the EMENS code

The EMENS code began in 1986 as an Euler code and continued being developed by implementing new capabilities every year at CASA. Viscous capabilities were implemented in 1996 and the results were validated during the AVTAC Project with the k-g TM (Kalitzin et al, 1996). EMENS is a multi-block structured grid code that uses a cell centred finite volume scheme. The multigrid scheme allows the code to decouple the blocks among them.

7.2.1 Implementation of the SALSA Turbulence Model

The SALSA turbulence model has been implemented into the EMENS during the Project, following Rung et al (2003). For this purpose, the TUB have provided support throughout the implementation and validation stages.

The SALSA (Strain-Adaptative Linear Spalart-Allmaras) model is an evolution of the Edwards variant of the Spalart-Allmaras model (1994) designed to tackle non-equilibrium situations. The main innovation put forward by the SALSA model is the modification of the parameter c_{b1} (a proportionality constant equal to 0.1355 in the original model, which appears in both the production and destruction terms of the transport equation), which becomes in the SALSA model a function of the flow variables through the strain rate S and the turbulent variable $\tilde{\nu}$:

$$c_{b1} = 0.1355\sqrt{\Gamma}, \qquad \Gamma = \min(1.25, \max(\gamma, 0.75)), \qquad \gamma = \max(\alpha_1, \alpha_2)$$

$$\alpha_1 = \left(1.01\frac{\tilde{\nu}}{\kappa^2 d^2 S^*}\right)^{0.65}, \qquad \alpha_2 = \max\left(0, 1 - \tanh\left(\frac{\rho\tilde{\nu}}{68\mu}\right)\right)^{0.65}$$

where d is the distance from the closest wall and

$$S^* = \sqrt{2S_{ij}S_{ij}}, \qquad S_{ij} = \frac{1}{2}\left(\frac{\partial u_i}{\partial x_j} + \frac{\partial u_j}{\partial x_i}\right) - \frac{1}{3}\frac{\partial u_k}{\partial x_k}\delta_{ij},$$

Additionally, and in order to obtain a more robust code a semi-implicit temporal integration scheme proposed by Eliasson and Wallin (2000) has been implemented. This scheme is devised to guarantee positivity of the turbulent variable $\tilde{\nu}$ throughout the computation. The basic scheme has been altered so as to guarantee that $\tilde{\nu}$ be always greater than the free-stream value $\tilde{\nu}_\infty$ as follows

$$\tilde{\nu}' = \frac{(\tilde{\nu} - \tilde{\nu}_\infty)\Delta\tilde{\nu}}{\tilde{\nu} - \tilde{\nu}_\infty + \Delta\tilde{\nu}}$$

7.2.2 Implementation of Scalable Wall Functions

Standard wall functions are based upon the logarithmic profile assumption. Thus, they are limited to sufficiently coarse grids for which y^+ of the first cell is greater than 30. This limitation requires as special a care in grid generation as in the case of full low-Reynolds number computations (with $y^+ \sim 1$). To overcome this constraint, Menter et al (1998) developed the so-called scalable wall functions, which allow users to apply arbitrarily fine grids (even with $y^+ \sim 1$) without running into inconsistencies.

Scalable wall functions have been implemented on the k-g model only. A limiter has been introduced so as to avoid that, for sufficiently fine grids ($y^+ < 11.06$), the next-to-wall cell switches within the computation towards the viscous sublayer (where the linear relation holds). The actual value of the limiter is given by the intersection between the logarithmic and the linear velocity profiles within the viscous sublayer ($y^+ = 11.06$). The turbulent kinetic energy is obtained by integrating the corresponding transport equation all the way to the wall but with modified source terms. The procedure is schematically as follows:

$$y^* = \frac{u^* y}{\nu}; \qquad u^* = \beta^{*1/4} k_p^{1/2}, \quad \tilde{y}^* = \max(y^*, Y^*_{\lim}), \quad u_\tau = \frac{\kappa U_p}{\log(E\tilde{y}^*)}$$

These values are used to compute the wall shear stress as $|\tau_w| = \rho u_\tau u^*$.

The source term for the k equation in the next-to-wall cells is modified:

$$\frac{\partial \rho k}{\partial t} + \frac{\partial \rho u_i k}{\partial x_i} = \tilde{P}_k - \tilde{D}_k + \frac{\partial}{\partial x_i}\left((\mu + \mu_t / \sigma_k) \frac{\partial k}{\partial x_i} \right)$$

where the modified source term is, according to Bredberg (2000), as follows:

$$\tilde{S}_k = \tilde{P}_k - \tilde{D}_k = \left(\frac{\rho u_* \kappa U_p}{Log(Ey_p^*)} \right) \frac{U_p}{Y_n} - \frac{\rho u_*^3 Log(Ey_p^*)}{\kappa Y_n}$$

where Yn is the height of the cell. To solve this equation a zero flow boundary condition on the wall ($k_w = 0$, $\left.\frac{\partial k}{\partial n}\right|_w = 0$) must be enforced on k.

The boundary condition on g is enforced as in standard wall functions. The g equation is not solved in the next-to-wall cell.

$$g = g_p = \frac{1}{\beta^{*1/4}} \sqrt{\frac{k y_p}{u_\tau}}$$

7.2.3 Implementation of Automatic Wall Functions

The idea behind automatic or hybrid wall functions (Vieser et al 2002) (Rung et al 2001) is to enforce boundary conditions on the mean flow and turbulent variables in such a way that, as the grid is refined, the original Low-Reynolds-Number (LRN) turbulence model is recovered. In this way, for grids with $y^+ \sim 30$ the law of the wall is at work, while for $y^+ \sim 1$ the LRN model is used. This allows the user to utilize, for a given problem, different grids with different sizes and near-wall resolution, such that coarse grids (with $y^+ \sim 30$) can be used in regions where the law of the wall is obeyed (attached flow, etc), while in regions where the original LRN turbulence model is required (regions with flow separation or in turbulent disequilibrium), finer grids (with $y^+ \sim 1$) can be used.

The implementation of automatic wall functions requires the use of blending functions to enforce the wall shear stress and the value of the turbulent variables in the next to wall cells. For the wall shear stress, irrespectively of the turbulent model under consideration, the following blending functions (due to Menter) are used:

$$\tau_w = \rho u_\tau u_\tau, \qquad u_\tau = \left((u_\tau^{vis})^4 + (u_\tau^{\log})^4 \right)^{1/4}, \qquad u_\tau^{vis} = \frac{U_p}{y^+} \;;\; u_\tau^{\log} = \frac{\kappa U_p}{Log(Ey^+)}$$

As for the treatment of the turbulent variables, it does depend on the particular turbulent model. For the k-g model, the turbulent kinetic energy is obtained in the next-to-wall cells by integrating the k transport equation

$$\frac{\partial \rho k}{\partial t}+\frac{\partial \rho u_i k}{\partial x_i}=(1-\phi)S_k+\phi\tilde{S}_k+\frac{\partial}{\partial x_i}\left((\mu+\mu_t/\sigma_k)\frac{\partial k}{\partial x_i}\right)$$

where S_k is the original source term of the k-g model and $\tilde{S}_k$ is a modified source term to account for the law of the wall, and ϕ is a blending function:

$$\phi=(1-e^{-0.09y^+})^2$$

designed to switch between the original transport equation when $y^+ \sim 1$and the modified transport equation when $y^+ > 30$.

As for the second turbulent variable, g, a Dirichlet boundary condition is used in the next-to-wall cell which incorporates the blending mechanism:

$$g=\left(\beta^*\omega\right)^{-1/2},\quad \omega=\left(\omega_{vis}^2(y^+)+\omega_{\log}^2(y^+)\right)^{1/2},\quad \frac{\omega_{vis}}{\rho u_\tau^2}=\frac{6}{\mu\beta}\frac{1}{\left(y^+\right)^2},\quad \frac{\omega_{\log}}{\rho u_\tau^2}=\frac{1}{\mu\left(\beta^*\right)^{1/2}\kappa}\left(\frac{1}{y^+}\right)$$

Finally, for the SALSA model the only modification is to enforce the value of $\tilde{\nu}$ in the next-to-wall cells, which takes advantage the linear behaviour of the variable near the wall, $\tilde{\nu}_p = ku_\tau y$.

7.3 Results Highlights

Among the different computations carried out during the FLOMANIA Project by EADS-CASA, the MRTT test case is reported separately. The results obtained for the Obi diffuser are reported here and not at the dedicated sub-chapter (which only deals with RSM models). Results for the NACA0012 airfoil with round tip and the RAE2822 airfoil are also shown in order to compare the results obtained using the different implementations performed.

7.3.1 Obi Diffuser

Computations have been carried out on the mandatory mesh, which has been supplemented with an entrance channel 50H long so as to guarantee a fully developed turbulent flow at the diffuser inlet. Flow conditions are Re_H = 20000 (based on the channel). A Mach number of M=0.2 (instead of M=0.05) has been used, keeping the compressibility effects at a low level at the same time that make the case more affordable in terms of convergence.

This test case is reported separately and is analysed there pursuing a homogeneous comparison of the different RSM implemented. The results presented here try to compare the performances of the k-g and SALSA models. The agreement with experimental data is not good for none of them, but the behaviour is different: while k-g could not to obtain separation at the inclined wall, the SALSA overpredicts the separation bubble. This result is in accordance with the results obtained for the MRTT test case.

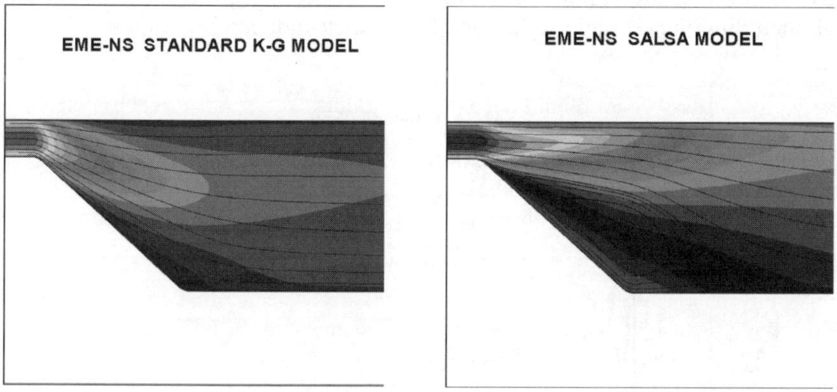

Figure 1 Streamlines and pressure maps for the Obi Diffuser

7.3.2 Naca0012 with round tip

As for the Obi diffuser, this test case is reported in a separate chapter. The results obtained by EADS-CASA are included there, the main conclusion being that the used mesh of 675000-cell (151 x 101 x 46) is not adequate to capture the wing tip vortex. Both (k-g and SALSA) models perform similarly but the differences between numerical and experimental results are very large due to the mesh resolution. Despite of that, we considered that it is worthwhile to compare the different behaviour of the k-g and SALSA throughout the test cases used.

The eddy viscosity predicted by both models does not decrease towards the centre of the core as should be expected on experimental (and physical) grounds (likely a consequence of the inability of the turbulence models used to predict the suppression of turbulence in regions where pure rotation is present). However, the SALSA model tends to predict significantly lower values for the eddy viscosity in the core region. This was also observed in the MRTT test case.

7.3.3 RAE2822

AGARD cases 9 and 10 have been chosen, as they are well-known benchmarks for testing model capabilities and performance under separation, especially the case 10. The case 10 was tested for comparison between SALSA and k-g. As for the MRTT test case, the k-g model fails to predict separation behind the shock wave, while the SALSA model does predicts detached flow, and it happens in the same way for the Obi diffuser (a sub-sonic case without shock wave)

The case 9 was computed with the k-g model and scalable wall-functions, on three grids of different densities (y^+ 30, 10 and 1). The results demonstrated that the Scalable wall-functions could tackle finer grids than standard wall functions without sensible degradation of the solution.

Figure 2 shows results for the case 10 using the SALSA model with automatic wall functions on two grids compared to the LRN model. It can be seen that, as the grid is refined, the results with wall functions converge to those of the LRN model, including the correct prediction of the shock-induced separation.

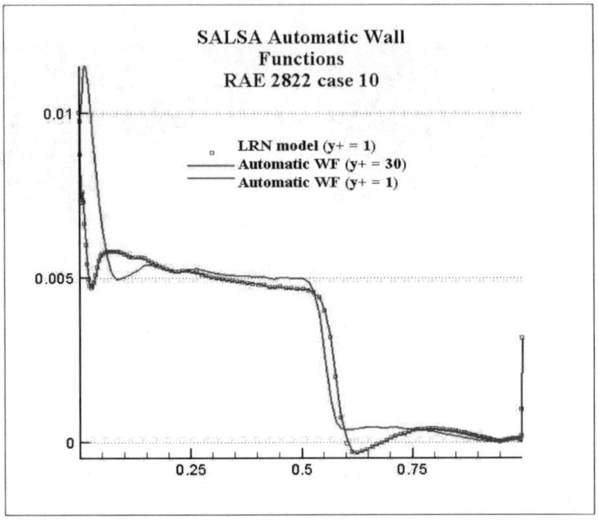

Figure 2 Friction coefficients for the SALSA model: automatic wall functions on various grids comparing to LRN computations

7.4 Conclusions

EADS-CASA carried out several implementations into the EMENS code that were tested in different test cases. In one side, the different approaches using wall functions show their ability to tackle different grid densities in the viscous layer and the expected convergence of their results with those of the LRN as the grid is refined, including the case of detached flows. This capability to predict separated flows seems to decrease as the y^+ moves away from $y^+ \sim 1$. Nevertheless, further computations will clarify better this point.

On the other hand, the SALSA TM was implemented and compared with the k-g model. A general trend was observed: SALSA predicted lower levels of eddy viscosity than k-g; or, conversely, k-g predicted higher levels. This may be part of the reason behind the general trend observed in all the tests performed: SALSA overpredicts detached flow regions, either behind shock waves (MRTT and RAE2822 case 10) or not (Obi diffuser), while k-g did not predict any flow separation for these test cases.

8 Methods and approaches used by EADS-M

W. Haase, S. Leicher, and W. Fritz, EADS-M

Abstract

The present chapter summarises effort at EADS-M on validation of turbulence models to bridge the gap between the most advanced research results obtained in recent years in the area of turbulence modelling and their application in daily industrial practice. To cope with the main aspects in flow physics, in particular in turbulence modelling, results and descriptions presented are concerned with reliability, robustness and accurate knowledge of the limitations of currently used eddy viscosity models. In addition, a process has been initiated to both investigating on full DSM models in industrial aerodynamic CFD codes and creating the industrial transition process towards DES and thus the next generation of turbulence modelling.

8.1 Method descriptions

8.1.1 2D Navier-Stokes method

For the A-airfoil investigation, in particular on the provision of a grid with the capability of producing grid independent solutions for all partners, a two-dimensional Navier-Stokes method (Haase et al, 1983) has been used in conjunction with the Johnson-King half-equation turbulence model. The latter is well known to produce accurate results for subsonic airfoil flow exhibiting separation. The method used is a Runge-Kutta time-stepping method with multi-grid acceleration. Blended second and fourth-order filtering is applied to prevent odd-even decoupling. For the three-stage scheme used, filtering is applied only once – which provides the best damping properties. Numerical dissipation is minimised by taking the eigenvalues in x- and y-direction independently and filtering was switched off in the boundary layer by scaling the filter value with the ratio of local to free stream Mach number. The Jameson-based scheme (Jameson et al, 1981) employs residual averaging and a variable time-step approach.

8.1.2 3D Navier-Stokes method

The DLR FLOWer code was used for all computations carried out with FLOMANIA, i.e. for the investigations on the delta wing and forward-swept wing cases. This code features parallel solutions of Euler as well as Navier-Stokes equations on structured meshes. FLOWer also bases on the Jameson et al (1981) approach with time-accurate computations using the dual time stepping method. Spatial discretisation uses a cell-vertex or cell-centred or AUSM finite volume formulation. Dissipative terms are explicitly added in order to damp high frequency oscillations and to allow sufficiently sharp resolution of shock waves. For steady state calculations, integration is accelerated by the techniques of local time stepping, enthalpy damping for inviscid flows and implicit residual smoothing. A sophisticated multigrid algorithm is employed allowing for standard single grid computations as well as successive grid refinement with the option of standard or full multigrid.

During the work carried out in FLOMANIA, the Speziale, Sarkar, Gatski (1991) differential Reynolds Stress turbulence model was implemented, which shows improved predictive accuracy for all applications (2D and 3D) that have been tested. In that respect, the "medium term" goals of FLOMANIA have been reached – although further investigations and validation work is needed to prove both accuracy and robustness for various applications.

8.2 A-airfoil – a grid dependence study

The main purpose was to provide a mesh that could be used for Navier-Stokes computations with advanced turbulence models in the fine-mesh version and in a coarser version to also serve for DES trials – where such fine meshes extended to 3D are "out of reach". Although a very fine mesh has been delivered to the partners involved in that test case, the application itself turned out to be a challenging case, compare chapter IV-18. The prediction of the laminar separation bubble on the upper surface at $\alpha=13.3^{o}$ and turbulent separation at about 85 to 90% chord remains a challenging task. Most of the turbulence models fail in predicting trailing edge pressure – although the mesh was constructed in a way that it contains about 130 to 140 points all along the surface across the (upper surface) boundary layer (from stagnation to trailing edge). The problem that still remained can be taken from the following Figure 1, showing a comparison between measured and computed momentum loss thickness.

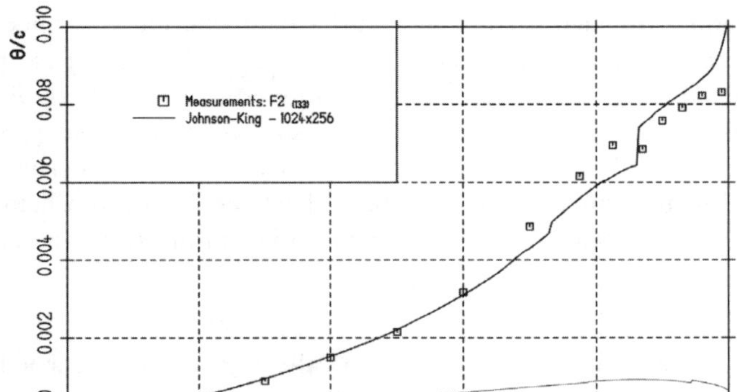

Figure 1 Momentum loss thickness for A-airfoil (finest mesh)

Apart from the non-smooth distribution – the Navier-Stokes velocity profiles are examined directly which sometimes causes this non-smooth behaviour – the agreement is excellent up to about 60%. When getting closer to the trailing edge separation, differences become more severe. When taking the smoothed values (like a mean value) up to 90% chord the results are in reasonable agreement, but the main problem shows up in the separated part where momentum loss thickness does not decrease as it should do but still increases. Further investigations might be needed – in particular a DRSM test. It should be mentioned additionally, that displacement thickness is slightly underpredicted beyond about 60% of chord.

8.3 Delta wing application

This test case served as a first application for testing rotation correction schemes on vertical flows. The test case, described in detail in chapter IV-11, was carried out on a structured mesh with 5 million mesh points and applicable for low-Re computations. With the EADS-M focus on the 25° case, the SAE and the Wilcox k-ω models were used with and without rotation correction according to Kok. Figure 2 provides a comparison between the measurements carried out by Breitsamter (1997) and the computations with and without rotation correction using the Spalart-Allmaras with Edwards correction (SAE) turbulence model.

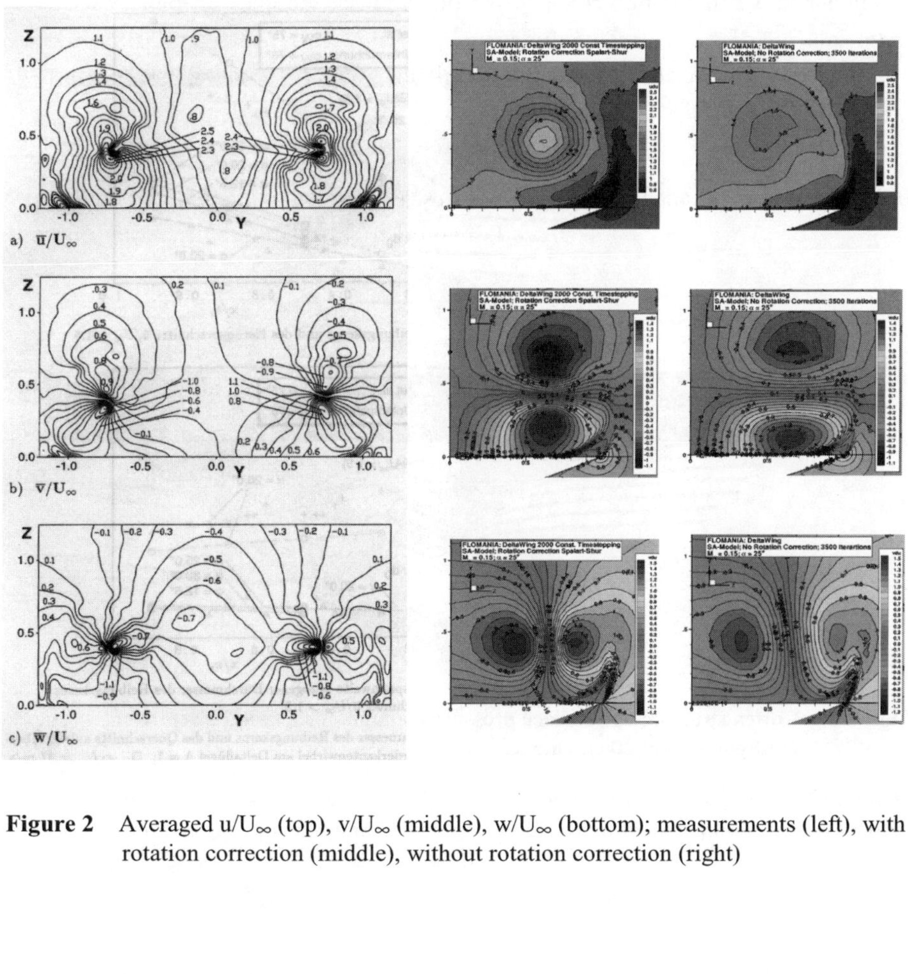

Figure 2 Averaged u/U∞ (top), v/U∞ (middle), w/U∞ (bottom); measurements (left), with rotation correction (middle), without rotation correction (right)

It can be easily detected that vortex or rotation correction applied is providing much steeper gradients and hence is clearly favourable for such types of flow. It must be added that for this case numerical viscosity was particularly kept at a minimum level.

8.4 Forward swept wing geometry

A test case that can be seen as an extension of the delta wing case briefly presented above has also been measured by Breitsamter et al (1998) at low Mach number (Ma=0.118), moderate Reynolds number (Re=434000) but high angle of attack. It serves as an industrial test case with vortices emanating from canard and main wing. Results shown in Fig. 3 have been computed on the medium grid with roughly 1.2 million mesh points (fine mesh with about 10 million points), using the Wilcox turbulence model with vortex correction. Vortices emanating from the tips of canard and main wing as well as from the canard- and wing-body junctions are well captured. Computations have been using (after initial cycles with local time steps) global (i.e. constant) time steps for reaching a steady state. For a more detailed analysis including velocity cross plots, please compare chapter IV-10.

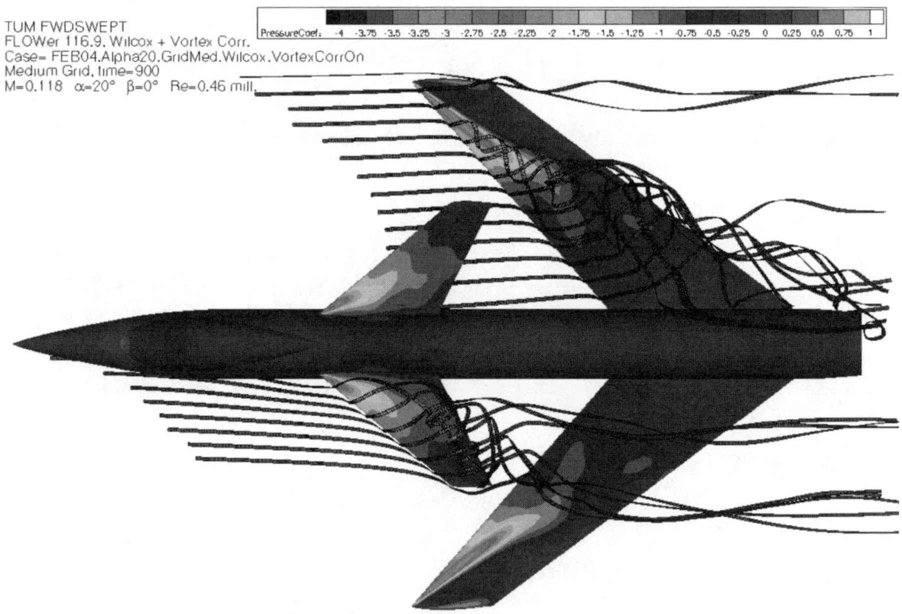

Figure 3 Vortex structure and surface pressure contours for forward-swept wing – canard configuration at 20^{0} incidence

9 Method Used and Highlight Results achieved with the *code_Saturne* Software at EDF

D. Laurence, S. Benhamadouche, and J. Uribe, Electricite de France

Abstract

The numerical method used in EDF's unstructured finite volume code is described, with an emphasis on boundary conditions. Through close collaboration with FLOMANA partners, UMIST in particular, implementation of SSG, SST, V2F models and scalable wall functions could be finalised. The wing-tip vortex and the 3D hill cases were computed in URANS mode, the latter also with LES using a synthetic turbulence method.

9.1 *Code_Saturne*: EDF's unstructured Finite Volume solver

9.1.1 Numerical method

The development of *Code_Saturne* was initiated in 1996 at EDF R&D to gradually replace the block-structured solver ESTET and the Finite Element code N3S. The new solver merges advantages of both methods: simplicity for coding complex models and ability to deal with complex geometries. The numerical method is described in detail in Archambeau et al. (2003). *Code_Saturne* is a collocated finite volume solver, all variables are collocated at the centres of gravity of the cells, which can be of any shape. Hanging nodes are treated as high order polygons. Gradient reconstruction methods as described in Ferziger & Perić, (1999) are used for non-orthogonal cells. Gradients at the cell centres are defined from the Gauss theorem. This requires interpolation of the variables on the cell faces. For structured grids the resulting scheme is similar to finite differencing along lines connecting cell centres. However, on non-rectangular grids, these lines are not orthogonal to cell faces and do not intersect the cell face centres. This interpolation is then corrected using 3D Taylor expansions which in turn involve the gradients of the variables. As these are not yet known, the deferred correction of Ferziger and Peric is introduced in the time-scheme: the implicit part of the fluxes (convection or diffusion) is written "as if" the cells were orthogonal, while "older", known values, of the gradients are used as a correction in the Taylor expansion correction.

The momentum equations are solved by considering an explicit mass flux. Velocity and pressure coupling is insured by the SIMPLEC algorithm. The Poisson equation is solved with a conjugate gradient method. The collocated discretisation requires a Rhie and Chow (1982) interpolation in the correction step to avoid oscillatory solutions, but is not essential for unstructured meshes.

9.1.2 Turbulence Models

Version 1.0 was released early 2001, with a standard k-epsilon and LRR differential stress model as basic turbulence models. FLOMANIA provided an opportunity to develop more advanced models, and all the while LES was developed on own resources. Switching from the LRR to the SSG has been a painless development, but the benefits of FLOMANIA lie in the demonstration of the superiority of the SSG model on a number of test cases. This model will become the default DSM at EDF. Furthermore, a number of near-wall models have been developed by the UMIST partner in EDF's code, and presented in the relevant section. However the Finite Volume numerical issues are presented here. Stability and positivity of turbulence variables are ensured by a combination of a fractional times-step method for and deferred correction. The latter allows balancing explicit source terms with convection-diffusion contributions.

9.1.3 Wall Functions

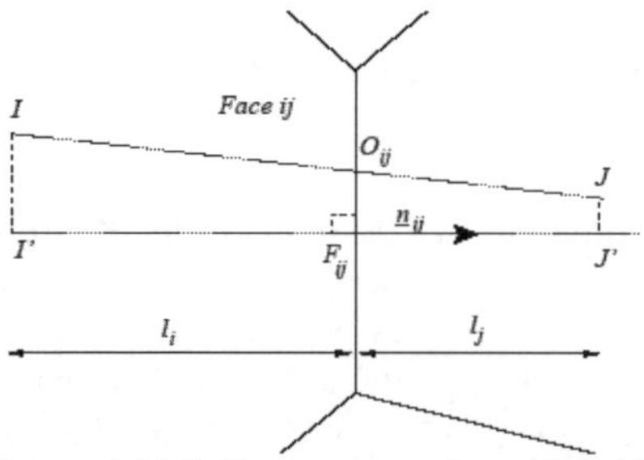

Figure 1 Cell face centre value obtained from cell centre values.

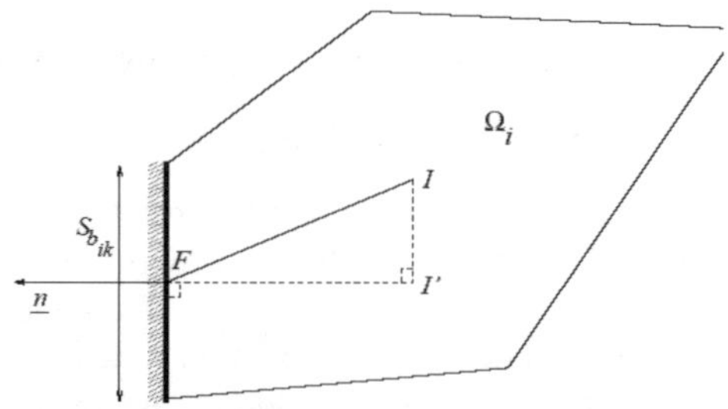

Figure 2 Face value on a solid boundary

Throughout the solver, gradients and cell face values are linked. Cell face values define the cell centre gradients through the Gauss theorem and gradients are used in turn for interpolations from centres to the faces, to correct the fact that the cell centre connecting lines do not intersect the faces at their centre:

$$\phi_{Fij} = \phi_{Oij} + O_{ij}F_{ij}(grad(\phi))_{Oij}$$

A value on a boundary cell face, F, can similarly be obtained by:

$$\phi_f = \phi_{I'} + I'F(grad(\phi).n)_f$$

Hence there is equivalence between specifying a Neumann condition or a face value on the border. Either way, both conditions are needed at some point in the calculation. In the standard wall function approach, the velocity at the border face *Uf* is calculated by equating the theoretical (or log-law) value of the velocity gradient and the calculated value obtained by the code:

$$G_{theo} = \frac{\partial U}{\partial y} = \frac{u*}{\kappa y}$$

$$G_{calc} = \frac{U_g - U_f}{2d} = \frac{(U_i + U_j)/2 - U_f}{2d}$$

A Cartesian mesh is assumed above for the purpose of simplicity: *Ug* , is the velocity at the upper face of the boundary cell, obtained by interpolating from *Ui* the velocity at the centre of the boundary element, and *Uj* is the velocity at the centre of the element above the boundary.

When introducing the scalable wall function approach recommended in FLOMANIA, the theoretical gradient is written as:

$$G_{theo} = \frac{\partial U}{\partial y} = \frac{u_*}{\kappa y} = \frac{u_*^{\,2}}{\kappa \tilde{y}^* \nu}$$

$$\tilde{y}^* = \min(y^*, Y^*); \qquad Y^* = 11.06$$

Previously, whenever the first cell fell below the buffer layer, G_{theo} was allowed to transition to the linear, viscous, velocity profile, while the turbulent variables retained their high-Re Neumann conditions. Although more accurate as concerns the velocity, this led to severe overestimation of the production of *k*, and too high friction. With the scalable wall functions, results were dramatically improved on the diffuser test case. The SWF reduces the velocity gradient at the wall, thus alleviating the need for damping functions.

9.1.4 Near wall models

As explained above, a segregated approach is used whereby convection-diffusion is solved successively for each variable, as a series of scalars (with the exception of destruction source terms inter-coupling). This works well when production and dissipation are in reasonable balance, but is more problematic with Near-Wall models where flow-physics dictates a balance between viscous diffusion and sink terms. Moreover, most applications are treated as time-dependant, which excludes stabilising measures such as under-relaxation and local time-stepping. Hence, early attempts with LRN models and necessarily very small CFL numbers had proved discouragingly expensive. On the other hand, EDF had supported early on the development of Durbin's brand of elliptic relaxation LRN models (Manceau et al 2001). During FLOMANIA, the V2F development was thus handed over to J. Uribe at UMIST who very successfully developed a code friendly version of the V2F model into *Code_Saturne* (Cf chapter II-17), by a consistent reformulation that decouples the boundary conditions so that they are compatible with the algorithm described above.

9.2 Highlight Results

9.2.1 Asymmetric Diffuser test case:

Results of case 3 are compiled in Chapter IV. What is analyzed here is the effect of reducing the tensorial diffusion (Daly Harlow) of stresses to an isotropic eddy viscosity, as suggested within the consortium to simplify the DSM model. Indeed divergence of a 3rd rank tensor is need in the full model. In a general purpose unstructured FV software which may include (as in *Code_Saturne*) angular periodicity, this may seem a daunting task. No differences between stress diffusion formulations were noticeable on the channel flow, but on the asymmetric diffuser, the simplification was observed to degrade somewhat the predictions. Nevertheless, the difference needs to be balanced against the coding complexity.

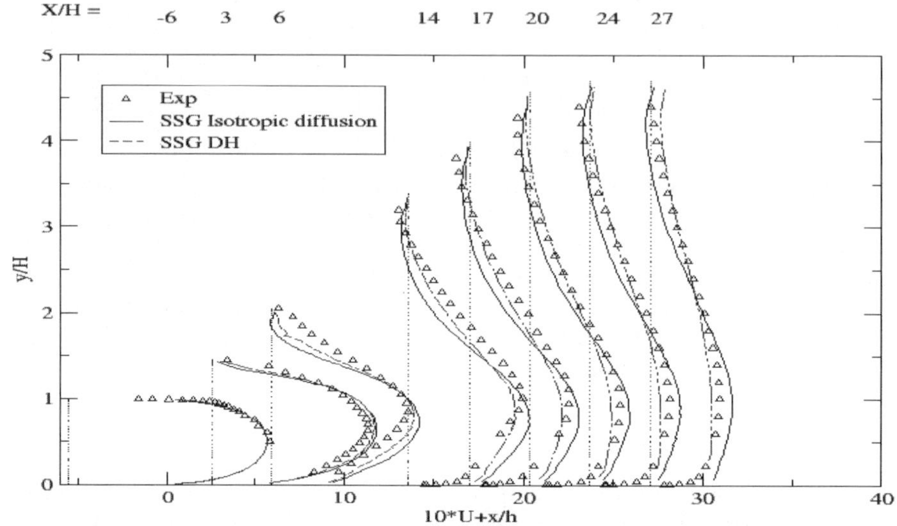

Figure 3 Effect of isotropic vs. tensorial diffusion in DSM model. Diffuser case.

9.2.2 Wing-tip

The wing-tip vortex has been computed with the DSM model and the initial mandatory grid. The initial stage of the formation of the vortex, with high momentum fluid (U/U_∞ =1.7) entrapped in the vortex core and surrounded by low momentum boundary layer fluid was well reproduced, but in later downstream stages, the too coarse grid lead to numerical diffusion and mixing of the 2 streams, as shown by subsequent simulations by C. Robinson on grids of up to 4 Million nodes.

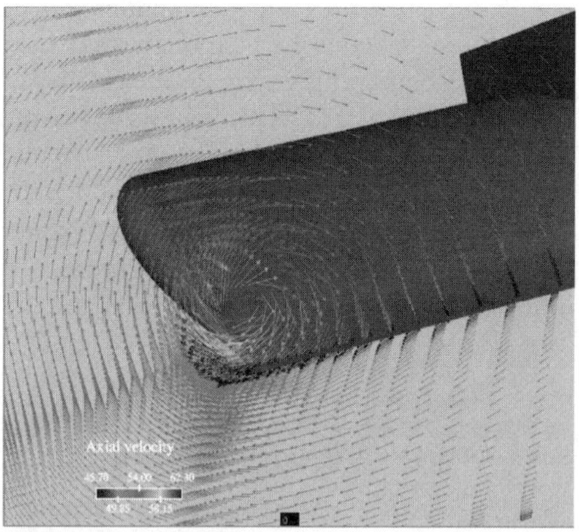

Figure 4 Wing-tip vortex formation, DSM.

9.2.3 The 3D hill

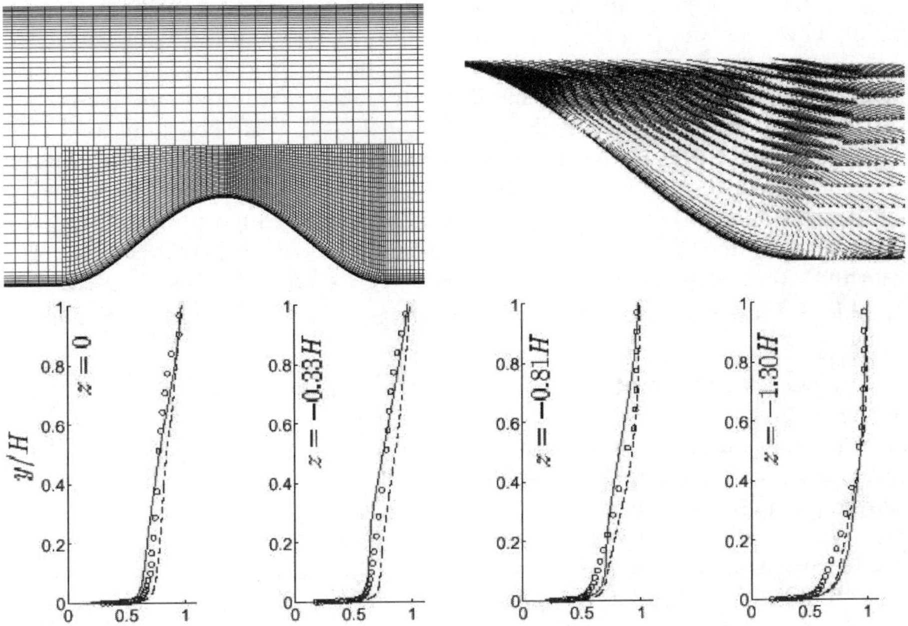

Figure 5 Flow over a 3D hill. LES results. Top: unstructured grid in symmetry plane, shallow separation. Bottom: Mean axial velocity in wake at x/H=3.69; dashed line - LES EDF, solid lines - LES Chalmers, symbols - experiments.

Figure 6 Non periodic channel flow with synthetic vortex inlet

The DSM model has been used to compute the 3D hill flow with a locally refined grid of 0.8 M cells (no symmetry plane), in time dependant mode, and produced an unsteady wake (although inlet was steady-state). These URANS results seem to be among the more realistic in the RANS group, yet not totally satisfactory.

An LES computation was run on the same grid, using the synthetic vortex method developed by Jarrin et al. (2003). This method generates realistic eddies which correspond to a prescribed distribution of Re stress (obtained e.g. from RANS) and are shown to be highly sustainable in channel and pipe flow tests without periodicity (Fig. 6). The hill-flow LES produced results in much better agreement with the experiments and refined RANS-LES of Chalmers. Therefore, this case shows no advantage of U-RANS compared to full LES which are able to reproduce accurately the experimental data in the wake, Fig 6. It can be concluded that none the various RANS or URANS approaches can reproduce this flow.

9.2.4 Near Wall models

Finally, the outcome from the FLOMANIA collaboration for EDF have been mainly new near wall modeling procedures: the scalable wall functions have been developed for two eqn. and DSM models and demonstrated indeed mesh independent behaviour as announced by F. Menter (despite the surprising simplicity of the suggested method). The Analytical Wall Functions of Craft, Gerasimov et al. have also been implemented, and mainly proved advantageous on buoyancy affected flows (as Fig 7), but not so much on Aerodynamic flows (hence not described herein). The SST model newly implemented has proved robust and more accurate that the standard Launder-Sharma model. In particular the code-friendly V2F model suggested by Uribe et al. (Cf. Chap II-17) also implemented in *Code_Saturne* was found to converge much faster than the original V2F model.

It is well known from V2F model publications that this model exhibits best its advantage in heat transfer cases, as illustrated figure 7 for natural convection in a tall cavity. The TU Delft group independently arrived at a similar formulation of the model and a joint publication highlights further benefits for heat transfer applications (Hanjalic et al. 2005).

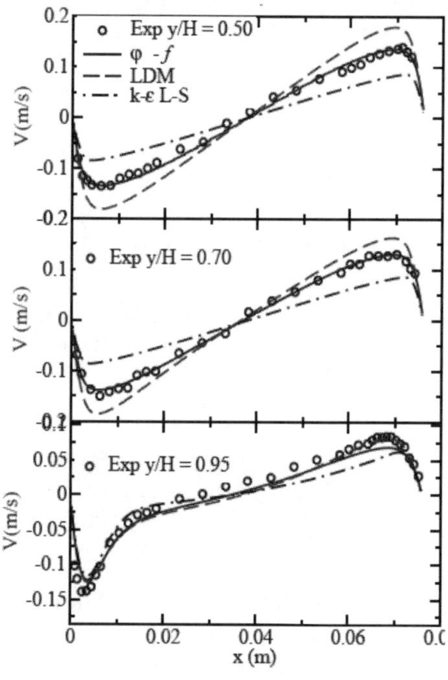

Figure 7 Natural convection in a tall cavity. Lines: new V2F model, dashed lines reference LDM (Lien-Durbin), dot-dashed Launder-Sharma k-epsilon. Symbols: exp. by Betts and Bokhari. 1995.

10 Contribution by FOI

S. Wallin, and P. Eliasson, FOI

Abstract

The activity at FOI was mainly focused around the implementation of a Reynolds stress transport model into our CFD code for unstructured grids. We showed that standard DRSMs may be solved down to the wall together with the standard ω equation without any near-wall corrections. Only minor recalibration was needed in order to fulfil the log law. We are working with an industrial CFD code so efforts have been made in making the implementation as efficient and robust as possible. 2D and 3D test cases have been computed and our results are in general close to other DRSM results. One interesting observation is that the results obtained by EARSMs are in general close to those of the corresponding full DRSMs.

10.1 The DRSM-omega model

Based on our experiences of EARSMs based on the Wilcox (1988) k-ω model (or derivation thereof, such as the BSL model, Menter 1994), we decided to build a DRSM that is as similar as possible as the corresponding EARSM. The main difference will, thus, be that the advection of the Reynolds stress anisotropy is accounted for in the DRSM. Moreover, the ω model makes it possible to formulate a model without any near-wall corrections or damping functions and still obtain the correct velocity profile. However, the Reynolds stresses will not be correctly predicted in the near-wall viscous and buffer regions, but we consider this as a secondary effect in high Reynolds number engineering flows. Moreover, most existing near-wall corrections are merely a curve fitting of DNS data and contain little of the flow physics.

The pressure-strain rate model is implemented as a generic model with terms corresponding to the SSG model (Speziale et al. 1991), and other DRSMs are obtained by changing the model coefficients. In this way the DRSMs that correspond to the Gatski & Speziale (1993) as well as the Wallin & Johansson (2000) are available in the FOI code.

The diffusion is modelled by eddy viscosity diffusivity with the Wilcox (1988) diffusivity coefficients. As an alternative, the Daly and Harlow model may be used, but in that case the usual values of the diffusion coefficients must be recalibrated in order to be consistent with the ω model since the magnitude of the diffusion in the k-ω model is about half of that in the standard k-ε model.

10.2 Numerical treatment and implementation guidelines

FOI co-ordinated task 5.4, the best practice recommendations for the implementation and calibration of turbulence models in FLOMANIA. The implementation guidelines are focused on the implementation of 2-equation turbulence models, the implementations of RSM and the implementation of DES. Also best practice guidelines for model validation were produced. FOIs contribution was focused around the numerical treatment of RANS models in explicit compressible solvers. The goal is to be able to safely and robustly iterate to steady state starting from any given initial solution.

10.2.1 2-equation models

For robust implementation of 2-equation turbulence models, maintaining positive values of the dependent turbulent variables are of critical importance. In an explicit local time stepping updating, the updating of the vector of turbulence (positive) variables, $\boldsymbol{q}$, is

$$\boldsymbol{q}^{+} = \boldsymbol{q} + \Delta\boldsymbol{q}\ .$$

In the case of explicit local time stepping $\Delta\boldsymbol{q} = \Delta t\boldsymbol{R}(\boldsymbol{q})$. By introducing an implicit approximation of the residual, $\boldsymbol{R}(\boldsymbol{q})$, active when the residual is negative (driving $\boldsymbol{q}$ towards zero) the following alternative scheme is obtained

$$\boldsymbol{q}^{+} = \boldsymbol{q} + \frac{\Delta\boldsymbol{q}}{1 - \min(\Delta q_\alpha / q_\alpha, 0)}\ .$$

It is important that the turbulence variables are strongly coupled such as the amount of under relaxation is the same for all turbulence equations. However, the mean flow equations do not have to be under relaxed because of the rather weak coupling between the turbulence and mean flow equations. This scheme is also used for the multigrid updating.

The spectral radius of the Jacobian of the negative source terms must be accounted for in the time step for stability. For $k-\omega$ it reads $\rho^{-} = -\max(\beta_S, 2\beta)\omega$.

10.2.2 Reynolds stress transport models

Realizability for DRSM is related to positivity for the two-equation turbulence models. The first requirement is that the model itself sustains realizable solutions. The second requirement is that the space discretisation is monotonic, which can be fulfilled by TVD schemes. The positivity treatment is extended to DRSM by requiring realizable values of the individual Reynolds stresses being equivalent to having positive eigenvalues of the Reynolds stress tensor. The eigenvalues are, however, not easily obtained and a more convenient measure is the determinant. Starting with realizable values, the Reynolds stress tensor will remain realizable if the determinant is not allowed to become zero in any instances.

The updating of the Reynolds stresses

$$\tau_{ij}^{+} = \tau_{ij} + c\Delta\tau_{ij}$$

where $c = 1$ gives the updating without any time-step reduction or under relaxation. If $\det(\tau_{ij}^{+}) = 0$ for any c between 0 and 1, the updating is restricted. The smallest value of c that gives zero $\det\tau_{ij}^{+}$ can now be determined by solving a third-degree equation and extracting the smallest real root c_0 for c between 0 and

1. In this work, an approximation of c_0 that is smaller than the exact root is used. The updating, used in both time stepping and multigrid, becomes

$$\tau_{ij}^{+} = \tau_{ij} + \frac{\Delta\tau_{ij}}{1-\min(-1/c_0,0)}.$$

The time step is determined by stability requirements from the convective and viscous terms from the coupled system of equation. The spectral radius of the Jacobian of the source terms must be accounted for, as in two-equation models. For the SSG-ω model it reads

$$\rho^{-} = -\max\left[2\beta, \beta^{*}, \beta^{*}\left(\frac{1}{2}C_1^0 + \frac{1}{6}C_5\sqrt{\frac{3}{2}II_a}\right)\right]\omega.$$

Because of the more advanced coupling between the momentum and the Reynolds stress equations, the solution of these equations must be strongly coupled. The same under relaxation must, thus, be used for all equations solved.

10.2.3 Wall function

The problems with wall function boundary condition are that (i) the grid must be sufficiently coarse and (ii) the boundary condition is a mixture of physics and numerical discretization. Problem (i) can be overcome with some kind of "adaptive" or "scalable" approach, but still a grid-convergence study is impossible because that the boundary value will change depending on the grid size.

A different approach to standard wall function has been considered by FOI. That is that the basic model equations and the "physical" boundary conditions are reformulated such as the near-wall viscous and buffer regions are eliminated. These new model equations are then discretizised and solved as usually, but the difference is that the requirement on the resolution is very much relaxed (similar to wall functions) and the same solution is obtained on both coarse and fine grids allowing grid refinement studies. There is no dependency on the grid size in the formulation of the equations or boundary conditions.

10.2.4 Description of the FOI solver Edge

The different developments have been implemented and validate in the FOI code Edge. It is an edge- and node-based compressible Navier-Stokes and RANS flow solver based on a finite volume formulation where a dual grid forms the control volumes. The governing equations are integrated explicitly to steady state with acceleration by multigrid. The solver supports several types of elements in 2D and 3D.

10.3 Test cases

The different DRSMs above have been implemented into the FOI code EDGE using the numerical considerations reported above. The Obi diffuser and the 3D hill (test cases 3 and 10) were computed using the SSG-ω model and the results were compared with results using two equation models. In particular, the comparison with the WJ-st-ω EARSM is interesting since the major difference between these models are the advection of the Reynolds stress anisotropy that is represented only by the DRSM. Computations using the DRSM that corresponds to the WJ-st-ω EARSM concerning the pressure-strain rate modelling have also been made. These results are close to the results obtained by the SSG-ω model and will not be shown here.

The asymmetry in the velocity profile for the Obi diffuser is well predicted both with EARSM and DRSM but the recirculation is under predicted, see Figure 1. One interesting observation here is that the EARSM and DRSM agree very well indicating that the algebraic assumption is reasonable for this case.

For the 3D hill flow, all models completely fail in representing the flow structure, see Figure 2, which is in line with the general conclusions concerning RANS models for this case (see the reporting of the case 16). Also, the difference between the EARSM and DRSM are large indicating that the advection of the anisotropy is of significant importance for this case. The EARSM and the SST results are fairly close to each other.

Finally, the DLR F6 cases were computed using different one- and two-equation models (see the reporting of case 20). These computations were made in the early stage of the project and no attempt of using DRSM on this case was made.

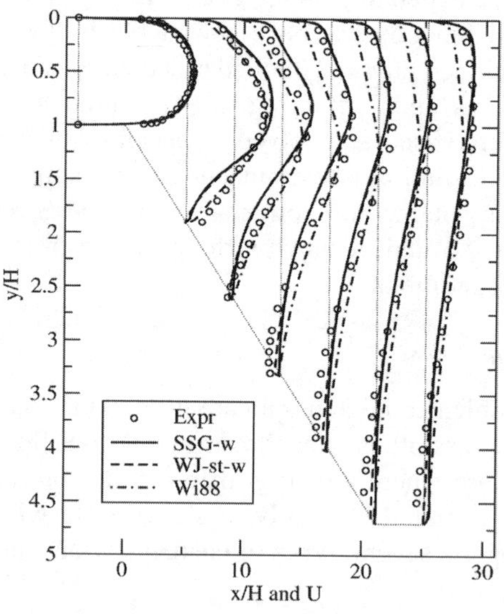

Figure 1 Computed velocity profiles for the Obi diffuser compared with experiments (Obi 1993)

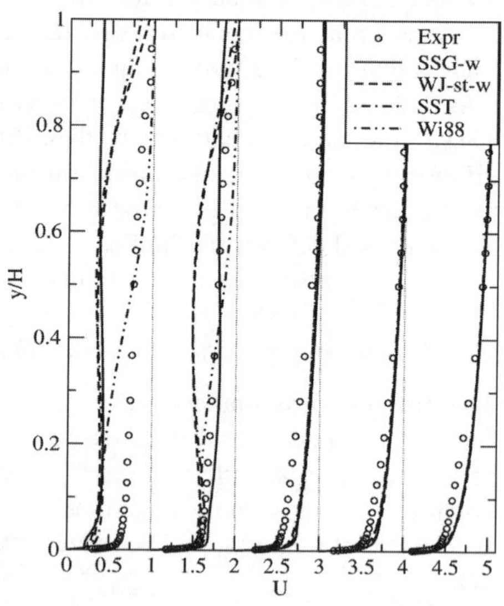

Figure 2 Computed streamwise velocity profiles for the 3D hill compared with experiments (Simpson et al. 2002). The profiles are taken at x/H=3.69 and from left to right z/H=0, -0.33, -0.81, -1.30, -1.79.

11 Contribution by ICSTM: Modelling generic 2d and 3d separated flows using anisotropy-resolving turbulence closures

M.A. Leschziner, C. Wang, and Y.-J. Jang, Imperial College London

Abstract

The contribution of Imperial College London has been targeted, principally, towards advanced anisotropy-resolving turbulence closures and their application to a range of physically complex "generic" flows in which 2d and 3d separation from continuous surfaces is the main linking feature. The closures included various non-linear eddy-viscosity models and explicit algebraic Reynolds-stress models as well as two full Reynolds-stress-transport models. The range of flows investigated extended from 2d separation in an asymmetric diffuser to the highly complex 3d flows around a 3d hill and a generic car body. For all geometries but one – the NACA 0012 aerofoil at moderate incidence – representatives from all three anisotropy-resolving model categories were investigated. One major conclusion derived from the studies is that observations made in statistically 2d flows often do not translate to much more complex 3d flows that feature massive separation and strong vortical transverse motion. While anisotropy-resolving closures may give a superior representation of the response of the stress field to different types of strain, they cannot account (in common with all other RANS models) for the complex dynamics associated with large-scale unsteadiness, and it appears that this remains a major obstacle to a satisfactory predictive performance of these models in highly separated conditions. In contrast, attached and mildly separated flows benefit from the much stronger fundamental rigour offered by elaborate anisotropy-resolving closures, but here too, the predictive performance varies greatly, and much depends on the precise details of the closure and its calibration.

11.1 Introduction

Although aerodynamic design generally strives to achieve attached flow over the external streamlined surfaces of an aircraft, separation and associated adverse consequences cannot be avoided in most practical circumstances. Examples in which separation almost invariably occurs are wing-body junctions, highly-curved intake ducts and nacelles, multi-element high-lift devices, afterbodies at transonic conditions and the rear of fuselage-type bodies. In the large majority of these cases, separation occurs as part of a complex and interactive 3d process. Thus, typically, a boundary layer develops over the curved surface, undergoes skewing, deflection and deceleration, eventually terminating in separation and a strongly rotational wake, with or without a closed recirculation zone. In reality, the flow is highly turbulent, of course, and the separation neither occurs along a continuous line, nor is steady. Indeed, separation may be intermittent and cover a major proportion of the curved surface.

The modelling of processes of this type with statistical closures is, without doubt, amongst the most challenging tasks faced by computational prediction procedures. There is, first, the ‘non-local’, highly dynamic character of turbulence, arising from the large, intermittent structures associated with separation. Second, the turbulence field may contain non-stochastic, periodic components associated with strong instabilities – e.g. shedding – which both ‘modulate’ the turbulence and also feed, through break-up, into stochastic turbulence. Even if the assumption is made that these processes can somehow be accommodated within statistical closures, there remain a whole host of other challenging aspects, including strong shear and skewness-induced anisotropy, wall effects on turbulence (both viscous and inviscid), and the effects on turbulence of normal straining ahead of separation and of strong curvature in organised vortices – e.g. in wakes. Thus, if statistical closures are at all expected to give an adequate representation of turbulence in separated aerodynamic flows, it is arguable that such closures must be able to resolve the anisotropy of turbulence and to return the correct response of the turbulent stresses to the very disparate types of strains that are involved in complex flows.

Against the above background, Imperial College’s contribution to FLOMANIA has been directed, principally, towards the formulation, improvement and the investigation of various anisotropy-resolving statistical closures for modelling 2d and 3d separation from curved surfaces. This contribution rests on many years of experience of the group, extending to the use of non-linear eddy-viscosity and Reynolds-stress models for both incompressible and compressible flows. Especially notable activities within FLOMANIA, some in conjunction with allied project, are the following:

- The formulation and validation of a new explicit algebraic Reynolds-stress model (EARSM) which correctly respects the asymptotic variation of all Reynolds-stress at the wall.
- The investigation of several non-linear eddy-viscosity models (NLEVMs), EARSMs and full Reynolds-stress models (RSMs) for several 2d and 3d test flows in which separation from curved surfaces is a key feature.
- The investigation of the importance of stress transport in 3d separated flows through a-priori testing and evaluation of the stress budgets.
- The coordination of the implementation and testing of two default RSM formulations for a reference test geometry – the Obi asymmetric diffuser (Obi, 1993).
- The extensive investigation of grid-sensitivity and the performance of several anisotropy-resolving models for separated flows from 2d and 3d hills (Wnd et al., 2004, Temmerman et al., 2004, Simpson et al., 2002).
- The use of RSMs and NLEVMs for a complex car-body flow (Lienhard and Becker, 2003).
- The investigation of the performance of NLEVMs for a stalled NACA0012 aerofoil which was used by others only to examine the effectiveness Detached Eddy Simulation (DES).

In what follows, a flavour is given of the range of studies undertaken. A relatively minor contribution was also made under the heading DES (Detached Eddy Simulation) and LES. This was in the form of the use of Imperial College's own LES and hybrid RANS/LES strategy (Temmerman et al., 2004) for two of the separated test flows. This contribution is reported under the appropriate Task headings in this book, but is not discussed herein.

11.2 Turbulence Modelling

A wide variety of turbulence models have been investigated as part of the FLOMANIA programme. These included linear two-equation models, used merely for reference against which to compare more elaborate models, NLEVMs, EARSM and full RSMs. A full list is given below – although not all models were applied to all flow configurations investigated:

(i) the linear low-Re k-ε model of Launder & Sharma (1974);
(ii) the linear k-ω model of Wilcox (1988, 1994);
(iii) the cubic k-ε EVM of Craft et al. (1996);
(iv) the cubic k-ε EVM of Apsley & Leschziner (1998);
(v) the quadratic k-ε EVM of Gatski and Speziale model (1993) with log-law wall functions;
(vi) the quadratic EVM/EARSM of Abe et al. (2003), both with ε and ω equations representing the length scale;
(vii) the Wallin-Johansson k-ω EARSM (2000);
(viii) the low-Re Reynolds-stress-transport model of Jakirlic & Hanjalic (1995);
(ix) the low-Re Reynolds-stress-transport model of Shima (1998);
(x) the high-Re Reynolds-stress-transport model of Speziale et al. (1991), both the version extended to apply to low-Re conditions by Chen et al. (2000) and in conjunction with wall functions (both scalable and log-law-based).

A complete mathematical description of all models is provided in Chapter III. Of these, the quadratic low-Re model of Abe et al. (AJL) deserves to be highlighted here, for it has been developed, calibrated and validated, in parts, within FLOMANIA. First, the model augments the basic quadratic constitutive stress-strain/vorticity equation by two additive fragments intended to account, respectively, for high normal straining and strong near-wall anisotropy. Second, it uses a form of the ω-equation (in one of two variants) that is much closer than Wilcox's form to the ε-equation. Specifically, it includes products of k and ω gradients and coefficients for the production and destruction terms that are directly equivalent to $C_{\varepsilon 1}$ and $C_{\varepsilon 2}$ normally used in the ε-equation.

An influential model fragment accounts specifically for strong near-wall anisotropy and for the correct decay towards two-component turbulence that is observed in DNS. This decay cannot be represented solely by the use of terms combining the strain and vorticity. The approach taken by Abe et al. is thus to add a tensorially correct wall-related term to the constitutive stress-strain/vorticity relation $a_{ij} \equiv \frac{\overline{u_i u_j}}{k} - \frac{2}{3}\delta_{ij} = f(S_{ij}, \Omega_{ij} ...)$, which takes into account the wall orientation. In the model variant used here, the wall-direction indicator is $d_i = N_i / \sqrt{N_k N_k}$, $N_i = \partial l_d / \partial x_i$, $l_d = y_n$ (wall distance) which is then used in the additive wall-anisotropy correction of the form:

$$^{w}a_{ij} = -f_w \left(d_i d_j - \frac{\delta_{ij}}{3} d_k d_k \right) \times f(S_{ik} S_{kj}, S_{ik} \Omega_{kj}, S_{kj} \Omega_{ik}, S^2, \Omega^2 ...)$$

where f_w is a viscosity-related damping function (see Jang et al (2002) for details). Alternative wall-orientation indicators that are independent of wall distance may readily be used. In the above damping function, a composite time scale is used, which combines the macro-scale k/ε with the Kolmogorov scale $\sqrt{\nu/\varepsilon}$. The damping function f_w then provides a smooth transition between the two scales across the near-wall layer. The model is fully described in Abe et al (2003) and Jang et al (2002), and the latter publication demonstrates, by way of results for the anisotropy and its invariants, that the model indeed returns the correct wall-asymptotic behaviour of the stresses for the separated flow in the 2-d constricted duct also considered in this report.

11.3 Computational Framework

The calculations were performed with a non-orthogonal, collocated, cell-centred finite-volume approach implemented in the code 'STREAM' (Lien and Leschziner (1994a), Apsley and Leschziner (1998). Convection of both mean-flow and turbulence quantities is approximated by the 'UMIST' scheme (Lien and Leschziner (1994b)) - a second-order TVD approximation of the QUICK scheme of Leonard. Mass conservation is enforced indirectly by way of a pressure-correction algorithm. Within this scheme, the transport and the pressure-correction equations are solved sequentially and iterated to convergence. Extensive grid-dependence tests were performed for all cases by the authors, except for the generic car body for which the grid was supplied by CFX, and this led to the final choice of grids for which results are included here as well as elsewhere in the present book. To provide an adequate resolution of the viscous sublayer and the buffer layer adjacent to the solid walls, when using low-Re models, the minimum wall-normal grid spacing at the wall was maintained in region y^+ of order 0.1-1. For the high-Re Gatski-Speziale model, the y^+ value, determined with the turbulence energy as the velocity scale, was maintained above 11.

11.4 The Flows Examined

A total of six 'generic' flows were investigated within the FLOMANIA programme, all featuring separation from curved or straight but continuous surfaces. The list in given below:

(i) TC9: flow in an asymmetric diffuser;
(ii) TC27: flow in a duct constricted by periodic 'hills' on one wall – this case was computed as a periodic segment as well as a sequence of three segments with inlet conditions taken from the reference LES databank;

(iii) TC26: flow over a 3d hill;
(iv) TC14: flow over the NACA 0012 aerofoil at low-to-medium (pre- and post-stall) incidence;
(v) TC24:flow around a generic car body;
(vi) flow around a wing-body junction (eventually discarded as a FLOMANIA test case).

Most of the results obtained are contained in test-case specific presentations and discussions in Chapter. A pictorial overview of the cases computed is conveyed in Figure 1.

The **asymmetric diffuser** was coordinated by the writers across the whole project, and cross-plotted results are presented and discussed in Chapter IV, Section 3. An exceptional aspect of this case is its purpose within the range of test flows: it was introduced to allow all participants who are engaged in second-moment modelling to verify the validity of their implementation of the default closure by reference to a simple and well-defined test geometry. Thus, Chapter IV contains cross-plots of results provided by partners, but only for the default closure. A number of partners have used other models, and this includes the present writers who used a range of NLEVMs and EARSMs in addition to RSMs.

The second geometry, a sequence of **2d hills**, is an especially sensitive indicator of the ability of models to capture (nominally 2d) separation from a curved surface, the structure of the separated shear layer, the response of reattachment to both and the behaviour in the post-reattachment recovery region. One particular attraction of this case is the availability of highly-regarded LES data (Temmerman et al., 2003), which satisfy, to an exceptional high degree, the constraints of spanwise homogeneity and streamwise periodicity that can only be achieved with a simulation. Extensive studies with several models are documented in Chapter IV, Section 17, in which either streamwise periodicity was imposed or the flow over a sequence of 3 hills was computed to examine the model-dependent trend towards the attainment of the periodic state, starting with the LES data upstream. Extensive investigations of this case are documented in Jang et al. (2002) and Wang et al (2004).

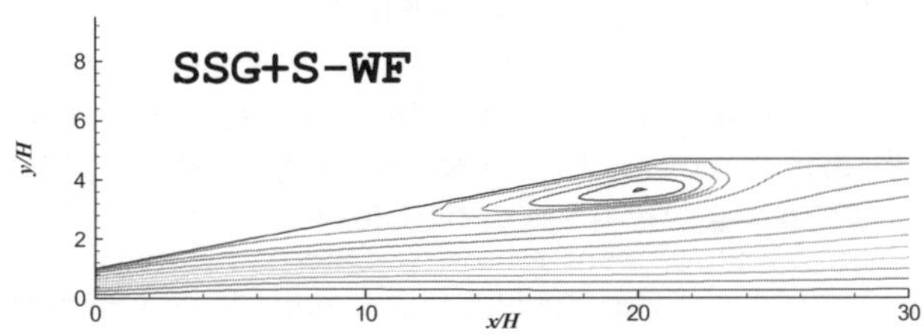

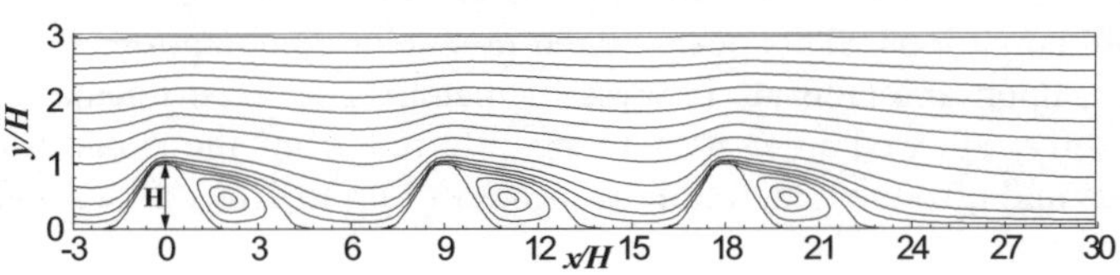

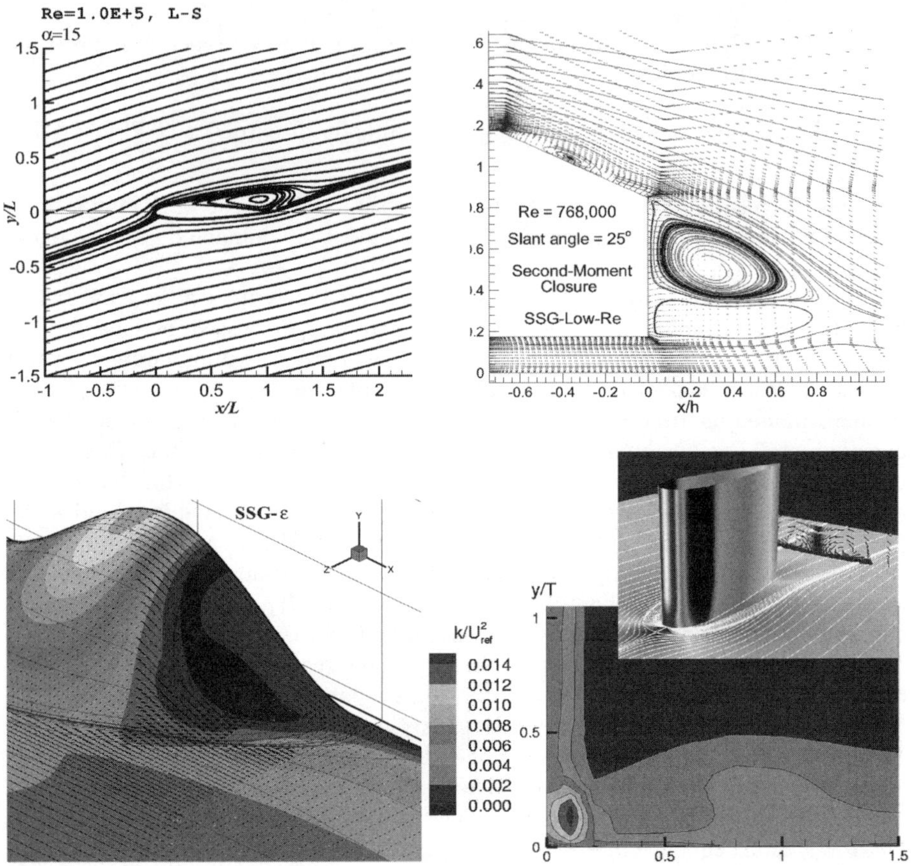

Figure 1 Test cases computed within FLOMANIA: asymmetric diffuser, sequence of 2d hills, NACA0012 aerofoil, generic car body, flow over 3d hill, flow in wing-body junction.

The flow over the 3d hill is, superficially, the 3d equivalent of the 2d configuration considered above. However, the flow here is, physically, much more complicated, involving extreme skewness upstream of the hill and a complex separation and 3d post-reattachment recovery pattern. Moreover, unlike the 2d configuration, the existence of a shedding-like, periodic component cannot be excluded and may coexist with the turbulent motion. One problem that presented itself in the present flow was the absence of measured upstream data that could be used as inlet conditions. This necessitated precursor computations for the hill-less duct, which, upon an elaborate process of matching to experimental data for the same geometry, allowed inlet data to be inferred from the numerical solution. This process and also the generation of meshes of up to 1.6M nodes that secured a grid-independent solution were both undertaken by the writers for the project as a whole. For this case, none of the models investigated yielded a solution even in fair agreement with the experiments. All model types – EVM, NLEVM, EARSM and RSM – gave similar (wrong) separation patterns, excessive separation length and too slow wake recovery. More information can be found in Chapter IV and in Wang et al (2004).

The **NACA0012** configuration had been chosen within FLOMANIA specifically to study the predictive capabilities of DES when the aerofoil is at the extreme incidence of 60 degrees. In this case, the aerofoil acts, effectively, as a bluff body, and a massive separation zone develops downstream of the aerofoil. While RANS methods are likely to perform poorly for such conditions, they are expected to do far better at low incidence, especially in attached and marginally-separated conditions. It is for such conditions that computations were performed by the writers. Only two models were applied to this geometry: the linear EVM of Launder and Sharma and the AJL NLEVM/EARSM developed as part of the FLOMANIA project. Results are given in Chapter IV, Section 6.

This **generic car body** is the most complex configuration considered by the writers within FLOMANIA. The principal challenge of this case arises from the need to capture the reattachment on the slanted rear portion of the body's roof, at 25 degrees of slant, following separation from the corner formed by the horizontal roof and the slanted surface. Steady-state computations were performed on a half-body, covered by a mesh of 1.8m points, which allowed low-Re models to be used. A total of five models were applied, two linear EVMs, two NLEVMs and the sixth the low-Re version of the RST of Speziale et al. (SSG). Only the last gave a broadly correct behaviour, with reattachment on the slanted surface (see Figure 1). Cross-plotted results can be found in Chapter IV, Section 15. An extensive exposition is also given by Jang et al. (2004).

11.5 Conclusions

As is evident from the foregoing account, the Imperial College group undertook wide-ranging modelling studies within FLOMANIA, covering a broad spectrum of advanced modelling strategies and separated flows. It also contributed materially to the coordination of efforts on three geometries across the FLOMANIA project. Additional modelling efforts on a wing-body junction and the application of LES and a hybrid RANS/LES method, whilst not included in the book, should at least be mentioned.

The principal objective of the work was to assess the ability of anisotropy-resolving models, including full second-moment closure, to represent the essential features of separation from curved surfaces, both in nominally 2d and 3d flows. A new NLEVM/EARSM has been introduced as part of the efforts, and this distinguishes itself by its adherence to the wall-asymptotic variation of all Reynolds-stress components at the wall. This model was found to perform well in 2d conditions. Studies on much more complex 3d flows that feature massive separation and strong vortical transverse motion demonstrate that conclusions derived in 2d do not, on the whole, translate to the 3d environment. While anisotropy-resolving closures give, in principle, a superior representation of the response of the stress field to different types of strain, they cannot account, as is the case with all other RANS models, for the complex dynamics associated with large-scale unsteadiness, and it appears that this remains a major obstacle to a satisfactory predictive performance of these models in highly separated conditions. However, attached and mildly separated flows benefit from the much stronger fundamental rigour offered by elaborate anisotropy-resolving closures, but here too, the predictive performance varies greatly, and much depends on the precise details of the closure and its calibration.

12 Advances in turbulence modelling for unsteady flows – IMFT

Y. Hoarau*, R. Perrin*, M. Braza*, D. Ruiz**, and G. Tzabiras***
*Institut de Mécanique des Fluides de Toulouse, Groupe EMT2,
**Computing Dept., ENSEEIHT, Toulouse,
***National Technical University of Athens

Abstract

The objective of the present study is to provide advanced statistical turbulence modelling closures for aerodynamic flows around bodies at high Reynolds number. This study has been carried out by means of theoretical analysis based on the IMFT's experimental data-base (see part 2 of the book). The two-equation modelling has been reconsidered in the context of the URANS/Organised Eddy Simulation, OES approach in respect of modifying the turbulence length scale involved in the eddy-diffusion coefficient. This has been done from second-order moment closures considerations in OES, from DNS and from the experimental sspectra for the turbulence fluctuating energy.

12.1 Second-order moment closures – improvement of two-equation modelling for unsteady flows

The LRR and SSG models has been implemented in the phase-averaged Navier-Stokes equations, by using the code ICARE of IMFT. The test-case of a NACA0012 airfoil at 20° and at Reynolds number 10^5 has been considered (Favier *et al* , (2002)). The turbulence anisotropy tensor has been evaluated in the unsteady boundary layer and in the detached region by means of the DNS study of IMFT (part 'NACA0012 – DNS approach'). Figure 1 shows that the diagonal term of the turbulent stress anisotropy, $b_{12}=-uv/k$ is not constant and much slower that in equilibrium boundary layers, (b_{12}~0.30 and practically constant according to (Bradshaw *et al* (1967)). The detailed analysis of this behavior is also provided by the DNS of (Hoarau *et al*, (2003)). Furthermore, the second-order moment closure (version of the LRR model with low-Reynolds damping by (Shima, (1988)) provides the variations of an equivalent eddy-diffusion coefficient, (table 1) that indicates a slight dispersion in the variation of the C_μ whose values are of an order of magnitude of 0.02, whereas the equilibrium turbulence value is 0.09.

The LRR and SSG models have been used for the flow prediction at $Re=10^5$. The drag and lift coefficients are shown in fig. 2. Their oscillation amplitudes are much more pronounced and their mean values are in good agreement with the experiment than in case of the standard URANS k-ε model (with Chien's damping). Furthermore, the LRR model is able to capture the leading-edge quasi-laminar separation bubble (fig.2), whereas the standard URANS k-ε modeling is unable to predict it. Both LLR and SSG models provide the von Kàrmàn mode and flow detachment in accordance with the physical experiment, fig. 2. A comparison with the turbulence quantities in the detached boundary layer is provided. The SSG model is found to provide a better behavior near the wall than the LLR model.

Table 1 Evaluation of an equivalent C_μ coefficient by means of the DSM (LRR); NACA0012 wing, α=20°, Re=10^5.

X/C	Y/C	C_μ équivalent
$0,3610^{-1}$	$0,3210^{-1}$	0,0158
0.45	$0,5710^{-1}$	0,01859
0,9064	$0,1510^{-1}$	0,01938
1,41	-0,678	0,0178
1,23	0,11	0,0172
0,73	0,19	0,024

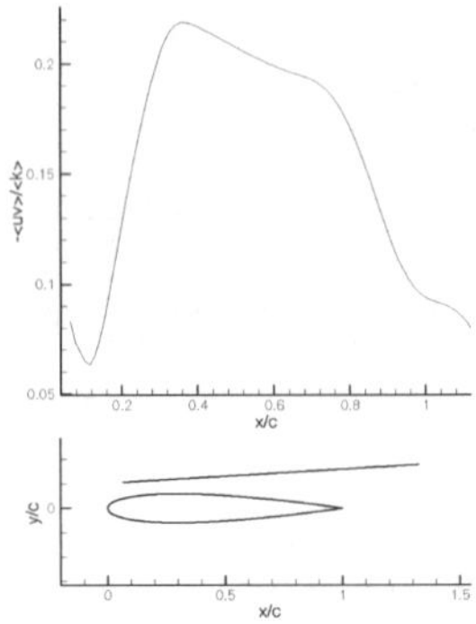

Figure 1 Mean velocity profile in the recirculation region - DSM model; exp. (Berton *et al* 2002).

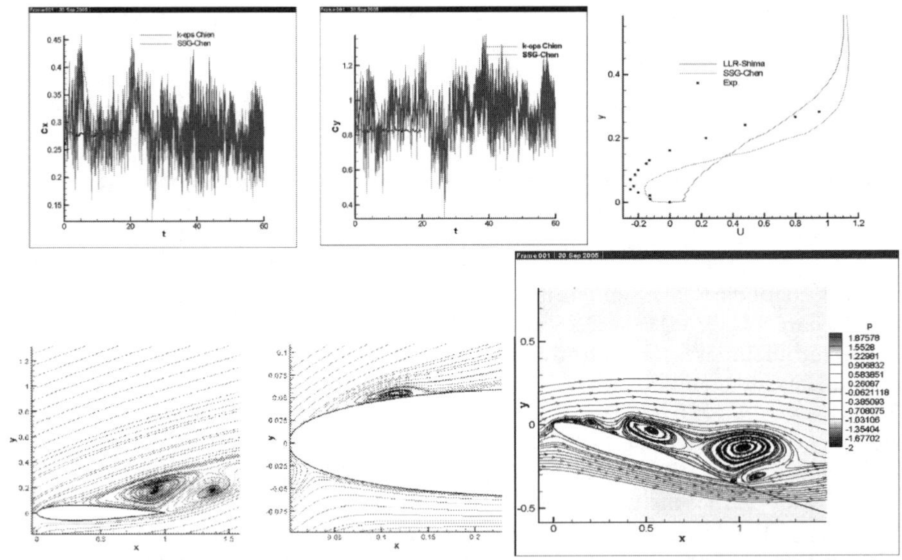

Figure 2 Time-dependent drag and lift coefficient; SSG model. Smaller amplitudes: k-ε-Chien model. Exp. C_D=0.32, exp. C_L=0.75

12.1.1 *Two-equation modelling improvement for detached unsteady flows*

IMFT has applied the modified OES/k-ε model derived by reconsidering the turbulence scales through the DSM model, in the context of the OES, organised Eddy Simulation approach, (Dervieux, Braza and Dussauge, (1998)), (Jin *et al*, 1994). The approach, described in detail by (Bouhadji *et al*, 2002a,b), (Hoarau *et al*, 2002) is schematically represented in fig.16, chapter 'Circular cylinder'). It consists of a dual spectrum splitting where the first part regroups the organised coherent flow structure and the second part, the random turbulence fluctuation.

This corresponds to the phase-averaged Navier-Stokes equations in the time-domain. The fact that the turbulence fluctuation energy spectrum is extended from the low to the high frequency range allows the use of the statistical approach for the modelling of this part, which is very efficient in respect of the high-Reynolds number. However, the turbulent scales *have to be reconsidered, in respect of the non-equilibrium turbulence*. This has been achieved by the IMFT's physical experiment around a circular cylinder. The solution of the time-marching phase-averaged Navier-Stokes equations ensures attendance of the steady-state equilibrium state in the regions where the flow is in statistical equilibrium, as well as the correct unsteady behaviour with an accurate prediction of the drag coefficient, in the strongly detached non-equilibrium flow regions. This is achieved inherently by the sum of $V(x,t)_{\text{resolved}} + V(x,t)_{\text{modelled}}$ in the physical domain, These elements had been detailed in the Flomania progress report by (Hoarau *et al* 2002). The improvement of two-equation turbulence modelling for unsteady flows is summarised as follows: an *optimum* C_μ *value of 0.02* has been suggested together with an optimum damping function, (Jin *et al* (1994)) re-scaled from DNS of IMFT (Hoarau, 2002) and originally based on boundary-layer DNS (Rodi *et al*, (1993)). The 3D test-case considered allows the investigation of the way the *inherent* unsteady three-dimensional motion is developed from a nominally 2D configuration. The prediction of the present inherent 3D unsteady turbulence motion is very well compared to the way of development of this kind of motion investigated in direct numerical simulations (Braza *et al,* 2001), fig.3. The Hopf bifurcation related to the development of the von Kármán mode is subjected to a spanwise predominant undulation, whose wavelength is evaluated to $\lambda z=0.64c$, in accordance with the DNS (Braza *et al*, (2001)) and experimental studies. The drag and lift coefficients versus time are compared with the experiment (Berton et al, (2002)). The mean drag coefficient is lower in the 3D case and the lift coefficient amplitudes are lower in 3D, as expected. A good comparison with this experiment is achieved (C_{D}_exp=0.320). An alternative in the context of OES modelling has been studied in collaboration UMIST-IMFT to derive a transport equation for the C_μ coefficient. This has been derived on the basis of the DSM and of the EARSM modelling of (Speziale *et al* 1996)) by A. Revell during his stay at IMFT, (Marie Curie Program, contract number HPMT-CT-2000-00079), (Revell *et al*, (2005a,b,c)). This model is sensitized in respect of the stress-strain misalignment in non-equilibrium turbulence. It provides C_μ values close to the previous approach. The C_{as} model evaluation and validation is in progress in the DESIDER program, by means of the IMFT's cylinder experimental data (fig.6), (Perrin, 2005)).

12.2 Conclusions

The 3D turbulent motion development in a strongly detached unsteady vortical flow around a wing is correctly predicted by the present OES/two-equation model. Therefore this closure has been applied by IMFT in the overall test-cases where IMFT contributes to in Flomania, as well as in DES modification of the l_{RANS} length scale, (chapter III_2). The provided two-equation model improvement (applicable either in k-ε *or in* k-ω), as well as in the RANS part of the DES (fig.3) is recommendable for use in industrial applications in aeronautics, concerning detached flows around bodies.

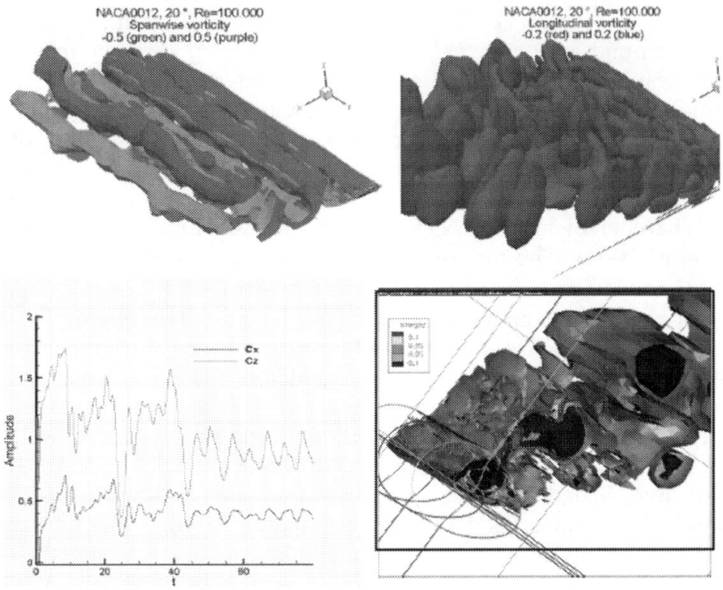

Figure 3 NACA0012 wing at Re=10^5 and α=20° (up), α=60° (down). k-ε_OES model. Up-left: spanwise vorticity ω_y :secondary-instability large-scale undulation. Up-right: longitudinal vorticity ω_x : smaller-scale counter-rotating cells. Down-left: Time-dependent drag and lift coefficients. Mean drag C_D_OES= 0.35, exp.C_D=0.32, standard k-ε_URANS C_D=0.44. Down-right: The OES_DES approach: iso-vorticity contours, NACA0012 beyond stall.

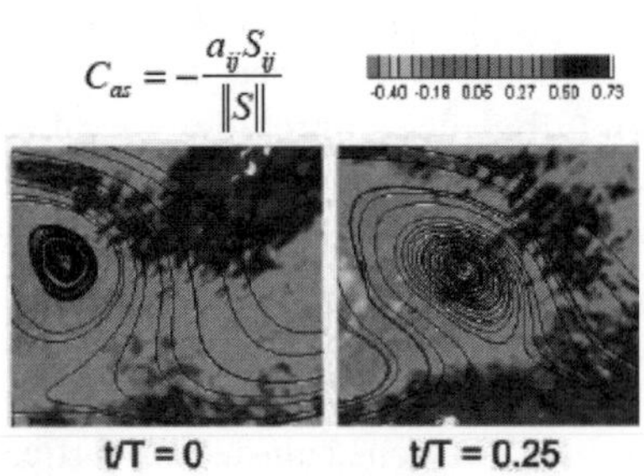

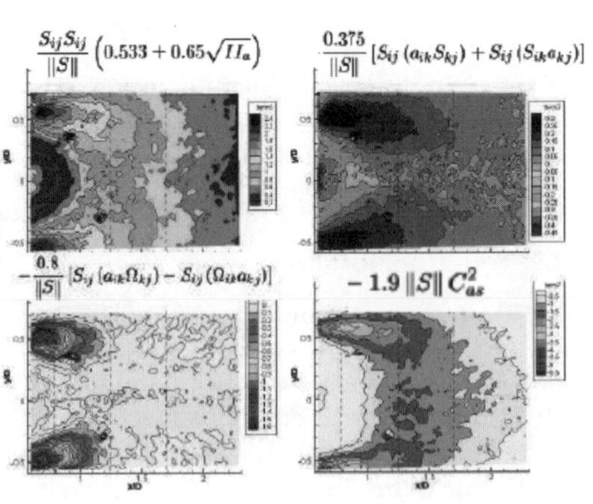

Figure 4 Variable $C\mu$ transport equation (the C_{as} model, Revell *et al*, 2005a,b). C_{as} term and C_{as} transport equation terms evaluation by means of the IMFT's cylinder experiment

13 Contribution of NUMECA: Main results achieved within the FLOMANIA project

A. Patel, B. Tartinville, L. Temmerman,
B. Leonard, and Ch. Hirsch, NUMECA

Abstract

The main focus of NUMECA during the FLOMANIA project was on the robust implementation of several established turbulence models and their validation on a range of generic and industrial test cases. NUMECA contribution strongly took benefit from interactions with other partners, UMIST in particular. In addition, grid adaptation criteria were developed. A last aspect concerned the implementation of the Detached Eddy Simulation (DES) in one of its solvers. During the course of this project, both NUMECA structured and unstructured solvers were put to use.

13.1 Description of numerical techniques

Part of NUMECA's contributions to FLOMANIA has been performed with the structured EURANUS flow solver. EURANUS is based on a collocated finite volume approach. The Jameson second order scheme with scalar artificial dissipation is used for the convective terms while the diffusive terms are discretised using second order central schemes. Temporal integration is based on the dual-time step method. At each time-step, a steady problem is thus solved using a 4th order Runge-Kutta scheme accelerated with the use of multi-gridding, local time-stepping and residual smoothing.

The other code, HEXSTREAM, used by NUMECA in this project is based on unstructured hexahedral grids and algorithms essentially similar to the EURANUS code previously described..

13.2 Turbulence modelling

13.2.1 In-house k-ε turbulence model

The in-house k-ε turbulence model implemented in the EURANUS structured code and in the HEXSTREAM unstructured code was improved with the assistance of UMIST in order to achieve robustness of the model on industrial cases.

The k-ε model is a low-Reynolds model due to Yang and Shih, and can also be used with extended wall functions. While modeling issues were not considered, all aspects of the integration of the model in the numerical frameworks of EURANUS and HEXSTREAM were investigated, with many modifications successfully applied: removal of ε relaxation, discretisation of the inviscid term with a first-order non-conservative upwind scheme, introduction of the Yap correction, removal of limiters on the production and the turbulent length scale, implementation of a linearized formulation for the production, point-implicit treatment of the negative part of the source terms.

These modifications actually helped to define a general framework for the implementation of other models (V_2f and RSM) which, to be successful, require a robust base model.

13.2.2 Other turbulence models

Based on the work performed on the in-house k-ε turbulence model, three additional models were implemented. The Lien's code friendly version of the V_2f model was integrated in the structured flow solver Euranus with the help of UMIST. The Menter SST model, the Wilcox k-ω model and the SARC model were implemented in the unstructured solver with, for the first two models, some support from UMIST again while, for the SARC model, the simplified version of SPTU was used. All models were then tested on the generic flow applications considered during this project.

13.2.3 The SSG model

The Sarkar Speziale Gatski (SSG) Reynold stress model with scalable wall function, chosen as common model by the FLOMANIA consortium, was implemented in the unstructured HEXSTREAM flow solver. In addition, several variants of the SSG model (SSG-ε variants, SSG-ω variants) were also implemented for evaluation. The major challenge was to achieve a robust implementation of the model in the numerical framework of the unstructured solver. All variants of the SSG model were treated numerically in the same way and shown to converge. Time-marching of the turbulent equations was done using a 4-stage Runge-Kutta scheme with under-relaxation. The convection terms were discretised using first order non-conservative upwind schemes while the diffusive terms used a simplified second-order centred scheme. Source terms were computed at cell centres with their negative contribution being treated as point-implicit as they strongly impact on the numerical stability of the code.

13.2.4 DES technique

Detached Eddy Simulation (DES) is a recently developped technique aiming at coupling the advantages of RANS and LES and is particularly well-suited for flows in which massive separation occurs. In this framework, the original version of DES, based on the Spalart-Allmaras turbulence model, was implemented in the HEXSTREAM code. The default numerical scheme (second order with scalar artificial dissipation) is however too dissipative as shown in preliminary tests with the decay of isotropic turbulence.

13.2.5 Adaptation sensor

A new adaptation sensor, based on the principle of reconstruction scheme, has been devised which allows to decrease the sensitivity of the adaptation criteria to user-defined parameters. Most adaptation criteria are based on flow features and often depend on user-defined parameters with values being strongly dependent of the flow considered.

The proposed sensor relies on the difference observed at a cell interface between the reconstructed values from left and right and the fact that this difference, on arbitrary meshes, is of the order of h^2. Refinement/coarsening is therefore triggered if the local value of the sensor is 4 times higher/lower than the averaged sensor value.

13.3 The asymmetric plane diffuser

The asymmetric plane OBI diffuser was computed on a grid made of 39000 cells with $y(1)^+ = 0.75$ for the Spalart-Allmaras model, the SARC model, the k-ε Yang-Shih model, the k-ω model, the SST model, the SSG model and the V_2f model and results where then compared with experimental data. Further details on this test-case as well as an extended set of results can be found in Section V.3 of this book. Figure 1 shows the influence of the choice of the turbulence model on the length of the recirculation zone.

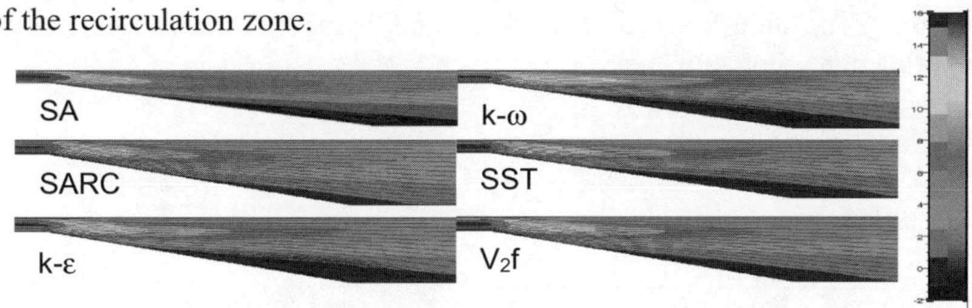

Figure 1 Prediction of the recirculation bubble in the Obi diffuser for different turbulence model

13.4 Grid adaptation on the Ahmed car body

The Ahmed car body with a slant angle of 25° was considered for testing the efficiency of the adaptation sensor developed in the course of Flomania. Starting from an unstructured mesh made of 417000 cells, 2 adaptations were performed leading to meshes of 756000 and 815000 cells. Figure 2 shows the evolution of velocity at mid-slant, as refinement progresses, the results converge towards a grid-independent solution.

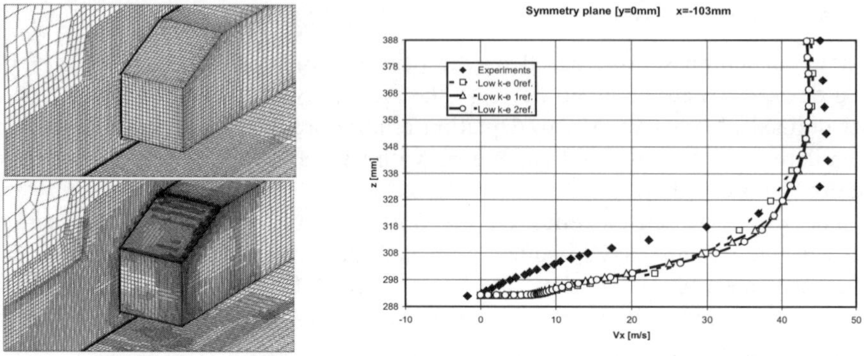

Figure 2 Initial and final adapted meshes around the Ahmed car body[Left] Streamwise velocity profiles for the three different grids in the slant region [Right]

13.5 The ONERA M6 wing

This case consists of a swept tapered wing with symmetric cross sections at an angle of incidence of 3.06° for a Mach number of 0.84 and a Reynolds number of 11.7 millions. Computations performed in the framework of FLOMANIA were done on the mandatory mesh (1.6 million cells) using the Spalart-Allmaras model and the Yang-Shih k-ε model. Both models agree well with each other and the experiment as shown in Figure 3.

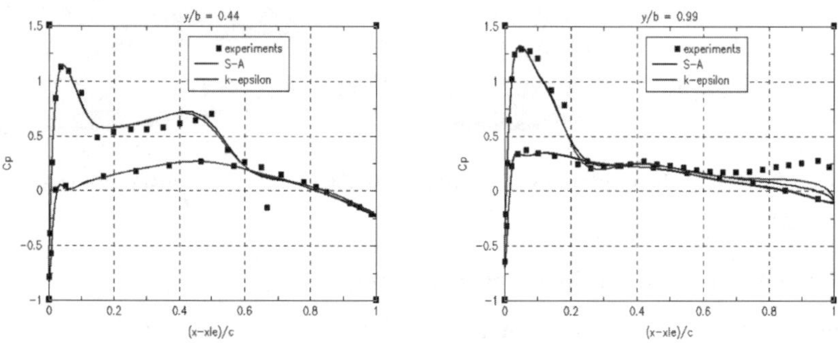

Figure 3 Pressure coefficient distribution at two different locations along the wing

13.6 The Ahmed car body

This test-case was used to assess the performance of the various turbulence models implemented during the course of FLOMANIA and the influence of mesh densities. A total of 5 turbulence models were thus tested (the original version of the in-house k-ε model and its improved version, the Spalart-Allmaras model, the SARC model and the SST model) on three different meshes: one structured (1,300,000 cells) and two unstructured (815,000 and 822,000 cells).

All models, except the SST model, failed to predict the recirculation on the slant. The improved k-ε model gave results very close to the Spalart-Allmaras model and the original k-ε model. The SARC model also predicted results similar to the Spalart-Allmaras model although more structures are visible. Profiles of the streamwise velocity at one location on the Ahmed body are shown on Figure 4.

Regarding the grid convergence studies, the Spalart-Allmaras model was used and results were in good agreement with each other and with the experimental data although it was again not possible to predict the recirculation region on the slant.

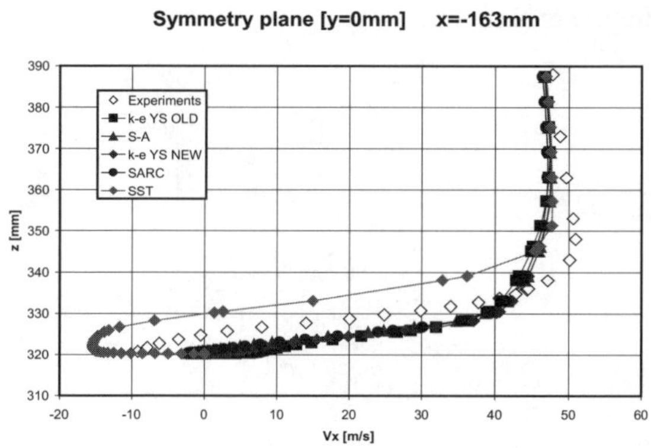

Figure 4 Streamwise velocity profiles for different turbulence models at one location on the Ahmed body: x = -163 mm (slant)

13.7 The DLR-F6 wide-body aircraft

The DLR-F6 configuration was also chosen for the 2nd AIAA drag prediction workshop held in 2003. It consists of wing-body configuration with or without nacelles at M = 0.6 and Re = 3,000,000 and experimental results are available for different angles of incidence. NUMECA's contribution for this test case is a series of computations at different incidence angle on a mesh made of 1,400,000 cells (coarse mesh for the previously mentioned workshop) using the Spalart-Allmaras model. Figure 5 shows the evolution of drag and lift coefficients for different angles of incidence.

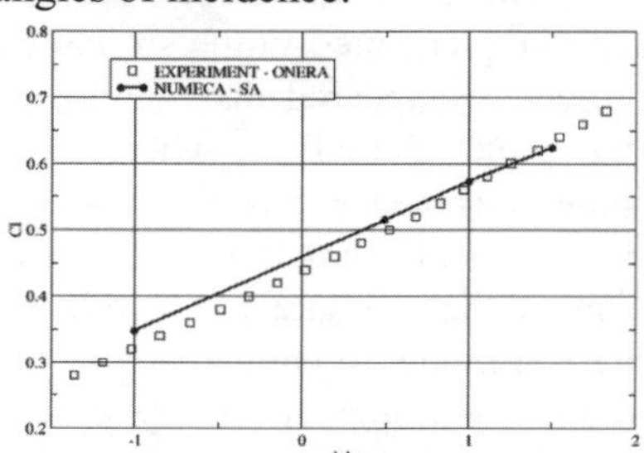

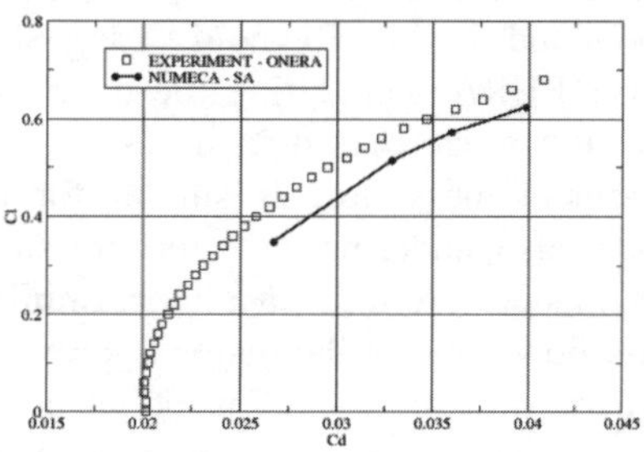

Figure 5 Evolution of lift coefficient against the angle of incidence [Left] and drag coefficient [Right]

13.8 The generic car mirror

The geometry considered is shown in Figure 8. Two DES computations on unstructured meshes were performed: a first one at a freestream velocity of 20 m/s with a mesh using 420000 cells and a second one at the reference freestream velocity of 40 m/s on a mesh of 1.1 millions cells. The computations were performed using the central JST scheme with low Mach number preconditioning and 50-100 iterations were required at each time-step to achieve two orders of convergence on the residuals. Overall, the unsteadiness of the flow was well captured and spectra, shown in Figure 6, agreed well with experimental data and did not show any dominant frequency.

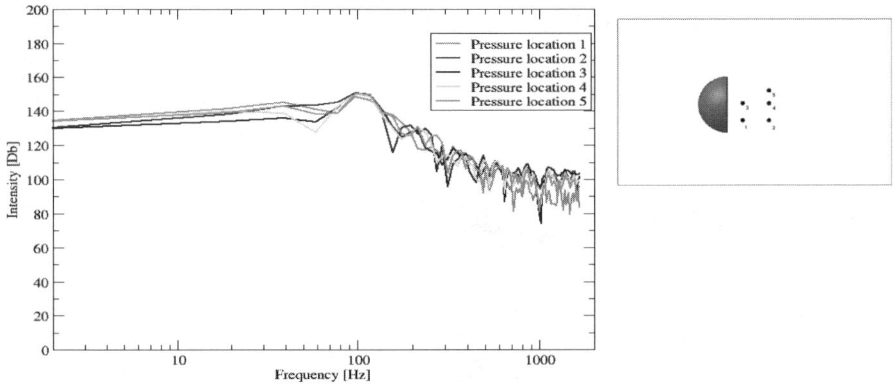

Figure 6 Noise spectra for the 5 different locations shown in the wake of the mirror

13.9 The rotor 37

The NASA rotor 37 is an isolated transonic axial compressor rotor with 36 blades and has been presented in the AGARD Advisory Report 355 entitled "CFD Validation for Propulsion System Components".

A multi-block low Reynolds mesh made of 615,000 grid points is used for this test case. Simulations were performed with the Spalart-Allmaras model, the v^2-f model and the low Reynolds Yang-Shih k-ε model implemented in the structured solver EURANUS during the course of this project. It appeared that the Yang-Shih k-ε model has difficulties in predicting the choke mass flow, whereas the maximum mass flow is similar for the two other turbulence models. The k-ε model also under-predicts the pressure ratio whereas both Spalart-Allmaras and v^2-f models produce an almost similar evolution of the pressure ratio with the mass flow. One of the major improvement of the v^2-f model in comparison to the Spalart-Allmaras model is in its prediction of the isentropic efficiency profile which is increased by 1% and so, closer to the experimental data. For more results, the reader is referred to Section IV.2 of the present book.

14 Contribution by ONERA

B. Aupoix, P. Ferrey, C. Gleyzes, R. Houdeville, and C. Marmignon, ONERA

14.1 Introduction

Within this project, ONERA activity was mainly focussed upon model implementation in elsA Navier-Stokes solver, model development as ONERA experience has shown that Reynolds stress models based upon a transport equation for ε are not well suited to predict separation and model validation. ONERA inherited test cases from AIRBUS-F after AIRBUS-F withdrawal. As the modelling work is not presented elsewhere, most of this section will be devoted to it.

14.2 Implementation of the common turbulence models in elsA software

The implementation of the common Reynolds stress models in the elsA software first required new developments of the code architecture, as it was initially restricted to two-equation models and eddy viscosity assumption. The Reynolds stress model implementation has unfortunately been delayed due to other severe revisions of the elsA software.

The high Reynolds number model form was implemented with elsA wall functions. First tests show the system to be very unstable. Use of a LUSSOR implicitation technique improved the model robustness. Converged results have been obtained on elsA validation test cases such as a flat plate and the RAE 2822 transonic airfoil. Application to the Obi diffuser evidenced problems in the model implementation as results departed from standard. Due to a lack of time, the cause of this discrepancy has not been identified.

The low Reynolds number form has also been coded, but not tested as the effort was first on the high Reynolds number version.

In parallel, work has been performed to improve turbulence model implementation and robustness in elsA. The emphasis has been put on the one hand upon the implicit treatment of the source terms and, on the other hand, upon the use of upwind schemes in the turbulence transport equations. These developments have shown to improve model robustness but are available only in a dedicated version of the code. They have to be transferred to the official version.

14.3 Turbulence model development

The goal of the ONERA activity was to develop a new Reynolds stress model which provides good predictions of flows up to separation. For that, the technique proposed by Catris and Aupoix (2000) has to be extended to Reynolds stress models.

14.3.1 Catris and Aupoix technique

Previous works at ONERA pointed out the importance of the length scale transport equation on the prediction of the flow response to pressure gradients. Catris and Aupoix (2000) proposed to write a length scale transport equation for an arbitrary length scale determining variable $\Phi = k^m \varepsilon^n$ in a generic form which, for Reynolds stress models, can be extended as:

$$\frac{D\Phi}{Dt} = (C_{\Phi 1} P_k - C_{\Phi 2}\varepsilon)\frac{\Phi}{k} + \frac{\partial}{\partial x_k}\left(D_{\Phi\Phi}\frac{k\overline{u_k' u_l'}}{\varepsilon}\frac{\partial \Phi}{\partial x_l}\right) + \frac{\partial}{\partial x_k}\left(D_{\Phi k}\frac{\Phi\overline{u_k' u_l'}}{\varepsilon}\frac{\partial k}{\partial x_l}\right)$$
$$+ C_{\Phi\Phi}\frac{k\overline{u_k' u_l'}}{\varepsilon\Phi}\frac{\partial \Phi}{\partial x_k}\frac{\partial \Phi}{\partial x_l} + C_{\Phi k}\frac{\overline{u_k' u_l'}}{\varepsilon}\frac{\partial k}{\partial x_k}\frac{\partial \Phi}{\partial x_l} + C_{kk}\frac{\Phi\overline{u_k' u_l'}}{k\varepsilon}\frac{\partial k}{\partial x_k}\frac{\partial k}{\partial x_l}$$

It has been checked that this equation is generic, i.e. that the transport equation for any other combination of k and ε can be deduced from this transport equation and the turbulent kinetic energy transport equation and has the same form.

The second point in the Catris and Aupoix approach is to determine the model coefficients so that the model exhibits typical behaviours which have been checked to be necessary if a good prediction of separation is desired whatever the flow conditions. These constraints are:

- consistency with homogeneous flow, including turbulence decay,
- prediction of the slope of the logarithmic region, whatever the pressure gradient (or at least for moderate pressure gradients),
- prediction of the square root region for strong adverse pressure gradients,
- good behaviour at turbulent/non-turbulent interfaces.

It must be pointed out that no existing model fulfils all these constraints. All the above behaviours can be expressed as relations between the constants in the length scale transport equation, but these relations also involve the associated Reynolds stress model which controls the anisotropy levels.

14.3.2 Model calibration

The constraints have been derived both for the Launder, Reece and Rodi (1976) and the Speziale, Sarkar and Gatski (1991) Reynolds stress models. A first range of model coefficients was guessed. In order to optimise these coefficients within the acceptable range, the complete model (SSG Reynolds stress model + length scale equation) has been implemented in a code computing self-similar solutions for wakes, plane and round jets, mixing layers and the outer region of the boundary layer.

However, it turned out that no reliable solution can be obtained, whatever the set of model constants. Generally, no convergence can be achieved. A lot of numerical tricks were tried to improve the convergence, some gave some hope but none really worked.

14.3.3 Analysis of the behaviour of turbulence models near a turbulent/non-turbulent interface

It appeared that all the numerical problems occurred near the outer edge of the turbulent region, although the model has been calibrated to give a good behaviour in the vicinity of a turbulent/non turbulent interface, following works by Cazalbou et al. (1994) and Cazalbou and Chassaing (2002). The classical analysis states that all variables evolve as power law solutions in the vicinity of the interface and constraints on the exponents ensure a fair behaviour, i.e. a smooth matching between the turbulent and non-turbulent regions, with turbulence propagation from the turbulent to the non-turbulent region.

A deeper analysis pointed out that they are not one but two power law solutions, and that the relevant solution is none of them but a third solution which bridges the two power laws. This lead to new constraints, more restrictive than the initial ones (Ferrey and Aupoix, 2005) and which were not fulfilled by the previously tested sets of model constants.

Moreover, it has been shown that any model with constant coefficients is unable to satisfy all the constraints.

14.3.4 Intermittency modelling

A model which satisfies all the constraints could be obtained by letting the model "constants" vary, at least near the turbulent/non-turbulent interface. In the framework of Reynolds stress models, this can easily be done which functions of the anisotropy invariants or of the production to dissipation ratio. However, it seemed more physical to account for the intermittency to better reproduce the physics near the interface. Once intermittency is accounted for, the behaviour near the turbulent/non-turbulent interface is completely decoupled from the other constraints so that model constants can be determined.

14.3.5 Final model

Accounting for the intermittency leads to three changes in the model. In the Reynolds stress transport equation, diffusion must account for the intermittency so that the diffusion term now reads:

$$Diff_{ij} = D_{kk} \frac{\partial}{\partial x_k}\left(\frac{k}{\varepsilon \gamma^{1-m_0}} \overline{u_k' u_l'} \frac{\partial \overline{u_i' u_j'}}{\partial x_l} \right)$$

A transport equation for the intermittency factor γ has to be added. In classical form, it reads:

$$\frac{D\gamma}{Dt} = \frac{\partial}{\partial x_k}\left(C_g (1-\gamma) \frac{k}{\varepsilon \gamma^{1-m_0}} \overline{u_k' u_l'} \frac{\partial \gamma}{\partial x_l} \right) + \left[C_g^{\mathrm{Prod}} \frac{P_k}{k} - C_g^{\mathrm{Diss}} \frac{\varepsilon}{k} \right] \gamma (1-\gamma)$$
$$+ C_g^2 \frac{k}{\varepsilon \gamma^{1-m_0}} \overline{u_k' u_l'} \frac{\partial \gamma}{\partial x_k} \frac{\partial \gamma}{\partial x_l}$$

Of course, on walls, the boundary condition is $\gamma = 1$ while γ is set to zero (or to a very small value) in non-turbulent regions.

At last, the length scale equation also has to be altered to account for intermittency, which leads to a more complex form. The final length scale equation has been written for the dissipation rate ε as:

$$\frac{D\varepsilon}{Dt} = (C_{\varepsilon 1} P_k - C_{\varepsilon 2}\varepsilon)\frac{\varepsilon}{k} + \frac{\partial}{\partial x_k}\left(D_{\varepsilon\varepsilon} \frac{k\overline{u_k' u_l'}}{\varepsilon \gamma^{1-m_0}} \frac{\partial \varepsilon}{\partial x_l} \right) + \frac{\partial}{\partial x_k}\left(D_{\varepsilon k} \frac{\overline{u_k' u_l'}}{\gamma^{1-m_0}} \frac{\partial k}{\partial x_l} \right)$$
$$+ C_{\varepsilon\varepsilon} \frac{k\overline{u_k' u_l'}}{\varepsilon^2 \gamma^{1-m_0}} \frac{\partial \varepsilon}{\partial x_k}\frac{\partial \varepsilon}{\partial x_l} + C_{\varepsilon k} \frac{\overline{u_k' u_l'}}{\varepsilon \gamma^{1-m_0}} \frac{\partial k}{\partial x_k}\frac{\partial \varepsilon}{\partial x_l} + C_{kk} \frac{\overline{u_k' u_l'}}{k \gamma^{1-m_0}} \frac{\partial k}{\partial x_k}\frac{\partial k}{\partial x_l}$$
$$+ \frac{\partial}{\partial x_k}\left(D_{\varepsilon\gamma} \frac{k\overline{u_k' u_l'}}{\gamma^{2-m_0}} \frac{\partial \gamma}{\partial x_l} \right) + C_{\varepsilon\gamma} \frac{k\overline{u_k' u_l'}}{\varepsilon \gamma^{2-m_0}} \frac{\partial k}{\partial x_l}\frac{\partial \gamma}{\partial x_k} + C_{k\gamma} \frac{\overline{u_k' u_l'}}{\gamma^{2-m_0}} \frac{\partial \varepsilon}{\partial x_l}\frac{\partial \gamma}{\partial x_k}$$

The final set of model constants is:

$$C_g = D_{kk} = 0.3 \quad D_{\varepsilon\varepsilon} = 0.4 \quad D_{\varepsilon k} = 0.08 \quad C_{\varepsilon\varepsilon} = -0.1533 \quad C_{\varepsilon k} = -0.8032$$
$$C_{kk} = 1.3581 \quad D_{\varepsilon\gamma} = -0.18 \quad C_{\varepsilon\gamma} = 0.6159 \quad C_{k\gamma} = -1.0175 \quad m_0 = \frac{2}{3}$$
$$C_g^{Prod} = 1.6 \quad C_g^{Diss} = 0.05 \quad C_g^2 = 0.1 \quad C_{\varepsilon 1} = 1.4 \quad C_{\varepsilon 2} = 1.92$$

The above model is only the high Reynolds number version. The near wall region corrections are those of the common model except the f_w damping function which has been altered as:

$$f_w = \exp\left[-\left(0.031 \frac{\sqrt{k}\, y}{\upsilon} \right)^4 \right]$$

14.3.6 Model validation

The above model constants have been determined by optimising the model for self-similar flows. Examples of the improvement obtained with the model are given in the figures for a self-similar, mixing layer flow. Fig. 1 shows examples of present model results for a mixing layer. The present model better predicts the velocity profile and provides a smooth matching with the external flow.

The model has then been introduced in a three-dimensional boundary layer solver to test its capabilities to predict boundary layers submitted to adverse pressure gradients. For moderate adverse pressure gradients, such as the Bradshaw 3300 test case (Fig. 2), as well as for strong adverse pressure gradients, such as the Skåre and Krogstadt experiment (Fig. 3), the proposed model is definitively superior to the common model, as it is better suited to predict separation. This is exemplified by the form of the velocity profile close to the wall, which is fuller with the common model. The present model also improves predictions compared to the SST model so that it should be superior to ω-based Reynolds stress models.

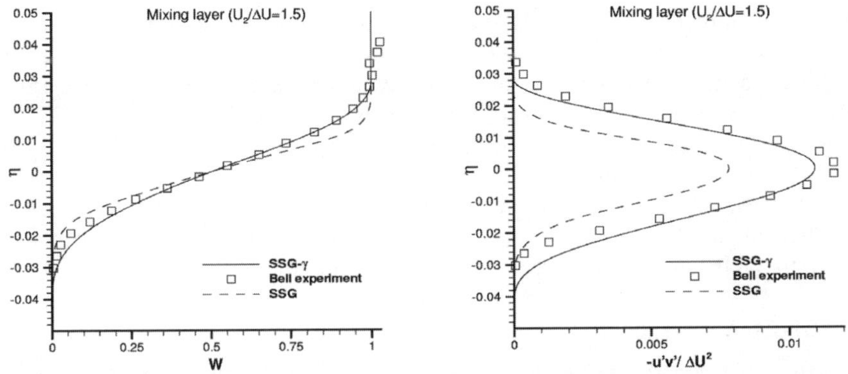

Figure 1 Prediction of the velocity and turbulent shear stress profiles for a mixing layer. Comparison of proposed (SSG-γ) model with the common model (SSG)

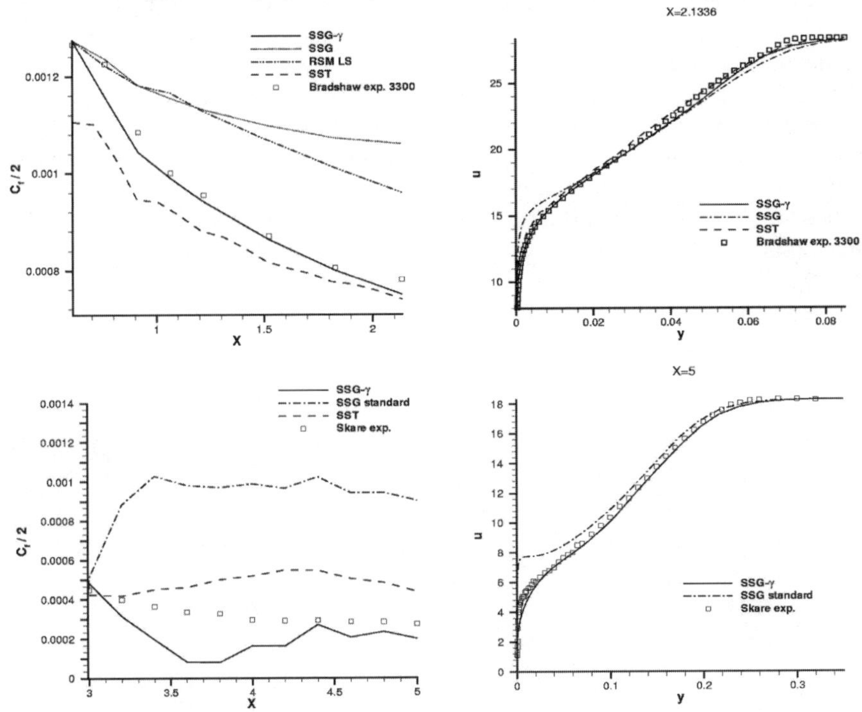

Figure 2 Prediction of the skin friction evolution and of the velocity profile. Comparison of proposed (SSG-γ) model with common (SSG) and SST models;
Top : Bradshaw mild adverse pressure gradient experiment;
Bottom : Skare and Krogstad strong adverse pressure gradient experiment

14.4 Validation test cases

ONERA contributed to four validation test cases: two academic (Obi diffusor and A-airfoil), two industrials (AS28 and generic air intake). These last two test cases were inherited from Airbus-F. As the A-airfoil experiment has been performed by C. Gleyzes, his expertise on this test case was particularly asked upon in the various discussions about separation bubble effects, flow two-dimensional character, effect of wind tunnel walls…

Due to problems in the common model implementation, ONERA contribution was restricted to eddy viscosity models, except for the Obi diffusor. The results obtained by ONERA are presented in the relevant sections.

14.5 ONERA's perspectives

The contribution to FLOMANIA has shown that Reynolds stress models can be implemented and used in Navier-Stokes solver at a reasonable cost. ONERA plans are first to debug the present implementation to have a reliable model. For accuracy as well as numerical reasons, a ω transport equation will be preferred to the common model.

Then, Reynolds stress model evaluation shall be pursued in the framework of different contracts and on house funding to evaluate the improvement that can be obtained for different applications, some of them not addressed by ONERA within FLOMANIA such as fighter aircrafts or turbomachinery.

Once an efficient Reynolds stress model architecture is available, other models such as the one presented above can easily be implemented in ONERA's software elsA.

15 Contribution by SPTU: Support of Partners' Efforts Directed to Implementation of DES Technology

A. Garbaruk, D. Magidov, M. Shur, M. Strelets, and A. Travin
St.-Petersburg State Technical University (SPTU)

Abstract

This chapter summarises the work performed by St.-Petersburg State Technical University as a major sub-contractor in FLOMANIA. Along with a support of Partners' work on implementation of the Detached-Eddy Simulation methodology (DES) being the main responsibility of SPTU within FLOMANIA, this includes some own DES and RANS studies carried out on requests of the Project coordinator.

15.1 Introduction

SPTU has occupied a central position in efforts to exploit DES by the other Partners within FLOMANIA that had been a long-term goal of the Project. This has resulted in a large number of SPTU interactions with different Partners, mostly concerned with different aspects of implementation and tools for analysis of LES and DES simulations. Other than that, SPTU provided meshes for DES of NACA 0012 airfoil beyond stall (Test Case 14) and 3D Circular Cylinder (Test Case 21) and performed DES of these and some other generic test cases that were used by the other Partners for validation of DES implementations in their in-house CFD codes.

Along with this, SPTU has provided a support to the interested Partners in implementation of the SARC turbulence model of Spalart and Shur, 1997 and performed RANS computations of several test cases (9, 14, 26, 28) with the use of the conventional linear eddy viscosity models (SA, SARC, SST). This activity has permitted to widen the comparison of different models capabilities as applied to complex turbulent flows.

Finally, an important outcome of SPTU participation in FLOMANIA the university has significantly benefited from, is an implementation and validation of the Common Differential Reynolds Stress Model (SSG-Ch) in the in-house code (NTS code).

Below we first briefly outline this code, which has been used by SPTU in all the numerical studies (Section 15.2). Then, in Section 15.3, we highlight some results of SPTU activity in DES area. Finally, in Section 15.4, results are presented of the 3D RANS computations of the A-airfoil aimed at the evaluation of 3D-effects caused by the sidewalls of the test section in the experiment.

15.2 NTS Code Description

The NTS computer code is solving 2D and 3D, compressible (arbitrary Mach number) and incompressible, steady and time-dependent Navier-Stokes equations with the use of a high-order implicit finite-volume formulation on the structured multi-block overlapping grids. Turbulence treatment is possible in the framework of the steady and unsteady RANS with a wide variety of conventional statistic turbulence models as well as with "turbulence resolving" approaches such as DES (based on SA and SST RANS models), LES, and DNS.

Following implicit flux-difference splitting numerical methods (based on MUSCL approach) are implemented in the code.

1. Method of Rogers and Kwak, 1988 (for incompressible flows).
2. Method of Roe, 1981 (for compressible flows).
3. Method of Weiss and Smith, 1995 with low Mach number preconditioning (for compressible flows at arbitrarily low Mach number).

Spatial approximation of the inviscid fluxes within all these methods is performed with the use of 3rd or 5th-order upwind-biased scheme or with the use of 4th-order centered scheme. Also, hybrid, weighted upwind/centered, schemes with the blending function dependent on the solution (Travin et al., 2002) or specified by the user are available and routinely used in DES, LES, or DNS. For the supersonic flows with shocks, flux limiting with a range of standard limiters can be turned on in order to provide smooth solutions in the vicinity of shock waves.

Viscous fluxes are approximated with the 2nd-order (default) or with the 4th–order (optionally) centered schemes.

Time integration in the code is implicit, with user-specified type of relaxation procedure including Gauss-Seidel relaxation by planes/lines, LU relaxation, or diagonally dominant approximate factorization (DDADI). Optionally, different implicit algorithms can be used in different grid blocks and, also, for solution of the gas-dynamic and turbulence-transport equations.

For the unsteady flows, time-derivatives are approximated with 2nd-order backward differences (three-layer scheme) with dual time-stepping (infinite default pseudo-time step) and sub-iterations. The number of sub-iterations at each time step depends on the problem being solved but usually is within the range from 5 up to 20 (this ensures reduction of the maximum residual by 3-4 orders of magnitude).

Parallelization of the solver is based on a "hybrid" conception that combines MPI and Open MP technologies. This permits to adjust the code to a specific hardware being used (shared, distributed or mixed memory structure).

It should be noted that the code permits computations with structured grid-blocks being not only simple rectangular parallelepipeds (in the computational coordinates) but also with the blocks containing arbitrary "holes" (cut-outs) inside such parallelepipeds. This feature is very helpful since it permits to minimize the number of the inter-block boundaries.

The inter-block interaction is implemented as follows. It is assumed that the grid-blocks used are the overlapping (not simply adjacent) ones. Coupled with the iterative procedure used for the solution of both steady-state and time-dependent problems, this permits to avoid imposing of any artificial boundary conditions at the inter-block boundaries. Namely, at each global iteration (for the steady problems) or at each subiteration of a time-step (for the time-accurate simulations), the values of all the primary variables (pressure, velocity components, and temperature) at the inter-block boundaries of a recipient block are computed by interpolation over the neighbouring cells of the donor grid-block. At the implicit stage of the solution procedure, zero values of all the residuals are imposed at the inter-block boundaries. In case of multiple blocks overlapping, a special system of priorities is applied to choose the block for interpolation. Other than that, an algorithm is developed that does not cause any lowering of the order of the spatial approximation in the case when the grids of the neighbouring blocks in the overlapping region are identical (a converged solution is the same as that obtained on a single-block grid). This feature is very important for parallel computations on computers with distributed memory, when the grid is being subdivided into several blocks artificially, i.e., only in order to provide a possibility for MPI-parallelization, since it permits to avoid any decrease of accuracy in this case.

The code has been intensively employed for many years now. A list of applications include a wide range of RANS, URANS, DES and DNS of aerodynamic flows and, also, LES-based computations of the jet noise. Accumulated experience shows that independently of the type of the flow through the boundary (subsonic or supersonic) the approach briefly outlined above does not virtually cause slowing down of the iterations convergence versus similar single-block computations, provided that the grid-steps in the overlapping regions are close to each other, and the overlapping is "deep" enough (more than one grid cell). Other than that, no noticeable defects of solutions are observed in the vicinity of the block interfaces. This is essential for the highly unsteady flows with vortices crossing the boundary, which is inevitable in any turbulence-resolving simulations (DES/LES/DNS).

15.3 Major Results of Activity in DES Area

15.3.1 Support of Partners in Implementation of DES

As mentioned in the Introduction, in accordance with its role in the Project, SPTU had a large number of interactions with different partners aimed at sharing experience in the area of DES.

First of all, software for calibration of the subgrid scale models for DES has been transferred to the partners with appropriate running and post-processing instructions. The software is a code generating a random divergence-free velocity field with a prescribed turbulence kinetic energy spectra needed to initiate simulation of decaying homogeneous isotropic turbulence in a cubic box with fully periodic boundary conditions. It also includes a post-processing routine for computing the energy spectrum by a given velocity field.

Further support was provided in a form of advices and comments on the implementation and verification of the DES technology in partners' in-house codes. Specifically, SPTU has provided statements and solutions for several test cases (backward facing step flow, the NACA0012 airfoil beyond stall (TC16)) to verify the implementation of the methodology. This includes a complete description of boundary conditions, meshes, pre- and postprocessing software, and supplementary information on numerical and implementation details.

Other than that, a current version of an empirical function designed to provide an automatic smooth switch from a high-order upwind-biased scheme in the RANS and irrotational regions of DES to central differencing in LES regions was distributed to the interested partners. On the basis of the suggested blending-function, AEA has developed a variant approach suited to CFX software.

Finally, in the course of a visit to TUB, Prof. M. Strelets gave a talk on DES fundamentals and applications. In the course of the visit to DLR, Dr. M. Shur provided assistance in implementation of the SA DES model in DLR's in-house CFD code TAU. In the course of a visit to IMFT, Prof. M. Strelets and Dr. M. Shur provided to IMFT background information on DES and some practical recommendation on its numerical implementation. Furthermore, SPTU interacted systematically with IMFT and provided some suggestions on the implementation of the DES technique in the IMFT codes. Much of this interaction was also devoted to an investigation of several aspects of the IMFT test case of flow around a circular cylinder (TC 21).

15.3.2 DES of 3D Circular Cylinder (IMFT Test Case 21)

The experimental study of this flow has been carried out by IMFT. A circular cylinder is placed in the 2,40m length test-section of S1/IMFT's subsonic wind tunnel between two end plates so that the cylinder aspect ratio, $L/D = 4.8$ (L is the distance between the two end-plates, D is the diameter of the cylinder) and the blockage ratio, D/H=0.208 (H is the distance between of the upper and lower walls of the section). The Reynolds number of the flow based on the cylinder diameter and inlet velocity was 140,000.

Preliminary, coarse grid, DES of the flow based on the assumption that the regime is sub-critical (laminar boundary layer separation and transition to turbulence in the separated shear layer) has shown that the boundary layers forming on the side-plates are very thin and do not affect the "core" of the flow any significantly. This is seen in Fig. 1, where we present an instantaneous swirl isosurface and the averaged pressure field from this simulation: the results visually reveal thin sidewall boundary layers and extended "2D" core. This finding has justified a much less expensive simulation with account of the upper and lower walls only, needed to represent their blocking effect, with the use of the periodic boundary conditions in the spanwise direction (two runs have been carried out with the span-period, L_z, equal to 2 and 4 cylinder diameters). A comparison of the results obtained in the simulations with the IMFT experimental data is presented in Figs.2-4. One can see that both mean flow velocity and turbulence statistics (Reynolds stresses) from the simulation agree with the data fairly well. It should be noted, that initially, due to some inconsistency in the pressure measurement technique, the experimental surface pressure significantly deviated from the predicted one. Exactly this circumstance gave a motivation to a double-checking of the experimental procedure by IMFT, which resulted in the correction of the original measurements (exactly these, corrected, data are presented in Fig. 4). Thus, this study provides an example of "validation of experiment by CFD" being far from typical today.

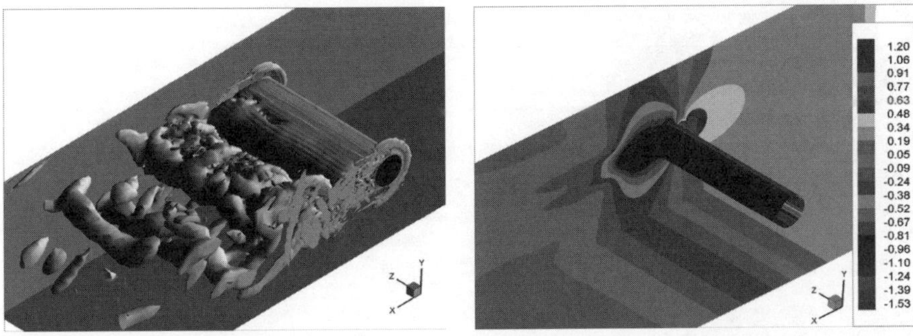

Figure 1 Instantaneous swirl (λ_2=0.5) isosurface (left) and time averaged pressure field (right) from SA DES of the 3D cylinder flow (Test Case 21)

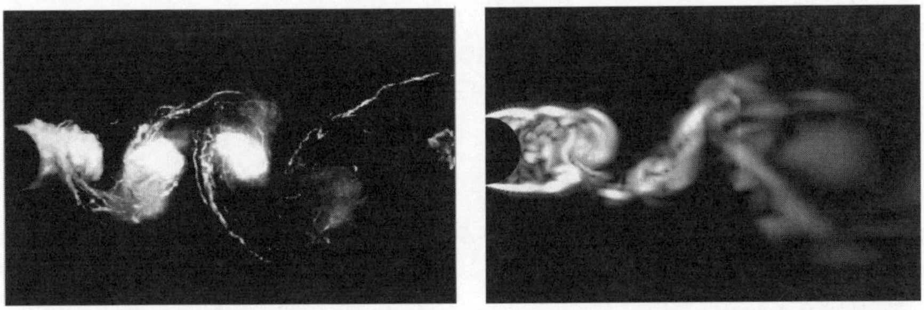

Figure 2 Comparison of flow visualisations from the SA DES (right) and IMFT experiment (left)

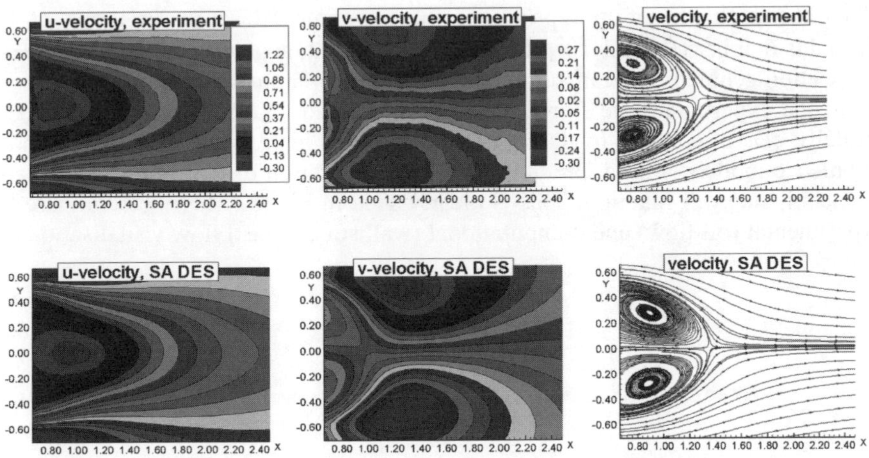

Figure 3 Comparison of SA DES mean flow prediction with IMFT experiment

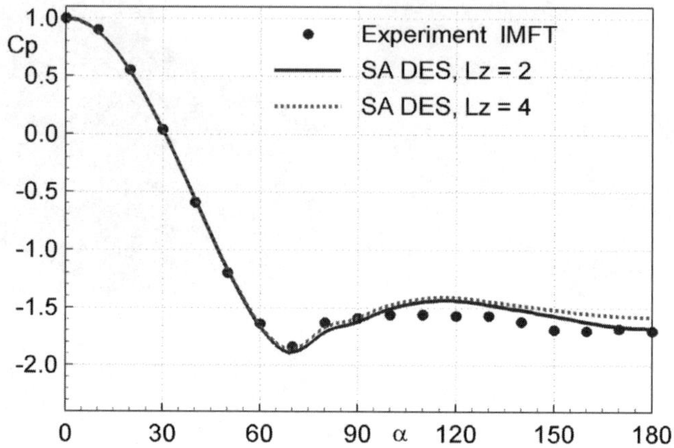

Figure 4 Comparison of SA DES prediction of mean pressure distribution over the cylinder with IMFT experiment

15.3 3D RANS of the ONERA A-Airfoil (TC 28): evaluation of sidewall effect

A goal of this study was getting an estimation of possible effect of the sidewalls on the flow past the A-airfoil studied in the experiments of Gleyzes and Capbern, 2003. In order to reach this objective we have performed 3D steady RANS computations of the flow past the airfoil with account of the sidewalls of the test section. The turbulence models used in these computations are: SA, SARC, and SST. The problem set-up reproduced the experimental set-up in terms of the Reynolds number ($Re=2.1\cdot10^6$), Mach number (0.15), angle of attack (13 degrees), test section geometry, and incoming boundary layer parameters.

In the experiment, the flow at the considered angle of attack turns out to be virtually 2D. In contrast to that, CFD, no matter which of the three listed above turbulence models is used, predicts a massive separation of the flow in the area of the airfoil/sidewall junction. This is clearly seen in Fig. 5, where we compare the experimental (oil flow) and computational (wall streamlines) flow visualisations.

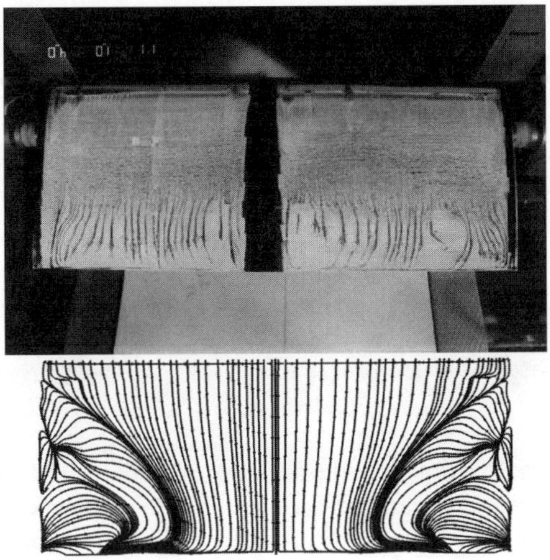

Figure 5 Comparison of experimental (up) and computational (down) flow visualisations

Figure 6 gives a more detailed idea on the flow topology predicted by 3D RANS, which clearly shows the large corner separation zones forming in the airfoil/sidewall corner due to the adverse pressure gradient on the suction side of the airfoil.

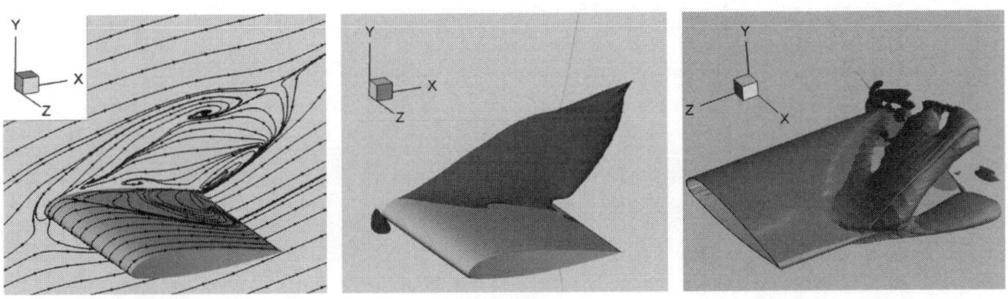

Figure 6 Streamlines at the airfoil surface and sidewall (left), surface of zero streamwise velocity (middle), and swirl isosurface λ_2=1.4 from 3D RANS solution

As a result, all the characteristics of the flow predicted by 3D RANS are far from those measured in the experiment and, also, from the similar predictions of 2D RANS (see Fig. 7).

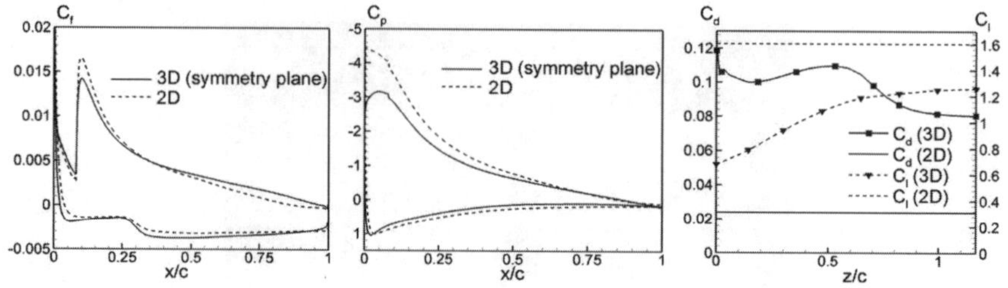

Figure 7 Comparison of the friction (left) and pressure (middle) coefficients distributions over the airfoil at symmetry plane from 3D RANS with 2D RANS predictions and spanwise distributions of drag and lift coefficients from 3D RANS (right)

As of today, we cannot find any quite definite explanation of this contradiction. It might be cased by an inadequate performance of the linear eddy-viscosity RANS models in this type of flow. A more optimistic explanation suggested by F. Menter in the course of the discussion of the results presented above is that we might be just near a bifurcation point, i.e., that with the same problem set-up but different initial conditions one could get different solution. A test would be to compute the flow at different angles of attack and see if there is a hysteresis when approaching the 13 degrees from below or from above, and at which angle the flow topology changes.

16 Method Used and Highlight Results Achieved in FLOMANIA

U. Bunge, C. Mockett, and F. Thiele, Technical University of Berlin (TUB)

Abstract

This chapter describes the method used by the Technical University of Berlin and it summarises the highlight results obtained with this method in the course of FLOMANIA. The main results are the successful implementation and calibration of DES for three different models and the improved representation of flow physics with these DES implementations.

16.1 Method used by TUB

16.1.1 Numerical method

The flow for all cases is computed numerically using an in-house finite-volume based code solving either the unsteady Reynolds-averaged or spatially filtered Navier-Stokes equations in case of a RANS or Large-Eddy simulation (LES), respectively. The procedure is implicit and of second order accuracy in space and time. All scalar quantities as well as the Cartesian components of tensorial quantities are stored in the cell centers of arbitrarily curvilinear, semi-structured grids. Diffusive terms are approximated with central schemes, whereas convective terms can be treated with central or upwind-biased limited schemes of higher order (Xue, 1998). A hybrid blending of both approaches for a Detached-Eddy Simulation (DES), as suggested by Travin et al., 2002 and discussed in chapter III-2 has been implemented, addressing the conflicting requirements posed by RANS and LES on the convection scheme. The linearized equations are solved sequentially and the pressure is iterated to convergence using a pressure-correction scheme of the SIMPLE type that assures mass conservation as the pressure equation is derived from the continuity equation (Karki and Patankar, 1989). Introducing apparent pressures and viscosities, a generalized Rhie & Chow interpolation is used to avoid an odd-even decoupling of pressure, velocity and Reynolds-stress components (Obi et al., 1991).

16.1.2 Turbulence Treatment

Turbulence is handled using three RANS turbulence models of different degrees of complexity, the details of which can be found in chapter III-2. The simplest is a modification of the SA model, the Strain-Adaptive Linear Spalart-Allmaras model (SALSA, Rung et al., 2003). Secondly, the linear local realizable (LLR) k-ω Model (Rung and Thiele, 1996), is a local linear two-parameter model derived from realizability and non-equilibrium turbulence constraints. Finally, the compact explicit algebraic stress model (CEASM, Lübcke et al., 2002), using the LL k-ε model as a background model (Lien and Leschziner, 1993) employs expressions for the Reynolds stresses chosen to yield physically correct results for highly 3D flows whilst preserving the simplicity of the quadratic formulation.

As long as grids are fine enough, a hybrid boundary condition (Rung et al., 2000) assures a low-Re formulation or solution down to the wall for all models.

All three models are also used for DES, where the constant C_{DES} is calibrated by computing the decay of isotropic turbulence as described by Bunge et al., 2003, and in the following section. To achieve a DES, the turbulence length scale in the model is replaced by the DES grid length scale, and the precise manner in which this is achieved for each of the TUB turbulence models is given in the DES section of chapter III-2. It is interesting to note that of the two length scales present in the background model of the CEASM, the wall-normal distance is left unchanged, allowing the construction of a shielding function to avoid grid-induced separation.

16.2 Highlight Results

16.2.1 Calibration of C_{DES}

As the DES modification (see chapter III-2) causes the underlying turbulence model to act as a subgrid-scale model in the LES zones, a test case to validate this LES capability is required. Furthermore, as the DES modification introduces a new model parameter C_{DES}, a reliable method must be found to determine a calibrated value for this. Both of these issues can be addressed simultaneously, employing the case of decaying isotropic (homogenous) turbulence (DIT, Comte-Bellot and Corrsin, 1971), as described by Shur et al., 1999. Furthermore, this procedure is to be followed for each of the DES implementations based on the three turbulence models mentioned above.

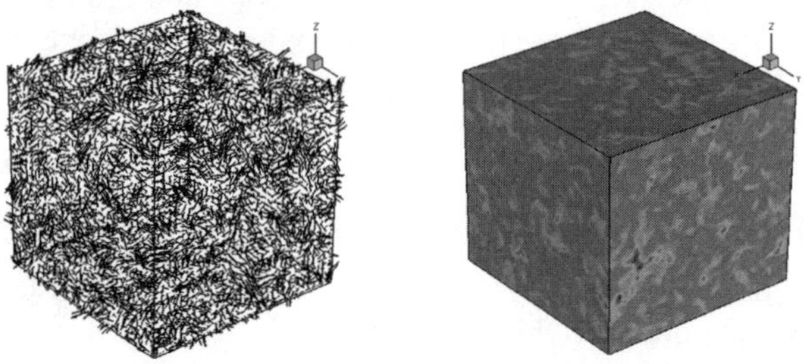

Figure 1 Initial velocity field and viscosity ratio for DIT

The decay of isotropic turbulence is simulated using three equidistant box grids of 16^3, 32^3 and 64^3 volumes, with periodic boundary conditions in all three directions. The flow field is initialised with a prescribed velocity field with zero mean flow, Fig. 1. This velocity distribution is obtained from the experimental energy distribution for t=0 using an inverse Fourier transformation from a tool provided by Prof. Strelets of SPTU. The same tool provides a means to extract an energy distribution from the computed velocity fields for comparison with experimental data at later time steps, as seen in Figs. 2 and 3.

Once an initial calibration investigation is complete, a deeper examination of the sensitivity of C_{DES} to variations of the underlying numerics is conducted.

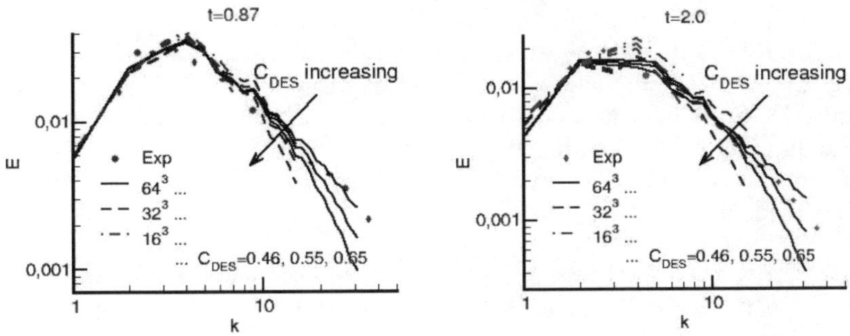

Figure 2 Calibration spectra for the SALSA model at t=0.87 (left) and t=2.0 (right)

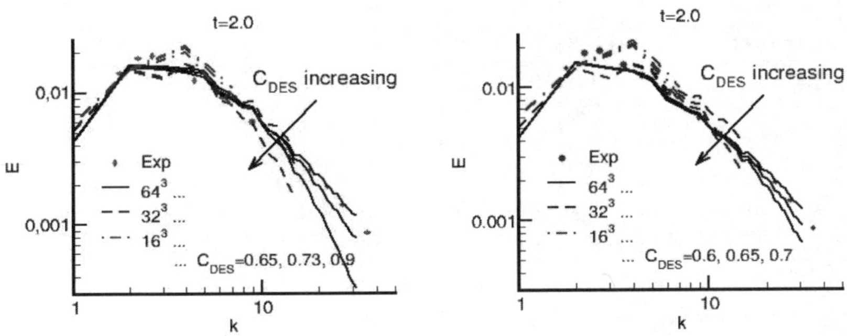

Figure 3 Calibration spectra for the LLR k-ω model (left) and the CEASM (right), at t=2.0

In Figs. 2 and 3, the resulting calibration spectra for each turbulence model are shown. The t=0.87 time point is shown only for one model (SALSA) as the long-term behaviour is more important. The results from the 32^3 grid were given precedence, as this was taken to be representative of the coarseness for practical applications. The level of dissipation in the underlying numerics can be varied using different flux blending parameters between upwind and central difference-based schemes. The effect of raising the dissipation is balanced by lower resulting C_{DES} values, Table 1, and is representative of realistic practical DES applications on coarser and less uniform grids.

Table 1 Range of applicable C_{DES} values for each model

Model:	Low dissipative:	Higher dissipative:
SALSA	0.65	0.55
LLR k-ω	0.78	0.73
CEASM	0.70	0.60

16.2.2 NACA0012 at high angle of attack

An application test case demonstrating the benefits of DES over URANS on the same grid is the computation of an airfoil in deep stall. In FLOMANIA, the NACA0012 profile was chosen at a range of angles of attack with Re=10^5, the complete set of results and a description of the grids for which can be found in chapter IV-6. Table 2 gives an overview for an angle of 60° of the mean lift (C_L) and drag (C_D) as well as the shedding frequency (St) for the coarse (c), medium (m) and fine (f) grids. The improvement of results in comparison to experimental data is evident.

Table 2 DES and URANS results on all the grids for 60° angle of attack in comparison to experiments for all models. Experimental force coefficient data as used by Strelets, 2001, Strouhal number from experiments of Swalwell et al., 2003

Model		CEASM			LLR k–ω			SALSA		
		$\overline{C_L}$	$\overline{C_D}$	St	$\overline{C_L}$	$\overline{C_D}$	St	$\overline{C_L}$	$\overline{C_D}$	St
Experiments		0.92	1.65	0.20	0.92	1.65	0.20	0.92	1.65	0.20
URANS	c	1.11	1.92	0.13	1.27	2.16	0.14	1.21	2.11	0.13
	m	1.08	1.83	0.13	1.19	1.99	0.16	1.12	1.93	0.12
	f	1.09	1.86	0.13	1.18	1.98	0.16	1.11	1.91	0.15
DES	c	0.97	1.63	0.17	0.93	1.57	0.18	0.91	1.53	0.17
	m	0.98	1.64	0.17	0.94	1.58	0.18	0.97	1.64	0.16
	f	0.92	1.56	0.18	0.94	1.58	0.20	0.95	1.59	0.18

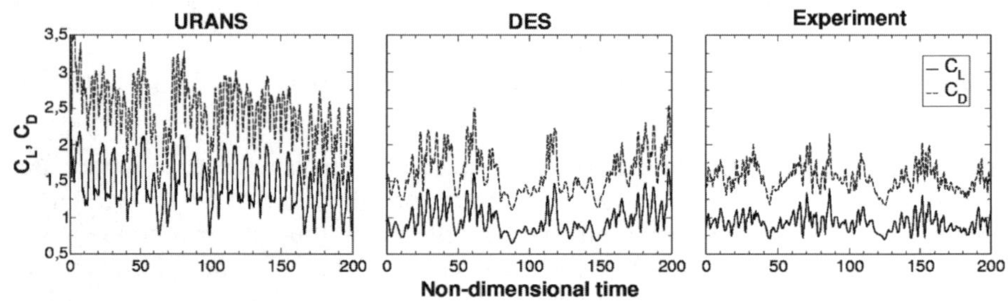

Figure 4 Lift vs. time for URANS, DES (CEASM) and experiments

Fig. 4 reveals that not only the frequency and mean value of lift and drag agree well with experimental data, but also the qualitative temporal behaviour is represented much better. The DES gives a more stochastic result with areas of reduced activity or "weak shedding cycles", which are also observed in the experiments and indicate a modulation between an organised flow pattern to a more chaotic one consisting of much finer structures.

17 Methods used and highlighted results from UMIST

J. Uribe, S. Utyuzhnikov, A. J. Revell, A. Gerasimov,
and D. Laurence, UMIST (The University of Manchester)

Abstract

The preferred models of the consortium (SSG, k-w, STT, SCW) have been implemented in a recent open source FV software and tested. Numerical behaviour of two popular versions of the V2F model have been compared and a new formulation proposed, that combines the accuracy of the original version and stability of the degraded "code friendly version". In the DES framework, the need for reduction of the EVM coefficient has been related to the stress-strain miss-alignment in rapidly changing time dependent mode and a single transport equation for this coefficient proposed. On the wall function front, the AWF of Craft and Gerasimov entailing an analytical integration of terms over the first cell (body force in particular) proved challenging to implement on unstructured grids, but was achieved. Towards the end of the project, a promising mathematical framework (Robin type conditions) was proposed, based on domain decomposition and mixed Dirichlet and Neumann conditions.

17.1 Introduction

The focus at UMIST has been set on "robust" formulations of the models and wall treatment, for implementation is professional software, and their suitability for industrial applications. Collaborations with Ansys-CFX and NUMECA, in addition to EDF, have been very fruitful in this respect. After fine tuning of the turbulence models in academic codes, UMIST implemented these models in the EDF unstructured finite volume software of EDF named *Code_Saturne*, described in Chapter II.9 (this software is available in full source code to academic partner-developers contributing to the *Code_Saturne* consortium). One exception is test-case 1 (wing-tip vortex) performed by C. Robinson with UMIST's STREAM code, but this as part of the MDAW project, and hence not described herein although those solutions will appear in chapter IV.

17.2 Two equations models improvements

17.2.1 Near-wall models

The viscous effects are not taken into account when using a wall function approach, therefore loosing information that can be important in some flows. In order to reproduce the near-wall effects, a low-Reynolds model should be used to compute all the sharp gradients present in the viscous sublayer. Many two equation models have a low-Reynolds counterpart but they have been developed (as if an afterthought) via damping functions in order to obtain the correct behaviour near the wall. These damping functions are usually design for a specific type of flow and are far from universal. Most of them contain the distance from the wall as a parameter, which can be ill-defined in certain geometries.

Durbin (1991) proposed a more physical approach to the near wall modelling by taking into account the effects of the wall via an elliptic equation. The formulation of the V2F model is based on the use of the correct velocity scale near the wall, $\overline{\upsilon^2}$ instead of k. Using this rationale, the turbulent viscosity is computed as $\nu_t = C_\mu \overline{\upsilon^2} T$, where T is a maximum between the turbulent and the Kolmogorov timescales. Additionally to the k and ε equations, a transport equation is solved for $\overline{\upsilon^2}$ and an elliptic equation is solved for f. The V2F model was proven to reproduce the wall effects accurately in many types of flows (Kalitzin 1999) but its major drawback is the stiffness present at the wall due to the boundary condition of the elliptic function f. At the wall, the boundary condition is given by:

$$f_w = \lim_{y \to 0} \frac{-20\nu\overline{\upsilon^2}}{\varepsilon_w y^4}$$

The fact that the denominator includes the distance to the wall to the power of four makes it very stiff and a coupled algorithm is necessary to obtain convergence (Lien and Kalitzin, 2001). It also imposes constrains on the mesh size near the wall.

During the course of the FLOMANIA project, UMIST developed a new formulation of the model that removes the stiffness while retaining the advantages of the original formulation (Laurence et al 2004). The new model solves for a variable $\varphi = \overline{\upsilon^2}/k$, and by adding a change on the definition of f, $\overline{f} = f + \frac{2\nu}{k}\frac{\partial k}{\partial x_k}\frac{\partial \varphi}{\partial x_k} + \nu \frac{\partial^2 \varphi}{\partial x_k^2}$. the equations can be solved uncoupled. The final equations for the two variables are:

$$\frac{D\varphi}{Dt} = \overline{f} - P\frac{\varphi}{k} + \frac{2}{k}\frac{\nu_t}{\sigma_k}\frac{\partial \varphi}{\partial x_j}\frac{\partial k}{\partial x_j} + \frac{\partial}{\partial x_j}\left[\frac{\nu_t}{\sigma_k}\frac{\partial \varphi}{\partial x_j}\right]$$

$$L^2\nabla^2\overline{f} - \overline{f} = \frac{1}{T}(C_1 - 1)\left[\varphi - \frac{2}{3}\right] - C_2\frac{P}{k} - 2\frac{\nu}{k}\frac{\partial \varphi}{\partial x_j}\frac{\partial k}{\partial x_j} - \nu\nabla^2\varphi$$

with the now simple, uncoupled, boundary conditions $\overline{f} = 0$ and $\varphi = 0$.

It is shown that the cross terms (products of gradients) are comparatively small and do not cause any stability problems.

The model was implemented in the unstructured FV *Code_Saturne*. The new formulation proposed presents an overall good performance, similar to the original model which required very small time steps (i.e. CFL of order 0.01, probably as a consequence of not being able to couple the boundary conditions in the professional FV code), whereas the above version convergence much more easily (CFL of order 1).

The φ model was tested on the asymmetric plane diffuser with other elliptic relaxation formulations as described (Laurence et al 2004). The velocity profiles predicted by the φ model, the LDM [V2F_Li] and the k-ω (dashed, solid and dotted lines, accordingly) are shown in Figure 2. The computational results are compared against the experimental data marked by triangles. The LDM under-predicts the recirculation length whereas the φ-model gives a larger recirculation zone, closer to the experiment. Overall, the φ -model performs better than the LDM. It should be noted that the pressure field is very sensitive to the recirculation bubble and affects the bulk of the flow as can be seen concerning the mean velocity in the region of the straight wall. The LDM model underestimates the velocity in this region, similarly to more standard $k-\varepsilon$ models.

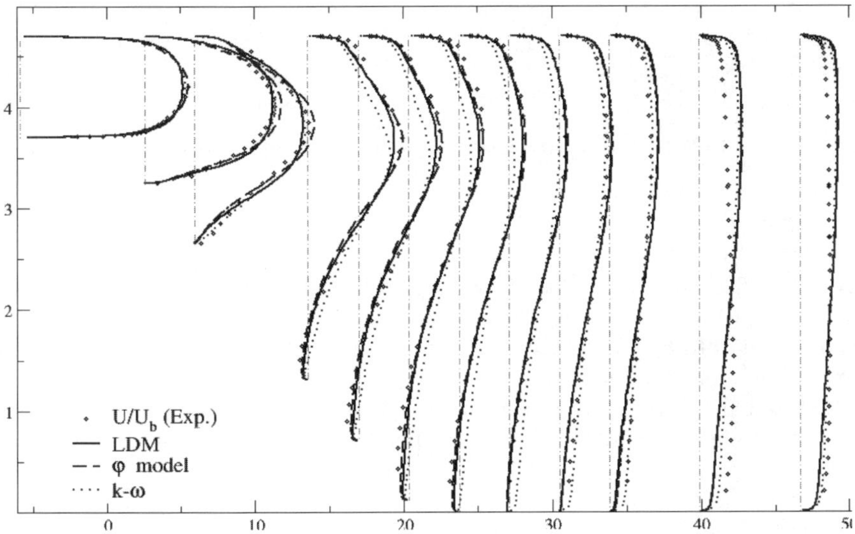

Figure 2 Asymmetric plane diffuser. Velocity profiles

The pressure coefficients are shown in Fig. 3. Here again, the φ-model produces better prediction than the LDM; while the both models are more accurate than the k-ω.

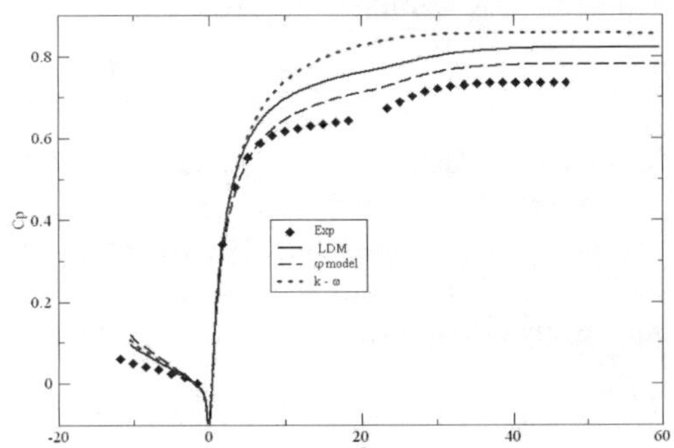

Figure 3 Pressure coefficient

Flow over periodic hills.
Velocity profiles for the flow over periodic hills can be seen in figure 4. Here the φ model performs better than the standard k-ε .

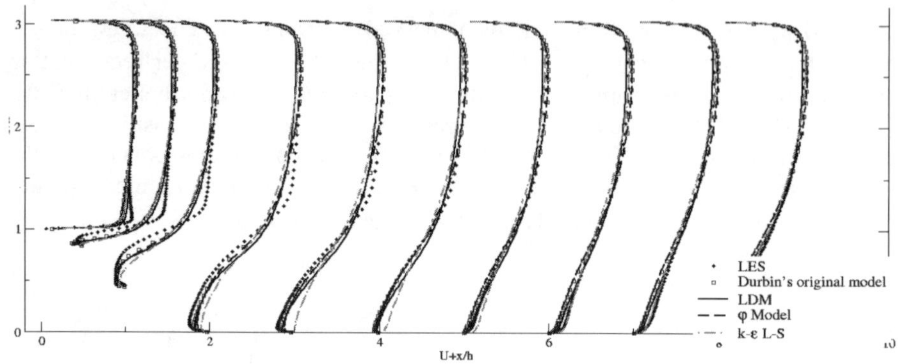

Figure 4 Flow over periodic hills. Velocity profile

17.2.2 New stress-strain eddy viscosity model: $k-\varepsilon-C_{as}$

A new eddy viscosity model was developed to include stress-strain lag effects in the modelling of unsteady mean flows. Standard EVMs significantly over-predict the production of turbulent kinetic energy in presence of strong strain. This has led to various limiters as in the V2F model (Durbin 1995), the SSTmodel (Menter et al. 2003) and the Linear Production model (Guimet et al 2002) among others. These corrections are not needed in DSMs, and Jakirlić (2004) solves in fact the full RSM to include stresses only in the k production term as a means of improving levels of k, while the standard EVM representation of the stresses is subsequently applied to the momentum equation. The results show some good improvements, but the computational cost is very high. The aim of this work was therefore to develop a model for the stress-strain lag, which appears to be a key parameter in rapidly evolving flows. The stress-strain lag parameter, hereby denoted C_{as} is written as follows.

$$C_{as} = -\frac{a_{ij}S_{ij}}{\|S\|} \qquad \|S\| = \sqrt{2S_{ij}S_{ij}}$$

A transport equation for *Cas* is derived from the full SSG DSM for the time derivative of the anisotropy tensor, a_{ij}, as well as terms including the rate of change of the strain tensor, S_{ij} (Revell et al). The stress-strain lag term is defined as follows. The strain rate parameter is $\eta= k\|S\|/\varepsilon$, and the second invariant of anisotropy is $A_2=a_{ij}a_{ij}$. The final transport equation is therefore as follows.

$$\frac{DC_{as}}{Dt} = \left(0.266+0.325\sqrt{A_2}\right)\|S\| - 2.7\frac{\varepsilon}{k}C_{as} + 0.1\|S\|C_{as}^2 + 1.05\frac{S_{ij}a_{ik}a_{kj}}{\eta}$$
$$+0.75\frac{S_{ij}a_{ik}S_{jk}}{\|S\|} + 1.6\frac{S_{ij}a_{ik}\Omega_{jk}}{\|S\|} - \frac{a_{ij}}{\|S\|}\frac{DS_{ij}}{Dt} - \frac{C_{as}}{\|S\|}\frac{D\|S\|}{Dt} + \mathrm{Diff}^{C_{as}}$$

The turbulent viscosity, ν_t, is redefined using an updated value of C_μ according to the following limit. The new diffusion term is of a standard gradient diffusion form with the constant σ_{Cas}=2.

$$\nu_t = C_\mu^{new} \frac{k^2}{\varepsilon} \qquad C_\mu^{new} = \min\left(0.09, \frac{C_{as}}{\eta}\right)$$

If the equation is to correctly model the log-layer region, then it is a requirement that it returns the standard value of C_μ in equilibrium conditions (i.e. when production to dissipation ratio is unity and h =3.33). The correctional effects of the C_{as} term on the parameter C_μ are expected to be even greater since normal anisotropy is non-zero and so more of the terms in the transport equation are in play. The motivation behind the model development stems from the observation that in rapidly varying mean flows, such as bluff body wakes or staggered tube bundles, RSMs can produce large unsteady structures similar to LES (Benhamadouche et al 2003). In homogenous turbulence subject to cyclic straining it was shown that the stresses increasingly lag behind the strains until production is shut off (Hadzic et al 2001). Therefore RSMs seem an attractive idea for DES, but are too computationally expensive. The present model is very economical since it is has almost no additional expense over a standard 2 equation model in terms of convergence time, and only slightly higher storage requirements. It is also very easy to implement in contrast with the RSM.

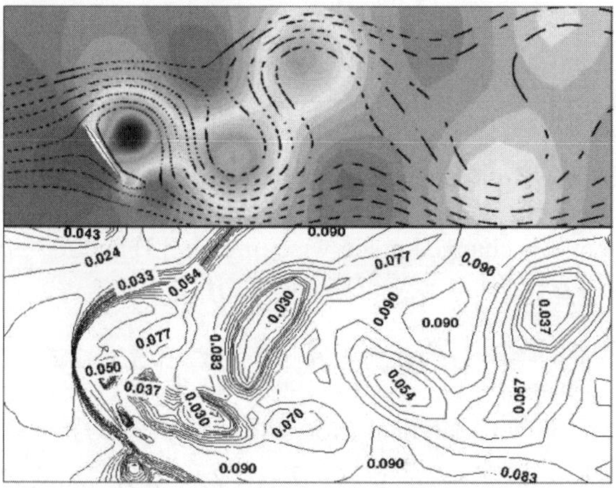

Figure 1. Flow around NACA0012 at 60° incidence using k-ε-C_{as} model. *top:* Pressure contours & streamlines. *Bottom:* modified C_μ isovalues

Figure 1 shows contour plots from some early calculations on the NACA0012 airfoil at an AoA of 60°. This is a massively separated flow and it can be seen that C_μ is reduced in the regions corresponding to the shedded structures (shown by the pressure contours and flow streamlines) while recovering its usual values at the boundaries of the unsteady wake. This behaviour is in agreement with the prediction of strongly detached flows by modified two-equation models that suggest the C_μ reduction based on the behaviour of the RSM modelling for the same class of flows (Hoarau et al 2002), in the context of Organised Eddy Simulation. It appears that the response of the model to regions of stress-strain lag could be used to automatically reduce *k* levels in the URANS-DES transition when unsteady structures start to emerge, instead of the ad-hoc switch based on grid characteristics as commonly used in DES; in which mesh dependency is one of the current breakdowns.

17.2.3 Generalized (Robin-type) wall functions

A method of boundary conditions transfer has been developed (Utyuzhnikov 2005). The method allows one to transfer a boundary condition from a wall to some intermediate surface. The boundary condition on the intermediate surface always becomes of Robin-type. In the vicinity of the wall, the boundary condition can be treated as the generalized wall function. Meanwhile, one should note the method can also be used far from the wall in the framework of a domain decomposition method.

If a Dirichlet boundary condition is set for a function u on the wall, the wall function (boundary condition) is formulated at point y^* (in the normal to the wall direction) as follows (Utyuzhnikov 2005)

$$u(y^*) = u_0 + y^* \frac{du}{dy}(y^*) f_1 - \frac{y^{*2}}{2\mu^*}\left(\int_0^1 R_h d\xi\right) f_2,$$

where $f_1 = \int_0^1 \frac{\mu(y^*)}{\mu(\xi)} d\xi$, $f_2 = 2\int_0^1 \mu(y^*)/\mu(\xi)\left(1 - \int_0^\xi R_h d\xi' / \int_0^1 R_h d\xi\right) d\xi$, $u(0) = u_0$, $R_h = R_h(y)$ is the right-hand side of the governing equation written in the boundary-layer-like form [1-3].

If the coefficient μ is changed piece-wise linearly [3] and $R_h = const$,

$$\mu = \begin{cases} \mu_w & \text{if } 0 \le y \le y_v \\ \dfrac{y - y_v}{y^* - y_v}(\mu^* - \mu_w) + \mu_w & \text{if } y_v \le y \le y^* \end{cases}$$

then it is possible to obtain analytical expressions for f_1 and f_2:

$$f_1 = \alpha_\mu \bar{y}_v (1 + \theta \ln \alpha_\mu), \quad f_2 = 2\alpha_\mu \bar{y}_v \left[1 - \theta - (1/2 - \theta)\bar{y}_v + \alpha_\mu \bar{y}_v \theta^2 \ln \alpha_\mu\right],$$

$$\text{where } \bar{y}_v = y_v / y^*, \alpha_\mu = \mu^* / \mu_w, \quad \theta^{-1} = \frac{\mu^* - \mu_w}{\mu_w} \frac{y_v}{y^* - y_v}.$$

The obtained general form of the wall function has been implemented for all independent variables (except dissipation ε) including the turbulent kinetic energy k and normal component of the velocity. These wall functions are formulated in the differential form, they are robust and can be applied to any kind of approximations of the governing equations. Boundary conditions are transferred to the point y^* which can be completely mesh independent.

III Presentation of turbulence models used by partners

The intention of the following chapter is to provide the reader with all necessary input related to the turbulence models used in the different methods and approaches of the FLOMANIA partners.

Models for the Reynolds averaged equations are presented first. These models are sorted according to the constitutive relation: linear eddy viscosity model are first addressed, followed by non-linear eddy viscosity models and explicit algebraic Reynolds stress models and finally differential Reynolds stress models. Acronyms for the model names are given, which are consistent throughout this book. Wall function treatment is then discussed. DES extensions of the models are detailed in the last section.

1 Models versus partners and model quick reference

B. Aupoix, ONERA, and all partners

The following matrices provide a complete list of models used by the FLOMANIA partners. This list is split into the main categories of turbulence models, eddy viscosity models (Table 1), EARSM & NLEVM (Table 2) and DRSM models (Table 3), and allows in insight into the diversity of models used. In addition, the letters one can find in the matrix refers to the treatment of models in the wall region, see Table 4.

Table 1 Eddy viscosity models versus partners

	Partner	**EADS-M**	**Bombardier**	**ANSYS**	**Alenia**	**EADS-CASA**	**Chalmers**	**Dassault**	**DLR**	**EDF**	**FOI**	**IMFT**	**ICSTM**	**NUMECA**	**ONERA**	**SPTU**	**TUB**	**UMIST**
Model Reference	**Acronym**																	
								0-equation model										
Baldwin78	**BL**														W			
								1-equation models										
Spalart92, 94	**SA**		H						W		W	W		W	W	W		
Edwards96	**SAE**	W																
+ Dacles95	**SAED**	W																
+ Shur00	**SARC**								W					W		W		
	SAERC	W																
+ Rung03	**SALSA**					W			W								H	

	Partner	**EADS-M**	**Bombardier**	**ANSYS**	**Alenia**	**EADS-CASA**	**Chalmers**	**Dassault**	**DLR**	**EDF**	**FOI**	**IMFT**	**ICSTM**	**NUMECA**	**ONERA**	**SPTU**	**TUB**	**UMIST**
Model Reference	**Acronym**																	
2-equation models: k-ε																		
Jones72	**JL**		H															
Launder74	**LS**			S				S V		W			W					W
Chen88	**CP**							W										
Jin94	**JB**											W						
Bouhadji02	**JB-BB**											W						
Yang92	**YS**													E W				
2-equation models: v2-f																		
Durbin95	**V2F_Du**							W		W								W
Lien01	**V2F_Li**							W		W				W				W
Laurence04	**V2F_La**							W		W								W
2-equation models: k-ω																		
Wilcox88	**Wi88**	W		A u					W		W	W	W		W			W
Peng97	**Peng**										W							
Menter94	**SST**	W		A u						W	W	W		W		W		W
+ Dol02	**SST-R**	W																
Kok00	**TNT**														W			
Rung96	**LLR**																H	
2-equation models: k-g																		
Kalitzin96	**k-g**					S												
2-equation models: k-L																		
Smith95	**Sm95**							S m							W			
Smith97	**Sm97**							S m										

It should be noted that EADS-M has used the Kok correction in all k-ω models, while Chalmers did not participate in work on RANS models.

Table 2 EARSM & NLEVM models versus partners

Model Reference	Partner / Acronym	EADS-M	Bombardier	ANSYS	Alenia	EADS-CASA	Chalmers	Dassault	DLR	EDF	FOI	IMFT	ICSTM	NUMECA	ONERA	SPTU	TUB	UMIST
Apsley98	**AL**												W					
Abe03	**AJL-ε**												W					
	AJL-ω												W					
Wallin00	**WJ**							W					W					
+ ω eq. (1994)	**WJ-ω**				W								W					
(1988)	**WJ-st-ω**										W							
+Arina96	**WJ-Rt**				W													
Wallin02	**WJ-C**																	
Hellsten04	**WJ-H**										W							
Grundestam	**WJ-G**				W													
Rung99	**LEA**	W							W									
Lubcke02	**CEASM**																H	

Table 3 DRSM models versus partners

Model Reference	Partner / Acronym	EADS-M	Bombardier	ANSYS	Alenia	EADS-CASA	Chalmers	Dassault	DLR	EDF	FOI	IMFT	ICSTM	NUMECA	ONERA	SPTU	TUB	UMIST
Speziale91	**SSG**			S						S			S	E S	S			S
+ Chen00	**SSG-Ch**									W		W	W	W		W		W
+ ω equation	**SSG-ω**										W							
Naot73	**IP**											W						
Hanjalic98	**HJ**												W					
Launder75	**LRR**									S								S
Shima98	**Sh**												W					
Wilcox98 + ω eq.	**Wi-ω**	W							W									
+ BSL eq	**Wi-BSL**			A u														

	Partner	**EADS-M**	**Bombardier**	**ANSYS**	**Alenia**	**EADS-CASA**	**Chalmers**	**Dassault**	**DLR**	**EDF**	**FOI**	**IMFT**	**ICSTM**	**NUMECA**	**ONERA**	**SPTU**	**TUB**	**UMIST**
Model Reference	**Acronym**																	
Hybrid SSG/LRR+ ω eq.	**SSG/ LRR-ω**	W							W									
Craft96	**TCL**																	A

Table 4 Wall treatment applied to models presented in Tables 1 to 3

Code	**Description**
A	Analytical wall functions (Craft et al., 2003)
Au	Automatic wall function
E	Extended wall functions (Hakimi et al., 2000)
H	Hybrid low/high Re (Rung et al., 2000)
S	Scalable wall functions (Grotjans and Menter, 1998)
Sm	Smith, 1990
V	Viegas and Rubesin, 1983
W	Solution down to the wall

In the tables to follow on the next pages, all turbulence models used are referenced for the sake of an easy search of models, names, and publications.

Acronym	Model Ref.	References
Eddy-viscosity models		
0-equation model		
BL	Baldwin, B.S., Lomax, H. (1978)	Thin layer approximation and algebraic model for separated turbulent flows. AIAA Paper 78-257, 16th Aerospace meeting, Huntsville, Alabama, January 16-18.
1-equation models		
SA	Spalart, P.R., Allmaras, S.R. (1992, 1994)	A one-equation turbulence model for aerodynamics flows. AIAA Paper 92-0349, 30th Aerospace Sciences Meeting and Exhibit, Reno, Nevada. A one-equation turbulence model for aerodynamics flows. La Recherche Aérospatiale, Volume 1, pp 5-21.
SAE	Edwards , Chandra (1996)	Comparison of eddy-viscosity turbulence models for three-dimensional, shock-separated flowfields. AIAA Journal, vol. 34, No. 4, pp 756-763.
SAED	SA + Dacles-Mariani, J., Zilliac, G.G., Crow, J.S., Bradshaw, P. (1995)	Numerical/experimental study of a wingtip vortex in the near field. AIAA Journal, Volume 3, No. 9, pp. 1561-1568.
SARC	SA + Shur, M., Strelets, M., Travin, A., Spalart, P.R. (2000)	Turbulence modelling in rotating and curved channels ; Assessing the Spalart-Shur correction term. AIAA Journal, Volume 38, No 5, pp 784-782.
SAERC	SAE + SARC	
SALSA	SA + Rung, T., Bunge, U., Schatz, M., Thiele, F. (2003)	Restatement of the Spalart-Allmaras eddy-viscosity model in a strain-adaptive formulation. AIAA Journal, Volume 51, No 7, pp. 1396-1399.
2-equation models: k-ε		
JL	Jones, W.P., Launder, B.E. (1972)	The prediction of relaminarization with a two-equation model of turbulence. International Journal of Heat and Mass Transfer, Volume 15, pp 301-314.
LS	Launder, B.E., Sharma, B.I. (1974)	Application of energy-dissipation model of turbulence to the calculation of flow near a spinning disc. Letters in Heat and Mass Transfer, Volume 1, pp 131-138.
CP	Chen, V.C., Patel, H.C. (1988)	Near-wall turbulence models for complex flows including separation. AIAA Journal, Volume 26, part 6, pp. 641-648.
JB	Jin, G., Braza, M. (1994)	A two-equation turbulence model for unsteady separated flows. AIAA Journal, Volume 32, No. 11, pp. 2316-2320.
JB-BB	A.Bouhadji, S. Bourdet, M. Braza, Y. Hoarau, P. Rodes, G.Tzabiras (2002)	Turbulence modelling of unsteady flows with a pronounced periodic character, NNFM, Vol 81, pp. 87-96, Eds. W. Haase et al, Springer.
YS	Yang, Z., Shih, T.H. (1992)	A k-ε calculation of transitional boundary layers. NASA TM ICOMP-92-08 CMOTT-92-05.

2-equation models: v2-f		
V2F_Du	Durbin, P.A. (1995)	Separated flow computations with the k-ω-v2 turbulence model. AIAA Journal, Volume 33, pp. 659-664.
V2F_Li	Lien, F.S., Kalitzin, G. (2001)	Computations of transonic flows with the v^2f turbulence model. International Journal of Heat and Fluid Flow, pp. 53-61.
V2F_La	Laurence, D.R., Uribe, J.C., Utyuzhnikov, S.V. (2004)	A robust formulation of the v^2f model. Journal of Flow, Turbulence and Combustion
2-equation models: k- ω		
Wi88	Wilcox, D.C. (1988)	Reassessment of the scale-determining equation for advanced turbulence models. AIAA Journal Volume 26, No 11, pp 1299-1310.
Peng	Peng, S-H., Davidsson, L., Holmberg, S. (1997)	A modified low-Reynolds number k-ω model for recirculating flows. ASME Journal of Fluids Engineering. Volume 119, pp. 867-875.
SST	Menter, F.R. (1994)	Two-equation eddy-viscosity turbulence models for engineering applications. AIAA Journal, Volume 32, No 8, pp 1598-1605.
SST-R	SST + Dol02	
TNT	Kok, J.C. (2000)	Resolving the dependence on freestream values of the k-ω turbulence model. AIAA Journal, Volume 38, pp. 1292-1295.
LLR	Rung, T., Thiele, F. (1996)	Computational modelling of complex boundary-layer flows. Proceedings of the 9th International Symposium on Transport Phenomena in Thermal-Fluid Engineering, Singapore, pp. 321-326.
2-equation model: k-g		
k-g	Kalitzin, G., Gould, A.R.B., Benton, J.J. (1996)	Application of two-equation turbulence models in aircraft design. AIAA Paper 96-0327, 34th Aerospace Sciences Meeting and Exhibit, Reno, Nevada.
2-equation models: k-L		
Sm95	Smith, B.R. (1995):	Prediction of hypersonic shock wave turbulent boundary layer interactions with the k-l two equation turbulence model. AIAA Paper 95-0232, 33rd Aerospace Sciences Meeting & Exhibit, Reno, Nevada.
Sm97	Smith, B.R. (1997):	A non-equilibrium turbulent viscosity function for the k-l two equation turbulence model. AIAA Paper 97-1959 Snowmass Village, Colorado. 28th AIAA Fluid Dynamics Conference.

EARSM & NLEVM		
AL	Apsley, D.D., Leschziner, M.A. (1998)	A new low-Reynolds-number non-linear two-equation turbulence model for complex flows. International Journal of Heat and Fluid Flows, Volume 19, pp. 209-222.
AJL- ω	Abe, K., Yang, J., Leschziner, M.A. (2003)	An investigation of wall-anisotropy expressions and length-scale equations for non-linear eddy-viscosity models. International Journal of Heat and Fluid Flows, Volume 24, pp. 181-198.
AJL- ω	Abe, K., Yang, J., Leschziner, M.A. (2003)	An investigation of wall-anisotropy expressions and length-scale equations for non-linear eddy-viscosity models. International Journal of Heat and Fluid Flows, Volume 24, pp. 181-198.
WJ	Wallin, S., Johansson, A.V. (2000)	An explicit algebraic Reynolds stress model for incompressible and compressible turbulent flows. Journal of Fluid Mechanics, Volume 403, pp. 89-132.
WJ-ω	WJ+ ω eq. (Wilcox, D.C. 1994)	
WJ-st-ω	WJ+ ω eq. (Wilcox, D.C. 1998)	Multiscale model for turbulent flows, AIAA Journal, Vol. 26, No. 11
WJ-Rt	WJ + Arina, R., Ceresola, N., Pianta, P.G. (1996)	Application of a two-equation turbulence model to the numerical prediction of the transonic buffet of an airfoil. Proceedings 20th ICAS Congress.
WJ-C	Wallin, S., Johansson, A.V. (2002)	Modelling streamline curvature effects in explicit algebraic Reynolds stress turbulence models. International Journal of Heat and Fluid Flows, Volume 23, No 5, pp. 721-730.
WJ-H	Hellsten, A. (2004)	New advanced k-ω turbulence model for high-lift aerodynamics. AIAA Paper 2004-1120, 42nd Aerospace Sciences Meeting and Exhibit, Reno, Nevada.
WJ-G	Grundestam, O., Wallin, S., Johansson, A.V. (2003)	A generalised EARSM based on a non-linear pressure strain model. Proceedings of the Third International Symposium on Turbulence and Shear Flow Phenomena, Sendai, Japan.
LEA	Rung, T., Lübcke, H., Franke, M., Xue, L., Thiele, F., Fu, S. (1999)	Assessment of explicit algebraic stress models in transonic flows. Engineering Turbulence Modelling and Experiments 4, pp. 659-668.
CEASM	Lübcke, H., Rung, T., Thiele, F. (2002)	Prediction of the spreading mechanism of 3D turbulent wall jets with explicit Reynolds-Stress closures. Engineering Turbulence Modelling and Experiments 5, pp. 127-145.

DRSM		
SSG	Speziale, C.G., Sarkar, S., Gatski, T.G. (1991)	Modelling the pressure-strain correlation of turbulence: An invariant dynamical systems approach. Journal of Fluid Mechanics, Volume 227, pp. 245-272.
SSG-Ch	SSG + Chen, H.C., Yang, Y.J., Han, J.C. (2000)	Computation of heat transfer in rotating two-pass square channels by a second moment closure model. International Journal of Heat and Mass Transfer, Volume 43, pp. 1603-1616.
SSG-ω	SSG + ω equation (Wilcox 1988)	
IP	Naot, D., Shavit, A., Wolfshtein, M. (1973)	Two-Point Correlation and the Redistribution of Reynolds Stresses, The Physics of Fluids, Vol.16, No 6, pp 738-743.
HJ	Hanjalic, K., Jakirlic, S. (1998)	Contribution towards the second-moment closure modelling of separated turbulent flows. Computers & Fluids, Volume 27, No. 2, pp. 137-156.
LRR	Launder, B.E., Reece, G.J., Rodi, W. (1975)	Progress in the development of a Reynolds-Stress turbulence closure. Journal of Fluid Mechanics, Volume 68, No 2, pp. 537-566.
Sh	Shima, N. (1998)	Low-Reynolds-number second-moment closure without wall-reflection redistribution terms. International Journal of Heat and Fluid Flows. Volume 19, pp. 549-555.
Wi-ω	Wilcox 88 + ω equation.	
Wi-BSL	Wilcox 88+ BSL equation (Menter 1994)	
SSG/ LRR-ω	Hybrid SSG/LRR+ ω equation (Wilcox 1988).	
TCL	Craft, T.J., Launder, B.E. (1996)	A Reynolds Stress Closure Designed for Complex Geometries, International Journal of Heat and Fluid Flows, Vol. 17, No. 3, pp. 245-254

2 Eddy-viscosity turbulence models

B. Aupoix, ONERA, and all partners

2.1 Introduction

As direct numerical solutions of the Navier-Stokes equations are beyond the reach of foreseeable computers, ensemble (or Reynolds) averaging has to be used, which only predicts the mean flow. For the sake of simplicity, as compressibility effects on turbulence models were not addressed in this project, only incompressible flows will be discussed. The Reynolds Averaged Navier-Stokes (RANS) equations thus read:

$$\frac{\partial U_i}{\partial x_i} = 0$$

$$\frac{\partial U_i}{\partial t} + U_j \frac{\partial U_i}{\partial x_j} = -\frac{1}{\rho}\frac{\partial p}{\partial x_i} + \frac{\partial}{\partial x_l}\left[\nu\left(\frac{\partial U_i}{\partial x_l} + \frac{\partial U_l}{\partial x_i}\right) - \overline{u_i' u_l'}\right]$$

where U_i, p, ρ and ν respectively designate the mean velocity component, the mean pressure, the density and the kinematic viscosity. u_i' designates a velocity fluctuation and the overbar an ensemble average. Averaging introduces new unknowns, the Reynolds stress tensor, $-\overline{u_i' u_j'}$, which represents the flux of momentum in the averaged flow, due to the turbulent motion, and must be modelled. The next sections will give a brief summary of the models used by partners. Models are sorted according first to the constitutive relation used to express the Reynolds stress tensor and then to the set of transport equations.

As they will be widely used in the following, let us define the rate of strain and rotation tensors:

$$S_{ij} = \frac{1}{2}\left(\frac{\partial U_i}{\partial x_j} + \frac{\partial U_j}{\partial x_i}\right) \quad \omega_{ij} = \frac{1}{2}\left(\frac{\partial U_i}{\partial x_j} - \frac{\partial U_j}{\partial x_i}\right)$$

together with their invariants:

$$S = \sqrt{2 S_{ij} S_{ij}} \quad W = \sqrt{2 \omega_{ij} \omega_{ij}}\,.$$

(Here, W is preferred to Ω to avoid confusion with reference frame rotation rate).

2.2 Eddy viscosity models

2.2.1 Boussinesq relation

These models are still widely used in the industry. They assume that, similarly to the viscous stress tensor, the Reynolds stress tensor can be related to the mean rate of strain as:

$$-\overline{u_i'u_j'} = 2\nu_t S_{ij} - \frac{2}{3}\delta_{ij}k$$

where $k = \frac{1}{2}\overline{u_i'u_i'}$ is the turbulent kinetic energy and ν_t the turbulent, or eddy, viscosity. This eddy viscosity has the dimension of the product of a turbulence velocity scale and a turbulence length scale, both to be determined.

2.2.2 Algebraic models

The algebraic models are rather old models in which all the turbulence scales are directly related to the mean flow, so that no extra transport equation is needed. Only the algebraic model by Baldwin and Lomax (1978) [BL] has been used in the framework of this project.

In the inner part of boundary layers, the eddy viscosity is given by a modified mixing length model as:

$$\nu_{t,inner} = l^2 F_{VD}^2 W$$

where $l = \kappa y$ is the mixing length, proportional to the distance y to the wall, $\kappa \approx 0.41$ being the von Kármán constant. $F_{VD} = 1 - \exp\left(-y^+/26\right)$ is the van Driest damping function based upon the dimensionless wall distance $y^+ = yu_\tau / \nu$ where the friction velocity u_τ is related to the wall shear stress τ_w as $u_\tau = \sqrt{\tau_w / \rho}$.

In the outer part of the boundary layer, the model has been formulated to be similar to the Cebeci and Smith (1974) model, but in a form more suitable for Navier-Stokes solvers. The outer eddy viscosity reads:

$$\nu_{t,outer} = KC_{CP}F_{wake}F_{Kleb}$$

where $K = 0.0168$, $C_{CP} = 1.6$. F_{wake} is derived from the maximum $F_{\max}$ of the function $F(y) = yWF_{VD}$ as $F_{wake} = \min\left(y_{\max}F_{\max}; 0.25 y_{\max}U_{ext}/F_{\max}\right)$, where U_{ext} is the velocity outside of the boundary layer and $y_{\max}$ the distance from the wall where F reaches its maximum. F_{Kleb} is derived from the Klebanoff's intermittency function as $F_{KLEB} = \left[1 + 5.5\left(0.3y/y_{\max}\right)^6\right]^{-1}$. Finally, the eddy viscosity reads:

$$\nu_t = \begin{cases} \nu_{t,inner} \text{ when } y \le y_{crossover} \\ \nu_{t,outer} \text{ when } y \ge y_{crossover} \end{cases}$$

where $y_{crossover}$ is the nearest location from the wall where both eddy viscosity expressions are equal.

2.2.3 One equation models

The Spalart and Allmaras (1992, 1994) model [SA] only solves one transport equation for a quantity $\tilde{\nu}$ which is equivalent to the eddy viscosity, except close to walls. This transport equation has been constructed by educating the model to reproduce flows of increasing complexity. It reads:

$$\frac{D\tilde{\nu}}{Dt} = c_{b1}\tilde{S}\tilde{\nu} - c_{w1}f_w\left(\frac{\tilde{\nu}}{d}\right)^2 + \frac{1}{\sigma}\left[\frac{\partial}{\partial x_k}\left((\nu+\tilde{\nu})\frac{\partial\tilde{\nu}}{\partial x_k}\right) + c_{b2}\frac{\partial\tilde{\nu}}{\partial x_k}\frac{\partial\tilde{\nu}}{\partial x_k}\right]$$

where d is the distance to the nearest surface. Far from walls, $\tilde{\nu} \equiv \nu_t$ and the source term is due to $S \equiv \tilde{S}$ which is the modulus of the vorticity or any other quantity which reduces to $|\partial u/\partial y|$ in thin shear layers. The model constants are

$$c_{b1} = 0.1355 \quad c_{b2} = 0.622 \quad \sigma = \frac{2}{3} \quad c_{w1} = \frac{c_{b1}}{\kappa^2} + \frac{1+c_{b2}}{\sigma}.$$

The function f_w damps the sink term in the outer part of the boundary layer and provides the sensitivity to pressure gradient. It reads:

$$f_w = g\left[\frac{1+c_{w3}^6}{g^6+c_{w3}^6}\right]^{1/6} \qquad g = r + c_{w2}\left(r^6 - r\right) \quad r = \frac{\tilde{\nu}}{\tilde{S}\kappa^2 d^2} \quad c_{w2} = 0.3 \quad c_{w3} = 2.$$

Finally, the wall treatment is made to preserve the same expression for $\tilde{\nu}$ as in the logarithmic region, i.e. $\tilde{\nu} = u_\tau \kappa d$. For that, the eddy viscosity is related to the transported quantity $\tilde{\nu}$ as:

$$\nu_t = f_{v1}\tilde{\nu} \quad f_{v1} = \frac{\chi^3}{\chi^3 + 7.1^3} \quad \chi = \frac{\tilde{\nu}}{\nu}$$

and the modified vorticity reads:

$$\tilde{S} = S + \frac{\tilde{\nu}}{\kappa^2 d^2} f_{v2} \quad f_{v2} = 1 - \frac{\chi}{1+\chi f_{v1}}.$$

The wall boundary condition is of course $\tilde{\nu} = 0$.

Edwards and Chandra (1996) [SAE] have proposed a modification of the near wall treatment of the Spalart and Allmaras model to improve numerical stability.

The definition of the vorticity in the wall region and the parameter r of the f_{v1} function are modified as:

$$\widetilde{S} = S\left[\frac{1}{\chi} + f_{v1}\right] \quad r = \frac{\tanh\left[\dfrac{\widetilde{\nu}}{\kappa^2 d^2 \widetilde{S}}\right]}{\tanh[1.0]}.$$

Dacles-Mariani et al. (1995) [SAD] reported a suggestion by Spalart to modify the production term to sensitise it to both strain and rotation rates and reduce the eddy viscosity production in regions where rotation is dominant. The revised form of the production term reads:

$$C_{b1}\widetilde{\nu}\left(W + 2\min(0, S - W)\right).$$

When this correction is used together with Edwards' correction, the model is labelled [SAED].

A more elaborate way to sensitise the Spalart and Allmaras model to curvature and rotation effects has been later proposed by Spalart and Shur (1997) and Shur et al. (2000). In a reference frame rotating at a rate Ω, the definition of the vorticity tensor is extended as:

$$\omega_{ij} = \frac{1}{2}\left(\frac{\partial U_i}{\partial x_j} - \frac{\partial U_i}{\partial x_j} + 2\varepsilon_{mji}\Omega_m\right)$$

where ε_{ijk} is the alternating or Ricci tensor. Introducing $D^2 = \frac{1}{2}\left(S^2 + W^2\right)$, the two following dimensionless quantities allow to quantify the curvature/rotation effects:

$$r^* = \frac{S}{W} \quad \widetilde{r} = \frac{2\omega_{ik} S_{jk}}{D^4}\left[\frac{DS_{ij}}{Dt} + \left(\varepsilon_{imn} S_{jn} + \varepsilon_{jmn} S_{in}\right)\Omega_m\right].$$

The production term $c_{b1}\widetilde{S}\widetilde{\nu}$ in the eddy viscosity transport equation is thus multiplied by the rotation function:

$$f_{r1} = (1 + c_{r1})\frac{2r^*}{1 + r^*}\left[1 - c_{r3}\tan^{-1}(c_{r2}\widetilde{r})\right] - c_{r1} \quad c_{r1} = c_{r3} = 1.0 \quad c_{r2} = 12.$$

This model is labelled [SARC] when the rotation correction is used with the basic [SA] model, [SAERC] when Edwards' correction is also used.

In the [SALSA] model, Rung et al. (2003) use Edwards' modification but with a different definition of the variable r:

$$r = 1.6\tanh\left[0.7\frac{\Psi}{\widetilde{S}}\right] \quad \Psi = \frac{\widetilde{\nu}}{\kappa^2 d^2}.$$

Moreover, the production term is modified to lower it in highly strained flows as:

$$c_{b1} = 0.1355\sqrt{\Gamma} \qquad \Gamma = \min(1.25;\max(\gamma;0.75)) \qquad \gamma = \max(\alpha_1;\alpha_2)$$

$$\alpha_1 = \left(1.01\frac{\tilde{\nu}}{\kappa^2 d^2 S}\right)^{0.65} \qquad \alpha_2 = \max\left(0.;1-\tanh\frac{\chi}{68}\right)^{0.65}.$$

2.2.4 *k-ε models*

Most two equation models solve a transport equation for the turbulent kinetic energy, which provides a turbulence velocity scale as $\sqrt{k}$. They are presented below according to the variable used to estimate the turbulence length or time scale.

$k-\varepsilon$ models solve the turbulent kinetic energy transport equation together with a transport equation for the turbulent kinetic energy dissipation rate ε. The eddy viscosity is related to these two quantities as:

$$\nu_t = C_\mu f_\mu \frac{k^2}{\varepsilon} \qquad C_\mu = 0.09$$

where f_μ is a damping function to decrease the eddy viscosity near walls.

The transport equation for the turbulent kinetic energy k can be derived from the Navier-Stokes equation. It requires little modelling. Its modelled form reads:

$$\frac{Dk}{Dt} = P_k - \varepsilon - D + \frac{\partial}{\partial x_k}\left[\left(\nu + \frac{\nu_t}{\sigma_k}\right)\frac{\partial k}{\partial x_k}\right] \qquad P_k = -\overline{u_i' u_j'}\frac{\partial U_i}{\partial x_j} = \nu_t S^2$$

where $\varepsilon + D$ is the turbulent kinetic energy dissipation rate, ε being the second transported turbulent quantity, which is equivalent to the dissipation in high turbulent Reynolds number regions and D an extra term to ensure the correct dissipation behaviour near solid surfaces. The no-slip condition imposes the turbulent kinetic energy being null at the wall.

The dissipation rate transport equation can also be deduced from the Navier-Stokes equations but no term by term modelling is possible. The general form of the modelled transport equation for the dissipation rate reads:

$$\frac{D\varepsilon}{Dt} = C_{\varepsilon 1}\frac{P_k \varepsilon}{k} - C_{\varepsilon 2} f_2 \frac{\varepsilon^2}{k} + E + \frac{\partial}{\partial x_k}\left[\left(\nu + \frac{\nu_t}{\sigma_k}\right)\frac{\partial \varepsilon}{\partial x_k}\right]$$

where f_2 represents both near wall and low Reynolds number dampings and E is a correction term to get the desired near wall behaviour. The real dissipation rate is finite at the wall but models use either an "isotropic" dissipation which is null at the wall, the real wall value being provided by the extra term D, or a non zero wall value, D being then generally null.

Jones and Launder (1972) [JL] and Launder and Sharma (1974) [LS] models are very similar. Both use an "isotropic" dissipation which is null at the wall. The near wall terms read:

$$D = 2\nu \frac{\partial\sqrt{k}}{\partial x_k}\frac{\partial\sqrt{k}}{\partial x_k} \qquad E = 2\nu\nu_t \frac{\partial^2 U_i}{\partial x_k \partial x_l}\frac{\partial^2 U_i}{\partial x_k \partial x_l}$$

and the two models just differ by the values of the model coefficients and the f_μ damping function, according to the following table:

	$C_{\varepsilon 1}$	$C_{\varepsilon 1}$	f_μ
Jones Launder	1.57	2.00	$\exp\dfrac{-2.5}{1+R_t/50}$
Launder Sharma	1.44	1.92	$\exp\dfrac{-3.4}{(1+R_t/50)^2}$

while in both models $\sigma_k = 1.0$, $\sigma_\varepsilon = 1.3$ and $f_2 = 1 - 0.3\exp\left(-R_t^2\right)$, $R_t = \dfrac{k^2}{\nu\varepsilon}$ being a turbulent Reynolds number.

A modified wall treatment has been proposed by Chen and Patel (1988) [CP]. They use the Launder and Sharma model [LS] without damping functions far from walls. In the wall region, only the turbulent kinetic energy transport equation is solved while the dissipation rate and the eddy viscosity are given, following Wolfshtein (1969), as:

$$\varepsilon = \frac{k^{3/2}}{l_\varepsilon} \qquad l_\varepsilon = C_l y\left[1 - \exp\left(-\frac{R_y}{A_\varepsilon}\right)\right] \qquad R_y = \frac{\sqrt{k}\,y}{\nu} \qquad C_l = \kappa C_\mu^{-3/4} \qquad A_\varepsilon = 2C_l$$

$$\nu_t = C_\mu \sqrt{k}\, l_\mu \qquad l_\mu = C_l y\left[1 - \exp\left(-\frac{R_y}{A_\mu}\right)\right] \qquad A_\mu = 70.$$

The matching between the two models is located for R_y about 250.

Jin and Braza (1994) [JB] have modified the Chien (1982) $k-\varepsilon$ model to sensitise it to vorticity. Chien's model differs from the Jones and Launder and Launder and Sharma models by a different choice of the coefficients in the dissipation equation $(C_{\varepsilon 1} = 1.35, C_{\varepsilon 2} = 1.80)$ and by a different near wall treatment. Chien's model still uses an "isotropic" dissipation but is sensitive to both the turbulence Reynolds number R_t and the reduced wall distance y^+. The extra terms in the transport equations to obtain the expected near wall behaviour read:

$$D = -2\nu\frac{k}{y^2} \qquad E = -2\nu\frac{\varepsilon}{y^2}\exp\left(-0.5y^+\right)$$

while the damping functions are now:

$$f_2 = 1 - \frac{2}{9}\exp\left(-\left(\frac{R_t}{6}\right)^2\right) \quad f_\mu = 1 - \exp\left(-0.0115 y^+\right).$$

The proposal from Jin and Braza is to make the model more sensitive to vorticity. For that, the production term is written as

$$P_k = \nu_t W^2$$

in both transport equations.

Bouhadji et al. (2002) [JB-BB] have reconsidered the eddy-diffusion coefficient in two-equation modelling in respect of non-equilibrium turbulent flows with massive separation, by means of the Launder, Reece and Rodi differential stress model, in the context of the OES approach. The recommended value of the C_μ coefficient (0.02) in association with the above mentioned damping function has been successfully applied in turbulent flows around lifting bodies.

Yang and Shih (1992) model [YS] is the only model presented here to deal with the real dissipation, which is not null at the wall. The wall condition for the dissipation rate thus reads:

$$\varepsilon_w = 2\nu\left(\frac{\partial\sqrt{k}}{\partial y}\right)^2.$$

Therefore, the extra term D in the turbulent kinetic energy transport equation is null. Another peculiarity of this model is to introduce a time scale T involving both turbulence and viscous effects as:

$$T = \frac{k}{\varepsilon} + \sqrt{\frac{\nu}{\varepsilon}}$$

so that the eddy viscosity definition and the dissipation transport equation are modified as:

$$\nu_t = C_\mu f_\mu k T$$

$$\frac{D\varepsilon}{Dt} = \left(C_{\varepsilon 1} P_k - C_{\varepsilon 2} f_2 \varepsilon\right)\frac{1}{T} + \nu\nu_t \frac{\partial^2 U_i}{\partial x_k \partial x_l}\frac{\partial^2 U_i}{\partial x_k \partial x_l} + \frac{\partial}{\partial x_k}\left[\left(\nu + \frac{\nu_t}{\sigma_k}\right)\frac{\partial\varepsilon}{\partial x_k}\right]$$

The f_2 damping function is borrowed from Chien's model while the eddy viscosity damping function f_μ reads:

$$f_\mu = 1 - \exp\left[\left(-a_1 R_y - a_3 R_y^3 - a_5 R_y^5\right)\right] \qquad R_y = \frac{\sqrt{k}\,y}{\nu}$$

$$a_1 = 1.5\,10^{-4} \qquad a_3 = 5.0\,10^{-7} \qquad a_5 = 1.0\,10^{-10}.$$

Finally, model constants are the same as for Launder and Sharma model except $C_{\varepsilon 1} = 1.45$.

2.2.5 v2-f models

Although general, the elliptic relaxation strategy proposed by Durbin (1995) to deal with near wall treatment has mainly been applied to $k-\varepsilon$ models. The idea is to solve one extra equation for $\overline{v^2}$ which represents the normal velocity fluctuation variance over a flat surface and otherwise reflects turbulence damping by walls. This damping is accounted for via a relaxation equation for the pressure strain-term. The basic model [V2F_Du] equations read:

$$\nu_t = C_\mu \overline{v^2} T$$

$$\frac{Dk}{Dt} = P_k - \varepsilon + \frac{\partial}{\partial x_k}\left[\left(\nu + \frac{\nu_t}{\sigma_k}\right)\frac{\partial k}{\partial x_k}\right]$$

$$\frac{D\varepsilon}{Dt} = \frac{C_{\varepsilon 1} P_k - C_{\varepsilon 2}\varepsilon}{T} + \frac{\partial}{\partial x_k}\left[\left(\nu + \frac{\nu_t}{\sigma_\varepsilon}\right)\frac{\partial \varepsilon}{\partial x_k}\right]$$

$$\frac{D\overline{v^2}}{Dt} = kf - \frac{\overline{v^2}}{k}\varepsilon + \frac{\partial}{\partial x_k}\left[\left(\nu + \frac{\nu_t}{\sigma_k}\right)\frac{\partial \overline{v^2}}{\partial x_k}\right]$$

$$L^2 \frac{\partial^2 f}{\partial x_k^2} - f = (1 - c_1)\frac{\frac{2}{3} - \frac{\overline{v^2}}{k}}{T} - c_2 \frac{P_k}{k}$$

The length and time scales appearing in the model equations are a blend between turbulent and Kolmogorov scalings as:

$$L = C_L \max\left(\frac{k^{3/2}}{\varepsilon}; C_\eta \left(\frac{\nu^3}{\varepsilon}\right)^{1/4}\right) \qquad T = \max\left(\frac{k}{\varepsilon}; 6\left(\frac{\nu}{\varepsilon}\right)^{1/2}\right).$$

The wall boundary conditions are rather unusual. Two boundary conditions are imposed for the turbulent kinetic energy:

$$k = 0 \qquad \frac{\partial k}{\partial y} = 0$$

to ensure a parabolic near wall behaviour. No near wall condition is then required upon the dissipation, but this is only possible when solving k and ε in a fully coupled manner. Another option is to impose the wall value of the dissipation to obtain the correct turbulent kinetic energy behaviour as:

$$\varepsilon = \lim_{y \to 0} 2\nu \frac{k}{y^2}.$$

The boundary conditions for $\overline{v^2}$ is to be null at the wall. Imposing a y^4 behaviour for $\overline{v^2}$ leads to the following boundary condition for f:

$$f(0) = \lim_{y \to 0} -\frac{20\nu^2 \overline{v^2}}{\varepsilon(0) y^4}.$$

Finally, model constants are

$$C_\mu = 0.19 \quad C_{\varepsilon 1} = 1.3 + \frac{0.25}{1 + \left(\frac{y}{2L}\right)^8} \quad C_{\varepsilon 2} = 1.9 \quad c_1 = 1.4$$

$$c_2 = 0.3 \quad C_L = 0.3 \quad C_\eta = 70 \quad \sigma_k = 1.0 \quad \sigma_\varepsilon = 1.3.$$

Lien and Kalitzin (2001) proposed a "code friendly" variant of the model [V2F_Li] to enhance numerical stability as the above wall condition for f can give rise to numerical oscillations. Therefore, they changed the definition of f to have a simpler wall boundary condition, i.e. $f = 0$. The transport equation for $\overline{v^2}$ and the elliptic relaxation equation become:

$$\frac{D\overline{v^2}}{Dt} = kf - 6\frac{\overline{v^2}}{k}\varepsilon + \frac{\partial}{\partial x_k}\left[\left(\nu + \frac{\nu_t}{\sigma_k}\right)\frac{\partial \overline{v^2}}{\partial x_k}\right]$$

$$L^2 \frac{\partial^2 f}{\partial x_k^2} - f = \frac{1}{T}\left((c_1 - 6)\frac{\overline{v^2}}{k} - \frac{2}{3}(c_1 - 1)\right) - c_2 \frac{P_k}{k}$$

Lastly, some model constants are revised to be:

$$C_\mu = 0.22 \quad C_{\varepsilon 1} = 1.4\left(1 + 0.050\sqrt{\frac{k}{\overline{v^2}}}\right) \quad C_L = 0.23$$

Laurence et al. (2004) [V2F_La] proposed to transport $\varphi = \frac{\overline{v^2}}{k}$ instead of $\overline{v^2}$ in order to make source terms numerically more stable. This led them to introduce a new function:

$$\overline{f} = f + \frac{2\nu}{k}\frac{\partial k}{\partial x_k}\frac{\partial \varphi}{\partial x_k} + \nu \frac{\partial^2 \varphi}{\partial x_k^2}.$$

The new transport equations for these two quantities read:

$$\frac{D\varphi}{Dt} = \overline{f} - P_k \frac{\varphi}{k} + \frac{2}{k}\frac{\nu_t}{\sigma_k}\frac{\partial \varphi}{\partial x_k}\frac{\partial k}{\partial x_k} + \frac{\partial}{\partial x_k}\left(\frac{\nu_t}{\sigma_k}\frac{\partial \varphi}{\partial x_k}\right)$$

$$L^2 \frac{\partial^2 \overline{f}}{\partial x_k^2} - \overline{f} = \frac{1}{T}(c_1 - 1)\left(\varphi - \frac{2}{3}\right) - c_2 \frac{P_k}{k} - \frac{2\nu}{k}\frac{\partial \varphi}{\partial x_k}\frac{\partial k}{\partial x_k} - \nu \frac{\partial^2 \varphi}{\partial x_k^2}$$

and some model constants are modified with respect to Durbin's original model as:

$$C_\mu = 0.22 \quad C_{\varepsilon 1} = 1.4\left(1 + \frac{0.05}{\sqrt{\varphi}}\right) \quad C_{\varepsilon 2} = 1.85 \quad C_L = 0.25 \quad C_\eta = 110$$

2.2.6 k-ω models

The use of the specific dissipation $\omega = \frac{\varepsilon}{k}$, which can also be viewed as the inverse of a turbulence time scale, to provide the second turbulent variable dates back to Kolmogorov. The Wilcox (1988) [Wi88] model uses a revised definition:

$$\omega = \frac{\varepsilon}{\beta^* k} \qquad \beta^* = 0.09$$

so that the eddy viscosity is given simply as:

$$\nu_t = \frac{k}{\omega}.$$

The transport equation for the turbulent kinetic energy is similar to the one in a $k-\varepsilon$ model while the specific dissipation transport equation is similar to the dissipation equation. They read:

$$\frac{Dk}{Dt} = P_k - \beta^* k\omega + \frac{\partial}{\partial x_k}\left[(\nu + \sigma \nu_t)\frac{\partial k}{\partial x_k}\right]$$

$$\frac{D\omega}{Dt} = \gamma \frac{P_k \omega}{k} - \beta \omega^2 + \frac{\partial}{\partial x_k}\left[(\nu + \sigma \nu_t)\frac{\partial \omega}{\partial x_k}\right]$$

with $\gamma = \frac{5}{9}, \beta = \frac{3}{40}, \sigma = 0.5$. This model requires no near-wall treatment but the near wall behaviour of ω deserves attention. Near the wall, the balance of the above equations shows that the turbulent kinetic energy behaves like $y^{3.23}$ and the specific dissipation as:

$$\omega = \frac{6\nu}{\beta y^2}.$$

Various strategies have been proposed to cope with this behaviour. Wilcox (1988) points out that an analytical value can be used to simulate rough walls and suggests to impose a large value for ω at the wall, which corresponds to very small roughnesses, in the hydrodynamically smooth regime. Wilcox (1993) recommends to force the above analytical behaviour of ω on several grid points above the surface. Menter (1994) imposes a large wall value, linked to the grid size, as $\omega_w = 10\frac{6\nu}{\beta \Delta y_1}$.

A slightly different expression can also be obtained by linear extrapolation of the analytical solution from the last grid point towards the wall.

Peng et al. (1997) developed a revised version of the $k-\omega$ model to ensure the proper behaviour of the turbulent kinetic energy and the turbulent viscosity in the wall region. An extra cross-diffusion term is added in the ω transport equation and damping functions are introduced to get the correct near wall behaviours. The model [Peng] reads:

$$\nu_t = f_\mu \frac{k}{\omega}$$

$$\frac{Dk}{Dt} = P_k - c_k f_k k\omega + \frac{\partial}{\partial x_k}\left[\left(\nu + \frac{\nu_t}{\sigma_k}\right)\frac{\partial k}{\partial x_k}\right]$$

$$\frac{D\omega}{Dt} = c_{\omega 1} f_\omega \frac{P_k \omega}{k} - c_{\omega 2}\omega^2 + c_\omega \frac{\nu_t}{k}\frac{\partial k}{\partial x_l}\frac{\partial \omega}{\partial x_l} + \frac{\partial}{\partial x_k}\left[\left(\nu + \frac{\nu_t}{\sigma_\omega}\right)\frac{\partial \omega}{\partial x_k}\right]$$

Model constants are re-established as:

$$c_k = 0.09 \quad \sigma_k = 0.8 \quad c_{\omega 1} = 0.42 \quad c_{\omega 2} = 0.075 \quad c_\omega = 0.75 \quad \sigma_\omega = 1.35$$

and the damping functions are:

$$f_\mu = 0.025 + \left(1 - \exp\left[-\left(\frac{R_t}{10}\right)^{3/4}\right]\right)\left(0.975 + \frac{0.001}{R_t}\exp\left[-\left(\frac{R_t}{200}\right)^2\right]\right)$$

$$f_k = 1 - 0.722\exp\left[-\left(\frac{R_t}{10}\right)^4\right] \quad R_t = \frac{k}{\nu\omega} \quad f_\omega = 1 + 4.3\exp\left[-\left(\frac{R_t}{1.5}\right)^{1/2}\right]$$

Menter (1992) pointed out the key drawback of the basic $k-\omega$ model: its predictions are sensitive to the free stream value for ω. Therefore, Menter (1994) proposed the BaseLine [BSL] model to cope with this problem. The $k-\omega$ model is blended with a $k-\varepsilon$ model so that it is equivalent to the $k-\varepsilon$ model in the outer part of the boundary layer and in free shear flows, while preserving the good performances of the $k-\omega$ model in the major part of the boundary layer. The k and ω transport equations are modified as:

$$\frac{Dk}{Dt} = P_k - \beta^* k\omega + \frac{\partial}{\partial x_k}\left[(\nu + \sigma_k \nu_t)\frac{\partial k}{\partial x_k}\right]$$

$$\frac{D\omega}{Dt} = \gamma\frac{P_k}{\nu_t} - \beta\omega^2 + 2(1-F_1)\sigma_{\omega 2}\frac{1}{\omega}\frac{\partial k}{\partial x_l}\frac{\partial \omega}{\partial x_l} + \frac{\partial}{\partial x_k}\left[(\nu + \sigma_\omega \nu_t)\frac{\partial \omega}{\partial x_k}\right]$$

The function F_1 controls the switching from a $k-\omega$ to a $k-\varepsilon$ model. It reads:

$$F_1 = \tanh\varsigma^4 \quad \varsigma = \min\left[\max\left(\frac{\sqrt{k}}{0.09\omega y};\frac{500\nu}{\omega y^2}\right);\frac{4\sigma_{\omega 2}k}{CD_{k\omega}y^2}\right]$$

$$CD_{k\omega} = \max\left(2\sigma_{\omega 2}\frac{1}{\omega}\frac{\partial k}{\partial x_k}\frac{\partial \omega}{\partial x_k};10^{-20}\right)$$

so that the switch occurs in the outer region of the boundary layer. All model coefficients are blended between $k-\omega$ constants when $F_1 = 1$ and $k-\varepsilon$ constants when $F_1 = 0$ as:

$$C = F_1 C_1 + (1-F_1)C_2$$

where the subscripts $_1$ and $_2$ respectively refer to the $k-\omega$ and $k-\varepsilon$ models with the following constants:

$$\begin{array}{llll} 1)k-\omega & \sigma_{k1} = 0.5 & \sigma_{\omega 1} = 0.5 & \beta_1 = 0.075 \\ 2)k-\varepsilon & \sigma_{k2} = 0.5 & \sigma_{\omega 2} = 0.856 & \beta_2 = 0.0828 \end{array}$$

and, for both models:

$$\gamma = \frac{\beta}{\beta^*} - \frac{\sigma_\omega \kappa^2}{\sqrt{\beta^*}} \quad \kappa = 0.41 \quad \beta^* = 0.09$$

Menter (1994) then proposed the Shear Stress Transport model [SST] to better predict the turbulent shear stress level. A limiter on the eddy viscosity is introduced as:

$$\nu_t = \frac{a_1 k}{\max(a_1\omega;\Omega F_2)} \quad F_2 = \tanh\left(\arg_2^2\right) \quad \arg_2 = \max\left(\frac{2\sqrt{k}}{0.09\omega y};\frac{500\nu}{\omega y^2}\right)$$

with $a_1 = 0.31$ to ensure that $-\overline{u'v'} = a_1 k$ in highly sheared regions. The constant σ_{k1} is changed to 0.85. Ω is the norm of the mean vorticity vector ($\Omega \equiv W$) but the norm of the rate of strain tensor is now preferred (Menter et al., 2003). Menter also recommends to limit the production in the turbulent kinetic energy equation to ten times the dissipation to avoid excessive turbulence production, e.g. near stagnation points.

This model has been sensitised to rotation effects using the rotation correction developed by Dol et al. (2002). For that, the production term in the specific dissipation transport equation is rewritten as:

$$\gamma \frac{P_k \omega}{k} = \gamma \max\left(S^2; W^2\right)$$

which yields the [SST-R] model.

Kok (2000) followed a similar route as Menter, adding a cross diffusion term to the $k-\omega$ model to overcome the free-stream sensitivity. The [TNT] model reads:

$$\frac{Dk}{Dt} = \max\left(P_k; 20\varepsilon\right) - \beta^* k\omega + \frac{\partial}{\partial x_k}\left[\left(\nu + \sigma_k \nu_t\right)\frac{\partial k}{\partial x_k}\right]$$

$$\frac{D\omega}{Dt} = \gamma \frac{P_k}{\nu_t} - \beta\omega^2 + \sigma_d \frac{1}{\omega}\max\left(0; \frac{\partial k}{\partial x_k}\frac{\partial \omega}{\partial x_k}\right) + \frac{\partial}{\partial x_k}\left[\left(\nu + \sigma_\omega \nu_t\right)\frac{\partial \omega}{\partial x_k}\right]$$

with the following model constants:

$$\beta^* = 0.09 \quad \sigma_k = \frac{2}{3} \quad \gamma = \frac{\beta}{\beta^*} - \frac{\sigma_\omega \kappa^2}{\sqrt{\beta^*}} \quad \sigma_d = \sigma_\omega = 0.5$$

The [LLR] (local linear realisable) model has been derived from ideas of non-linear models, in order to sensitise the $k-\omega$ model to strain and rotation effects and to obtain a proper behaviour in the wall region. Therefore, the eddy viscosity relation depends upon the strain and rotation levels while model coefficients depend upon the turbulence Reynolds number. The model reads:

$$\nu_t = c_\mu \frac{k}{\omega}$$

$$\frac{Dk}{Dt} = P_k - \beta_k k\omega + \frac{\partial}{\partial x_k}\left[\left(\nu + 0.5\nu_t\right)\frac{\partial k}{\partial x_k}\right]$$

$$\frac{D\omega}{Dt} = P_\omega - \beta_\omega \omega^2 + \frac{\partial}{\partial x_k}\left[\left(\nu + 0.5\nu_t\right)\frac{\partial \omega}{\partial x_k}\right]$$

$$c_\mu = \frac{f_\mu}{4 + A_s \tilde{U}} \quad A_s = 3\cos\left(0.5\arccos\left(\frac{\sqrt{6}S_{ij}S_{jk}S_{ki}}{\left(S_{ij}S_{ij}\right)^{1.5}}\right)\right) \quad \tilde{U} = \frac{\sqrt{0.5\left(S^2 + W^2\right)}}{\omega}$$

$$f_\mu = \frac{\frac{1}{80} + R_\mu}{1 + R_\mu} \quad R_\mu = \left(\frac{R_t}{70}\right)^\alpha \quad R_t = \frac{k}{\nu\omega} \quad \alpha = 1 + 0.9 sign\left(1; \frac{R_t}{70} - 0.9\right)$$

$$P_k = \nu_t S^2$$

$$P_\omega = S^2 \sqrt{c_\mu} f_1\left(c_1 - \frac{c_\mu S}{\omega}\right) \quad c_1 = \max\left(0.43; \frac{S}{S + 5\omega}\right) \quad f_1 = \frac{\frac{1}{90} + \left(\frac{R_t}{70}\right)^2}{1 + \left(\frac{R_t}{70}\right)^2}$$

$$\beta_k = \frac{\frac{0.83}{3} + R_k}{1 + R_k} \quad R_k = A^*\left(\frac{R_t}{100}\right)^{2.5} + \left(1 - A^*\right)\left(\frac{R_t}{100}\right)^{0.5} \quad A^* = \tanh\left(0.5\frac{R_t}{100}\right)$$

$$\beta_\omega = \frac{1.83}{1 + \sqrt{\frac{c_\mu}{1 + \frac{\upsilon_t}{\upsilon}}}} - 1$$

2.2.7 k-g model

Kalitzin et al. (1996) proposed to reformulate the $k - \omega$ model following three key ideas: to get rid of the wall distance, to have simple source terms and to have straightforward boundary conditions. They first proposed to use $\tau \propto \frac{1}{\omega} \propto \frac{k}{\varepsilon}$ as new transported quantity, introduced a limiter R to avoid high levels of τ in the outer part of turbulent regions and finally introduced $g = \sqrt{\tau}$ as transported quantity to get a better numerical behaviour. The final [k-g] model reads:

$$\nu_t = \beta^* k g^2$$

$$\frac{Dk}{Dt} = P_k - \frac{\beta^* k^2}{R} + \frac{\partial}{\partial x_k}\left[\left(\nu + \frac{\nu_t}{\sigma_k}\right)\frac{\partial k}{\partial x_k}\right]$$

$$\frac{Dg}{Dt} = -\alpha\frac{\beta^* g^3}{2R} P_k + \frac{\beta k g}{2R} - \left(\nu + \frac{\nu_t}{\sigma_g}\right)\frac{3\beta^* k g}{R}\frac{\partial k}{\partial x_k}\frac{\partial g}{\partial x_k} + \frac{\partial}{\partial x_k}\left[\left(\nu + \frac{\nu_t}{\sigma_g}\right)\frac{\partial g}{\partial x_k}\right]$$

$$R = \max\left(0.01\nu; \beta^* k \tau\right)$$

and the model constants are:

$$\beta^* = 0.09 \quad \sigma_k = \sigma_g = \sigma_\omega = 2.0 \quad \alpha = \frac{5}{9} \quad \beta = 0.075.$$

From its definition, $g=\sqrt{\dfrac{k}{\varepsilon}}$ is null at the wall.

2.2.8 k-l models

From previous works by Rotta, Wolfshtein and Mellor and Herring, Smith derived a $k-kL$ model he later transformed into a $k-L$ model. The final version of the model (Smith, 1995) [Sm95] solves a transport equation for $q^2=2k$ together with a l transport equation. The model has been initially derived for compressible flows. Its incompressible form reads:

$$\frac{Dq^2}{Dt}=2P_k-\frac{2q^3}{B_1 l}-2\nu\frac{\partial q}{\partial x_k}\frac{\partial q}{\partial x_k}+\frac{\partial}{\partial x_k}\left[\left(\nu+S_q\nu_t\right)\frac{\partial q^2}{\partial x_k}\right]$$

$$\frac{Dl}{Dt}=\left(2-E_2\right)\frac{q}{B_1}\left[1-\left(\frac{l}{\kappa L}\right)^2\right]-S_q\frac{\nu_t}{l}\frac{\partial l}{\partial x_k}\frac{\partial l}{\partial x_k}\left(\frac{l}{\kappa L}\right)^2+2S_q\frac{\nu_t}{q^2}\frac{\partial l}{\partial x_k}\frac{\partial q^2}{\partial x_k}$$
$$+\frac{\partial}{\partial x_k}\left[\left(\nu+S_q\nu_t\right)\frac{\partial l}{\partial x_k}\right]$$

where L is the distance to the nearest surface. The eddy viscosity is related to the transported quantities as

$$\nu_t=\nu\chi\Phi \quad \chi=\frac{ql}{\upsilon B_1^{1/3}} \quad \Phi=\left(\frac{C_1^4 f_1+C_2^2\chi^2+\chi^4}{C_1^4+C_2^2\chi^2+\chi^4}\right)^{1/4} \quad f_1=\exp\left[-50\left(\frac{l}{\kappa L}\right)^2\right]$$

and the model constants are:

$$B_1=18.0 \quad E_2=1.2 \quad S_q=0.7 \quad \kappa=0.41 \quad C_1=25.5 \quad C_2=2.0.$$

Here again, the boundary condition is that the turbulence length scale l be null at the wall. Smith (1997) [Sm97] proposed a revision of this model with a sort of SST limiter. For that, he considers the production to dissipation ratio. Introducing

$$P'=\max\left(0.;\nu\chi\Phi\left(\frac{\partial U_i}{\partial x_j}+\frac{\partial U_j}{\partial x_i}\right)\frac{\partial U_i}{\partial x_j}\right) \quad \varepsilon=\frac{q^3}{B_1 l} \quad \varepsilon_\nu=\nu\frac{\partial q}{\partial x_k}\frac{\partial q}{\partial x_k},$$

the eddy viscosity is multiplied by a function σ of the production to dissipation ratio as:

$$\nu_t=\sigma\nu\chi\Phi \quad \sigma=\frac{8\alpha-2\sqrt{\alpha}+7}{8\alpha^{3/2}+5} \quad \alpha=\frac{P'}{\varepsilon+\varepsilon_\nu}$$

The constant C_1 has been reduced to 23.5 to better predict the viscous sublayer.

3 Non Linear Eddy Viscosity Models and Explicit Algebraic Reynolds Stress Models

B. Aupoix, ONERA, and all partners

3.1 Introduction

Non Linear Eddy Viscosity Models (NLEVM) and Explicit Algebraic Reynolds Stress Models (EARSM) provide an alternative to the linear Boussinesq relation. The terminology NLEVM designates models which are derived a priori while EARSM models are obtained as simplifications of Reynolds stress transport equations, under equilibrium assumptions.

From the pioneering work of Pope (1975), the Reynolds stress tensor is developed in terms of the strain rate and vorticity tensors. More precisely, introducing the anisotropy tensor:

$$a_{ij} = \frac{\overline{u_i' u_j'}}{k} - \frac{2}{3}\delta_{ij},$$

the dimensionless strain rate and vorticity tensors:

$$S_{ij}^* = \tau S_{ij} \qquad \omega_{ij}^* = \tau \omega_{ij}$$

where τ is a turbulence time scale, and their invariants:

$$II_S = S_{lm}^* S_{ml}^* \qquad II_\Omega = \omega_{lm}^* \omega_{ml}^* \qquad III_S = S_{lm}^* S_{mn}^* S_{nl}^*$$
$$IV = S_{lm}^* \omega_{mn}^* \omega_{nl}^* \qquad V = S_{lm}^* S_{mn}^* \omega_{np}^* \omega_{pl}^*$$

the anisotropy tensor is expressed as:

$$\begin{aligned}
a_{ij} = {} & \beta_1 S_{ij}^* + \beta_2 \left(S_{ik}^* S_{kj}^* - \frac{\delta_{ij}}{3} II_S \right) + \beta_3 \left(\omega_{ik}^* \omega_{kj}^* - \frac{\delta_{ij}}{3} II_\Omega \right) + \beta_4 \left(S_{ik}^* \omega_{kj}^* - \omega_{ik}^* S_{kj}^* \right) \\
& + \beta_5 \left(S_{ik}^* S_{kl}^* \omega_{lj}^* - \omega_{ik}^* S_{kl}^* S_{lj}^* \right) + \beta_6 \left(\omega_{ik}^* \omega_{kl}^* S_{lj}^* + S_{ik}^* \omega_{kl}^* \omega_{lj}^* - \frac{2\delta_{ij}}{3} IV \right) \\
& + \beta_7 \left(\omega_{ik}^* \omega_{kl}^* S_{lm}^* S_{mj}^* + S_{ik}^* S_{kl}^* \omega_{lm}^* \omega_{mj}^* - \frac{2\delta_{ij}}{3} V \right) \\
& + \beta_8 \left(S_{ik}^* \omega_{kl}^* S_{lm}^* S_{mj}^* - S_{ik}^* S_{kl}^* \omega_{lm}^* S_{mj}^* \right) + \beta_9 \left(\omega_{ik}^* S_{kl}^* \omega_{lm}^* \omega_{mj}^* - \omega_{ik}^* \omega_{kl}^* S_{lm}^* \omega_{mj}^* \right) \\
& + \beta_{10} \left(\omega_{ik}^* S_{kl}^* S_{lm}^* \omega_{mn}^* \omega_{nj}^* - \omega_{ik}^* \omega_{kl}^* S_{lm}^* S_{mn}^* \omega_{nj}^* \right)
\end{aligned}$$

where the β coefficients can be functions of the invariants formed with the dimensionless rate of strain and vorticity tensors. This provides a more accurate representation of all the components of the Reynolds stress tensor and allows to capture rotation and curvature effects.

All the turbulence models presented in this section couple such a representation of the Reynolds stress tensor with two transport equations for the turbulent kinetic energy and a length scale determining variable. Boundary conditions for these equations are similar to those previously discussed and will not be detailed.

3.2 Apsley and Leschziner model

Apsley and Leschziner model [AL] is based upon a truncation of the above expansion which is written in a slightly different form as:

$$a_{ij} = -2CS^*_{ij} + q_1\left(S^*_{ik}S^*_{kj} - \frac{1}{3}S^*_{kl}S^*_{kl}\delta_{ij}\right) + q_2\left(\omega^*_{ik}S^*_{kj} - S^*_{ik}\omega^*_{kj}\right) + q_3\left(\omega^*_{ik}\omega^*_{kj} - \frac{1}{3}\omega^*_{kl}\omega^*_{lk}\delta_{ij}\right)$$
$$-\gamma_1\left(S^*_{kl}S^*_{kl}\right)S^*_{ij} - \gamma_2\left(\omega^*_{kl}\omega^*_{lk}\right)S^*_{ij} - \gamma_3\left(\omega^*_{ik}\omega^*_{kl}S^*_{lj} + S^*_{ik}\omega^*_{kl}\omega^*_{lj} - \left(\omega^*_{kl}\omega^*_{lk}\right)S^*_{ij} - \frac{2}{3}S^*_{km}\omega^*_{mn}\omega^*_{nk}\delta_{ij}\right)$$
$$-\gamma_4\left(\omega^*_{ik}S^*_{kl}S^*_{lj} - S^*_{ik}S^*_{kl}\omega^*_{lj}\right)$$

The relations between the model constants are first derived from successive approximations of an algebraic model based upon a linear Reynolds stress model. The constants read:

$$C = \frac{-\tilde{a}_{12}}{\tilde{\sigma}}\left(1 + \frac{\beta^2}{3} - \gamma^2\right)^{-1} f_p \quad q_1 = \frac{6\left(\tilde{a}_{11} + \tilde{a}_{22}\right)}{\tilde{\sigma}^2} \quad q_2 = \frac{\tilde{a}_{11} - \tilde{a}_{22}}{\tilde{\sigma}^2} \quad q_3 = 0$$

$$\gamma_1 = \frac{4}{3}C\left(\frac{\beta}{\tilde{\sigma}}\right)^2 f_p^2 \quad \gamma_2 = 4C\left(\frac{\gamma}{\tilde{\sigma}}\right)^2 f_p^2 \quad \gamma_3 = 6C\left(\frac{\gamma}{\tilde{\sigma}}\right)^2 f_p^2 \quad \gamma_4 = 6C\frac{\beta}{\tilde{\sigma}}\frac{\gamma}{\tilde{\sigma}} f_p^2$$
$$\beta = 0.22 \quad \gamma = 0.62$$

The reference values of the anisotropy tensor and the reduced strain rate, denoted by tildes, correspond to simple shear flow, including near-wall damping, and are represented, from direct numerical simulation data, as:

$$\tilde{a}_{11} = \frac{1}{3} + 0.42\exp\left(0.296\sqrt{y^*} - 0.040y^*\right) \qquad y^* = \frac{\sqrt{k}\,y}{\nu}$$
$$\tilde{a}_{22} = 0.404\left[1 - \exp\left(-0.001y^* - 0.000147y^{*2}\right)\right] - \frac{2}{3}$$
$$\tilde{a}_{12} = 0.3\left[1 - \exp\left(-0.00443\sqrt{y^*} - 0.0189y^*\right)\right]$$
$$\tilde{\sigma} = 3.33\left[1 - \exp\left(-0.45y^*\right)\right]\left[1 + 0.277y^{*3/2}\exp\left(-0.088y^*\right)\right]$$

Departure from simple shear flow is accounted for via the correction factor:

$$f_p = \frac{2f_0}{1 + \sqrt{1 + 4f_0\left(f_0 - 1\right)\left(\frac{\sigma}{\tilde{\sigma}}\right)}} \qquad f_0 = 1 + 1.25\max\left(0.09\tilde{\sigma}^2; 1.0\right) \quad \sigma = \frac{k}{\varepsilon}\sqrt{\left(\frac{\partial U_i}{\partial x_j}\right)^2}$$

3.3 Abe, Jang and Leschziner models

These models are improvements of a previous model developed by Abe et al. (1977), which uses specific definitions for the anisotropy, rate of strain and vorticity tensors:

$$b_{ij}^* = C_D b_{ij} \quad b_{ij} = \frac{\overline{u_i' u_j'}}{2k} - \frac{\delta_{ij}}{3} \quad S_{ij}^* = C_D \tau S_{ij} \quad \omega_{ij}^* = C_D \tau \omega_{ij} \quad C_D = 0.8$$

where the time scale is defined as:

$$\tau = \frac{\nu_t}{k} \quad \nu_t = C_\mu f_\mu \frac{k^2}{\varepsilon} \quad C_\mu = 0.12 \quad f_\mu = \left[1 - \frac{35}{R_t^{\frac{3}{4}}} \exp\left(-\left(\frac{R_t}{30}\right)^{\frac{3}{4}}\right)\right] [1 - f_w(26)]$$

$$R_t = \frac{k^2}{\nu\varepsilon} \quad f_w(\xi) = \exp\left(-\left(\frac{n^+}{\xi}\right)^2\right) \quad n^+ = \frac{(\varepsilon\nu)^{\frac{1}{4}} n}{\nu}$$

and n is the distance to the nearest surface.

The basic constitutive relation by Abe et al; (1997) reads:

$$b_{ij}^* = {}^1b_{ij}^* + {}^2b_{ij}^*$$

$${}^1b_{ij}^* = -C_B S_{ij}^* \quad {}^2b_{ij}^* = C_B \left[-2\left(S_{ik}^* \omega_{kj}^* - \omega_{ik}^* S_{kj}^*\right) + 2\left(S_{ik}^* S_{kj}^* - \frac{\delta_{ij}}{3} S^{*2}\right)\right]$$

where

$$C_B = \frac{1}{1 + \frac{22}{3}\omega^{*2} + \frac{2}{3}\left(\omega^{*2} - S^{*2}\right) f_B} \quad S^{*2} = S_{ij}^* S_{ij}^* \quad \omega^{*2} = \omega_{ij}^* \omega_{ij}^*$$

$$f_B = 1 + C_\eta \left(\omega^* - S^*\right) \quad C_\eta = 100.$$

A first term is added to the anisotropy tensor to improve predictions for strong normal strain as:

$${}^s b_{ij}^* = -C_B f_{s1} S_{ij}^* + 2 C_B f_{s2} \left(S_{ik}^* S_{kj}^* - \frac{\delta_{ij}}{3} S_{lk}^* S_{kl}^*\right)$$

$$f_{s1} = f_{r1} f_{r2} C_{s1} \left(\omega^{*2} - S^{*2}\right) \quad f_{s2} = -f_{r1} f_{r2} \left[1 + C_{s2}\left(\omega^{*2} - S^{*2}\right)\right]$$

$$f_{r1} = \frac{W^2 - S^2}{W^2 + S^2} \quad f_{r2} = \frac{S^2}{W^2 + S^2} \quad C_{s1} = 0.15 C_\eta \quad C_{s2} = 0.07 C_\eta$$

Another extra term is added to cope with the strong anisotropy in the near-wall region as:

$${}^{w}b_{ij}^{*} = C_D f_w(26)^{w}b_{ij}$$

$${}^{w}b_{ij} = -\frac{\alpha_w}{2}\left(d_i d_j - \frac{\delta_{ij}}{3} d_k d_k\right) + \left(1 - f_{r1}^2\right)\left[-\frac{\beta_w C_w}{1 + C_w\sqrt{S^{**2}\omega^{**2}}}\left(S_{ik}^{**}\omega_{kj}^{**} - \omega_{ik}^{**}S_{kj}^{**}\right)\right.$$

$$\left. + \frac{\gamma_w C_w}{1 + C_w S^{**2}}\left(S_{ik}^{**}S_{kj}^{**} - \frac{\delta_{ij}}{3}S^{**2}\right)\right]$$

where

$$S_{ij}^{**} = \tau_d S_{ij} \quad \omega_{ij}^{**} = \tau_d \omega_{ij} \quad S^{**2} = S_{mn}^{**}S_{mn}^{**} \quad \omega^{**2} = \omega_{mn}^{**}\omega_{mn}^{**}.$$

The unit vector representing the wall normal direction reads:

$$d_i = \frac{N_i}{\sqrt{N_k N_k}} \quad N_i = \frac{\partial d}{\partial x_i}$$

where d designates the distance to the nearest wall. The time scale τ_D is a blend between the turbulence and Kolmogorov time scales as:

$$\tau_D = \left(1 - f_w(15)\right)\frac{k}{\varepsilon} + f_w(15)\sqrt{\frac{\varepsilon}{\nu}}$$

and the model coefficients read:

$$\alpha_w = 1. \quad \beta_w = \frac{1}{4} \quad \gamma_w = 1.5 \quad C_w = 0.5$$

Finally, the constitutive relation is extended in the near-wall region as:

$$b_{ij}^{*} = {}^{1}b_{ij}^{*} + \left(1 - f_w(26)\right)\left({}^{2}b_{ij}^{*} + {}^{s}b_{ij}^{*}\right) + {}^{w}b_{ij}^{*}$$

and the Reynolds stress tensor reads:

$$\overline{u_i' u_j'} = \frac{2}{3}k\delta_{ij} + 2k\frac{b_{ij}^{*}}{C_D}.$$

The above constitutive relation can be coupled either with a $k-\varepsilon$ model or with a $k-\omega$ model. Both models use the same transport equation for the turbulent kinetic energy:

$$\frac{Dk}{Dt} = -\overline{u_i' u_j'}\frac{\partial U_i}{\partial x_j} - \varepsilon + \frac{\partial}{\partial x_l}\left[\left(\nu + \frac{\nu_t}{\sigma_k}\right)\frac{\partial k}{\partial x_l}\right]$$

where

$$\nu_t = C_\mu f_\mu \frac{k^2}{\varepsilon} \quad \sigma_k = \frac{1.2}{f_t} \quad f_t = 1 + 5 f_w(5.0)$$

The [AJL-ε] model uses the following, rather classical, form for the dissipation rate transport equation:

$$\frac{D\varepsilon}{Dt} = -C_{\varepsilon 1}\overline{u_i' u_j'}\frac{\partial U_i}{\partial x_j}\frac{\varepsilon}{k} - C_{\varepsilon 2} f_\varepsilon \frac{\varepsilon^2}{k} + \frac{\partial}{\partial x_l}\left[\left(\nu + \frac{\nu_t}{\sigma_\varepsilon}\right)\frac{\partial \varepsilon}{\partial x_l}\right]$$

$$C_{\varepsilon 1} = 1.45 \quad C_{\varepsilon 2} = 1.83 \quad f_\varepsilon = \left[1 - 0.3\exp\left(-\left(\frac{R_t}{65}\right)^2\right)\right]\left[1 - f_w(3.3)\right] \quad \sigma_\varepsilon = \frac{1.5}{f_t}$$

The [AJL-ω] model is based upon a revised version of the specific dissipation equation, with a mixed derivative term and near-wall damping term:

$$\frac{D\omega}{Dt} = -\alpha\overline{u_i' u_j'}\frac{\partial U_i}{\partial x_j}\frac{\omega}{k} - \beta\omega^2 + \left(1 - f_w(600)\right)C_{\omega 1}\tau\frac{\partial k}{\partial x_l}\frac{\partial \omega}{\partial x_l} + \frac{\partial}{\partial x_l}\left[\left(\nu + \frac{\nu_t}{\sigma_\omega}\right)\frac{\partial \omega}{\partial x_l}\right]$$

$$\alpha = 0.45 \quad \beta = 0.0747 \quad C_{\omega 1} = 1.5 \quad \sigma_\omega = \frac{1.5}{f_t}$$

$$\omega = \frac{\varepsilon}{\beta^* k} \quad \beta^* = \beta_0 \frac{\dfrac{\beta}{3\beta_0} + \left(\dfrac{R_{tw}}{9}\right)^4}{1 + \left(\dfrac{R_{tw}}{9}\right)^4} \quad \beta_0 = 0.09 \quad R_{tw} = \frac{k}{\nu\omega}$$

3.4 Wallin and Johansson model

Wallin and Johansson (2000) [WJ] derived an explicit algebraic Reynolds stress model as an equilibrium solution for the Launder, Reece and Rodi (1975) model. In the formulation of the dimensionless rate of strain and vorticity tensors S_{ij}^* and ω_{ij}^*, the time scale is chosen as:

$$\tau = \frac{k}{\varepsilon}.$$

For two-dimensional flows, only two terms of Pope's expansion are needed. The coefficients are thus obtained as:

$$\beta_1 = -\frac{6}{5}\frac{N}{N^2 - 2II_\Omega} \quad \beta_4 = -\frac{6}{5}\frac{1}{N^2 - 2II_\Omega}$$

where N is linked to the production to dissipation ratio as:

$$N = c_1' + \frac{9}{4}\frac{P_k}{\varepsilon} \quad c_1' = \frac{9}{4}(c_1 - 1) \quad c_1 = 1.8$$

and is obtained through a cubic equation as:

$$N = \begin{cases} \dfrac{c_1'}{3} + \left(P_1 + \sqrt{P_2}\right)^{1/3} + sign\left[\left(P_1 - \sqrt{P_2}\right)\left|P_1 - \sqrt{P_2}\right|^{1/3}\right] & P_2 \geq 0 \\ \dfrac{c_1'}{3} + 2\left(P_1^2 - P_2\right)^{1/6}\cos\left(\dfrac{1}{3}\arccos\left(\dfrac{P_1}{\sqrt{P_1^2 - P_2}}\right)\right) & P_2 < 0 \end{cases}$$

$$P_1 = \left(\frac{1}{27}c_1^{'2} + \frac{9}{20}II_S - \frac{2}{3}II_\Omega\right)c_1^{'}$$

$$P_2 = P_1^2 - \left(\frac{1}{9}c_1^{'2} + \frac{9}{10}II_S + \frac{2}{3}II_\Omega\right)^3$$

For three-dimensional flows, five terms must be considered. The related coefficients now read:

$$\beta_1 = -\frac{N\left(2N^2 - 7II_\Omega\right)}{Q} \quad \beta_3 = -\frac{12IV}{NQ} \quad \beta_4 = -\frac{2\left(N^2 - 2II_\Omega\right)}{Q}$$

$$\beta_6 = -\frac{6N}{Q} \quad \beta_9 = \frac{6}{Q} \quad Q = \frac{5}{6}\left(N^2 - 2II_\Omega\right)\left(2N^2 - II_\Omega\right)$$

The production to dissipation ratio is now obtained via a sixth order equation. An approximate solution reads:

$$N = N_c + \frac{162\left(\phi_1 + \phi_2 N_c^2\right)}{20N_c^4\left(N_c - \frac{c_1^{'}}{2}\right) - II_\Omega\left(10N_c^3 + 15c_1^{'}N_c^2\right) + 10c_1^{'}II_\Omega^2}$$

$$\phi_1 = IV^2 \quad \phi_2 = V - \frac{II_S II_\Omega}{2}$$

where N_c is the above solution for two-dimensional flows.

Near-wall treatment first leads to introduce a modified time scale borrowed from Durbin (1993):

$$\tau = \max\left(\frac{k}{\varepsilon}; C_\tau\sqrt{\frac{\nu}{\varepsilon}}\right) \quad C_\tau = 6.0$$

and to modify the constitutive relation as:

$$a_{ij} = f_1\beta_1 S_{ij}^* + \left(1 - f_1^2\right)\frac{3B_2 - 4}{\max\left(II_S; II_S^{eq}\right)}\left(S_{ik}^* S_{kj}^* - \frac{\delta_{ij}}{3}II_S\right)$$

$$+ f_1^2\beta_3\left(\omega_{ik}^*\omega_{kj}^* - \frac{\delta_{ij}}{3}II_\Omega\right) + \left(f_1^2\beta_4 - \left(1 - f_1^2\right)\frac{3B_2}{2\max\left(II_S; II_S^{eq}\right)}\right)\left(S_{ik}^*\omega_{kj}^* - \omega_{ik}^* S_{kj}^*\right)$$

$$+ f_1\beta_6\left(S_{ik}^*\omega_{kl}^*\omega_{lj}^* + \omega_{ik}^*\omega_{kl}^* S_{lj}^* - \frac{2\delta_{ij}}{3}IV\right) + f_1^2\beta_9\left(\omega_{ik}^* S_{kl}^*\omega_{lm}^*\omega_{mj}^* - \omega_{ik}^*\omega_{kl}^* S_{lm}^*\omega_{mj}^*\right)$$

$$B_2 = 1.8 \quad II_S^{eq} = 5.74 \quad f_1 = 1 - \exp\left(-\frac{y^*}{26}\right) \quad y^* = 2.4\sqrt{R_y} + 0.003R_y^2 \quad R_y = \frac{y\sqrt{k}}{\nu}$$

This model is coupled with the Chien's $k-\varepsilon$ model presented in section 3.1.5. Only one near-wall term is modified in the turbulent kinetic energy transport equation as:

$$D = -\frac{2\nu k}{y^2}\exp\left(-C_k y^*\right) \quad C_k = 0.04$$

Similarly, the [WJ-ω] model couples the above EARSM model with Wilcox' $k-\omega$ model. The turbulence time scale used to make velocity gradient dimensionless is then

$$\tau = \frac{1}{\beta^* \omega}$$

In order to predict the correct near-wall behaviour of the Reynolds stress tensor, the 1994 version of the $k-\omega$ model, which ensures $k \approx y^2$, is preferred. The turbulent kinetic energy and specific dissipation rate transport equations read:

$$\frac{Dk}{Dt} = P_k - \beta^* k\omega + \frac{\partial}{\partial x_k}\left[\left(\upsilon + \frac{\upsilon_t}{\sigma_k}\right)\frac{\partial k}{\partial x_k}\right]$$

$$\frac{D\omega}{Dt} = \alpha\frac{P_k \omega}{k} - \beta\omega^2 + \frac{\partial}{\partial x_k}\left[\left(\upsilon + \frac{\upsilon_t}{\sigma_\omega}\right)\frac{\partial \omega}{\partial x_k}\right]$$

$$\beta = \frac{3}{40} \qquad \sigma_k = 2 \qquad \sigma_\omega = 2$$

$$\alpha^* = \frac{\frac{1}{40} + \frac{R_{eT}}{R_k}}{1 + \frac{R_{eT}}{R_k}} \qquad \alpha = \frac{5}{9}\frac{\frac{1}{10} + \frac{R_{eT}}{R_\omega}}{1 + \frac{R_{eT}}{R_\omega}}\frac{1}{\alpha^*} \qquad \beta^* = \frac{9}{100}\frac{\frac{5}{18} + \left(\frac{R_{eT}}{R_\beta}\right)^4}{1 + \left(\frac{R_{eT}}{R_\beta}\right)^4}$$

$$R_k = 6 \qquad R_\omega = 2.7 \qquad R_\beta = 8 \qquad R_{eT} = \frac{\rho k}{\omega \mu}$$

An alternative is the [WJ-std-ω] version which couples the above Wallin and Johansson EARSM model with the standard $k-\omega$ model [Wi88], without any wall damping. This Wilcox (1988) $k-\omega$ model does not reproduce the near wall asymptotics nor the near wall peak in turbulent kinetic energy only the mean velocity profile is correct. Therefore, the EARSM model is thus used without damping functions i.e. with $f_1 = 1$.

At last, the [WJ-Rt] model couples the above EARSM model with a $k-R$ model, where $R=k^2/\varepsilon$. This model is mainly a reformulation of standard $k-\varepsilon$ models to improve numerical behaviour and reads:

$$\frac{Dk}{Dt} = P_k - \frac{k^2}{R} + \frac{\partial}{\partial x_k}\left[\left(\nu + \frac{\nu_t}{\sigma_k}\right)\frac{\partial k}{\partial x_k}\right]$$

$$\frac{DR}{Dt} = (2 - C_{\varepsilon 1})\frac{P_k R}{k} - (2 - C_{\varepsilon 2} f_2)k + \left(\nu + \frac{\nu_t}{\sigma_\varepsilon}\right)\frac{\partial^2 R}{\partial x_k^2} - \frac{1}{\sigma_\varepsilon}\frac{\partial \nu_t}{\partial x_k}\frac{\partial R}{\partial x_k}$$

$$\nu_t = C_\mu f_\mu R$$

$$C_\mu = 0.09 \quad C_{\varepsilon 1} = 1.44 \quad C_{\varepsilon 2} = 1.92 \quad \sigma_k = 1.0 \quad \sigma_\varepsilon = 0.8566$$

$$f_2 = 1 - 0.22\exp\left[-\left(\frac{R_t}{6}\right)^2\right] \quad f_\mu = \exp\left(-\frac{3.4}{\left(1+\frac{R_t}{50}\right)^2}\right) \quad R_t = \frac{k^2}{\nu\varepsilon}$$

3.5 Wallin and Johansson curvature correction

Wallin and Johansson (2002) improved the above presented EARSM model to sensitise it to rotation and curvature effects [WJ-C]. As EARSM models are based upon the invariance of the anisotropy tensor in a given reference frame, this frame has to be prescribed. They proposed to use a strain-rate based co-ordinate system, i.e. a reference frame linked to the evolution of the rate of strain tensor. The co-ordinate system rotation rate tensor $\Omega_{ij}^{(r)}$, which is related to the reference frame transformation matrix T_{ij} as:

$$\Omega_{ij}^{(r)} = \frac{DT_{ik}}{Dt} T_{kj} = -T_{ik}\frac{DT_{kj}}{Dt}$$

is linked to the evolution of the strain rate tensor as:

$$\Omega_{ij}^{(r)} = -\varepsilon_{ijk}\omega_k^{(r)} \quad \omega_i^{(r)} = \varepsilon_{pqj}\frac{II_S^2\delta_{ij} + 12III_S S_{ij} + 6II_S S_{ik}S_{kj}}{2II_S^3 - 12III_S^2} S_{pl}\frac{DS_{lq}}{Dt}.$$

The reference frame rotation is thus accounted for in the EARSM model by modifying the dimensionless vorticity tensor as:

$$\omega_{ij}^* = \tau\omega_{ij} - \frac{\tau}{A_0}\Omega_{ij}^{(r)}.$$

Moreover, while the above presented EARSM model was based upon a specific Reynolds stress model which allowed drastic simplifications, the underlying Reynolds stress model constants had to be optimised to correctly capture rotation effects, i.e. mainly bifurcation points for rotating homogeneous shear flows. The recalibration was made such as the equilibrium solution in the non-rotating case was kept, resulting in $A_0 = -0.72$. The above modification of ω_{ij}^* is thus the only modification of the EARSM model to account for curvature effects.

3.6 Hellsten model

Hellsten 2004 [WJ-H] developed a carefully tailored $k-\omega$ model. The model reads:

$$\frac{Dk}{Dt} = P_k - \beta^* k\omega + \frac{\partial}{\partial x_k}\left[(\upsilon + \sigma_k \upsilon_t)\frac{\partial k}{\partial x_k}\right]$$

$$\frac{D\omega}{Dt} = \gamma\frac{P_k\omega}{k} - \beta\omega^2 + \frac{\partial}{\partial x_k}\left[(\upsilon + \sigma_\omega \upsilon_t)\frac{\partial \omega}{\partial x_k}\right] + \sigma_d \frac{1}{\omega}\max\left(\frac{\partial k}{\partial x_l}\frac{\partial \omega}{\partial x_l};0\right)$$

where, as in Menter's model, all the coefficients are blends between inner and outer values as:

$$C = f_{mix} C_1 + (1 - f_{mix}) C_2 .$$

The model coefficients read:

	γ	β	σ_k	σ_ω	σ_d
C_1	0.518	0.0747	1.1	0.53	1
C_2	0.44	0.0828	1.1	1.0	0.4

The mixing function f_{mix} is defined as:

$$f_{mix} = \tanh\left(C_{mix}\Gamma^4\right) \quad C_{mix} = 1.5 \quad \Gamma = \min\left[\max(\Gamma_1;\Gamma_2);\Gamma_3\right]$$

$$\Gamma_1 = \frac{\sqrt{k}}{\beta^* \omega d} \quad \Gamma_2 = \frac{500\nu}{\omega d^2} \quad \Gamma_3 = \frac{20k}{\max\left[\frac{d^2}{\omega}\frac{\partial k}{\partial x_l}\frac{\partial \omega}{\partial x_l};200k_\infty\right]}$$

where d is the distance to the nearest surface. The ω wall value is prescribed using rough wall formula, in the limit of the smooth wall regime.

Here again, as the $k-\omega$ model reduces to the standard (1988) version in the wall region and thus does not provide the right near wall behaviours, the EARSM model is used without wall damping functions.

In Hellsten's works, the EARSM model includes the above presented curvature correction but, in the framework of this project, it was used without it.

3.7 Grundestam, Wallin and Johansson model

The above EARSM models have been derived from a linear model for the pressure-strain term. Grundestam et al. (2003) extended the EARSM approach to deal with pressure-strain models which are non-linear with respect to the mean velocity gradients. This model is labelled [WJ-G].

Curvature effects are accounted using the above presented Wallin and Johansson curvature correction:

$$\overset{*}{\omega}_{ij} = \tau\omega_{ij} - \frac{\tau}{A_0}\Omega_{ij}^{(r)} .$$

and the associated invariant:

$$\overset{*}{II}_\Omega = \overset{*}{\omega}_{lm}\overset{*}{\omega}_{ml} .$$

As in the [WJ] model, the model uses a restricted tensorial basis with only five tensors. The associated coefficients now read:

$$\beta_1 = A_1\left(N^* c^2 II_\Omega + 8cII_\Omega^*\sqrt{-II_\Omega} - 14N^* II_\Omega^* + 4N^{*3}\right)/(Q_1Q_2)$$

$$\beta_3 = 4IVA_1\left(2c^2 II_\Omega N^{*2} + c^3 N^*(-II_\Omega)^{3/2} - 18II_\Omega^{*2} + 18cN^*\sqrt{-II_\Omega}\, II_\Omega^* + 7c^2 II_\Omega II_\Omega^*\right)/$$
$$\left[II_\Omega^{*2}\left(2c\sqrt{-II_\Omega} - 3N^*\right)Q_1Q_2\right]$$

$$\beta_4 = 4A_1/Q_1$$

$$\beta_6 = -2A_1\left(N^* c^2 II_\Omega + 4cII_\Omega^*\sqrt{-II_\Omega} + 2N^{*2}c\sqrt{-II_\Omega} - 6N^* II_\Omega^*\right)/\left(II_\Omega^* Q_1Q_2\right)$$

$$\beta_9 = 2A_1\left(4cN^*\sqrt{-II_\Omega} + c^2 II_\Omega - 6II_\Omega^*\right)/\left(II_\Omega^* Q_1Q_2\right)$$

$$Q_1 = -\left(2N^* - c\sqrt{-II_\Omega}\right)^2 + 2II_\Omega^*$$

$$Q_2 = N^{*2} - 2II_\Omega^*$$

The parameter N^* is linked to the previously defined parameter N as:

$$N^* = N - c\left(\sqrt{II_S} - \sqrt{II_\Omega}\right)$$

and is given by a third rank equation the solution of which reads:

$$N^* = \begin{cases} \dfrac{A_3^*}{3} + \left(P_1 + \sqrt{P_2}\right)^{1/3} + sign\left(P_1 - \sqrt{P_2}\right)\left|P_1 - \sqrt{P_2}\right|^{1/3} & P_2 \geq 0 \\ \dfrac{A_3^*}{3} + 2\left(P_1^2 - P_2\right)^{1/6}\cos\left(\dfrac{1}{3}\arccos\left(\dfrac{P_1}{\sqrt{P_1^2 - P_2}}\right)\right) & P_2 < 0 \end{cases}$$

$$P_1 = \left(\frac{A_3^{*2}}{27} + \frac{A_1A_4}{6} II_S - \frac{2}{3} II_\Omega^*\right)A_3^*$$

$$P_2 = P_1^2 - \left(\frac{A_3^{*2}}{9} + \frac{A_1A_4}{3} II_S + \frac{2}{3} II_\Omega^*\right)^3$$

This solution is applied even for three-dimensional flows. The model constants are now optimised as:

$$A_0 = -0.9 \quad A_1 = 1.20 \quad A_2 = 0. \quad A_3 = 1.80 \quad A_4 = 2.25 \qquad c = 0.56$$

where c is the coefficient which governs the new non-linear terms, with respect to the mean velocity gradients. The coefficient A_3^* is given by:

$$A_3^* = A_3 - c\left(\sqrt{II_S} - \sqrt{-II_\Omega}\right)$$

This model is also coupled with the $k-R$ turbulence model.

3.8 Linear explicit algebraic model

Rung et al. (1999) proposed the linear explicit algebraic model [LEA] as a modification of the initial algebraic Reynolds stress model derived by Pope (1975) and Gatski and Speziale (1993). For two-dimensional flows, they express the anisotropy tensor as:

$$a_{ij}^{2D} = -2\frac{\nu_t}{k}\left[S_{ij} + \beta_2\frac{k}{\varepsilon}\left(S_{ik}\omega_{kj} + S_{jk}\omega_{ki}\right) - \beta_3\frac{k}{\varepsilon}\left(S_{ik}S_{kj} - S_{lk}S_{kl}\delta_{ij}\right)\right]$$

$$\nu_t = c_\mu^*\frac{k^2}{\varepsilon} \qquad c_\mu^* = \frac{\beta_1}{1 - \frac{2}{3}\eta^2 + 2\xi^2} \qquad \eta^2 = \frac{(\beta_3 S)^2}{8} \qquad \xi^2 = \frac{(\beta_2 W)^2}{2}$$

$$S = \frac{k}{\varepsilon}\sqrt{2S_{kl}S_{kl}} \qquad W = \frac{k}{\varepsilon}\sqrt{2\omega_{kl}\omega_{kl}}$$

$$\beta_1 = \frac{\frac{2}{3} - \frac{C_2}{2}}{g} \qquad \beta_2 = \frac{1 - \frac{C_4}{2}}{g} \qquad \beta_3 = \frac{2 - C_3}{g}$$

where the model coefficients correspond to a modified version of the linearized SSG model:

$$C_1 = 2.5 \quad C_2 = 0.39 \quad C_3 = 1.25 \quad C_4 = 0.45.$$

For three-dimensional flows, the above expression is complemented as:

$$a_{ij} = a_{ij}^{2D} - \frac{6\beta_3}{2 - \eta^2 + \xi^2}\frac{\nu_t}{\varepsilon}\left[\xi^2 S_{ik}S_{kj} - \left(\frac{2\eta\beta_2}{\beta_3}\right)^2\omega_{ik}\omega_{kj} - \frac{\beta_2 k}{\varepsilon}\left(1 - \frac{\xi}{\eta}\right)\left(S_{ik}S_{kl}\omega_{lj} + S_{jk}S_{kl}\omega_{li}\right)\right].$$

Compared to previous Pope and Gatski and Speziale models, the improvement is the proposed model for the g factor, which is linked to the production to dissipation ratio, as:

$$g = \left(1 + 0.95\left[1 - \tanh\left(\frac{S}{2.15}\right)^2\right]\right)(C_1 - 1) + \frac{S^2}{4 + 1.83\sqrt{0.2S^2 + 0.8W^2}}.$$

This model is coupled with the standard Wilcox (1988) $k - \omega$ model.

3.9 Compact explicit algebraic stress model

Lübcke et al. (2002) developed the compact explicit algebraic stress model [CEASM] for applications to strongly three-dimensional flows. This model uses a restricted set of tensors in Pope's representation, with the following coefficients:

$$\beta_1 = -2c_\mu \quad \beta_2 = -\frac{4A_3c_\mu}{g} \quad \beta_4 = \frac{A_2c_\mu}{g} - \frac{II_S}{2}\beta_8 \quad \beta_5 = \frac{12A_2c_\mu\left(A_3 - A_2\sqrt{-\frac{II_\Omega}{II_S}}\right)}{-2g^2 + A_3^2 II_S + A_2^2 II_\Omega}$$

$$\beta_8 = -\frac{12A_2c_\mu}{II_S g} \quad A_1 = -0.472 \quad A_2 = -0.775 \quad A_3 = -0.375$$

$$c_\mu = \frac{-A_1 g}{g^2 - \frac{2}{3}A_3^2 II_S - 2A_2^2 II_\Omega}$$

and g is given by :

$$g = C_1 - 1 + \frac{II_S}{4 + 1.83\sqrt{0.4II_S - 1.6II_\Omega}}\frac{2k^2}{\varepsilon^2}$$

This model is coupled with Lien and Leschziner (1993) $k-\varepsilon$ model which reads:

$$\frac{Dk}{Dt} = P_k - \varepsilon + \frac{\partial}{\partial x_k}\left[\left(\nu + \frac{\nu_t}{\sigma_k}\right)\frac{\partial k}{\partial x_k}\right]$$

$$\frac{D\varepsilon}{Dt} = (C_{\varepsilon 1}P_k - C_{\varepsilon 2}\varepsilon)\frac{\varepsilon}{k} + \frac{\partial}{\partial x_k}\left[\left(\nu + \frac{\nu_t}{\sigma_\varepsilon}\right)\frac{\partial \varepsilon}{\partial x_k}\right]$$

$$C_{\varepsilon 1} = 1.44 f_1 \quad C_{\varepsilon 2} = 1.92 f_2 \quad \sigma_k = 1.0 \quad \sigma_\varepsilon = 1.3 \quad C_1 = 2.5$$

$$f_1 = 1 + \frac{P_k^*}{P_k} \quad P_k = -\overline{u_i' u_j'}S_{ij} \quad P_k^* = \frac{C_{\varepsilon 2} f_2 k^{3/2}}{C_{\varepsilon 1} L_\varepsilon}\exp\left(-\alpha_d R_k^2\right) \quad \nu_t = 0.09 f_\mu \frac{k^2}{\varepsilon}$$

$$f_2 = 1 - 0.3\exp\left(-R_t^2\right) \quad L_\varepsilon = \kappa c_\mu^{-3/4} n(1 - \exp(-\alpha_\varepsilon R_k)) \quad f_\mu = \frac{1 - \exp\left(-\alpha_\mu R_k\right)}{1 - \exp\left(-\alpha_\varepsilon R_k\right)}$$

$$c_\mu = 0.09 \quad \alpha_d = 0.0022 \quad \alpha_\varepsilon = 0.263 \quad \alpha_\mu = 0.016 \quad R_t = \frac{k^2}{\nu\varepsilon} \quad R_k = \frac{\sqrt{k}n}{\nu}$$

This is the correct definition for f_1 as there is a misprint in the original reference).

4 Differential Reynolds stress turbulence models

B. Aupoix, ONERA, and all partners

4.1 Introduction

Solving the transport equation for the Reynolds stress tensor allows to account for Reynolds stress history effects and to dispose of any direct link between the Reynolds stress tensor and the local velocity gradients. The exact transport equation for the Reynolds stress tensor can be deduced from the Navier-Stokes equations and reads:

$$\frac{D\overline{u_i'u_j'}}{Dt} = P_{ij} + \phi_{ij} - \varepsilon_{ij} + V_{ij} + T_{ij}$$

where $P_{ij} = -\overline{u_i'u_l'}\frac{\partial U_j}{\partial x_l} - \overline{u_l'u_j'}\frac{\partial U_i}{\partial x_l}$ is the production term, due to the action of the mean velocity gradients, $\phi_{ij} = \overline{\frac{p'}{\rho}\left(\frac{\partial u_i'}{\partial x_j} + \frac{\partial u_i'}{\partial x_j}\right)}$ the pressure-strain term or redistribution term as it is traceless and exchanges energy between the normal stresses, $\varepsilon_{ij} = 2\nu\overline{\frac{\partial u_i'}{\partial x_l}\frac{\partial u_j'}{\partial x_l}}$ the "dissipation" term due to viscous effects, although it can be shown that it is a real sink term only for the normal stresses, $V_{ij} = \frac{\partial}{\partial x_l}\left[\nu\frac{\partial \overline{u_i'u_j'}}{\partial x_l}\right]$ and $T_{ij} = \frac{\partial}{\partial x_l}\left[-\overline{u_i'u_j'u_l'} - \overline{\frac{p'}{\rho}\left(u_i'\delta_{jl} + u_j'\delta_{il}\right)}\right]$ are respectively the viscous and turbulent diffusion terms. Only the production and viscous diffusion terms are exact while all the other terms need to be modelled.

Models used in this project mainly differ in the closure of the pressure-strain term. In the following, they will therefore be sorted first according to the pressure-strain term model.

As they will be widely used in the following, the anisotropy tensors for both the Reynolds stress tensor and the dissipation tensor are introduced here as:

$$a_{ij} = \frac{\overline{u_i'u_j'}}{k} - \frac{2\delta_{ij}}{3} \qquad e_{ij} = \frac{\varepsilon_{ij}}{\varepsilon} - \frac{2\delta_{ij}}{3}$$

together with their invariants:

$$A_2 = a_{ij}a_{ij} \quad A_3 = a_{ij}a_{jk}a_{ki} \quad A = 1 - \frac{9}{8}(A_2 - A_3)$$

$$E_2 = e_{ij}e_{ij} \quad E_3 = e_{ij}e_{jk}e_{ki} \quad E = 1 - \frac{9}{8}(E_2 - E_3).$$

The problem of the boundary condition can be discussed once for all the models. The no-slip condition imposes the velocity fluctuations and hence the Reynolds stresses to be null at the wall.

4.2 Speziale, Sarkar and Gatski model

Although being among the most recent pressure-strain term models used in this project, the Speziale et al. model (1991) is presented first as it is the basis for the common Reynolds stress models most participants implemented in their codes.

Speziale et al. (1991) proposed a redistribution term model which is tuned to reproduce the return to isotropy in absence of mean velocity gradient, highly sheared flows and flows subjected to both shear and rotation. It reads:

$$\phi_{ij} = -\left(c_1\varepsilon + c_1^* P_k\right)a_{ij} + c_2\varepsilon\left(a_{ik}a_{kj} - \frac{\delta_{ij}}{3}A_2\right) + \left(c_3 - c_3^*\sqrt{A_2}\right)ks_{ij}$$

$$+ c_4k\left(a_{ik}s_{kj} + a_{jk}s_{ki} - \frac{2}{3}a_{nm}s_{nm}\delta_{ij}\right) + c_5k\left(a_{ik}\omega_{jk} + a_{jk}\omega_{ik}\right)$$

$$c_1 = 1.7 \quad c_1^* = 0.90 \quad c_2 = 1.05$$

$$c_3 = 0.8 \quad c_3^* = 0.65 \quad c_4 = 0.625 \quad c_5 = 0.20$$

The "dissipation" tensor is classically modelled assuming isotropy of the small scales as:

$$\varepsilon_{ij} = \frac{2}{3}\delta_{ij}\varepsilon.$$

The turbulent diffusion is modelled, with a first gradient hypothesis, as:

$$T_{ij} = \frac{\partial}{\partial x_l}\left[\frac{2c_s}{3}\frac{k^2}{\varepsilon}\frac{\partial \overline{u_i'u_j'}}{\partial x_l}\right] \qquad c_s = 0.22\,.$$

To solve high turbulent Reynolds number flows, the above Reynolds stress model is coupled with the standard form of the dissipation equation:

$$\frac{D\varepsilon}{Dt} = C_{\varepsilon 1}\frac{P_k\varepsilon}{k} - C_{\varepsilon 2}\frac{\varepsilon^2}{k} + \frac{\partial}{\partial x_l}\left(\frac{C_\mu k^2}{\sigma_\varepsilon \varepsilon}\frac{\partial \varepsilon}{\partial x_l}\right)$$

with

$$C_{\varepsilon 1} = 1.44 \quad C_{\varepsilon 2} = 1.92 \quad C_\mu = 0.09 \quad \sigma_\varepsilon = 1.3\,.$$

This forms the basis for the high Reynolds number common model, to be used with wall functions, and hereafter designed as [SSG].

Chen (1995) proposed an extension [SSG-Ch] of the [SSG] model to solve it down to the wall. As concerns the Reynolds stress transport equation, the viscous diffusion term is now considered while the turbulent diffusion is modelled, following Daly and Harlow (1970), as:

$$T_{ij} = \frac{\partial}{\partial x_k}\left[c_s \frac{k}{\varepsilon} \overline{u_k' u_l'} \frac{\partial \overline{u_i' u_j'}}{\partial x_l} \right] \qquad c_s = 0.22\,.$$

To account for near-wall effects, the redistribution term is altered as:

$$\phi_{ij} = -\left[\left(c_1\varepsilon + c_1^* P_k\right)\left(1 - f_w\right) + f_w \varepsilon\right] a_{ij} - c_2 \left(1 - f_w\right)\varepsilon\left(a_{ik}a_{kj} - \frac{\delta_{ij}}{3} A_2 \right)$$
$$+ \left(c_3 - c_3^* \sqrt{A_2}\right) k s_{ij} + c_4 k\left(a_{ik}s_{kj} + a_{jk}s_{ki} - \frac{2}{3} a_{nm}s_{nm}\delta_{ij} \right) + c_5 k\left(a_{ik}\omega_{jk} + b_{jk}\omega_{ik}\right).$$

and a wall reflection term is added as:

$$\phi_{ij}^{w} = f_w \left[0.45\left(P_{ij} - \frac{2\delta_{ij}}{3} P_k \right) - 0.03\left(Q_{ij} - \frac{2\delta_{ij}}{3} P_k \right) + 0.16 k S_{ij} \right]$$

where $Q_{ij} = -\overline{u_i' u_l'} \frac{\partial U_l}{\partial x_j} - \overline{u_l' u_j'} \frac{\partial U_l}{\partial x_i}$. The wall damping function is defined as:

$$f_w = \exp\left[-\left(0.0184 \frac{\sqrt{k}\, y}{\nu} \right)^4 \right].$$

The transport equation for the dissipation rate is also revised to provide the correct near wall behaviour as:

$$\frac{D\varepsilon}{Dt} = C_{\varepsilon 1}\left(1 + f_w\right)\frac{P_k \varepsilon}{k} - C_{\varepsilon 2} f_\varepsilon \frac{\varepsilon \varepsilon^*}{k} + \left[\left(\frac{7}{9} C_{\varepsilon 2} - 2 \right) \frac{\varepsilon \varepsilon^*}{k} - \frac{\varepsilon^{*2}}{2k} \right] f_w$$
$$+ \frac{\partial}{\partial x_k}\left[\nu \frac{\partial \varepsilon}{\partial x_k} + C_\varepsilon \frac{k}{\varepsilon} \overline{u_k' u_l'} \frac{\partial \varepsilon}{\partial x_l} \right]$$

with $\varepsilon^* = \varepsilon - 2\frac{\nu k}{y^2}$, which is null at the wall, and the following constants and damping function:

$$C_{\varepsilon 1}=1.35 \quad C_{\varepsilon 2}=1.80 \quad C_{\varepsilon}=0.15 \quad f_{\varepsilon}=1-\frac{2}{9}\exp\left[-\left(\frac{R_t}{6}\right)^2\right] \quad R_t=\frac{k^2}{\nu\varepsilon}.$$

Another extension of the [SSG] model, referred to as [SSG-ω], has been devised in this program by combining the high Reynolds number form of the Reynolds stress equation, without wall damping nor reflection terms, with the standard version [Wi88] of the ω equation. Diffusion is modelled using the eddy viscosity diffusivity with the Wilcox (1988) diffusion coefficients. The model is solved down to the wall, without any near-wall corrections. This is numerically robust but the near-wall asymptotic behaviour of the turbulence quantities is not correct.

4.3 Isotropisation of production model

The isotropisation of production model is the simplest model for the pressure-strain term, initially proposed by Naot et al. (1973). One of its advantages is its simple form, which may improve its numerical robustness. In this kind of model, the pressure strain term is represented as:

$$\phi_{ij}=-c_2\left(P_{ij}-\frac{2}{3}P_k\delta_{ij}\right)-c_1\frac{\varepsilon}{k}\left(\overline{u_i' u_j'}-\frac{2}{3}k\delta_{ij}\right).$$

The first term represents the linear part of the pressure-strain term, modelled as a reduction of the production, and the second the return to isotropy, borrowed from Rotta. This model, with $c_1=0.6$ and $c_2=1.8$, has been coupled with a wall reflection term to account for near wall treatment as:

$$\phi_{ij}^{w}=\left[0.45\left(P_{ij}-\frac{2}{3}P_k\delta_{ij}\right)+0.08kS_{ij}\right]f_w \quad f_w=\exp\left[-\left(0.015\frac{\sqrt{k}\,y}{\nu}\right)^4\right]$$

and with the above presented dissipation and turbulent diffusion models to form the model [IP]. The associated length scale equation is a classical dissipation equation, based upon Chien's (1982) model:

$$\frac{D\widetilde{\varepsilon}}{Dt}=C_{\varepsilon 1}\frac{P_k\widetilde{\varepsilon}}{k}-C_{\varepsilon 2}f_2\frac{\widetilde{\varepsilon}^2}{k}-2\nu\frac{\widetilde{\varepsilon}}{y^2}\exp\left(-0.5y^+\right)+\frac{\partial}{\partial x_k}\left[\nu\frac{\partial\widetilde{\varepsilon}}{\partial x_k}+C_{\varepsilon}\frac{k}{\varepsilon}\overline{u_k' u_l'}\frac{\partial\widetilde{\varepsilon}}{\partial x_l}\right]$$

$$C_{\varepsilon 1}=1.44 \quad C_{\varepsilon 2}=1.9 \quad f_2=1-\frac{2}{9}\exp\left(-\left(\frac{R_t}{6}\right)^2\right) \quad C_{\varepsilon}=0.18.$$

This equation provides the "isotropic" dissipation which is null at the wall (" " are added to avoid confusion with the isotropic model for the dissipation tensor). The real dissipation, which appears in the "dissipation" term is corrected as:

$$\varepsilon=\widetilde{\varepsilon}+2\nu\frac{k}{y^2}.$$

Hanjalić and Jakirlić (1998) [HJ] proposed another model based upon the isotropisation of production approach and applicable down to the wall. Here again, wall effects on the pressure-strain term are accounted for via two reflection terms:

$$\phi^{w}_{ij,1} = c_1^w f_w \frac{\varepsilon}{k}\left(\overline{u_k' u_m'} n_k n_m \delta_{ij} - \frac{3}{2}\overline{u_i' u_k'} n_k n_j - \frac{3}{2}\overline{u_k' u_j'} n_k n_i\right)$$

$$\phi^{w}_{ij,2} = c_2^w f_w\left(\phi_{km,2} n_k n_m \delta_{ij} - \frac{3}{2}\phi_{ik,2} n_k n_j - \frac{3}{2}\phi_{kj,2} n_k n_i\right) \quad \phi_{ij,2} = -c_2\left(P_{ij} - \frac{2}{3}P_k \delta_{ij}\right)$$

where n_k represents the components of the wall normal. The turbulent diffusion model is again the Daly and Harlow one.

The model is sensitised to the strong anisotropy induced by the wall through the use of the anisotropy tensors for both the Reynolds stress tensor and the dissipation tensor so that model constants read:

$$c_1 = c + \sqrt{A}E \qquad c = 2.5AF^{1/4} f \qquad c_2 = 0.8\sqrt{A}$$

$$c_1^w = \max(1 - 0.7c; 0.3) \qquad c_2^w = \min(A; 0.3) \qquad F = \min(0.6; A_2)$$

$$f = \min\left(1; \left(\frac{R_t}{150}\right)^{3/2}\right) \qquad f_w = \min\left(1.4; \frac{k^{3/2}}{2.5\varepsilon x_n}\right) \qquad R_t = \frac{k^2}{\upsilon\varepsilon}$$

The dissipation rate tensor is modelled to reproduce DNS data as:

$$\varepsilon_{ij} = f_s \varepsilon_{ij}^* + (1 - f_s)\frac{2\delta_{ij}}{3}\varepsilon$$

$$\varepsilon_{ij}^* = \frac{\varepsilon}{k}\frac{\overline{u_i' u_j'} + f_d\left(\overline{u_i' u_k'} n_j n_k + \overline{u_j' u_k'} n_i n_k + \overline{u_k' u_l'} n_k n_l n_i n_j\right)}{1 + 1.5\frac{\overline{u_p' u_q'}}{k} n_p n_q f_d}$$

$$f_s = 1 - \sqrt{A}E^2 \qquad f_d = \left(1 + \frac{R_t}{10}\right)^{-1}$$

The Reynolds stress transport equations are coupled with a transport equation for the dissipation rate which reads:

$$\frac{D\varepsilon}{Dt} = C_{\varepsilon 1}\frac{P_k \varepsilon}{k} - C_{\varepsilon 2} f_\varepsilon \frac{\varepsilon\tilde{\varepsilon}}{k} + C_{\varepsilon 3}\nu \frac{k}{\varepsilon}\overline{u_j' u_k'}\frac{\partial^2 U_i}{\partial x_j \partial x_l}\frac{\partial^2 U_i}{\partial x_k \partial x_l}$$

$$+ \max\left(\left[\left(\frac{1}{C_L}\frac{\partial l}{\partial x_j}\right)^2 - 1\right]\left(\frac{1}{C_L}\frac{\partial l}{\partial x_j}\right)^2; 0\right)\frac{\varepsilon\tilde{\varepsilon}}{k}A + \frac{\partial}{\partial x_k}\left[\nu\frac{\partial \varepsilon}{\partial x_k} + C_\varepsilon \frac{k}{\varepsilon}\overline{u_k' u_l'}\frac{\partial \varepsilon}{\partial x_l}\right]$$

where $l = \dfrac{k^{3/2}}{\varepsilon}$ is the turbulence length scale and the model constants and damping functions are:

$$C_{\varepsilon 1} = 2.6 \quad C_{\varepsilon 2} = 1.92 \quad C_{\varepsilon 3} = 0.25 \quad C_{\varepsilon} = 0.18 \quad C_L = 2.5$$

$$f_{\varepsilon} = 1 - \frac{C_{\varepsilon 2} - 1.4}{C_{\varepsilon 2}} \exp\left[-\left(\frac{R_t}{6} \right)^2 \right]$$

4.4 Launder, Reece and Rodi model

Launder et al. (1975) developed a model for the redistribution term as a linear function of the Reynolds stress tensor:

$$\phi_{ij} = -\frac{c_2 + 8}{11}\left(P_{ij} - \frac{2}{3}\delta_{ij} P_k \right) - \frac{30c_2 - 2}{55} k \left(\frac{\partial U_i}{\partial x_j} + \frac{\partial U_j}{\partial x_i} \right)$$

$$-\frac{8c_2 - 2}{11}\left(D_{ij} - \frac{2}{3}\delta_{ij} P_k \right) - c_1 \frac{\varepsilon}{k}\left(\overline{u_i' u_j'} - \frac{2}{3}\delta_{ij} k \right)$$

$$P_k = \frac{P_{ii}}{2} \quad D_{ij} = -\overline{u_i' u_l'} \frac{\partial U_l}{\partial x_j} - \overline{u_l' u_j'} \frac{\partial U_l}{\partial x_j}$$

The model constants have been calibrated with respect to experiments available at that time as $c_1 = 1.5$, $c_2 = 0.4$. The dissipation tensor is here again modelled assuming small scale isotropy.

In this project, the [LRR] model has been used with $c_1 = 1.8$ and the Daly and Harlow diffusion law. It was coupled with the following dissipation equation:

$$\frac{D\varepsilon}{Dt} = C_{\varepsilon 1} \frac{P_k \varepsilon}{k} - C_{\varepsilon 2} f_2 \frac{\varepsilon^2}{k} + \frac{\partial}{\partial x_k}\left[\nu \frac{\partial \varepsilon}{\partial x_k} + C_{\varepsilon} \frac{k}{\varepsilon} \overline{u_k' u_l'} \frac{\partial \varepsilon}{\partial x_l} \right]$$

$$C_{\varepsilon 1} = 1.44 \quad C_{\varepsilon 2} = 1.92 \quad C_{\varepsilon} = 0.18.$$

This model is only valid far from walls and has to be coupled with a wall law.

Shima's model (1998) [Sh] is based upon the Launder, Reece and Rodi form of the pressure strain term, written as:

$$\phi_{ij} = -c_1 \frac{\varepsilon}{k}\left(\overline{u_i' u_j'} - \frac{2\delta_{ij}}{3} k \right) - c_2 \left(P_{ij} - \frac{2\delta_{ij}}{3} P_k \right) - c_3 \left(D_{ij} - \frac{2\delta_{ij}}{3} P_k \right)$$

$$-c_4 k \left(\frac{\partial U_i}{\partial x_j} + \frac{\partial U_j}{\partial x_i} \right)$$

To be applicable down to the wall, this model avoids any wall reflection term by making model constants dependant upon the anisotropy invariants as:

$$c_1 = 1 + 2.45 A_2^{1/4} A^{3/4}\left[1 - \exp\left(-\left(7A^2\right)\right)\right]\left[1 - \exp\left(-\left(\frac{R_t}{60}\right)^2\right)\right]$$

$$c_2 = 0.7A \quad c_3 = 0.3\sqrt{A}$$

$$c_4 = 0.65A\left(0.23c_1 + c_2 - 1\right) + 1.3A_2^{1/4} c_3$$

The isotropic dissipation tensor model is still used together with the Daly and Harlow model for the turbulent diffusion.

The model is coupled with the following transport equation for the dissipation rate:

$$\frac{D\varepsilon}{Dt} = C_{\varepsilon 1}\frac{P_k \varepsilon}{k} - C_{\varepsilon 2}\frac{\varepsilon\tilde{\varepsilon}}{k} + \frac{\partial}{\partial x_k}\left(\nu\frac{\partial \varepsilon}{\partial x_k} + C_\varepsilon \frac{k}{\varepsilon}\overline{u_k' u_l'}\frac{\partial \varepsilon}{\partial x_l}\right)$$

where $\tilde{\varepsilon} = \varepsilon - 2\nu\left(\frac{\partial\sqrt{k}}{\partial x_l}\right)^2$ is the "isotropic" dissipation which is imposed to be null at the wall. The model constants are:

$$C_{\varepsilon 1} = 1.44 + \beta_1 + \beta_2 \quad C_{\varepsilon 2} = 1.92 \quad C_\varepsilon = 0.15$$

$$\beta_1 = 0.25A\min\left(\frac{\lambda}{25} - 1;0\right) - 1.4A\min\left(\frac{P_k}{\varepsilon} - 1;0\right) \quad \beta_2 = A\lambda^2 \max\left(\frac{\lambda}{2.5} - 1;0\right)$$

$$\lambda = \min\left(\lambda^*;4\right) \quad \lambda^* = \sqrt{\frac{\partial}{\partial x_l}\left(\frac{k^{3/2}}{\varepsilon}\right)\frac{\partial}{\partial x_l}\left(\frac{k^{3/2}}{\varepsilon}\right)}$$

Wilcox (1998) has proposed a coupling of a revised Launder, Reece and Rodi model for the Reynolds stress model with a transport equation for the specific dissipation. A first gradient diffusion model is used so that the Reynolds stress transport equation reads:

$$\frac{D\overline{u_i' u_j'}}{Dt} = P_{ij} + \phi_{ij} - \frac{2}{3}\beta^* \omega k \delta_{ij} + \frac{\partial}{\partial x_l}\left[\left(\nu + \sigma^* \nu_t\right)\frac{\partial \overline{u_i' u_j'}}{\partial x_l}\right]$$

where the pressure-strain term is adapted from the Launder et al. model as:

$$\phi_{ij} = -\frac{c_2 + 8}{11}\left(P_{ij} - \frac{2}{3}\delta_{ij}P_k\right) - \frac{30c_2 - 2}{55}k\left(\frac{\partial U_i}{\partial x_j} + \frac{\partial U_j}{\partial x_i}\right)$$
$$- \frac{8c_2 - 2}{11}\left(D_{ij} - \frac{2}{3}\delta_{ij}P_k\right) - c_1\beta^*\omega\left(\overline{u_i' u_j'} - \frac{2}{3}\delta_{ij}k\right),$$

with $c_1 = 1.8, \quad c_2 = 0.52$.

In the framework of this project, this model has not been coupled with the elaborate ω equation proposed by Wilcox (1998) but with either the BSL equation [Wi-BSL] or the basic [Wi88] equation [Wi-ω]. In this last case, a value of $c_2 = \frac{5}{9}$, as recommended by Taulbee (1992), has been used.

Another model has been built during this project using a blend of the above Wilcox' extension of the Launder et al. model to treat the inner region of the boundary layers with the high Reynolds SSG model for the outer region. The blending of the two models, which is indeed a blending of the coefficients of the pressure-strain term, is done as in the [BSL] model, using the F_1 function. The length scale equation is again the [BSL] model one. Therefore, the model will be designated as [SSG/LRR-ω].

4.5 Non-linear models

All the above models use expressions for the pressure-/strain correlation which are linear with respect to the anisotropy tensor. This forbids the model to satisfy realisability conditions. A non-linear model, proposed by Craft et al. (1996), and which is consistent with the two-component limit at the wall, has been used. It is labelled [TCL]. The pressure-strain term is modelled as:

$$\phi_{ij} = -c_1\varepsilon\left[a_{ij} + c_1'\left(a_{ik}a_{kj} - \frac{\delta_{ij}}{3}A_2\right)\right] - \varepsilon a_{ij}\sqrt{A}$$

$$-0.6\left(P_{ij} - \frac{\delta_{ij}}{3}P_{kk}\right) + 0.3a_{ij}P_{kk}$$

$$-0.2\left[\frac{\overline{u_k'u_j'}\,\overline{u_l'u_i'}}{k}\left(\frac{\partial U_k}{\partial x_l} + \frac{\partial U_l}{\partial x_k}\right) - \frac{\overline{u_l'u_k'}}{k}\left(\overline{u_i'u_k'}\frac{\partial U_j}{\partial x_l} + \overline{u_j'u_k'}\frac{\partial U_i}{\partial x_l}\right)\right]$$

$$-c_2\left[A_2\left(P_{ij} - D_{ij}\right) + 3a_{mi}a_{nj}\left(P_{mn} - D_{mn}\right)\right]$$

$$+c_2'\left\{\left(\frac{7}{15} - \frac{A_2}{4}\right)\left(P_{ij} - \frac{\delta_{ij}}{3}P_{kk}\right) + 0.1\left(a_{ij} - \frac{1}{2}\left(a_{ik}a_{kj} - \frac{\delta_{ij}}{3}A_2\right)\right)P_{kk} - 0.05a_{ij}a_{kl}P_{kl}\right.$$

$$+0.1\left[\left(\frac{\overline{u_i'u_m'}}{k}P_{mj} + \frac{\overline{u_j'u_m'}}{k}P_{mi}\right) - \frac{2\delta_{ij}}{3}\frac{\overline{u_l'u_m'}}{k}P_{ml}\right] + 0.2\frac{\overline{u_l'u_i'}\,\overline{u_k'u_j'}}{k^2}\left(D_{lk} - P_{lk}\right)$$

$$\left.+0.1\left[\frac{\overline{u_l'u_i'}\,\overline{u_k'u_j'}}{k^2} - \frac{\delta_{ij}}{3}\frac{\overline{u_l'u_m'}\,\overline{u_k'u_m'}}{k^2}\right]\left[6D_{lk} + 13k\left(\frac{\partial U_l}{\partial x_k} + \frac{\partial U_k}{\partial x_l}\right)\right]\right\}$$

where D_{ij} is the same as in the LRR model and the coefficients read:

$$c_1 = 3.1 f_A \sqrt{A_2} \quad c_1^{'} = 1.1 \quad c_2 = \min\left(0.55; \frac{3.2A}{1+S^*}\right)$$

$$c_2^{'} = \min(0.6; A) + 3.5\frac{S^* - W^*}{3+S^* + W^*} - 2S_I$$

$$f_A = \begin{cases} \sqrt{\frac{A}{14}} & A < 0.05 \\ \frac{A}{\sqrt{0.7}} & 0.05 < A < 0.7 \\ \sqrt{A} & 0.7 < A \end{cases}$$

$$S^* = \frac{k}{\varepsilon}\sqrt{\frac{S_{ij}S_{ij}}{2}} \quad W^* = \frac{k}{\varepsilon}\sqrt{\frac{\omega_{ij}\omega_{ij}}{2}} \quad S_I = \frac{S_{ij}S_{jk}S_{ki}}{\left(S_{nm}S_{nm}/2\right)^{1.5}}$$

This form, consistent with the two-component limit at the wall, no longer requires the use of wall reflection terms.

The dissipation tensor is modelled as a blend between isotropic dissipation and Reynolds stress controlled anisotropy as:

$$\varepsilon_{ij} = (1 - f_\varepsilon)\frac{\overline{u_i^{'}u_j^{'}}}{k}\varepsilon + \frac{2}{3}\delta_{ij} f_\varepsilon \varepsilon \quad f_\varepsilon = A^{1.5}$$

Daly and Harlow diffusion model is used, as for the common SSG model.
The model is coupled with a transport equation for the "isotropic" dissipation which reads:

$$\frac{D\varepsilon}{Dt} = C_{\varepsilon 1}\frac{P_{kk}\varepsilon}{2k} - C_{\varepsilon 2}\frac{\varepsilon^2}{k} + Y_E + \frac{\partial}{\partial x_l}\left(\nu\frac{\partial \varepsilon}{\partial x_l} + C_\varepsilon \frac{k}{\varepsilon}\overline{u_k^{'}u_l^{'}}\frac{\partial \varepsilon}{\partial x_k}\right)$$

$$C_{\varepsilon 1} = 1.0 \quad C_{\varepsilon 1} = \frac{1.92}{1 + 0.7 A_d \sqrt{A_2}} \quad C_\varepsilon = 0.15 \quad A_d = \max(A; 0.25)$$

$$Y_E = C_{\varepsilon l}\frac{\varepsilon^2}{k}\max\left(F(F+1)^2; 0\right) \quad F = \frac{\partial l}{\partial x_k}\frac{\partial l}{\partial x_k} - c_l\left[1 - (1 - B_\varepsilon R_t)\exp(-B_\varepsilon R_t)\right]$$

$$C_{\varepsilon l} = 0.5 \quad l = \frac{k^{3/2}}{\varepsilon} \quad c_l = 2.55 \quad B_\varepsilon = 0.1069 \quad R_t = \frac{k^2}{\nu\varepsilon}$$

This model has been used, in this project, only with wall functions.

5 Wall functions

B. Aupoix, ONERA, and all partners

5.1 Introduction

Turbulence models can be solved down to the wall. However, this is usually costly since the strong gradients of the computed quantities in the wall region require a large number of grid points. Assuming that the near-wall behaviour is more or less universal, wall functions can be used. They allow larger cells in the wall region and CPU time savings. The various wall functions used by partners are described in this chapter and acronyms are given. In all this section, y designates the distance to the nearest surface and the superscript $^+$ a wall scaling, i.e. a scaling with a reference wall velocity, generally the friction velocity $u_\tau = \sqrt{\frac{\tau_w}{\rho}}$, and the viscosity ν.

5.2 Scalable wall functions [S]

These wall functions (Grotjans and Menter, 1998) are presented first as they are the ones used with the common Reynolds stress model [SSG].

The basic idea is to assume that the mean velocity profile obeys a logarithmic law. To avoid problems near separation, the wall scaling involves the turbulent kinetic energy at the first grid point. Denoting $_1$ the values at the center of the wall-adjacent cells in a cell-centered scheme, or estimated values at one quarter of the cell height in a cell-vertex scheme, the friction velocity u_τ is obtained as:

$$u_\tau = \frac{U_1}{\frac{1}{\kappa}\log \tilde{y}^* + c} \qquad \tilde{y}^* = \max\left(\frac{y_1 u^*}{\nu};11.06\right) \qquad u^* = C_\mu^{1/4}\sqrt{k_1} \qquad C_\mu = 0.09$$

where the use of the max function forces the point to be considered above the viscous, linear region and to fulfil the logarithmic region behaviour. This ensures that the approach can be used, whatever the grid resolution.

In order to solve the momentum equation in the wall-adjacent cells, the momentum flux, i.e. the wall friction, is estimated as $\tau_w = \rho u_\tau u^*$. The turbulent kinetic energy and Reynolds stress equations are also solved in the wall-adjacent cells. The turbulent kinetic energy and Reynolds stress wall fluxes are null. When using coarse grids, since the mean velocity gradients cannot be accurately determined, approximations for the source terms in these equations are also required; source terms are assumed to be null as convection and diffusion terms should be null in the logarithmic region.

Concerning the dissipation equation, it is simpler not to solve it but to impose the dissipation rate value at the first grid point above the wall as $\varepsilon_1 = \dfrac{u^{*4}}{\kappa \widetilde{y}^* \nu}$.

5.3 Automatic near-wall treatment [Au]

To avoid the simplifications made above when the first point above the wall lies below the logarithmic region, Esch et al. (2003) proposed the automatic wall treatment which ensures a smooth blending between the logarithmic and viscous region solutions. In both regions, the friction velocity can be deduced from the velocity respectively as:

$$u_\tau^{vis} = \frac{U_1}{\widetilde{y}^*} \quad u_\tau^{\log} = \frac{U_1}{\frac{1}{\kappa}\ln \widetilde{y}^* + c}$$

and the friction velocity is finally reconstructed as:

$$u_\tau = \sqrt[4]{\left(u_\tau^{vis}\right)^4 + \left(u_\tau^{\log}\right)^4}.$$

This technique is used with ω -based models, taking advantage of the analytical behaviours of ω in the viscous and logarithmic regions. The specific dissipation is thus reconstructed as:

$$\omega_{vis} = \frac{6\nu}{0.075 y^2} \quad \omega_{\log} = \frac{u_\tau}{0.3\kappa y} \quad \omega_1 = \sqrt{\omega_{vis}^2 + \omega_{\log}^2}.$$

5.4 Hybrid wall function [H]

A similar technique has been proposed by Rung et al. (2000) to blend viscous and logarithmic regions treatment. This is based upon an approximation for the velocity profile in the wall region, using wall scaling, as:

$$y^+ = U^+ + \frac{1}{E}\left[\exp\left(\kappa U^+\right) - \left(1 - \kappa U^+ + \frac{\left(\kappa U^+\right)^2}{2} + \ldots - \frac{\left(\kappa U^+\right)^4}{4!}\right)\right]$$

with $\kappa = 0.41$ and $E = 8.43$. The above velocity law, valid for $0 \le y^+ \le 100$, allows to determine the blending function between the "low Reynolds number" expression for the wall shear stress $\tau_w^{LR} = \mu \dfrac{U_1}{y_1}$ and the "high Reynolds number" expression $\tau_w^{HR} = \rho\left(\dfrac{\kappa U_1}{\ln\left(E y_1^+\right)}\right)^2$ as:

$$\tau_w = (1-\varsigma)\tau_w^{LR} + \varsigma \tau_w^{HR} \quad \varsigma = \left(1 - \exp\left[-0.09 y_1^+\right]\right)^2.$$

The turbulent kinetic energy is solved in the wall-adjacent cells by redefining the production as:

$$P_k = \frac{C_\mu^{1/4} |\tau_w| \sqrt{k_1}}{\kappa y_1}.$$

The kinetic energy dissipation rate or the specific dissipation are not computed in the wall-adjacent cells but imposed, using prescribed length scales which match the near-wall asymptotic behaviours, as:

$$\varepsilon_1 = \frac{k_1^{3/2}}{L_\varepsilon} \quad L_\varepsilon = \kappa C_\mu^{-3/4} y_1 \left(1 - \exp(-\gamma_\varepsilon Re_k)\right) \quad \gamma_\varepsilon = 0.2 \quad Re_k = \frac{\sqrt{k_1}\, y_1}{\nu}$$

$$\omega_1 = \frac{k_1^{1/2}}{L_\omega} \quad L_\omega = \kappa C_\mu^{-3/4} y_1 \left(1 - \exp(-\gamma_\omega Re_k)\right) \quad \gamma_\omega = 0.057$$

When coupled with the Spalart and Allmaras model, the transported quantity $\tilde{\nu}$ is imposed at the first grid point as:

$$\frac{\tilde{\nu}_1}{\nu} = \frac{\kappa}{E}\left(\exp\left(\kappa U_1^+\right) - 1 - \kappa U_1^+ - \frac{\left(\kappa U_1^+\right)^2}{2} \right).$$

5.5 Extended wall function [E]

The extended wall functions by Hakimi et al. (2000) are just an extension of classical wall functions to improve reproduction of turbulent kinetic energy and dissipation rate profiles. The wall shear stress is still determined by inverting the wall law for the mean velocity profile, which is written as:

$$u^+ = \min\left(y^+ ; \frac{1}{\kappa} \ln y^+ + C \right) \quad \kappa = 0.41 \quad C = 5.5$$

The turbulent kinetic energy and its dissipation rate are thus imposed at the first point above the surface as:

$$k^+ = 0.23 \left(\frac{y^+}{1.4} \right)^2 \qquad y^+ \le 1.4$$

$$k^+ = \min\left(P_1\left(y^+\right); P_2\left(y^+\right)\right) \quad 1.4 \le y^+ \le 400$$

$$k^+ = P_2(400) \qquad y^+ \ge 400$$

$$\varepsilon^+ = \min\left(\frac{1}{\kappa y^+} ; P_3\left(y^+\right) \right)$$

where the P_i are fourth order polynomials.

5.6 Analytical wall function [A]

A major weakness of the log-law type wall functions described so far is the assumed logarithmic velocity profile in the inner turbulent region. This condition is based on the assumption that the boundary layer is two-dimensional, there is zero pressure-gradient and the flow is in local equilibrium. For many applications these assumptions are not valid. For example if the boundary layer is being accelerated or decelerated near a flow reattachment or separation, or if there is an external force applied to the fluid such as a buoyancy force from a heated wall.

In the analytical wall function neither the log-law velocity profile nor the constant or linear variation in shear stress is assumed. A simplified momentum equation is specified in the near-wall cell:

$$\rho\frac{\partial UU}{\partial x}+\rho\frac{\partial UV}{\partial x}=-\rho\frac{\partial p}{\partial x}+\frac{\partial}{\partial x}\left[(\mu+\mu_t)\frac{\partial U}{\partial y}\right]$$

A prescribed viscosity profile is also adopted (near wall node denoted $_p$, sublayer thickness denoted $_v$):

$$\mu_t=0 \qquad for \qquad y<y_v$$

$$\mu_t=\rho c_\mu c_l k_p^{1/2}(y-y_v) \qquad for \qquad y\geq y_v$$

Which is the Prandtl-Kolmogorov condition and in which $c_l=2.55$. The simplified momentum equation is integrated to find analytical values of velocity gradient, wall shear stress and average production of turbulent kinetic energy for the near-wall cell. The derivation of these terms for an isothermal wall function is summarised below, where the prime suffix denotes non-dimensional variables and the subscript 'ref' denotes far-field reference values used to normalise the variables.

Wall Shear Stress:

$$\tau'_w=-\frac{k_p^{'1/2}}{\rho_{ref}}A_1$$

Average Production of Turbulent Kinetic Energy (north cell face denoted $_n$):

$$\overline{P}'_k=\frac{\rho\alpha k_p^{1/2}}{y'_n}\int_{y_v^*}^{y_n^*}\left(y^*-y_v^*\right)\left[\frac{A_1+C_1y_v^*+C_2\left(y^*-y_v^*\right)}{1+\alpha\left(y^*-y_v^*\right)}\right]^2dy^*$$

Constant Terms:

$$A_1=\frac{\alpha U'_n-C_2\left(y_n^*-y_v^*\right)+\left[C_1y_v^*-\frac{C_2}{\alpha}\right]\ln\left[1+\alpha\left(y_n^*-y_v^*\right)\right]+\frac{\alpha C_1y_v^{*2}}{2}}{\alpha y_v^*+\ln\left[1+\alpha\left(y_n^*-y_v^*\right)\right]}$$

$$C_1 = \frac{\nu'^2}{k_p'}\left[\frac{\partial p'}{\partial x'} + \gamma U' \frac{\partial U'}{\partial x'} + \gamma V' \frac{\partial V'}{\partial y'}\right]$$

$$C_2 = \frac{\nu'^2}{k_p'}\left[\frac{\partial p'}{\partial x'} + U' \frac{\partial U'}{\partial x'} + V' \frac{\partial V'}{\partial y'}\right]$$

Average Dissipation Rate:

$$\bar{\varepsilon}' = \frac{k_p'^{3/2}}{y_n'}\left[\frac{k_p'^{3/2}}{y_\varepsilon^*} + \frac{1}{c_l}\ln\left(\frac{y_n^*}{y_\varepsilon^*}\right)\right]$$

γ is an empirical constant which is used to control the influence of convection inside the laminar sub-layer, $\alpha = c_\mu c_l$. The turbulence and dissipation viscous sublayer thickness y_v^* and y_ε^* are 10.8 and 5.1 respectively.

5.7 Viegas and Rubesin wall law [V]

The Viegas and Rubesin (1983, 1985) wall function is dedicated to compressible flows. The extension to compressible flows is beyond the scope of this presentation. The wall shear stress is determined assuming that the velocity profile satisfies either a linear or a logarithmic relation as:

$$U^+ = y^+ \qquad y^+ \le 11.13$$

$$U^+ = \frac{1}{\kappa}\ln y^+ + 5.25 \qquad y^+ \ge 11.13$$

The peculiarity of this model is the way source terms are treated in the turbulent kinetic energy equation. The wall adjacent cell is split into two parts: the viscous sublayer for $y^+ \le 5$ and a turbulent region above. The turbulent kinetic energy production in the viscous sublayer is neglected while the production is estimated in the rest of the cell assuming that the turbulent stress is equal to the wall shear stress and that the velocity gradient is given by the logarithmic law.

Similarly, the dissipation rate is averaged over the wall adjacent cell assuming two different behaviours. In the viscous sublayer, the turbulent kinetic energy is quadratic, i.e. $k = k_v\left(\frac{y}{y_v}\right)^2$ where the subscript $_v$ designates values at the edge of the viscous sublayer, so that the dissipation reads $\varepsilon = 2\nu\frac{k_v}{y_v^2}$. In the logarithmic region, the turbulent kinetic energy is assume to vary linearly, the linear law being given by the two control points above the surface. The local value of the dissipation is given here by the equilibrium relation $\varepsilon = \frac{\left(\sqrt{C_\mu}k\right)^{3/2}}{\kappa y}$.

5.8 Smith wall layer model [Sm]

Smith wall layer model (1990) is specific to his $k-L$ turbulence models [Sm95,Sm97]. Here again, the model has been initially derived for compressible flows but only incompressible flows will be addressed.
The basic idea is to rewrite the turbulent kinetic energy equation in the wall region, neglecting advection terms, as a equation for $q^4(=2k^2)$ for which an analytical solution is obtained, assuming the length scale profile. Once the turbulent kinetic energy and the length scale profiles are known, the eddy viscosity can be computed and the velocity profile can be obtained by integrating the momentum equation in which advection terms have been neglected. This is an iterative process as the turbulent kinetic energy and velocity profiles can be determined only by prescribing the wall shear stress, the correct value of which is obtained when the correct velocity value at the first grid point above the surface is retrieved. One of the advantages of this approach is that pressure gradient effects are accounted for, both in the analytical expression for the turbulent kinetic energy and in the integration of the momentum equation. However, it is a rather CPU expensive method compared to the previous ones.

6 Coupled RANS/LES methods

U. Bunge, TU Berlin, and all partners

Abstract

The present chapter concisely describes the hybrid methods as applied in FLOMANIA by different partners. In the first chapter the non-zonal DES method is described, whereas in the second chapter the zonal approaches are presented. Beginning with a description of the general idea of the method in the chapter on DES an overview of existing versions is presented.

6.1 DES

At first the DES idea and part of its history is outlined to motivate its invention before a definition of the idea in words is given.

6.1.1 Description and definition of the DES approach

The Detached-Eddy Simulation of turbulence was suggested in 1997 (Spalart et al., 1997) as a response to the conflict between the inability of the conventional RANS method to treat properly a wide class of high Reynolds number industrial flows with massive separation, on one hand, and an unaffordable computational cost of a fully resolved LES of such flows due to the presence of turbulent boundary layers populated with small (in this context: attached) eddies, on the other.

The recognition of this conflict mounted into the creation of an approach that combines the fine-tuned RANS technology in the attached boundary layers with LES in the separated regions. A general idea of such an approach belongs to P.Spalart. It implies that the attached, or boundary layer eddies should be modelled in the classic RANS mode, while the larger, or detached ones populating the separation regions and wakes would be simulated. Small subgrid eddies in these regions are also modelled, but have much less influence than the boundary-layer eddies. For this reason the proposed approach that claims to reach this goal was given the name Detached-Eddy Simulation or DES in short.

The core DES idea was expounded in 1997 (Spalart et al., 1997) together with its formulation based on the Spalart-Allmaras (SA) turbulence model (Spalart et al., 1994). A more thoughtful definition of the technique, not linked with any specific turbulence model was given later (Travin et al., 2000), including the mentioning of the grey area problem between the RANS and LES region in a DES. The LES region in the standard version of DES is calibrated to act like the simple Smagorinsky SGS model.

In accordance with this definition, DES is a three-dimensional unsteady numerical solution using a single turbulence model, which functions as a subgrid-scale model in regions where the grid density is fine enough for an LES, and as a RANS model in regions where it is not (Travin et al., 2000).

A grid fine enough for a LES is the one whose maximum spatial step in all three dimensions, Δ , is much smaller than the flow turbulence length-scale, l_{turb}, constituting an integral length scale of the turbulence - much larger than the Kolmogorov scale, of course. Thus, in the LES regions little control is left to the model, and the larger, most geometry-sensitive, eddies are directly resolved. As a result, the range of scales in the solution and, therefore, the accuracy of the non-linear interactions available to the largest eddies are expanding when the grid spacing is decreasing. In other words, the model adjusts itself to a lower level of mixing, relative to RANS, in order to unlock the large-scale instabilities of the flow and to let the energy cascade extend to length scales close to the grid spacing.

In contrast, in the RANS regions where Δ is larger than l_{turb}, the model has strong control over the solution, which, however, remains unsteady and 3D even with a 2D geometry. This situation is typical for thin shear layers (either boundary layers or free mixing layers), which are precisely the flows where the RANS approach is known to be most adequate in terms of computational cost, robustness, and credibility.

As a LES requires a rather low-diffusive numerical scheme, and RANS usually goes together with upwind based, limited schemes, a hybrid numerical scheme is seen to be an important ingredient of a DES implementation (Travin et al., 2002). This aims at an upwind scheme in the RANS regions or regions of irrotational flow in order to avoid stability problems, whereas in the LES region or regions of higher vorticity central differencing is applied to avoid too much numerical diffusion that could deteriorate any LES results containing physical small-scale oscillations.

An important feature of DES is that the approach is non-zonal and, as such, provides a single velocity and eddy viscosity field, and there is no issue of smoothness and complex conditions at the RANS and LES interface. Note also that if the boundary layers remain attached and the steady RANS solution is stable, DES finds that solution. On the other hand, with a grid fine enough for an LES everywhere, including the attached boundary layer, DES is equivalent to a Smagorinsky LES, although no explicit filtering of the Navier-Stokes equations is performed, and with a further grid refinement it is gradually transforming to DNS.

Another important feature of DES is that, in general, the user can apply fine resolution only in regions of special interest, which is similar to tagging these regions for LES treatment, but in an implicit manner.

6.1.2 General principles of DES implementation

On the basis of the ideas outlined above, it is possible to suggest some general principles of designing DES based on different RANS models. Indeed, due to the general DES definition given in the previous section, a DES model can be obtained from a RANS model by an appropriate modification of the length scale, which is, explicitly or implicitly, involved in any RANS turbulence model.

Denoting this length scale by $l_{RANS}(\underline{r})$ ($\underline{r}$ is the radius-vector of the considered field point) an appropriate DES length scale, l_{DES}, may be defined as:

$$l_{DES} = \min(l_{RANS}, C_{DES}\Delta), \tag{1.1}$$

where C_{DES} is the only new adjustable DES model constant and, and Δ can based on the largest dimension of the local grid cell, e.g.:

$$\Delta = \max(\Delta_x \Delta_y \Delta_z). \tag{1.2}$$

However, other methods of computing a characteristic cell dimension are possible as well. Applied to a wall-bounded flow, where $l_{RANS}(\underline{r})=O(d_w)$, with d_w representing the distance to the wall, this definition of l_{DES} results in a hybrid model that functions as a standard RANS model in the whole or, at least, in the major part of the attached boundary layer, and as its subgrid-scale version in the remaining flow region including the separated regions and the near wake. Indeed, inside the boundary layer, due to the significant grid anisotropy in common grids with high cell aspect ratios, and thus very large cells compared to the wall distance, one usually gets $l_{RANS}(\underline{r}) < C_{DES}\,\Delta$. Therefore, in accordance with equation (1.1), $l_{DES}=l_{RANS}$, and the model reduces to the unmodified RANS model.

Otherwise, once a field point is outside the RANS area ($l_{RANS}(\underline{r}) > C_{DES}\,\Delta$), the length scale of the model transforms to $l_{RANS}(\underline{r})=C_{DES}\,\Delta$, i.e., the eddy viscosity becomes grid-dependent, and the model performs as a subgrid-scale version of the RANS model.

It should be noted that inherent to the above mentioned definition a possible error can arise when a too fine grid is generated such that the length scale is replaced inside the boundary layer. This will lead to a unphysical reduction of eddy viscosity due to an early switch into LES mode which may lead to "grid-induced separation" (Menter et al., 2002 and 2003). The result is a premature (with the grid still much too coarse for the well-resolved LES) switching to LES mode while in the boundary layer, which, in turn, would cause a "corruption" of the major boundary layer characteristics (Spalart et al., 1997). Such situations can be avoided by a careful control of the grid structure, especially, at moderately high Reynolds numbers. In high Reynolds number flows, where DES is aimed to be used preferably, this situation initially seemed to be not very probable with available computers, especially regarding the - with respect to their geometry – simple cases treated with DES in the beginning. For these cases the grid dimensions could easily be controlled.

However, it has been shown that in practical applications employing complex grid structures these situations can indeed occur. Moreover, in order to enhance the general applicability of DES, a remedy has to be found. So far, only model-dependent solutions to that problem exist (Menter, et.al., 2003 b), as well as for the problem that low-Re terms in a model that contain the wall distance have to be treated with care. Their activation in LES mode of a DES has to be avoided. These topics are also dealt with in the ongoing project DESider.

6.1.3 Numerical principles

Hybrid numerical scheme

As DES is essentially a hybrid approach both RANS and LES areas exist. Considering the rather different demands for RANS and LES with respect to numerics, this inevitably results in some DES-specific numerical issues. In some of the first DES publications (Travin et al., 2000; Shur et al. 1999; Forsythe et al., 2000; Constantinescu et al., 2000) a priority has been given to satisfying the RANS demands. In particular, these DES studies were performed with the use of implicit upwind schemes dictated by the lack of stability of the less-dissipative central schemes in the flow regions where DES is operating in the RANS mode. The causes of the instability are well known. They are: high values of the cell Reynolds number based either on molecular or on eddy viscosity, dispersion especially where the flow is not aligned to the grid lines, non-uniform grid spacing and coefficients, and non-linearity in general.

On the other hand, in the LES regions of DES upwind schemes seem to be sub-optimal since they are commonly considered as "too dissipative" for LES (Moin, 1998). Thus, though a statement such as "upwind schemes are unacceptable for LES" cannot be correct as any numerical method, if consistent, results in an acceptable accuracy with a grid fine enough, numerical dissipation in DES requires special attention: although excessive dissipation does not result in an unstable or meaningless solution, it prevents the solution from taking full advantage provided by the grid and so makes DES more "grid-demanding". In other words, it stops the energy cascade before the SGS eddy viscosity does, or in collaboration with the eddy viscosity, but still at scales that are larger than the best possible with a given grid.

An approach that allows to resolve this issue or, at least, to weaken significantly the harm caused by the use of upwind methods in DES is suggested by Travin et al., 2002. It is based again on the hybrid DES nature and uses the following hybrid, central/upwind, approximation of the inviscid fluxes, F_{inv}, in the governing equations:

$$F_{inv} = (1-\sigma)F_{ctr} + \sigma F_{upw}, \tag{1.3}$$

where F_{ctr} and F_{upw} denote respectively the central and upwind approximations of F_{inv}, and σ is a blending function. The function should be designed so that in regions treated by DES in RANS mode, σ is close to its maximum value σ_{max}=1.0, resulting in an almost full upwind scheme, while in the LES regions σ is close to zero resulting in an almost full centered scheme. In addition, the blending function has to ensure the choice of the upwind scheme in the inviscid or irrotational areas of the flow. This is needed to guarantee stability of the scheme with coarser grids typically used in such regions. A possible specific form of the blending function on the basis of numerical experiments with SA-DES of cylinder flow is as follows:

$$\sigma = \sigma_{\max} \tanh(A^{C_{H1}}). \tag{1.4}$$

Here the function A is defined as

$$A = C_{H2} \max\{[(C_{DES}\Delta / l_t)/g - 0.5], 0\}, \tag{1.5}$$

the turbulence length scale, l_t, is defined via the eddy viscosity and a combination of the magnitudes of the mean strain, S, and vorticity, Ω,

$$l_t = [(\nu_t + \nu)/(C_\mu^{3/2} K)]^{1/2}, \tag{1.6}$$

where $K = \max\{[(S^2 + \Omega^2)/2]^{1/2}, 0.1\tau_0^{-1}\}$ and $\tau_0 = L_0 / U_0$ is the characteristic convective time, and the parameter g is introduced to ensure the dominance of the upwind scheme in the disturbed irrotational flow regions where $\Omega << 1$ while S remains finite:

$$g = \tanh(B^4),$$

$$B = C_{H3}\Omega \max\{S, \Omega\} / \max\{(S^2 + \Omega^2)/2,\ \Omega_{\min}^2\},\quad \Omega_{\min} = 10^{-3}\tau_0^{-1}. \tag{1.7}$$

To avoid the possibility of a singularity in equation (1.5) the product of l_t and g is limited by $10^{-15} L_0$. The constants of the blending functions are:

$$\sigma_{\max} = 1.0,\ C_{H1} = 3.0,\ C_{H2} = 1.0,\ C_{H3} = 2.0.$$

In Figure 1 a snapshot of the blending function σ is presented from a DES of a circular cylinder (Travin et al., 2002) together with a simultaneous snapshot of the eddy viscosity. It gives a clear idea of the performance of the suggested approximation. In the vortical area of the cylinder wake, where DES performs in LES mode, σ is close to zero and the scheme is a centered scheme. Conversely, in the near-wall RANS regions and also in the irrotational outer part of the flow, σ is close to 1.0 and the scheme works an upwind one.

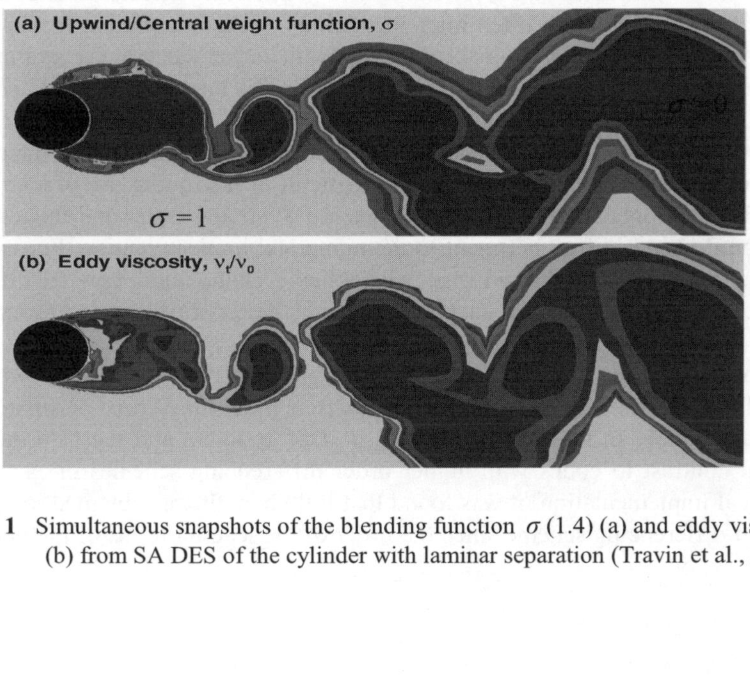

Figure 1 Simultaneous snapshots of the blending function σ (1.4) (a) and eddy viscosity (b) from SA DES of the cylinder with laminar separation (Travin et al., 2002).

Available examples of DES performance with a purely upwind scheme and with the hybrid scheme outlined above convincingly support better capabilities of the hybrid scheme. Note however, that the hybrid scheme defined by the relations (1.4)-(1.7) is not the only possible one, and other examples or enhancements of the blending function are available (Menter et al., 2002 and 2005).

The switch in the CFX code is performed between the second order upwind scheme (NAC – numerical advection correction) and the central difference scheme (CDS) in the following way:

$$\Theta_{ip} = \sigma_{CFX} \cdot \beta_1 \Theta_{NAC} + (1 - \sigma_{CFX}) \beta_2 \Theta_{CDS}$$

where Θ_{ip} is a correction to the first order upwind scheme of the integration point variable. The variables β_1 and β_2 have been introduced as an additional degree of freedom (e.g. β_2=0.95 would result in a 95% CDS +5%upwind scheme). Otherwise, the present implementation is carried out according to the model of Strelets (2001). In order to stay consistent with the DES limiter, the only modification made to the F_{inv} function is to include the SST zonal formulation also in the numerical switch:

$$\sigma_{CFX} = \max\left(\sigma, F_{SST}\right),$$

where F_{SST} is either F_1 oder F_2 as given in chapter 3.2.6 [BSL and SST Model].

In cases where the general flow structure is known in advance, a more suitable method to the above presents itself. Rather than the non-linear relations (1.4)-(1.7), designing an explicit geometry-specified or zonal blending function is more effective.

As noted earlier in this section, hybrid numerical schemes are considered standard for DES and are often implemented in a DES calculation. However, this should not be considered as an absolute prerequisite for success. For example, BT computes DES with a pure central scheme while the unstructured TAU solver of the DLR can either use a central or upwind scheme. Since the motivation for a hybrid differencing scheme is related to the numerical diffusion and to the order of the combined differencing methods, it is difficult to justify the use of a combined upwind-central scheme when both schemes are of similar order and the numerical diffusion does not differ, pertinent to the numerical implementation. For example, the Tau code on a structured grid will achieve central and upwind difference schemes of O(2), whereas for unstructured grids the order decreases to somewhere between first and second order.

Experience with the TAU code to date suggests that the upwind scheme is generally more robust, however the central scheme with Matrix dissipation was used successfully in the computation of the DIT problem and the NACA0012 at stall. In contrast to codes with higher order differencing schemes or a different numerical implementation, it was found that little benefit was obtained in applying a hybrid differencing scheme since the order of the leading terms of the numerical error (and thus the numerical dissipation) are similar. Implementation of a higher order scheme in a unstructured solver is, however, highly nontrivial, particularly when the discretisation is based on a dual grid metric (for details of Tau code see (Gerhold et al.,1999),(Schwamborn et al.,1999)).

6.1.4 Existing DES implementations

Based on SA or SALSA model

The implementation of DES for these two models is straightforward and achieved simply by replacing the wall normal distance d by l_{DES} according to equation (1.1) in the dissipation terms and all damping functions where it occurs, cf. chapter 3.2.3.

Based on SST model

Though a two equation model and therefore more complex in nature, the implementation turns out to be as easy as for the one-equation SA model. Only the dissipation term $\beta^* k\omega$ in the k-transport equation as given in chapter 3.2.6 is manipulated by replacing the RANS turbulent length scale $k^{1/2}/\beta^*\omega$ by l_{DES}. This results in the following modification of the dissipation term

$$\beta^* k\omega \rightarrow k^{3/2} / l_{DES} \ . \tag{1.8}$$

The main practical problem with the DES formulation (both for the Spalart Allmaras and the SST-DES model) is that there is no mechanism for preventing the limiter from becoming active in the attached portion of the boundary layer, as mentioned before. This will happen in regions where the local surface grid spacing is less than the boundary layer thickness $\Delta_S > c\delta$ with c of the order of one. This is not a situation unlikely to occur, especially when unstructured grids are used in the simulation. In the case where the limiter is activated in the boundary layer, the result will in most cases be grid induced separation (GIS). In other words, the separation onset and therefore the flow topology can be altered by the grid provided by the user.

It can be argued that the grid induced separation could be avoided by a larger grid spacing in the lateral direction, but that implies that the flow direction is known at the grid generation stage, which is not the case in most complex three-dimensional simulations. Furthermore, unstructured prism/tetrahedron based grids are typically isotropic on the surface, eliminating this option. It should also be noted that a grid-spacing in the lateral direction (spanwise for a wing) which is solidly on the "safe" side of the DES limiter would prevent the DES mode in the region downstream of separation and thereby limit the effectiveness of the DES model to produce unsteady structures in the separating shear layer.

One way of alleviating the grid induced separation problem is to "shield" the boundary layer from the DES limiter, thereby avoiding/reducing the problem. As the SST model is based on a zonal formulation, differentiating between the boundary layer and the rest of the flowfield, the blending functions of the model can also be used to formulate a zonal DES limiter. The following modification is therefore proposed for the SST-DES model:

$$F_{DES-CFX} = \max\left(\frac{L_t}{C_{DES}\Delta}(1-F_{SST}),1\right); \quad with \quad F_{SST} = 0, F_1, F_2$$

$F_{SST}=0$ recovers the Strelets et al. model. F_1 and F_2 are the two blending functions of the SST. F_2 shields more of the boundary layer and is therefore the preferred default. This technique is introduced as 'GIS-shield' or 'shielding' (Menter, et.al., 2003 b),. It should however be noted that even F_2 does not completely eliminate the problem, but reduces it by an order of magnitude, $\Delta_S > c\delta$ where c is now of the order of 0.1.

Based on LLR model
No other replacement than the one described above for the SST model in equation (1.8) above is required as the wall distance is not an ingredient of this model. Only the coefficient in this case is β_k instead of β^* , cf. chapter 3.2.6.

Based on CEASM model
For this model which is based on a k-ε model ε simply has to be replaced by $k^{3/2}/l_{DES}$. The wall distance is included in the background model, yet it is explicitly not to be replaced by l_{DES}. This guarantees that the near-wall behaviour of the model is retained and the near-wall mode cannot become active in areas far away from the wall which are usually treated in LES modus in a DES. Moreover, a method to introduce a GIS-shield is then readily implemented based on the wall distance n by:

$$\varepsilon \to k^{\frac{3}{2}} \max\left(\frac{1}{L_{RANS}};\frac{1}{C_{DES}\Delta}\left(1-e^{-\alpha_d \mathrm{Re}_k^2}\right)\right)$$

$$\mathrm{Re}_k = \frac{\sqrt{k}n}{\nu}$$

Taking the constant α_d directly from the background model is not mandatory, one can easily extend the range of shielded near-wall region by decreasing the factor below the given value of α_d.

OES and DES
In OES (Organised Eddy Simulation) method, (Dervieux, Braza, Dussauge, 1998), applicable in strongly detached non-equilibrium flows, the reconsideration of the turbulence length scale and of the eddy-diffusion coefficient C_μ, concerning the two-equation modelling had been performed as mentioned in the chapter 3.2 and in the D4.2_30. This was achieved by means of the energy spectrum splitting in a first resolved part regrouping the coherent wave numbers (frequency peaks) and in a second part to be modelled, regrouping the continuous spectrum due to the chaotic motion. The equations of motion in the time-domain are the phase-averaged (or ensemble – averaged) Navier-Stokes equations. The second part of the energy spectrum has a modified shape comparing to the classic equilibrium spectrum with the standard slope value of -5/3. The experimental studies of IMFT concerning the data-base (chapter 2) have demonstrated the modification of the shape and of the slope value of the spectrum due to the coherent structures formation from the wall. Taking into account these modifications, the nonlinear

transfer of energy has been quantified in an analogous way as adopted from the Kovasznay hypothesis (Hinze, 1975) and the turbulence kinetic energy spectrum that in case of equilibrium takes the form

$$E\left(\kappa(m)\right)=\left(\gamma_k\alpha_m\right)^{-\frac{2}{3}}\frac{\left[1-\dfrac{\kappa(m-1)}{\kappa(m)}\right]\dfrac{5}{3}e^m}{\kappa(m)-\kappa(m-1)},$$

has been modelled by taking into account these modifications, m being the number of multi-component spectrum splitting (Hoarau *et al*, 2002). In case of one periodicity, m=1 and κ is the wave number. It has been shown that $k^{0.5}$ (k is the turbulence kinetic energy) is reduced by a factor of 4.5 that corresponds to a reduction of the length scale l_{OES} by 0.022 in the non-equilibrium regions, comparing with the equilibrium turbulence length scale $l = k^{3/2}/\varepsilon$.

Furthermore, in the time-domain, the same facts have been shown by means of modelling the second part of the energy spectrum by the DSM model (the LRR version). The equivalent eddy-diffusion coefficient C_μ takes the value of 0.02, according to the relation $\nu_t= C_\mu k^{0.5} l$ (Bouhadji et al, 2002). Therefore, in DES implementation of the two-equation modelling, the eddy-diffusion coefficient is taken to be 0.02 and consequently the dissipation term in the k- transport equation, $D^k{}_{DES}=\rho k^{3/2}/l_{DES}$ is modified according to l_{DES}=min(l_{OES}, $C_{DES}\Delta$).

Based on RS model

The DES based on RSM has been initiated in the course of FLOMANIA, yet with mixed success and no clear description can be provided on which RSM has to be used. Pertaining to the general idea of DES and the increased insensitivity to the background model with increasing spatial resolution this development branch is highly debatable anyway.

6.1.5 Calibration of DES constant

The constant C_{DES} must be calibrated. This is done analogously to the Smagorinsky constant in LES, employing the case of decaying isotropic (homogenous) turbulence (DIT) (Comte-Bellot and Corrsin, 1971), as described by (Shur, et al., 1999).

Fig. 2 shows an example of such a calibration performed by three different partners yielding similar results for the constant CDES . Here, only the final results are shown, as one usually starts with a given value and iterates until the curves for energy versus wave number fit. Therefore, computations are performed on three different grid levels. The Cartesian and equidistant grids represent a box with periodic boundary conditions in all three directions consisting of 163, 323 and 643 volumes, respectively. No more details are given and the reader is referred to the DESider project for ongoing work on this calibration procedure.

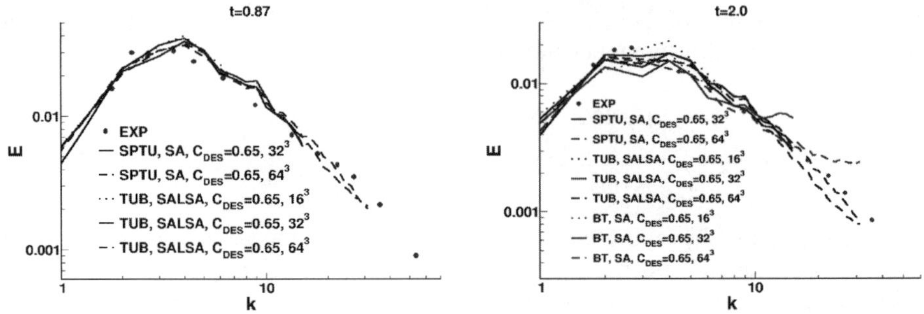

Figure 2 Decay of isotropic turbulence: Comparison of experimental data (Comte-Bellot and Corrsin, 1971), and numerical results by different partners, Energy (E) vs. wave number (k) for two different times (t). Computations on two different grids.

It has to be mentioned, that the optimal value of C_{DES} is in fact not constant pertaining to the inherent numerical diffusion that influences the calibration and more or less strongly depends on the grid.

6.2 Hybrid methods

Apart from DES, other hybrid approaches were used in FLOMANIA that are described in this chapter.

6.2.1 Description of methods

Hybrid RANS/LES employing Interface Condition

As in DES, the object of hybrid LES-RANS is to get rid of the requirement of high near-wall resolution in wall-parallel planes. In the near-wall region (the URANS region), a low-Re number RANS turbulence model (usually an eddy-viscosity model) is used. In the outer region (the LES region), the usual LES is used, see Fig. 3. The idea is that the effect of near-wall turbulent structures should be predicted by the RANS turbulence model rather than being resolved. The matching between the RANS region and the LES region for this method usually takes place in the inner part of the logarithmic region (i.e. around 30 to 60 wall units away from the wall). In the LES region, coarser grid spacing in wall-parallel planes can be used where the grid resolution is presumably dictated by the requirement of resolving the largest turbulent scales in the flow (which are related to the outer length scales, e.g. the boundary layer thickness), rather than the near-wall turbulent processes. The unsteady momentum equations are solved throughout the computational domain. The turbulent RANS viscosity is used in the RANS region, and the turbulent SGS viscosity is used in the LES region.

Although good results have been presented with hybrid LES-RANS, it has been found that the treatment of the interface between the RANS region and the LES region is crucial for the success of the method. The resolved turbulence supplied by the RANS region to the LES region does not have any reasonable turbulent characteristics and is not representative of turbulence at all. This results in poorly resolved stresses on the LES side of the interface and this gives a hack -- also referred to as a shift -- in the velocity profile approximately at the location of the matching plane. Because the interface is located in the 'constant-shear-stress-layer', the too low-resolved stresses in the LES return a too small wall shear stress.

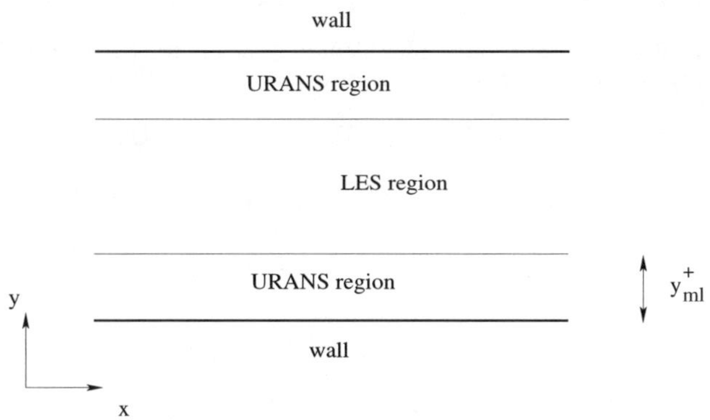

Figure 3 The LES and RANS region

Here we propose to add fluctuations to the momentum equations at the LES side of the interface RANS (Dahlström, 2003; Dahlström & Davidson, 2003; Davidson & Dahlström, 2005). The turbulent fluctuations u'_{DNS}, v'_{DNS}, w'_{DNS} are taken from a DNS of a generic boundary layer. The aim is to create resolved turbulence with reasonable structural information of relevant time and length scales.

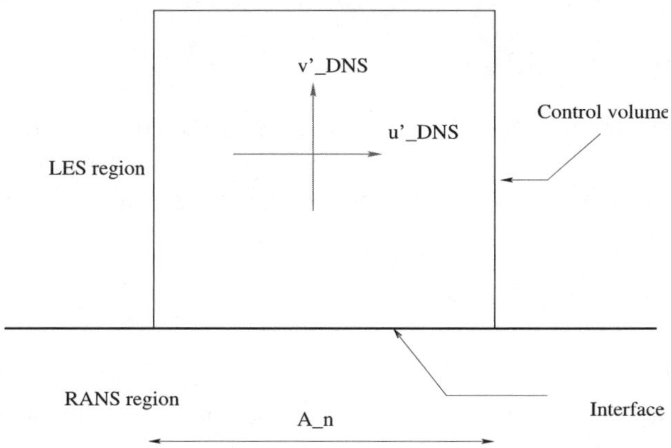

Figure 4 Added fluctuations (u'_{DNS}, v'_{DNS} and w'_{DNS}) in the cells in the LES region adjacent to the interface.

A one-equation model is employed in both the RANS region and LES region,

$$\frac{\partial k_T}{\partial t}+\frac{\partial}{\partial x_j}(\bar{u}_j k_T)=\frac{\partial}{\partial x_j}\left[(\nu+\nu_t)\frac{\partial k_T}{\partial x_j}\right]+P_{k_T}-C_\varepsilon\frac{k_T^{3/2}}{\ell}$$

$$P_{k_T}=2\nu_t\bar{s}_{ij}\bar{s}_{ij}.$$

No special treatment is used in the equations at the matching plane except that the form of the turbulent viscosity and the turbulent length scale are different in the two regions, cf. Table 1.

Table 1 Turbulent viscosities and turbulent length scales in the URANS and LES regions. n denotes the distance to the nearest wall.

	URANS region	LES region
l	$2.5n\left[1-e^{-0.2k^{1/2}n/\nu}\right]$	$l=\Delta=(\delta V)^{1/3}$
ν_t	$2.5k^{1/2}n\left[1-e^{-0.014k^{1/2}n/\nu}\right]$	$0.07k^{1/2}l$
C_ε	1.0	1.07

In the present work an improved hybrid LES-RANS (Dahlström, 2003; Dahlström & Davidson, 2003; Davidson & Dahlström, 2005) is employed in which forcing is applied at the interface. Sources including turbulent velocity fluctuations are added to the three momentum equations in the LES region in the cells adjacent to the interface. The fluctuations $u'_{DNS}, v'_{DNS}, w'_{DNS}$ are obtained from a channel DNS ($\mathrm{Re}_\tau=500$), in which the velocity field along a line $x=x_0$, $y=y_{ml}$ was stored on disk. Fluctuations $u'_{i,DNS}(x_0,y_{ml},z,t)$ along lines are used to compute the source

$$S_i=-\gamma\rho u'_{i,DNS}u'_{j,DNS}\delta_{j2}A_n \tag{1.9}$$

in the $\bar{u}_i$ momentum equation. A_n is the area of the control volume, see Fig. 4 and γ is a scaling function which is the ratio of the local modelled turbulent kinetic energy and the turbulent kinetic energy of the superimposed DNS fluctuations, i.e.

$$\gamma(x,y_{ml},z,t)=c_\gamma k(x,y_{ml},z,t)/\left[0.5\left(u'^2_{DNS,rms}+v'^2_{DNS,rms}+w'^2_{DNS,rms}\right)\right]$$

with $c_\gamma=0.4$.

Taylor's hypothesis is used to achieve streamwise variation of the DNS fluctuations in equation (1.9), i.e.

$$u'_{DNS}(x, y_{ml}, z, t) = u'_{DNS}(x_0, y_{ml}, z, \tau)$$
$$\tau = M_{DNS}(t - (x - x_0)/V_S), \quad V_S = \langle \bar{u}_{y_{ml}} \rangle$$

With M_{DNS} the streamwise turbulent length scale of the added fluctuations can be controlled. A low M_{DNS} value gives a large streamwise turbulent length scale. In (Davidson and Billson, 2004) different values of M_{DNS} were evaluated, and it was found that $M_{DNS} = 0.25$ gives the best result. This M_{DNS} -value gives a streamwise integral length scale L_x of the added DNS fluctuations which is approximately four times larger than the physical value. However, the magnitude of L_x is similar to the streamwise grid spacing. It makes no sense adding fluctuations whose integral length scale is much smaller than the streamwise grid spacing (i.e. with $L_x < \Delta x$), as the equations cannot respond to a forcing with such small length scale. This is probably the reason why the results are improved when the streamwise integral length scale of the added fluctuations is increased by a factor of four.

Combined RANS-LES strategy with arbitrary interface location

The objective of all LES/RANS hybrid formulations is to diminish the unsteady activity near the wall, delegate the representation of a large proportion of the turbulence activity to the RANS model and allow much higher grid aspect ratios. The first and simplest realisations of the DES method uses a one-equation model near the wall, with the length-scale proportional to the distance from the wall, and that same formulation as a subgrid-scale model in the LES region, except that the length scale is replaced by a grid-distance norm. The switch between the two regions is dictated by the intersection of the two length-scales, therefore, explicit control over the location of the interface is only possible via the spatial resolution. Given grid constraints in the LES domain outside the near-wall region, this location can be very close to the wall at relatively low Reynolds numbers. Another disadvantage is that this simple DES realisation tends to over-estimate the turbulence activity, especially in the buffer region, as a consequence of the 'double-counting' arising from the RANS turbulence stresses being added to the non-negligible resolved stresses.

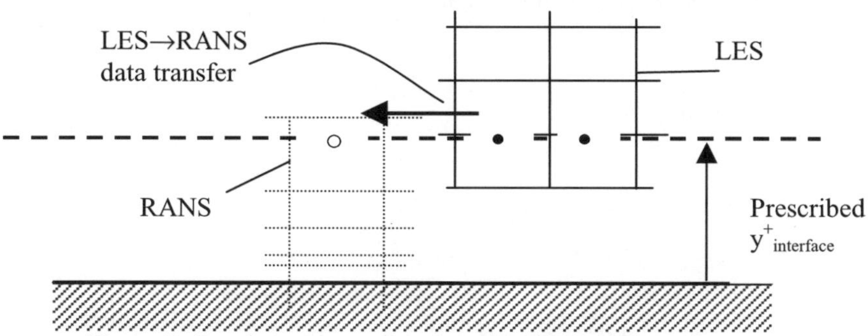

Figure 5 Two-layer near-wall region with LES-RANS overlap

In a method proposed recently (Temmerman et. al. 2004, Hanjalic et. al., 2004), the LES/RANS interface may be specified and the coupling conditions at the interface are dynamically adjusted on the basis of an examination of the turbulence activity predicted by the LES, as indicated in Fig. 5. At the interface, the requirement is imposed that the sum of the two viscosity contributions arising, respectively, from the turbulence model and the resolved motion should be continuous:

$$\nu_{RANS}^{mod} + \nu_{RANS}^{res} = \nu_{LES}^{mod} + \nu_{LES}^{res} \tag{1.9}$$

Continuity in the resolved (suitably averaged) contributions across the interface then imply the need for equality of the modelled contributions in equation. (1.9). If, for the sake of simplicity, a one-equation model is used in the RANS layer (two-equation modeling has also been used),

$$\nu_t = C_\mu k^{0.5} l_\mu ,$$

continuity can be secured by modifying the interface value of C_μ as follows:

$$C_{\mu,int} = \frac{\nu_{LES}^{mod}}{l_\mu k_{RANS}^{0.5}}$$

This value is determined from the subgrid-scale model at LES nodal positions shown as solid dots in Fig. 5 and used by the RANS scheme at the node indicated by an open circle. It now needs to be 'relaxed' in the RANS region towards the value 0.09 normally used in the turbulence model. The choice of the relaxation function is necessarily arbitrary. One function that has been investigated is:

$$C_\mu = 0.09 + \left(C_{\mu,int} - 0.09\right) \frac{(1 - \exp(-y/\Delta)}{(1 - \exp(-y_{int}/\Delta_{int}))}$$

(Temmerman et al, 2004) show, on the basis of a-priori studies for channel flow, that the absence of the above dynamic approach leads to a serious over-estimation of the total turbulence activity, because the resolved motion is insufficiently attenuated by the modelled component.

IV Applications – test cases

1 NACA0012 wing with rounded tip

P. Albrecht, D. Wu, A. Martin, U. Bunge, and F. Thiele, TUB, and partners ALENIA, EADS-CASA, UMIST

Abstract

This chapter describes the common effort of UMIST, EADS-CASA, ALENIA and TUB to compute the steady vortex evolving at the round tip of a wing at 10° angle of attack. Results for different numerical grids obtained by the aforementioned four partners using a variety of codes and turbulence models are presented in comparison to experimental data. The main parameters of influence and suggestions for future computations of that case are identified.

1.1 Introduction

The examined test case is the flow around a NACA0012 airfoil with a round wing tip, as shown in Fig. 1 (left). For the flow around an airfoil of finite length, three dimensional effects have to be taken into account, as the pressure difference between upper and bottom surfaces levels out at the wing tip. The streamlines are thereby deflected inwards at the upper and outwards at the underside of the airfoil, Fig. 1 (right). This causes the flow to separate at the wing tip and generate a free vortex that rolls up more and more of the wing wake as it moves downwards and whose properties depend on a variety of parameters such as the Reynolds number, the angle of attack or the geometry of the wing tip. Dependent on these conditions, a pressure distribution develops which determines the flow around the wing tip and the point of separation. These parameters have to be taken into account by the numerical simulations.

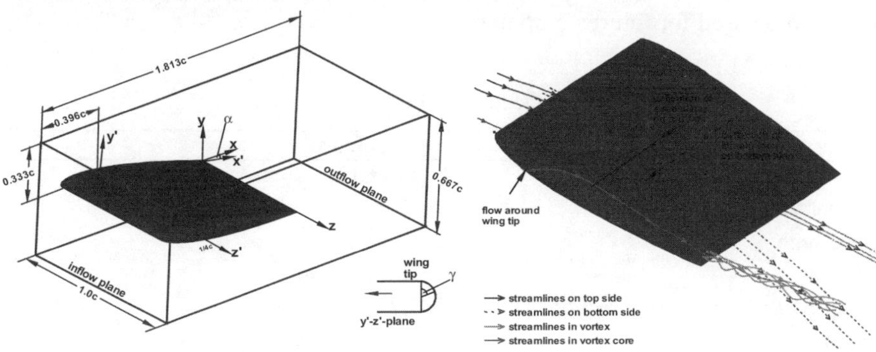

Figure 1 Wing model in wind tunnel (left) and sketch of developing vortex (right)

Steady numerical simulations for this case (Dacles-Mariani, 1995) have been conducted employing a modified version of the Baldwin-Barth one-equation model prescribing the experimental inflow data in the inflow plane very close to the airfoil representing the test section shown in Fig. 1 (left). The grid consisted of roughly $1.5 \cdot 10^6$ volumes with resolved boundary layers even on the wind-tunnel walls. However, for the wake alone, additional uniform grids were generated constituting the wake case which serves to simulate and describe the vortex alone. Among the essential findings for the given numerical procedure were the necessity of fifth order schemes together with a cell size in the wake of not more than $5 \cdot 10^{-3}$ times the chord length c and boundary conditions representing experimental data, especially at the inflow plane close to the leading edge of the wing. The need to modify the Baldwin-Barth model in order to prevent excessive diffusion of the vortex was also outlined, which essentially corresponds to a curvature correction.

1.2 Test case setup

1.2.1 Geometry

The configuration of the test case is presented in Fig. 1 (left). The domain represents the test section of the open-loop wind tunnel at the Fluid Mechanics Laboratory (FML) at the NASA Ames Research Centre, where the experiments were conducted (Dacles-Mariani, 1995; Chow et al., 1997a and 1997b). Investigated is an airfoil with a NACA0012 profile and a rounded, i.e. in this case semi-circular, wing tip which is mounted on one wind tunnel wall and the angle of attack is adjusted by rotating the airfoil about its quarter chord. The relevant data is summarised in Tab. 1.

The ambiguous Reynolds number stems from a evaluation of the experimental data resulting in unphysical conditions using the given $4.6 \cdot 10^6$. Therefore, implying standard conditions a Reynolds number of $4.35 \cdot 10^6$ is more realistic. This number is used by UMIST, whereas the other partners used the higher value as given in the cited references. However, Prof. Bradshaw, who was involved in the experimental investigations, agreed that the lower number might be correct and should be used for future computations.

Table 1: Geometry of test case setup

Quantity	Corresponding value
Cross section	812.8mm x 1219.2mm
Length of test section	2209.8mm
Chord length	c=1219.2mm
Semi-span	b=3/4 c=914.4mm
Semi-span of constant chord section	b_0=841.248mm
Position of quarter chord	482.6mm downstream of the inlet, centred between upper and lower wall
Inlet velocity	U_∞=51.8167m/s
Reynolds number	$Re_c = U_\infty c/\nu = 4.35 \cdot 10^6$ or $4.6 \cdot 10^6$
Turbulence level	Tu=0.15%
Angle of attack	α=10°

1.2.2 Experimental data available

Several measurements have been conducted in order to fully capture and describe the wing-tip vortex. The complete description can be found in a NASA Technical Memorandum (Chow et al., 1997), and the complete set of experimental data in electronic format has been provided by Prof. Bradshaw. Next to the measurements for flow visualization, the mean pressure and flow field as well as turbulent quantities such as the shear stresses have been recorded. An example of the shear-stress contours is given in Fig. 2 which was taken from the memorandum. The velocity profile at the inflow plane is also available.

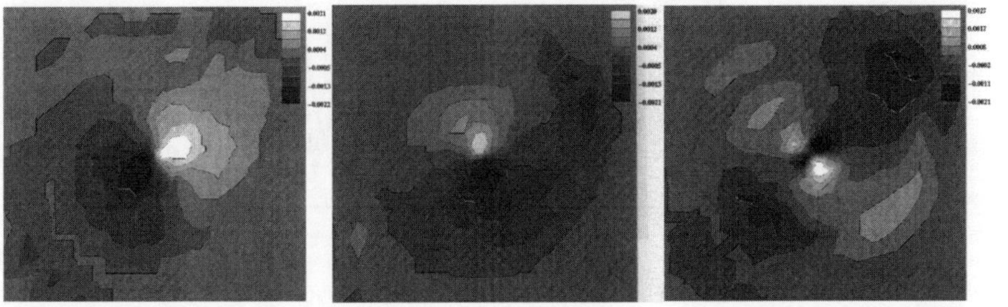

Figure 2 Example of experimentally obtained shear stress contours for u'v' (left), u'w' (middle) and v'w' (right) at x/c=0.452 (Chow et al., 1997b)

Natural transition was measured and made visible, however, for the measurements of mean and turbulent quantities, transition was fixed by roughness elements near the leading edge to enhance comparability with numerical simulations.

222 static pressure taps were located on the upper half of the wing surface in 12 chordwise rows to measure the surface pressure for ±10° angle of attack, and a seven-hole pressure probe was used to determine pressure and velocity in different crossflow planes as seen in Fig. 5 employing 21x29 data points. The vortex core was defined as the point where crossflow velocity or pressure reaches a minimum value. Hot-wire measurements provided the turbulence quantities. Fig. 3 gives an example of the measured Reynolds shear stresses in the vortex above the wing before the trailing edge in a form that can readily used for comparison.

A summary is provided for FLOMANIA on the project's web page (Summary of NASA TM 110418; description of test case, experimental setup and results).

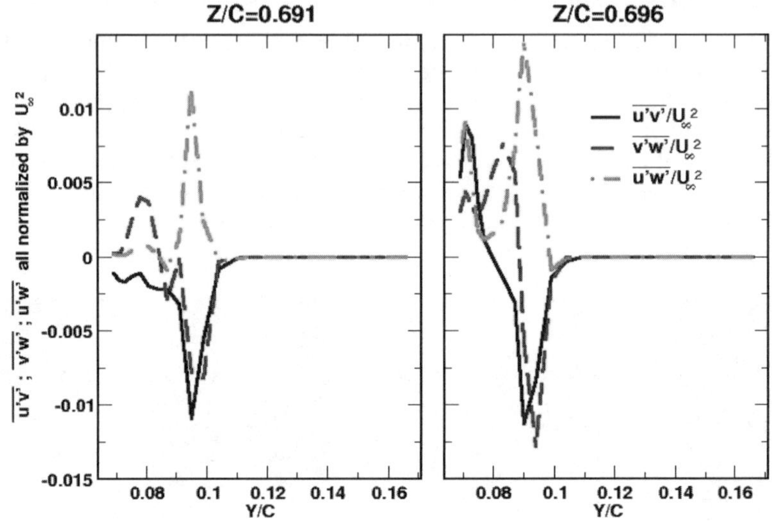

Figure 3 Reynolds stresses at X/C=-0.197

1.2.3 Numerical Grids

In the course of FLOMANIA, it has been shown that the spatial resolution of the computational domain is of utmost importance. Dependent on the order of a given numerical scheme, the $1.5{\cdot}10^6$ volumes used by Dacles-Mariani, 1995, might be taken as a lower limit. This shows how much computational effort will have to be used to get reasonable results for this test case. Different computational meshes were provided by UMIST ranging from roughly 250,000 to 5,400,000 for the full configuration and even 4,000,000 volumes in the wake alone, cf. Tab. 2. These grids, shown in Fig. 4, were generated for the M-DAW project (Contract No. G4RD-CT-2002-00837) and are used in FLOMANIA with permission.

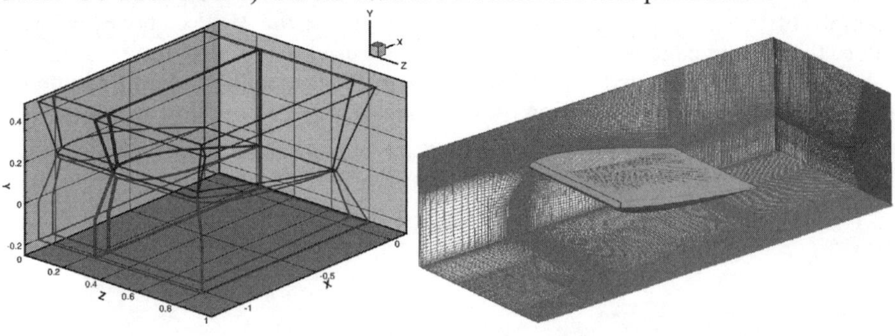

Figure 4 Block structure immediately around the wing for grid G III and G IV (left) and distribution of control volumes for G III on the root wall, outlet and wing surface (right)

Computations on the coarse grids G I and G II performed by many partners were all in vain and a lot of time was lost, yet, the computations were comparatively short and raised the awareness as to the problems of this case. Results and conclusions on these computations can be found in another FLOMANIA added deliverable for this case (Report on test case 01 by TUB).

Table 2: Numerical grids provided by UMIST

Grid	Total number of volumes
G I (single block)	340,000
G II (single block)	700,000
G III (new multiblock topology)	4,200,000
G IV (new multiblock topology)	5,400,000
G V DS (wake grid only)	4,000,000

It has to be mentioned that the inflow plane of grids G III and G IV is not as close to the leading edge of the wing as for grids G I and G II, which exactly represent the experimental domain shown in Fig. 1. This eliminates to some extent the influence of the inflow velocity profile on the results, and a block profile can be applied. This issue is also discussed in the aforementioned deliverables, where it is shown that the flow at the beginning of the measurement area is strongly affected by the airfoil, and an inflow plane further upstream for the numerical simulation is suggested. This has been realised in grids G III and G IV, which were generated following the initial investigations. Although the magnitude of the influence measured was of minor significance, it was decided that since this can easily be avoided, it should be done. The block structure of grids G III and G IV can be seen in Fig. 4 (left) and the distribution of control volumes for G III is shown on the right.

1.3 Results

In Fig. 5 the pressure distribution in all experimental data planes is shown. This result is obtained on grid G III but is qualitatively similar on all grids and for all partners. However, the quantitative agreement with the experimental data and downstream sustainability of the pressure minimum in the centre of the vortex strongly depends on the grid resolution. With unrealistically high dissipation, the flow recovers too far upstream, closer to the trailing edge.
It can be seen that the path of the vortex core (characterised by a minimum of pressure) is bent slightly upwards and towards the wall upon which the wing is mounted.

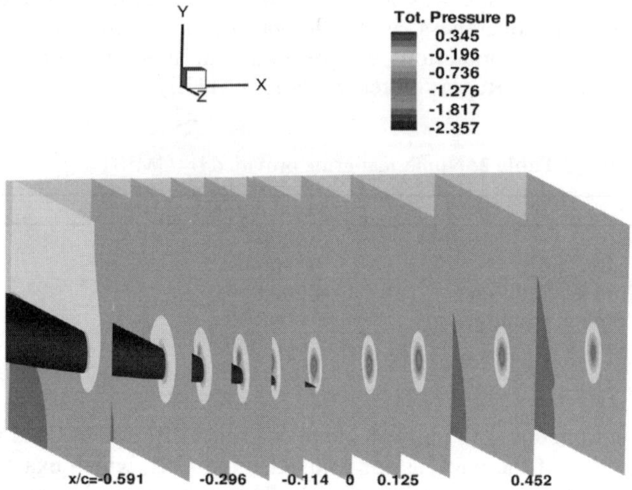

Figure 5: Pressure in all experimental cut locations

Fig. 6 shows a comparison of the C_P distribution around the wing at z/c=0.666. The experimental data are shifted somewhat upwards, and the experimental data in the area of x/c=0.75 are rather coarse, so that only a qualitative comparison with experiments can be considered. This is further compounded as the reference pressure location, or alternatively, the value for the free stream pressure was not agreed upon by the various partners. Nonetheless, the correct tendency for all results to yield a vortex with a corresponding suction peak at the rear end of the suction side can be seen. However, taking into account the possibility that the reference pressure for the experimental data could be too low, the experimental curve would shift downwards giving a better agreement with all numerical data.

The SALSA and CEASM results on grid G III over-predict this suction whereas the coarse grid (G II) underestimates the suction except for WJ-G results, which hint at the improvements one can obtain by employing a curvature correction.

In Figs. 7 and 8, the location of the vortex centreline computed numerically is compared to the experimental results. Deviations occur mainly for the coarse grid G II where the spatial resolution especially in z-direction in the wake is simply too coarse to give reasonable results.

In the vortex core, an excess of flow velocity with a maximum of up to 1.77 close to the trailing edge is measured. The magnitude of flow velocity in the x-direction of all computations together with experimental data is depicted in Fig. 9. Obviously, only computations on the finer grids G III and G IV achieve this amplification of velocity.

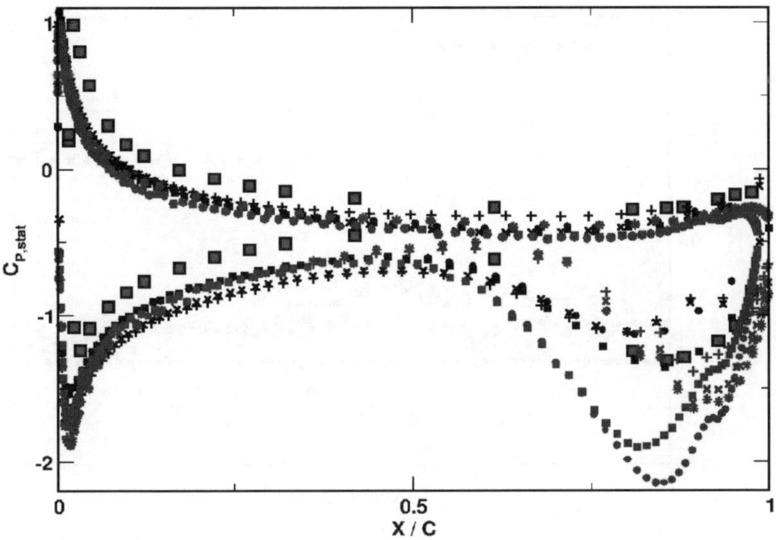

Figure 6 Pressure distribution at z/c=0.666 (legend cf. Fig. 9)

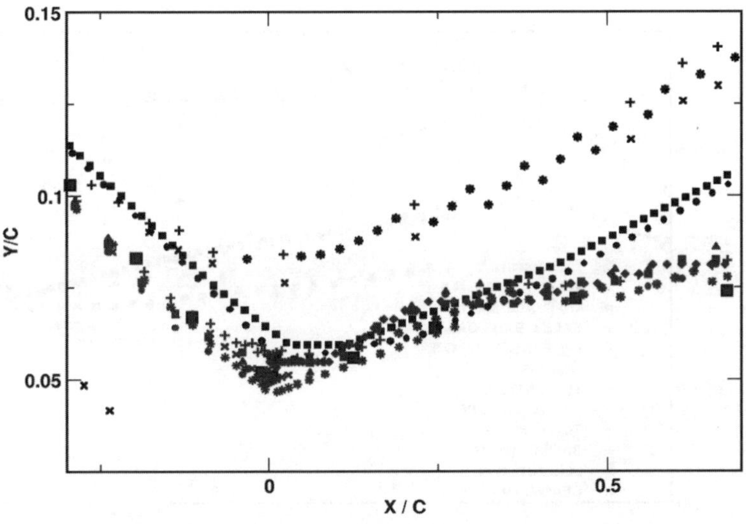

Figure 7 vortex core location (legend cf. Fig. 9)

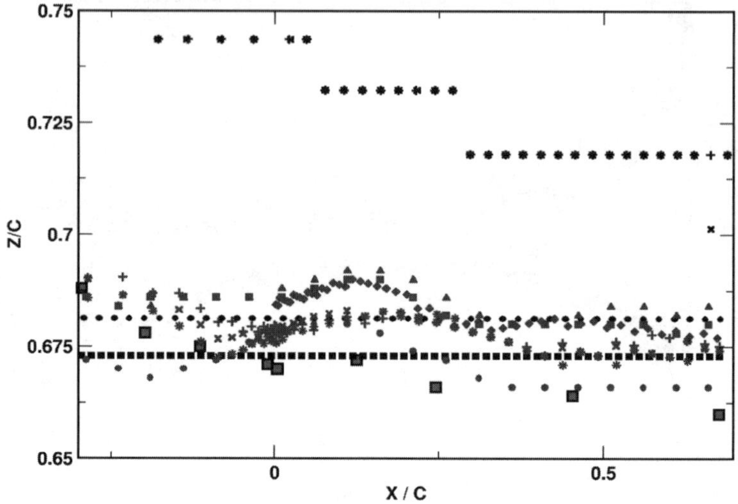

Figure 8 Vortex core location (legend cf. Fig. 9)

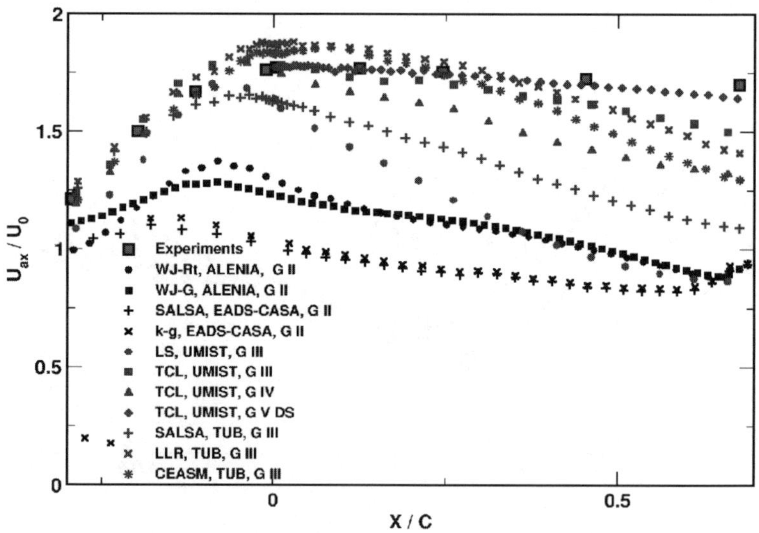

Figure 9 x-component of flow velocity in the vortex core

This velocity excess decays rather slowly in the wake and dependent on the numerics and turbulence model this is represented to differing extents in the simulations. In general, the linear models, e.g. SALSA and LS exhibit excessive dissipation. The LLR-k-w and CEASM even overestimate the peak velocity and the best result, in general, is obtained by the second moment closure TCL on grid G III, however, no comparison in required computational effort was made. For the finer G IV grid the results deteriorate, which is due to the fact that the refinement of the grid in the boundary layer around the wing is achieved partially at the expense of grid points in the wake, so that the results cannot be as good. This, again, demonstrates the high number of grid points required to achieve good results. Therefore, the best result in the wake is obtained on grid G V DS with 4,000,000 uniformly distributed nodes in the wake alone, which can be compared to the wake case of Dacles-Mariani, 1995.

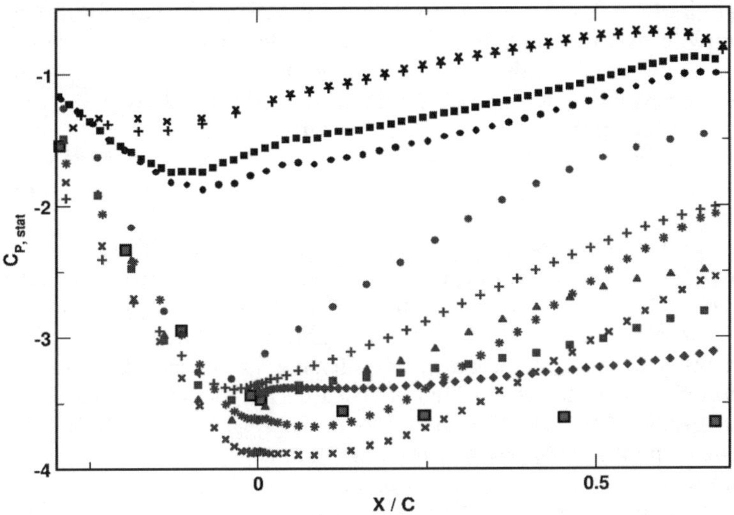

Figure 10 Static pressure coefficient in the wake (legend cf. Fig. 9)

Along with the increase in velocity comes a drop in static pressure in the core of the wake which is shown in Fig. 10. The best result is again obtained on the fine wake grid G V DS, and the overestimation of flow velocity in the vortex core at the trailing edge by the LLR-k-w and the compact explicit algebraic stress model CEASM is accompanied by a pressure that is lower than the experimental data. The best result is again achieved by the TCL second moment closure.

The results obtained on the coarse grid G II again cannot be compared and it is not possible to draw reliable conclusions, except that the spatial resolution is insufficient. Unfortunately, computations on the fine grid were so far performed only by two partners.

1.4 Conclusions and outlook

The computations and examinations have not come to an end and will be continued in the M-DAW project (Contract No. G4RD-CT-2002-00837). This is due to the fact that to much time was spent initially on incorrect grids (representing an erroneous geometry) and coarse-grid computations where even the flow around the wing, and therefore the pressure distribution deviated strongly. Against this background, the quantities in the wake such as the Reynolds stresses, could not have been expected to be correct and were therefore not compared. However, an important lesson learnt is that grids of high quality with a large number of cells are required, enabling the numerical scheme to develop and sustain the vortex in the wake. This is clearly demonstrated by the early coarse grid computations and can be taken as a starting point for future computations together with the resolved obscurities in the Reynolds number and inflow location. The experimental data are complete and very good, so that this case is worthwhile to be taken as a validation case, especially for anisotropy-resolving models or the application of curvature corrections. We thank Prof. Bradshaw for the provision of the complete experimental data. Some debatable inconsistencies remain, such as the pressure-reference value and the Reynolds number, but these can be easily resolved if partners agree upon a common procedure and common values which, was not entirely the case in FLOMANIA. It is recommended to use the lower Reynolds number, to use a common free stream pressure and to correct the experimental pressure coefficients accordingly by fixing the value in the stagnation point.

The results so far show a supremacy of the second moment closure approach TCL over the linear or nonlinear EVM, although no information about the higher computational effort is available. The LLR and CEASM tend to over-predict the pressure drop in the vortex and yield too rapid a breakdown of the vortex due to their inability to correctly represent the physics, in this case the strong streamline curvature present in the flow. This is also true for the SALSA model, yet it does not over-predict the velocity in the wake. Not surprisingly, the LS model performs even worse and gives the fastest vortex breakdown.

Unfortunately, results by ALENIA and CASA are only available on the coarse grid G II, and the results by TUB on this grid also revealed that it is not possible to obtain reasonable results on such a coarse grid, based on a procedure of 2nd order accuracy.

The considerable amount of work done, although only partially successful, has delivered some important ‘lessons learnt’ about the numerical requirements and the treatment of the experimental data. Furthermore, the completeness of the experimental and now-available numerical data, together with the practically relevant and challenging flow physics render this case very interesting, and as such is recommended for future study.

2 Rotor 37

B. Tartinville, C. Hirsch, NUMECA

2.1 Introduction

The NASA rotor 37 is an isolated transonic axial compressor rotor with 36 blades. This case was initially included in a wider research program intended to cover a range of design parameters typical of aircraft turbine engine high-pressure compressor inlet stages. It has been presented in the AGARD Advisory Report 355 entitled "CFD Validation for Propulsion System Components" and was used for CFD validation and comparison exercises including a wide range of numerical models.

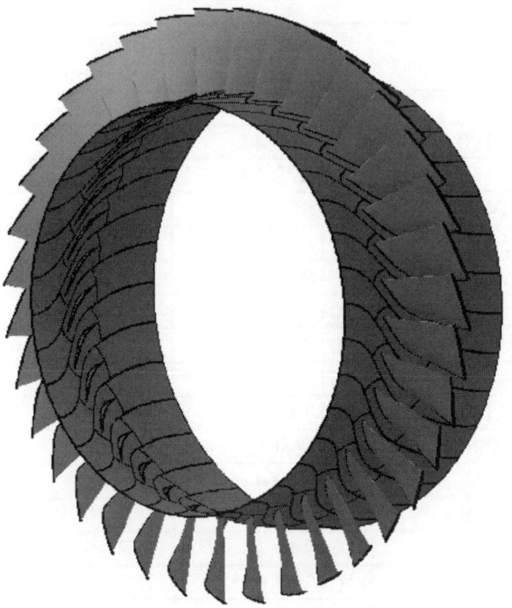

Figure 1 View of the rotor 37

2.2 Physical conditions

All the data available for this case were measured with the NASA rotor 37 operating at the equivalent design rotational speed of 17188.7 rpm. To establish reference points for detailed flow field comparisons, overall performance was determined at equivalent mass flow rates from the choke mass flow to a minimum slightly above the rotor stall flow. The experimental choke mass flow as determined by NASA was m_{choke} = 20.93 kg/s whereas the near stall flow rate was

experimentally determined to be $m = 0.925\ m_{choke}$=19.36 kg/s. A total of 13 sets of overall performance data were measured over this range.
Overall performance includes average total pressure ratio and adiabatic efficiency. Laser anemometer velocity data were acquired in four hub-to-tip measuring planes but only at mass flow rates of 0.98 m_{choke} and 0.925 m_{choke} (i.e., at near stall flow rate).

Inlet boundary conditions could be inferred from experimental data available at a first measurement plane located 4.19 cm upstream of the blade leading edge. Radial profiles of absolute total pressure and total temperature are available at this entrance station and are provided in Table 1 below. Furthermore, a 3% turbulent intensity was also measured in this plane.

Table 1 Total pressure and total temperature inlet profile

Radius (cm)	Absolute total pressure (Pa)	Absolute total temperature (K)
17.9222	99946.98	288.265
18.3490	101740.43	287.862
18.7452	101872.15	287.775
19.1414	101872.15	287.804
19.5682	101872.15	287.804
19.9644	101872.15	287.804
20.5435	101872.15	287.862
21.0922	101872.15	287.977
21.6713	101811.36	288.035
22.2504	101740.43	288.265
22.7990	101669.50	288.381
23.2258	101740.43	288.323
23.6220	101811.36	288.208
24.0182	101740.43	288.150
24.4450	101527.65	288.208
24.8412	100909.56	288.265
25.1765	98913.46	288.265
25.4203	95600.13	288.381

Both absolute total pressure and temperature profiles are imposed at the inlet of the computational domain. The velocity was assumed axial. The turbulent quantities at the entrance are defined in order to have a 3% turbulent intensity and a turbulent viscosity ratio of the order of 10. Therefore, the values ok k and ε are 44 $m^2.s^{-2}$ (assuming that the inlet axial velocity is of the order of 180 m/s) and 10^6 $m^2.s^3$, respectively. In order to use the same turbulent time and length scales at the inlet for k-ε and v^2-f models, these inlet values for turbulent quantities are used in all the calculations. The walls are assumed adiabatic. Furthermore, only a fraction of the hub (z within [–0.005 m, 0.05 m]) is rotating together with the blade. The outlet condition (imposed static pressure with radial equilibrium) varies with the operating point considered.

The fluid is air perfect gas with the following properties:

- Specific heat C_p: 1004.5 J/Kg.
- Gamma: 1.
- Prandtl number: 0.70
- Kinematic viscosity following the Sutherland Law.

2.3 Grid and computation settings

A multi-block low Reynolds mesh is used for this test case. A H-O-H topology is used with about 615,000 grid points. 73 grid points are defined in the span-wise direction. The mesh clustering close to solid walls was defined in order to have y^+ close to unity at the first inner cell. Four grid levels are used for this computation.

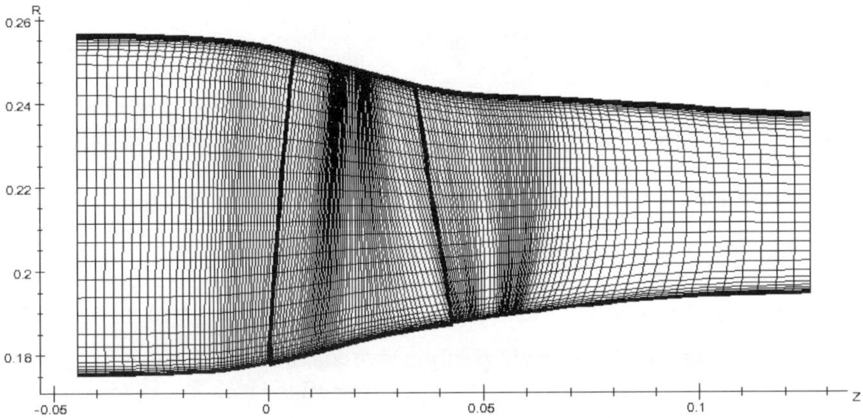

Figure 2 Meridional representation of the mesh used for the rotor 37

Three types of turbulence models are tested on this case: the Spalart-Allmaras model, the low Reynolds Yang-Shih version of the k-ε model and the code friendly version of the v^2-f model. The full working line has been computed for these three models on the finest grid level. Furthermore, and in order to investigate the effect of grid coarsening, the full working line has also been computed on a coarser mesh (number of grid points divided by two in all directions, i.e., about 77,000 grid points) for both the v^2-f and the Spalart-Allmaras models.

2.4 Results

2.4.1 Verification of the mesh clustering

The resolution of the boundary layer depends on the mesh clustering. Figure 3shows the value of y+ in the first inner cell around the blade at mid-span. It confirms that the mesh clustering is relevant for low Reynolds calculations.

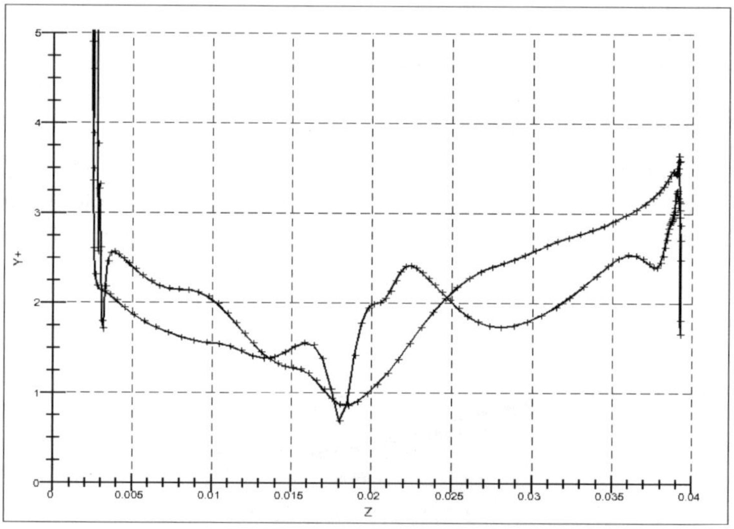

Figure 3 Axial distribution of y+ along the blade at mid span

2.4.2 Overall performance

Experimental data on the overall performance includes average total pressure ratio, P_{ratio}, and adiabatic efficiency, η_{is}. They are defined as:

$$P_{ratio} = \frac{P_{t1}}{P_{t2}}, \qquad \eta_{is} = \frac{\left(\frac{P_{t1}}{P_{t2}}\right)^{\frac{\gamma-1}{\gamma}} - 1}{\frac{T_{t2}}{T_{t1}} - 1}.$$

Subscripts 1 and 2 stand for inlet and outlet of the computational domain, respectively. Computed and experimental results for both quantities are displayed in Figure 4 and Figure 5.

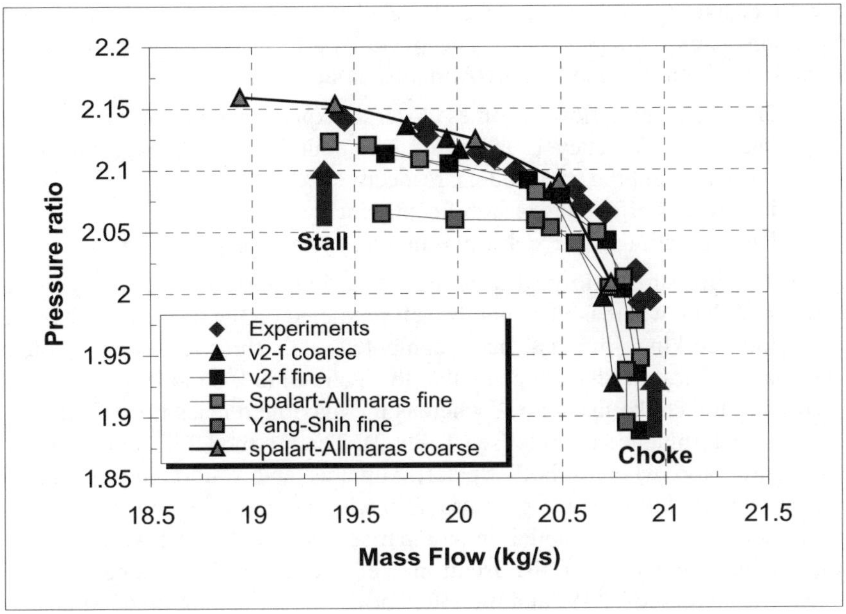

Figure 4 Compressor map for the rotor 37 – Pressure ratio

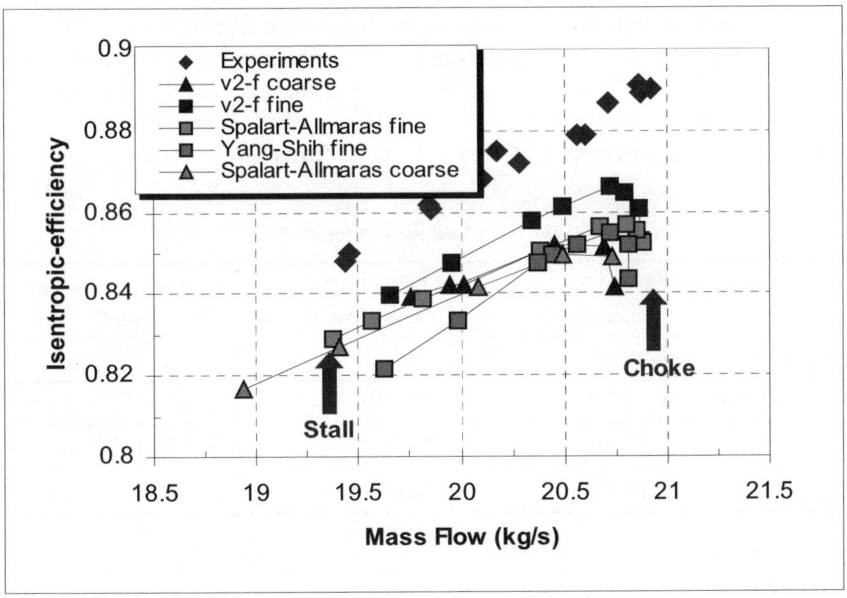

Figure 5 Compressor map for the rotor 37 – Isentropic efficiency

For both coarse mesh computations the choke mass flow is underpredicted in comparison to experimental data. The overall performances are improved on a finer mesh for both v^2-f and Spalart-Allmaras models:

- The choke mass flow becomes closer to experimental data.
- The peak efficiency is increased by about 0.5 and 1% for the Spalart-Allmaras and the v^2-f models, respectively.
- The overall efficiency is significantly increased.
- The prediction of the stability limit close to stall is improved.

However the pressure ratio profile is too flat at low mass flow in comparison to experimental data, whereas the slope is well predicted on the coarser mesh.

The full working line has been computed with three different turbulence models on the finest mesh. It appears that the Yang-Shih k-ε model has difficulties in predicting the choke mass flow, whereas the maximum mass flow is similar for the two other turbulence models (see Table 2). The k-ε model also under-predicts the pressure ratio whereas both Spalart-Allmaras and v^2-f models produce an almost similar evolution of the pressure ratio with the mass flow. One of the major improvement of the v^2-f model in comparison to the Spalart-Allmaras model resides in its prediction of the isentropic efficiency profile. Indeed, the peak efficiency is increased by 1% and the efficiency is closer to experimental data.

2.4.3 Results at 98% of the choke mass flow

The choke mass flow is different for all the turbulence models used and different from the experimental choke mass flow. Therefore the experimental results obtained at 98% of the choke mass flow will be compared to numerical results obtained at a slightly different mass flow. The numerical results obtained on the fine mesh are summarized in the Table 2 (the number in parenthesis in the left column is the exact percentage of the choke mass flow).

Table 2 Mass flow comparisons

	Choke mass flow [kg/s]	98% choke mass flow [kg/s]	Closest numerical result [kg/s]
Experiment	20.93	20.51	-
Spalart-Allmaras	20.878	20.460	20.379 (97.6% choke)
k-ε	20.808	20.392	20.376 (97.9% choke)
v^2-f	20.875	20.458	20.498 (98.2% choke)

The AGARD advisory report has outlined that the performances (in term of total pressure or temperature ratios) could greatly vary with the mass flow (see for instance their figure 3.18). As far as our selected numerical experiments are not strictly at 98% of the choke mass flow, this can be a potential source of differences. However, the error in the mass flow is less than 0.4%, and it is more than likely that the differences between the numerical experiments can be mainly attributed to difference in the turbulence models used.

2.4.4 Relative Mach Number at 90% span

Figure 6 shows the relative Mach numbers at 90% span-wise location for the three different computations performed on the fine mesh. It appears that all the models produce a relatively similar pattern. A strong shock attached at the blade trailing edge interacts with the suction side boundary layer. Whereas the shock positions are nearly the same for the k-ε and v^2-f models, the Spalart-Allmaras model produces a shock slightly upstream.

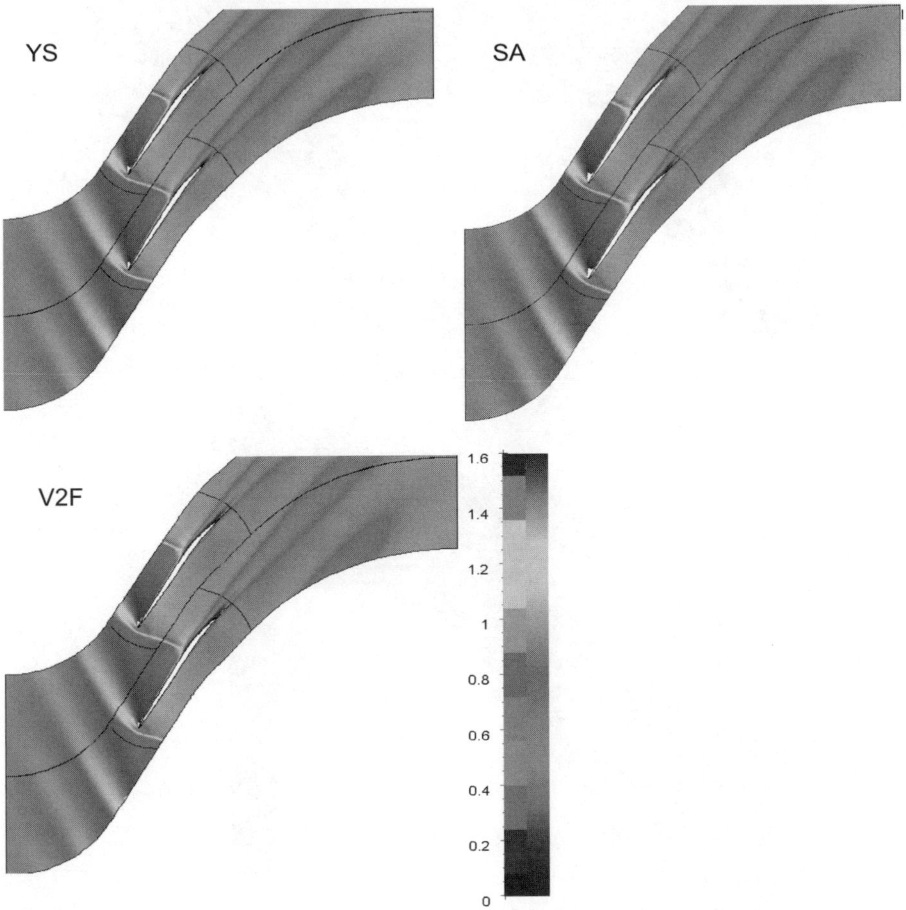

Figure 6 Relative Mach number at 90% span-wise position for the three different turbulence models

2.4.5 Relative Mach Number at 95% span

Figure 7 shows the relative Mach number distribution at 95% span-wise location for the three different computations performed on the fine mesh. At this span-wise location the flow pattern is affected by the interaction between the tip leakage flow and the shocks. This induces a strong vorticity close to the blade corner on the suction side. All the turbulence models behave in a similar way. The blade-to-blade gradient in the relative Mach number within the passage is representative of the influence of the tip vortex. This tip vortex is slightly more important when the v^2-f model is used than with the k-ε or Spalart-Allmaras models.

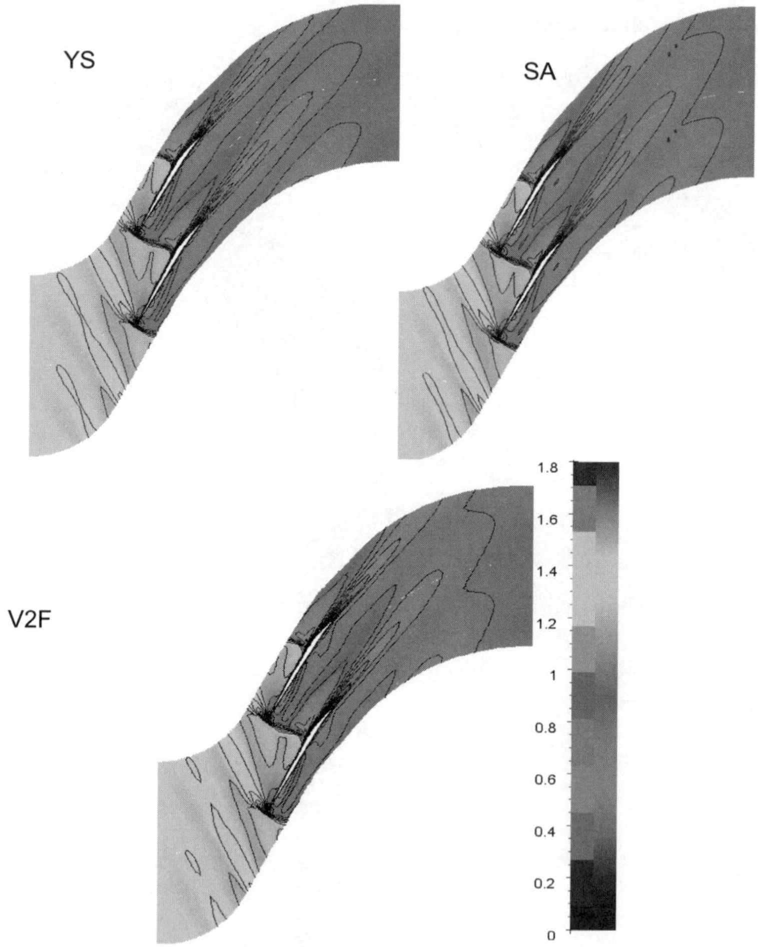

Figure 7 Relative Mach number at 95% span-wise position for the three different turbulence models

2.4.6 Downstream radial profiles

Radial profiles of pitch-wise averaged total pressure ratio, isentropic efficiency, total temperature ratio, and absolute flow angle are available at a downstream station located at z = 0.1067 m, and shown in Figure 8 and Figure 9.

The prediction in the total pressure ratio is improved by using the more complex v^2-f model compared to the Yang-Shih version of the k-ε model. Both v^2-f and Spalart-Allmaras models produce almost identical pressure ratio profiles. Both results are in close agreement with experimental data from mid-span to the shroud. However, none of the models is able to reproduce the pressure deficit close to the hub. Indeed, a leakage flow exists through the axial gap between rotating and non-rotating parts of the hub. It has been proved that such a leakage could explain the observed pressure deficit near the hub (Shabbir et al. 1997).

The numerical prediction in the efficiency is almost similar for all the models. As in the AGARD advisory report or in the simulation reported by Gerolymos and Vallet (2001), there is a strong deficit of efficiency from 80% span-wise to the tip. None the less, the v^2-f model produces a slightly higher efficiency than the two other models for relatively high radius. Since the tip wall region is an important origin of the global loss over-prediction in the simulations, this can explain the differences already observed in the operating map.

All the models produce an absolute flow angle that is about 2 degrees too small in comparison to experimental data. For all the models the radial variations in total temperature ratio are under-predicted. This was also observed in the AGARD report.

On the one-hand, the v^2-f model does satisfactorily predict the two local minima of stagnation temperature at 15 and 70% span-wise positions. However, it fails to predict the maximum at 40%. On the other hand, the Spalart-Allmaras model does reproduce this local maximum but at mid-span.

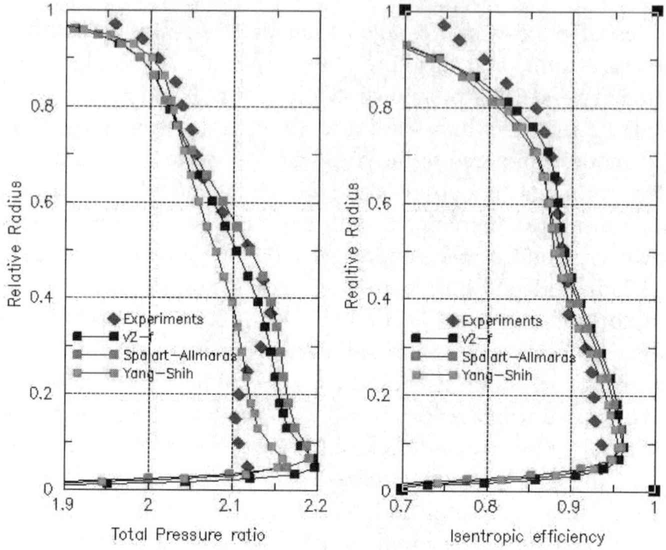

Figure 8 Total pressure ration and isentropic efficiency

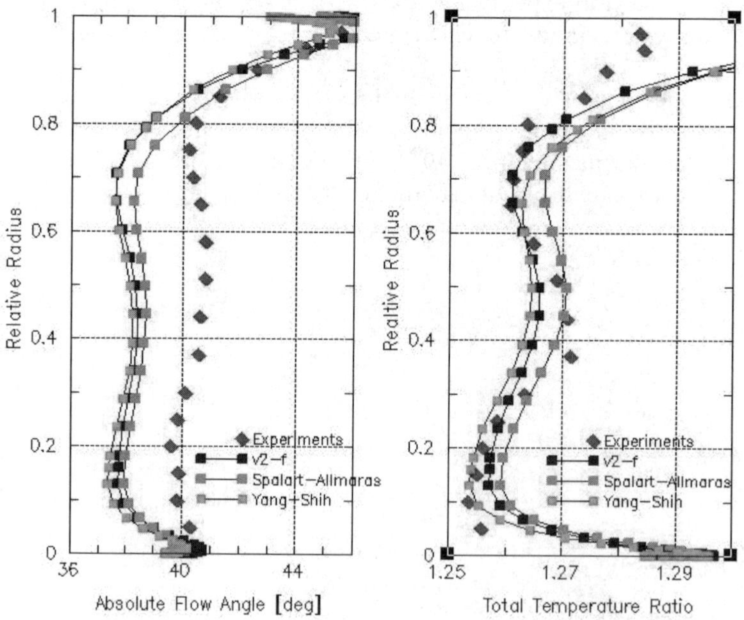

Figure 9 Absolute flow angle and total pressure ratio

3 The Asymmetric Plane Diffuser

M.A. Leschziner and C. Wang, Imperial College London

3.1 Introduction

It has been one of the major objectives of FLOMANIA to support the uptake, especially by industrial organisations, of advanced turbulence closures, in general, and Reynolds-stress-transport models, in particular. The rationale of doing so, in the context of efforts to predict ever more physically and geometrically complex flows, has been enunciated for many years by academic groups engaged in turbulence-modelling research. It has been argued, in particular, that such models account, in a physically meaningful manner, for the disparate interactions between the Reynolds stresses and different types of strain. This disparity is rooted, essentially, in the strong sensitivity of the anisotropy in the Reynolds stresses to the different types of strains and the strong effects on the generation rates of stresses that spring from it.

Second-moment models are not only mathematically complex, but also pose a whole host of numerical problems. The most obvious one is the absence of viscosity-related second-order derivatives, the presence of which can be exploited very effectively in efforts to construct stable and rapidly-converging solution algorithms. The incorporation of these models into general turbulent-flow algorithms is not trivial, therefore, and the process 'offers' numerous 'opportunities' for both errors of coding and for significant predictive differences associated with minor (or not so minor) variations in the numerical practices applied in the course of implementation. A particularly potent source of discrepancies is the near-wall region and the way in which the interaction between the wall-boundary conditions and relevant terms in the model equations is represented numerically.

Against this background, the decision was taken, at the outset, to define a 'mandatory' test flow for which all partners electing to implement and/or apply second-moment closure would provide solutions for one particular second-moment model and a prescribed numerical mesh to verify the validity of their implementation practices. Desirable major attributes of the test case were judged to be: statistical two-dimensionality, geometric simplicity, the presence of separation, reattachment and recirculation, demonstrable high sensitivity to modelling that comes with an internal flow dominated by shear throughout its extent, the existence of previous computational experience with the flow and the availability of well-regarded experimental data. The asymmetric diffuser of Obi (1993), shown schematically in Fig. 1, was eventually chosen to be that case.

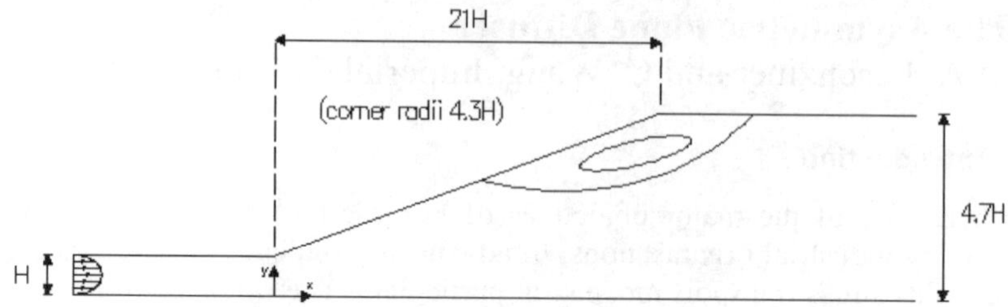

Figure 1: Geometry of asymmetric diffuser

While the Obi diffuser satisfies most of the desiderata stated above, it is not free from problems and uncertainties. It is certainly close to being two-dimensional, but is necessarily contaminated, to some extent, however minor at the large experimental aspect ratio, by 3d features arising from side walls. Moreover, the experimental database is incomplete, lacking skin-friction information and the pressure distribution on one wall, and there is some ambiguity in respect of the precise inflow data and associated non-dimensionalisation of the data further downstream. However, the essential validity of the data is supported by further experiments by Buice and Eaton (1997) for an essentially identical geometry and also by large eddy simulations by Kaltenbach (2001), which impose spanwise homogeneity. In any event, an important argument as to why the accuracy of the experimental data was not regarded as a critical issue here is that this case was not chosen primarily to study physical issues – although this is of some interest – but, rather, to act as a simple and transparent means for demonstrating agreement between solutions arising from (what should be) a specific, fully-defined turbulence model. The uncertainty in the inlet conditions, in particular, was addressed by requiring that all computors should generate, as a precursor, fully-developed channel-flow distributions to be prescribed at some distance upstream of the diffuser section. This was in harmony with the manner in which the inlet flow was generated in the experimental configuration, and this is also in line with the approach adopted in previous computational studies of this case – e.g. Apsley and Leschziner (1999).

Discussions on the choice of default second-moment closure were intensive and occupied a substantial amount of time. Specifically in relation to the objectives of the present test case, this choice could be argued to have been of marginal importance. However, it was felt that the closure selected should be accessible to the variety of numerical methodologies used by the partners, not only for the diffuser case, but also for other, more complex applications to be addressed in FLOMANIA. The decision was eventually taken to adopt the model of Speziale et al (1991), denoted SSG henceforth. One important consideration, apart from the absence of wall-reflection corrections to the pressure-strain model, was that, alongside the high-Re baseline form, a low-Re extension was available (see Chapter III). This allowed solutions from two groups of codes to be included, one that would use the "scalable wall functions" (see Chapter III) and the other the low-Re version allowing integration to the wall. Both versions include a length-scale-related equation for the dissipation rate, ε.

In total, seven groups contributed to comparisons to follow. Ahead of the discussion of the results, attention is directed towards the process of coordinating the exercise.

3.2 Description of geometry, grid and flow conditions

The geometry has already been introduced – Fig. 1. Also, as noted, contributors were asked to generate their own inlet conditions through pre-cursor channel computations. These, on their own, can be regarded as a test of the implementation of the model. At the exit, zero-gradient conditions were recommended, while at walls, the no-slip conditions or the specified wall functions provide an unambiguous prescription. The two SSG model variants were fully documented (see Chapter III) and distributed. Moreover, the experimental data were obtained in digital form and made available to partners. This left the task of generating mandatory grids. The grids eventually prescribed are shown in Figs. 2 and 3, while Fig. 4 gives the distance of the nodes closest to the wall for both grids, obtained with the Launder-Sharma, linear k-ε eddy-viscosity model.

The high-Re grid, Fig. 2, chosen following grid-dependence tests, was generated from its low-Re-grid counterpart, Fig. 3, by simply removing a number of near-wall grid lines, creating a void layer of a desired thickness. This might appear to be an inappropriate approach, a perception perhaps strengthened by the appearance of the grid in Fig. 2. However, there are good arguments for taking this route. First, it is noted that the Reynolds number of the flow is quite low – 22,000, based on flow rate. An inevitable consequence is that a physically fairly thick wall-function layer is required to achieve a y^+-distribution appropriate to the use of wall functions: as seen from Fig. 4, the y^+-values are still quite low, especially at the inclined wall, although the layer thickness, in dimensional terms, is substantial. Thus, to maintain this layer and to arrange for a much more gradual variation of the wall-normal grid above the layers would have required a substantial coarsening of the grid in this region and consequent loss of numerical accuracy in influential (highly sheared) parts of the flow domain. Second, it can be argued that the rather non-uniform high-Re grid so created would be useful in bringing to the fore, more than would a graduated grid, errors and numerical peculiarities in the implementations of the wall-function version of the SSG model. Thus, this practice and its outcome were deemed acceptable.

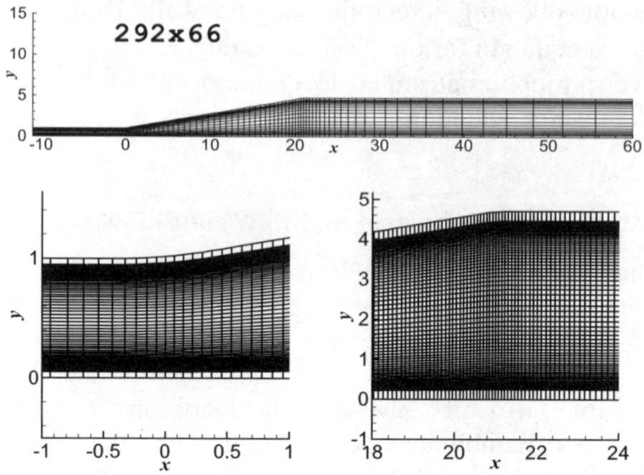

Figure 2 High-Re grid (only every fifth line shown in upper plot)

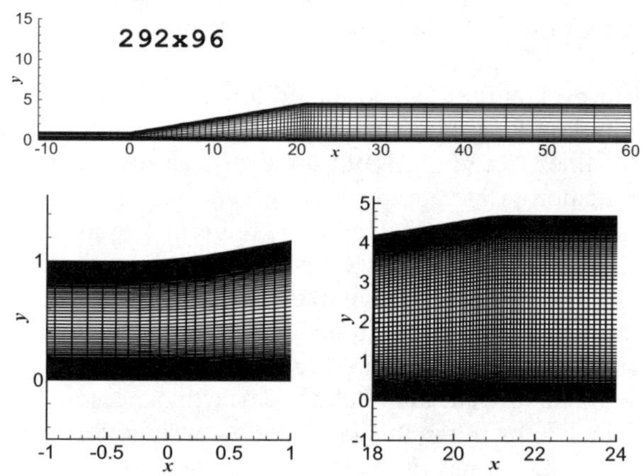

Figure 3 Low-Re grid (only every fifth line shown on upper plot)

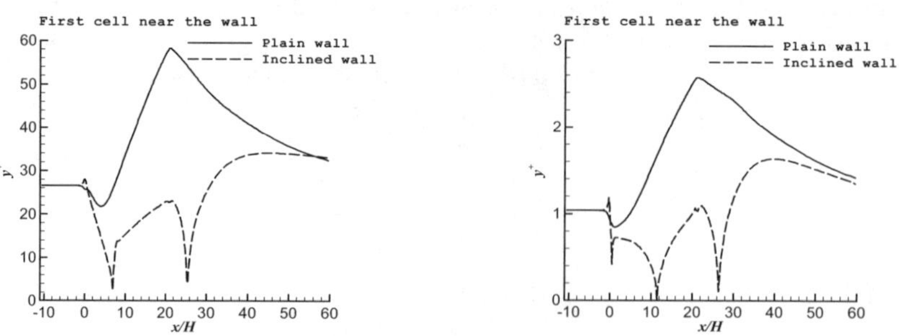

Figure 4 variations of y^+ of the wall-nearest lines of nodes (cell centres). Left: high-Re; right: low-Re

3.3 Contributors and their contributions

Contributed computations are included herein for 8 partners. Table 1 summarises the principal elements of their numerical practices, while Table 2 lists the modelling practices. As seen from the latter, there are 3 categories of implementations, one involving different non-standard forms of the length-scale equation, namely for ω rather than for ε. This necessitates the presentation of cross-plotted results in three parts, relating to the respective variations in modelling practice. A further complication arises from the fact that the implementations involving the ω-equation differ. FOI uses the Wilcox (1988) model, while NUMECA and DLR uses the BSL (blending) practice of Menter (1994). DLR's implementation also blends the pressure-strain models of Speziale et al (SSG) and Launder et al (LRR) in the BSL manner. The equations are integrated to the wall despite the absence of *Re*-dependent damping functions. Thus, the near wall anisotropy will not be correctly represented.

Table 1 Numerical practices

CFX	Implicit	Compressible and incompressible	Unstructured	Finite volume 2nd in space
DLR	Implicit: turbulence, Explicit: mean	Compressible	Structured	Central with artificial dissipation
FOI	Explicit, local-time steping, MG	Compressible	Unstructured	Finite volume 2nd central/upwind in space
ICSTM	Implicit	Incompressible	Structured	Finite volume, 2nd in space
Numeca	Explicit	Compressible & Incompressible	Unstructured	Finite volume, 2nd in space
SPTU	Implicit	Incompressible	Structured	Finite volume, 2nd in space
UMIST	Implicit	Incompressible	Unstructured	Finite volume, 2nd in space

Table 2 Modelling practices

	Turbulence model	Length - scale equation	Wall treatment
AEA	SSG	ε	Scalable wall function
DLR	SSG+LRR[1]	ω–BSL Menter	Integration to the wall
FOI	SSG	ω	Integration to the wall
ICSTM	SSG SSG+Chen low-Re	ε ε	Scalable wall function Integration to the wall
Numeca	SSG SSG[2]	ε ω–BSL Menter	Scalable wall function Integration to the wall
ONERA	SSG	ε	Scalable wall function
SPTU	SSG+Chen low-Re	ε	Integration to the wall
UMIST	SSG	ε	Scalable wall function

[1] Blending of SSG and LRR models, the latter effective near the wall, with blending controlled by the equivalent of Menter's BSL approach.

[1,2] Blending of ω equation near the wall with ε away from the wall with the BSL approach.

3.4 Results

3.4.1 General description

Prior to the presentation of cross-plotted results, it is instructive to consider some overall flow characteristics and illustrate the extreme sensitivity of this flow to different modelling practices. Thus, Figure 5 shows 8 computations performed by the writers: one with a linear EVM, 3 with NLEVMs/EARSMs and 4 with two second-moment-closure models. Figure 5 shows some major differences in respect of the ability of the models to resolve the separated layer. It must be pointed out again that this feature is quite weak, however, and may be regarded as a 'second-order' (i.e. refined) indicator of predictive capabilities. In contrast, a 'first-order' indicator is the ability of the model to return the substantial flow asymmetry (between the upper and lower wall) that arises as a consequence of the inclined wall. Thus, the solutions for LS and SSG+Chen may look close, but they are not, the latter giving a much greater level of asymmetry reflecting near-separation.

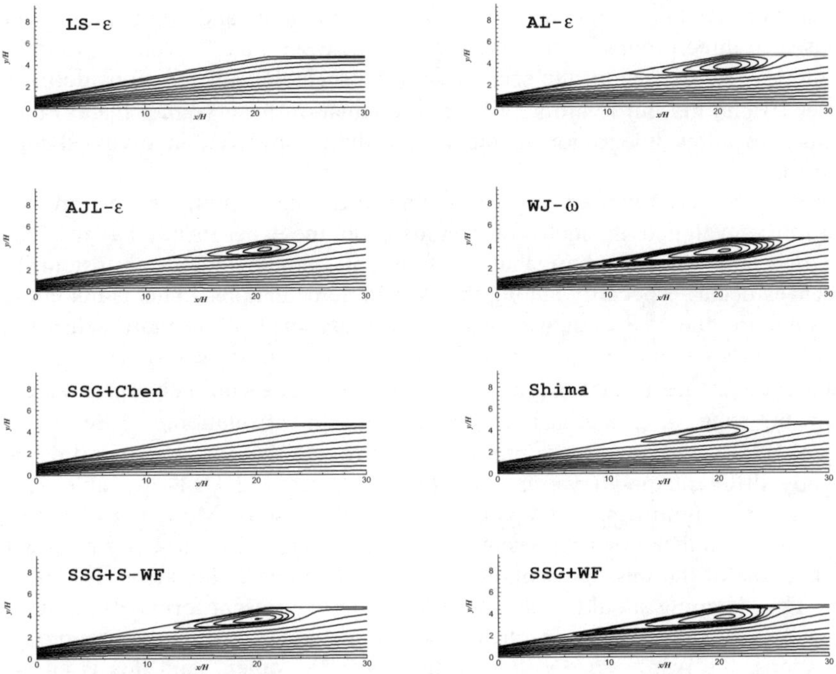

Figure 5 Stream function calculated based on 8 models

An important fact to highlight in relation to the SSG model is that the use of this model with a low-*Re* extension gives a substantially different behaviour to that achieved with wall function, the precise nature of the latter being of subordinate importance. It thus seems that the resolution of the near-wall region is extremely influential. While the low-Re treatment might be expected to be superior, it evidently gives worse results (as the real flow is separated), and this points to defects in the near-wall modelling elements. All three NLEVMs are seen to resolve separation, but the WJ model seriously over-estimates its extent. Thus, here again, there is a substantial sensitivity to modelling. A final point to make is that the nature of the basic second-moment closure is also of great importance. This is reflected by the differences between the Shima and SSG models. While the latter gives a pronounced recirculation zone, the writers have observed serious anomalies in the near-wall flow, including a weak forward flow between the recirculation zone and the inclined wall. More generally, second-moment models tend to return anomalous reattachment behaviour with a characteristic 'doubling-up' of the separation streamline at reattachment.

3.4.2 Cross-plotted results

This section contains comparisons among predictions and with experiments, organised in three groups, corresponding to three SSG model variations used: one with the low-Re extension, the second with the scalable wall functions defined in Chapter III and the third with a Wilcox-type equation for ω (rather than ε) that is presumed to allow integration to the wall without, however, involving damping functions.

Computors were required to generate their own inlet conditions, corresponding to the fully-developed channel flow established in the experiment. The results are given in Figure 6, and show that, even at this simple level, not insignificant differences occur, especially among the WF implementations. This is not entirely surprising, for the WF computations must be assumed to be more sensitive to numerical issues as a consequence of the highly non-uniform wall-normal grid. An encouraging feature is that the group of low-Re SSG-ε computations give close agreement, suggesting a consistent numerical implementation. Assessing the group of SSG-ω results is more problematic, however, because every partner used a slightly different model variant (see FOI, Numeca and DLR in Table 2). As seen, there are significant discrepancies among the results. Moreover, this model group gives a rather poor representation of the streamwise normal stress at the wall, because of the misrepresentation of stress damping in the semi-viscous wall layer. This damping should be anisotropic, but in the present formulation, without near-wall corrections, such a mechanism is not included. Within the group of WF calculations, the result by ONERA at odds with the others, and this is an issue which will be seen to persist and worsen in plots to follow.

Comparisons for the friction factor are given in Figure 7. The experimental data do not originate from Obi's study, however, but from follow-up experiments by Buice and Eaton (1997) for the same geometry. Here, the inability of the low-Re form of SSG to predict the separation process is clearly brought out. In contrast, the SSG-ω formulations appear to give an excessive tendency towards separation, but results must be viewed as inconclusive, in the light of the significant discrepancies between Numeca's solutions and the other two sets. In particular, the poor performance of Numeca's implementation at the plane wall is curious, indicating an excessive asymmetry in the streamwise flow. Again, the results of the WF implementations show significant differences from one to others, with the ONERA implementation displaying distinctly anomalous features. Finally, ONERA's and Numeca's skin-friction level upstream of the diffuser is excessive. This could be due to an error in non-dimensionalisation. However, reference to Fig. 6 suggests that a more likely reason is a defect in the implementation, for both are seen to give normal-stress levels which exceed those of other solutions in the respective model categories, and also the experimental level.

Distributions of the pressure coefficient are given in Figure 8. Experimental data are only available for the plane wall, but the cross-flow variations are very minor (at least when judged on the basis of most computations). This set of plots gives rise to the greatest concern about the variability of the results obtained with the wall functions and the ω-based formulations. The pressure distribution is, of course, a primary quantity, and a failure to predict it correctly indicates some major defects in the velocity field. Both ONERA's and, to a lesser extent, UMIST's and Numeca's data suggest an excessively large recirculation region and/or a much too extensive region of slow flow that is accompanied by excessive lateral displacement, preventing recovery at the requisite rate. As will be seen by reference to the velocity profiles given below, this is indeed the case. Specifically, UMIST's and especially ONERA's and Numeca's solutions show an excessive fluid displacement away from the inclined wall, evidently leading to too slow a pressure recovery.

Profiles of streamwise velocity and Reynolds stresses at two streamwise positions are presented in Figures 9-11. It is not possible (or, indeed, appropriate, in the present context), to provide here a searching analysis of the physical issues implied by the plots. The emphasis is, rather, on the consistency of and the differences among the predictions.

The velocity profiles show most implementations to under-estimate the near-wall separation region. Yet, the velocity predicted below the separated zone, further away from the wall, tends to be depressed below that measured. Most models return the strong velocity asymmetry, in contrast to standard linear EVMs. In fact, the SSG-ω model (DLR's and FOI's implementation) and most WF implementations exaggerate this asymmetry, despite missing the upper separated region. The WF implementations are fairly closely bunched, except for the ONERA's results which must be regarded with some concern. Most solutions (ONERA, UMIST, Numeca) show excessively large regions of slow flow close to the inclined wall, which then leads to the slow pressure recovery noted earlier. Figure 9 thus demonstrates, as did results already discussed, how sensitive model performance is in the present flow to the details of the near-wall representation. This sensitivity extends well beyond the near-wall region, clearly affecting the gross characteristics of the flow.

Profiles of the shear stress and wall-normal stress are contained in Figures 10 and 11, respectively. These are sensitive indicators of model performance. The inclusion of the wall-normal stress component, in particular, is motivated by the fact that this component derives its level, principally, from a transfer of energy from the streamwise stress by the pressure-strain process. This stress thus reflects sensitively the performance of the model as well as the validity of its implementation. The two results of the wall-resolving SSG-ε models are close, with differences in maximum stress being of order 5%. Differences among the profiles in the other two groups are significantly larger, with the ONERA and Numeca profiles again being anomalous relative to the general tend. All models underestimate this stress (and also over-estimate the normal-stress anisotropy), suggesting defects in the pressure-strain model. The WF-results are, again, bunched together, which is an encouraging outcome suggesting an essentially correct implementation of all models, despite the differences.

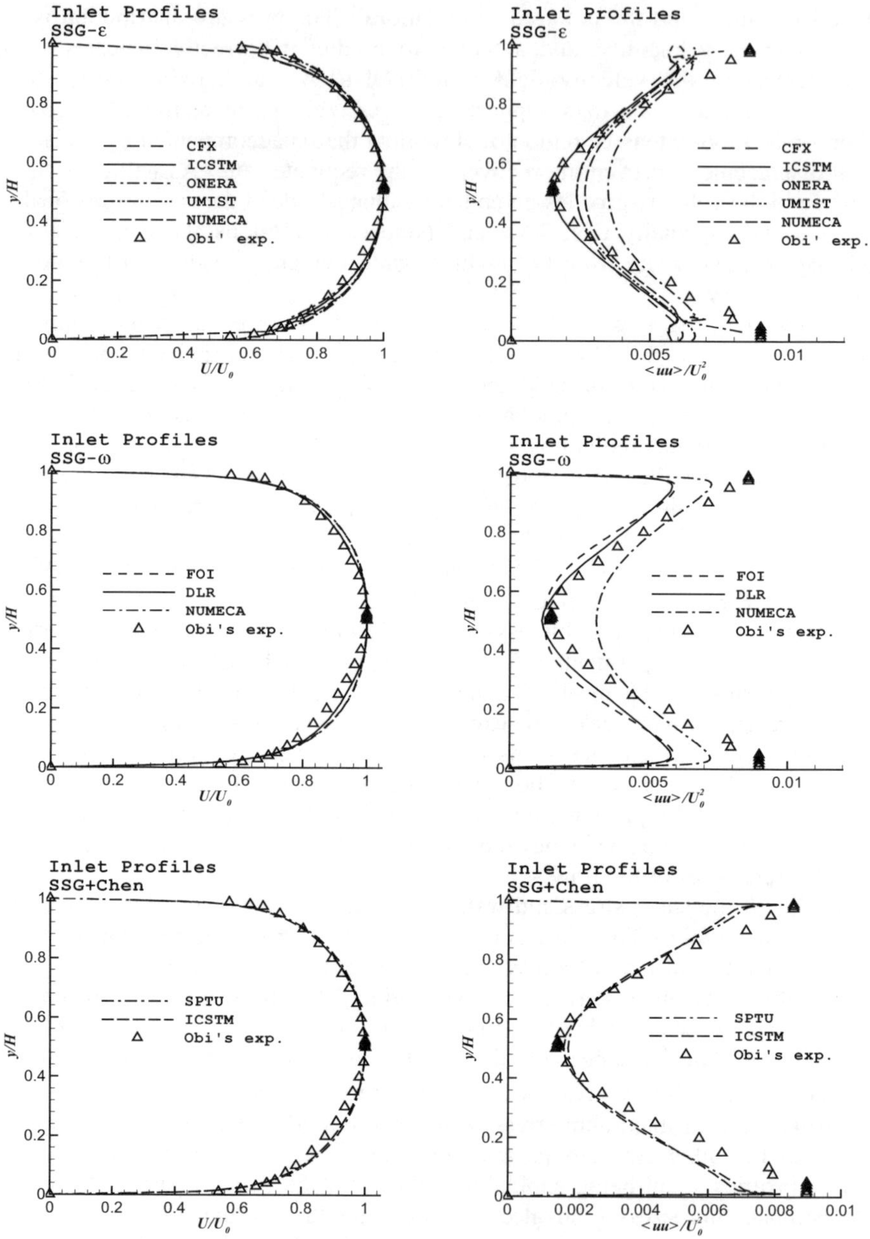

Figure 6 Profiles of velocity and streamwise normal stress at inlet

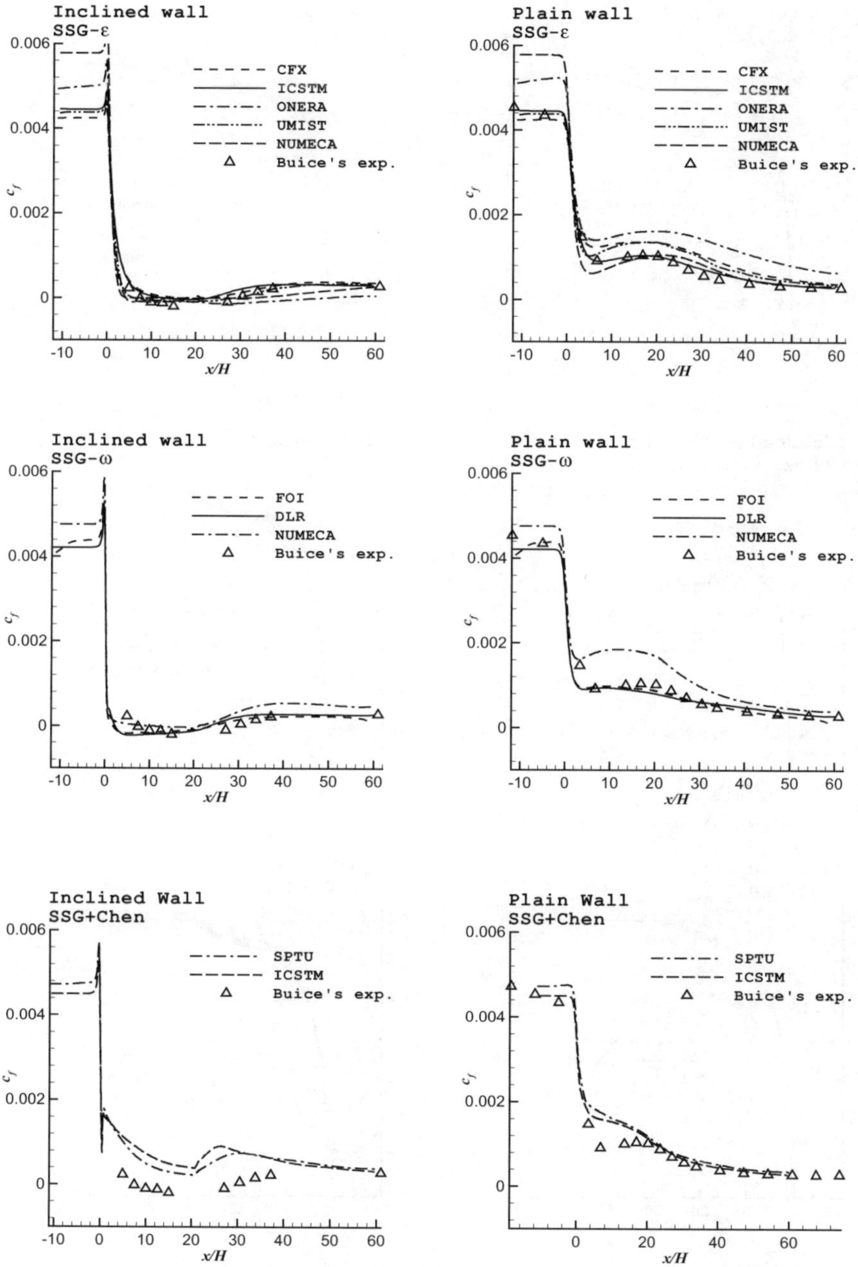

Figure 7 Distributions of skin friction on both walls

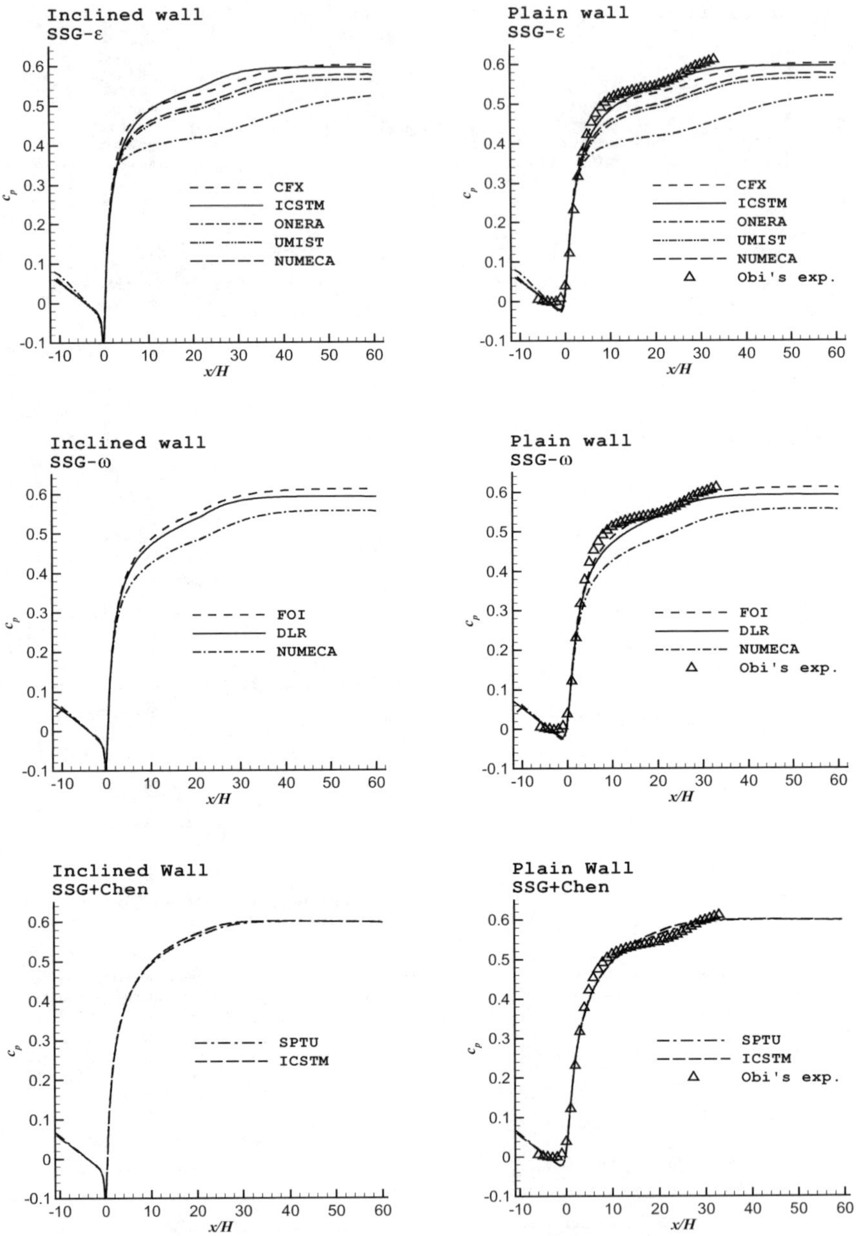

Figure 8 Distributions of pressure coefficients on both walls

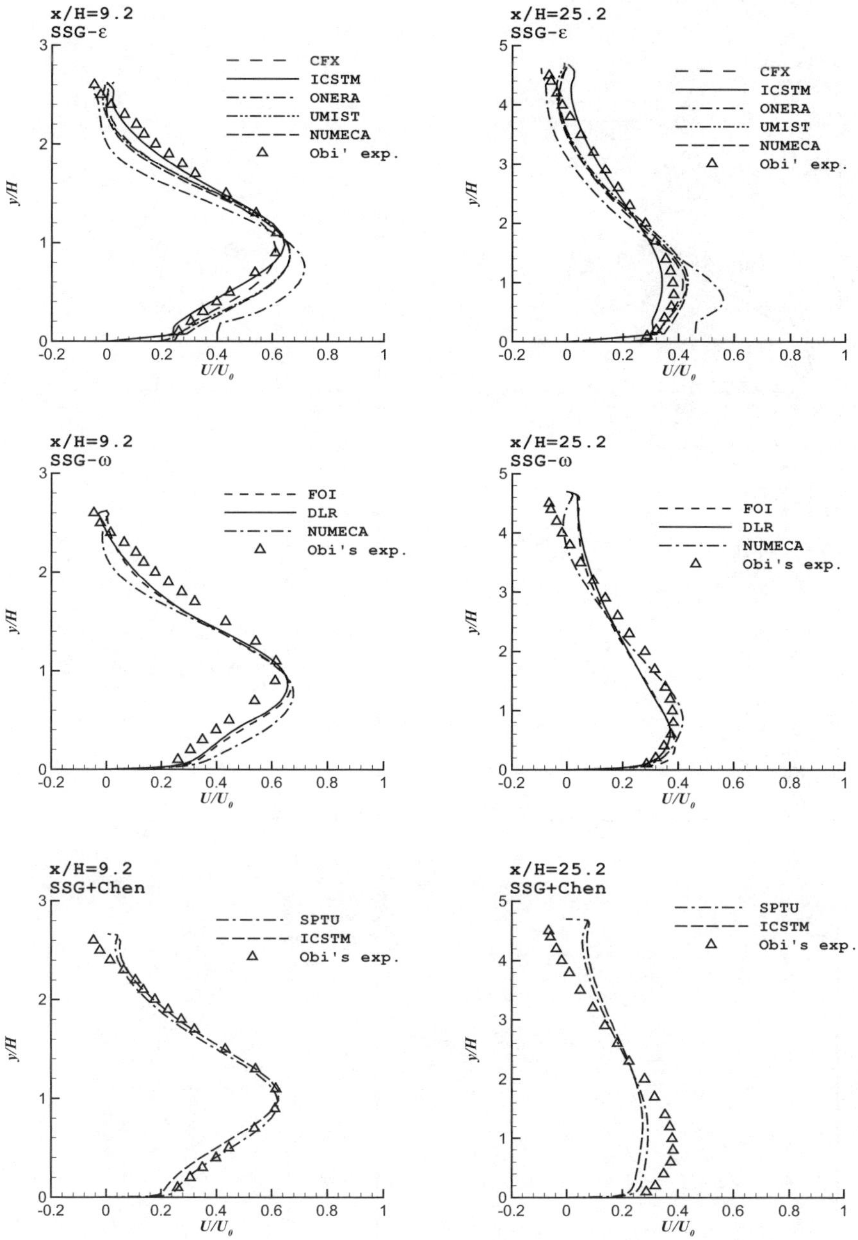

Figure 9 Profiles of velocity and streamwise stress at two streamwise locations

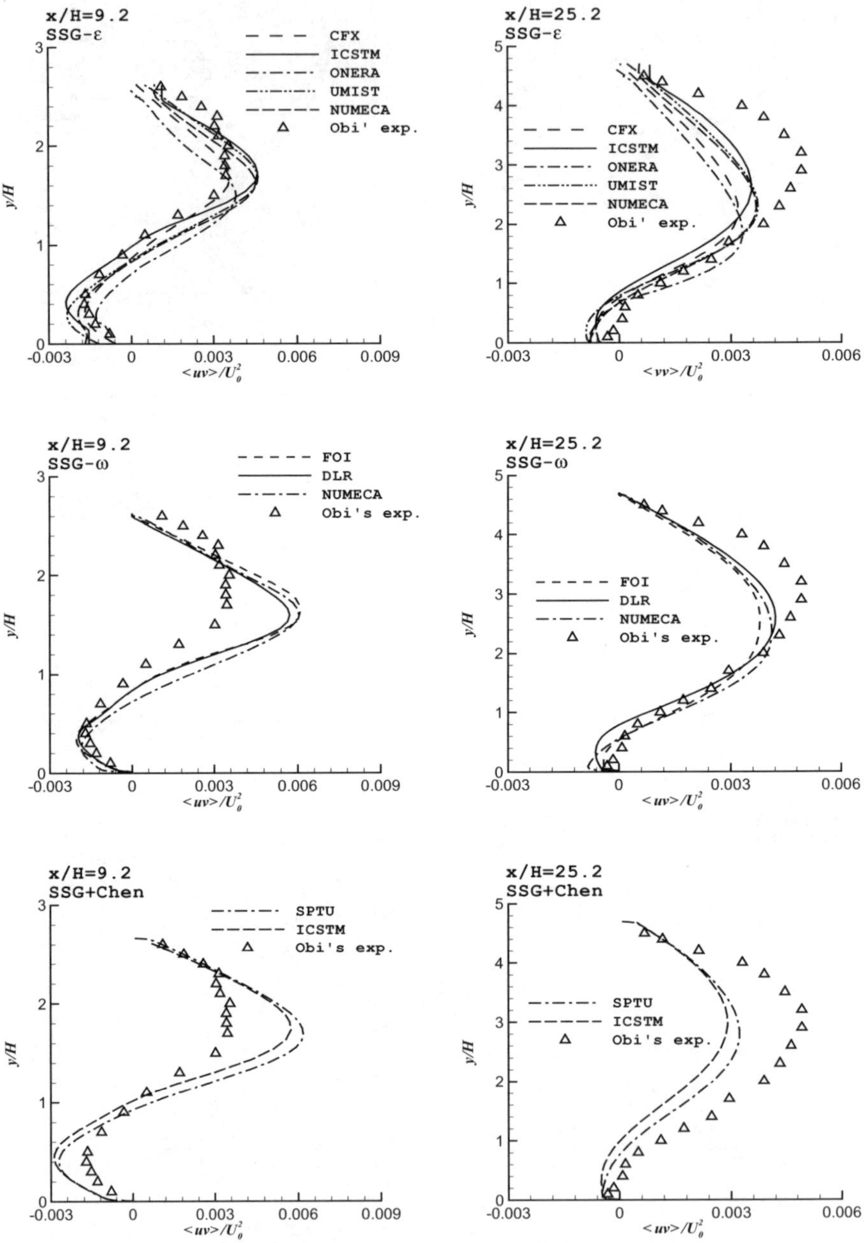

Figure 10 Profiles of shear stress at two streamwise locations

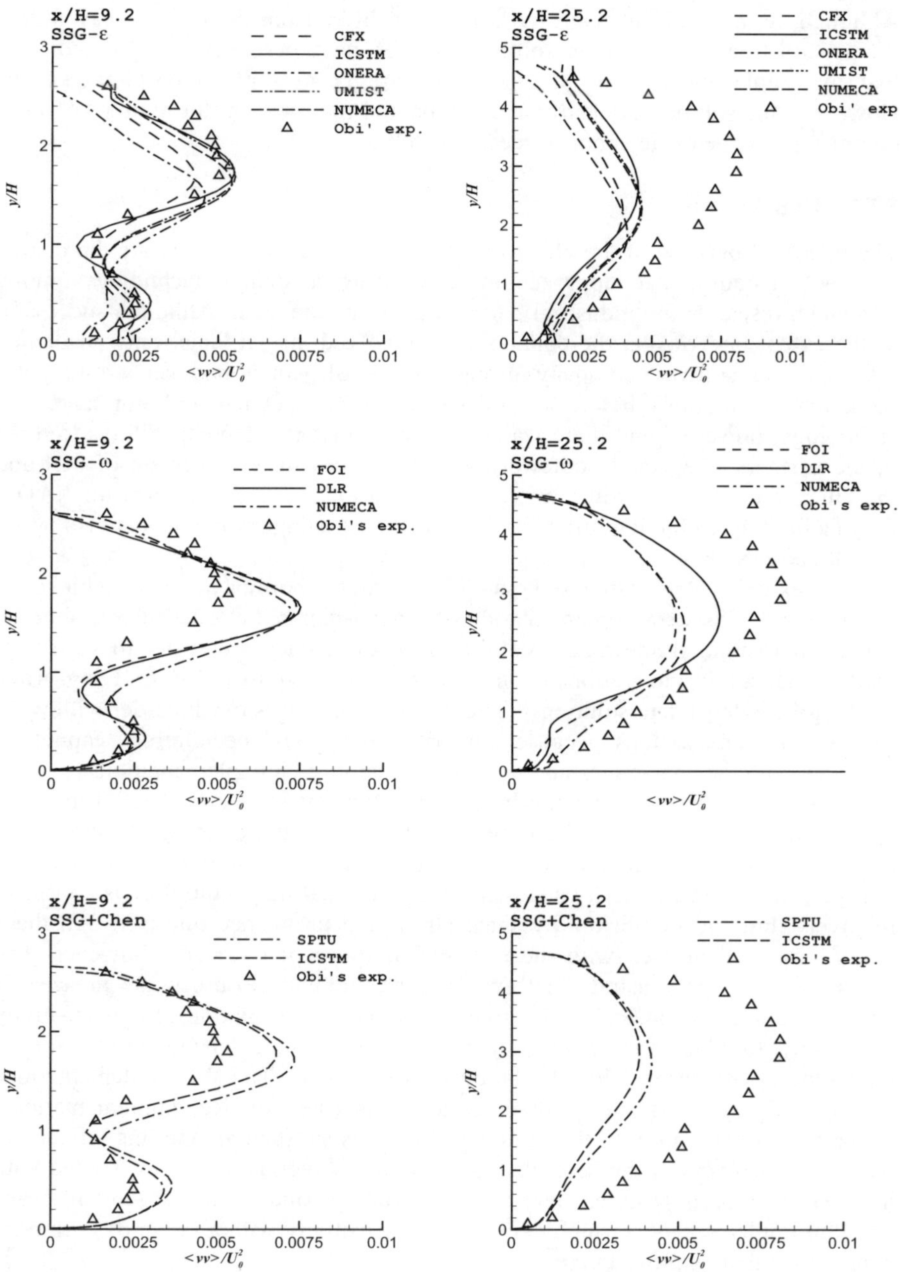

Figure 11 Profiles of transverse normal stress at two streamwise locations

Although, as pointed out, ONERA's profiles differ from the others, they follow the general trends of the other solutions and of the experiment, and this indicates that the implementation of the model is correct in some important respects. However, this solution is clearly anomalous and requires careful review. Similar comments may be made about Numeca's results.

3.5 Conclusions

The principal purpose of the effort reported herein was to demonstrate the degree to which a coordinated approach to transferring modelling technology among academic, research and industrial partners could be pursued. Although successful to the extent undertaken, the collaboration involved, prima facie, only academic, research and software-company groups, some of which had second-moment-modelling capabilities before FLOMANIA started. Others did not have this technology, however, and there was a genuine transfer of the capabilities to three of the partners. Especially noteworthy is the successful involvement of FOI and the DLR, the latter providing a major computational-design platform for EADS, thus facilitating technology transfer to industry. Analogous comments also apply to FOI and ONERA.

The exercise demonstrated how difficult and laborious it is to achieve an accurate and consistent implementation of second-moment closure in an elaborate, general numerical framework. Validity and consistency are easier to verify for models that allow integration to the wall and thus require fully-resolving grids, giving grid-independent solutions. The task of doing so is much harder with wall-function implementations because numerical errors and peculiarities cannot be excluded with the coarse meshing inevitably used in such implementations.

As regards physical aspects, the present flow distinguishes itself from most others examined in FLOMANIA by its extremely high sensitivity to modelling details. In particular, separation is a tenuous feature that can easily be missed. To a large extent, this sensitivity stems from the fact that the whole flow is confined, relatively thin and dominated by shear strain and turbulence transport. All these feature are well aligned with the principal purpose of the study, however. The results included demonstrate that even second-moment closure is no panacea. It does, in contrast to linear EVMs, give a credible representation of the sensitivity of the flow to adverse pressure gradient, but performance depends greatly on the modelling details, especially in relation to wall effects. The WF implementations return fairly good results, often better than the low-Re implementations. However, the variability of the implementations is substantial, and this reflects, in part, the coarseness of the near-wall grid and the numerical errors associated with it. What perhaps is required next is to undertake a study with the scalable wall-function implementation which explores this sensitivity with a range of grid more benign than that employed herein.

4 ONERA M6 wing

B. Eisfeld, DLR Braunschweig

Abstract

The present chapter presents the results obtained by Dassault, DLR and NUMECA for the ONERA M6 wing. Solutions by structured and unstructured methods are compared, where different turbulence models have been applied ranging from one-equation to full Reynolds stress models.

4.1 Test case description

The ONERA M6 wing has been established as a standard 3D test case for transonic flow by the AGARD Specialist Working Group WG 04, seeking for the highest quality experimental data available at that time (Barche, 1979). The ONERA M6 is a low aspect ratio swept, untwisted tapered half-wing model generated from the symmetric ONERA D airfoil. It has been tested in the ONERA S2MA wind tunnel at different Mach numbers and angles of attack at a Reynolds number of $Re_c = 11.72 \cdot 10^6$ based on the mean chord length (Schmitt et al, 1979). For FLOMANIA the conditions $Ma = 0.84$ and $\alpha = 3.06^0$ have been chosen from the data base, where a pronounced lambda-shock occurs on the suction side.

In the experiments a boundary layer diverter has been used, i. e. the half-wing model has been mounted on a plate positioned at some distance off the wind tunnel wall. In contrast in the computations symmetry conditions have been applied at the plane aligned with the inner wing section.

4.2 Computations

4.2.1 Computational grids

The codes applied by Dassault, DLR and NUMECA require different types of computational grids. Therefore Dassault generated a fully tetrahedral mesh consisting of 270000 nodes with a near wall stretching factor of $f = 1.15$ normal to the wall (Figure 1). DLR generated a structured one-block mesh with 800000 cells and a near wall stretching factor of $f = 1.4$, which has also been used by NUMECA (Figure 2). Its maximum aspect ratio in streamwise direction is $a = 1.4 \cdot 10^6$.

Both grids were found to resolve the flow field sufficiently down to the wall, i.e. the wall nearest grid points have a distance to the wall of $y_1^+ \approx 1$. Nevertheless the DLR mesh showed a too low resolution in spanwise direction for capturing the shock junction on the suction side correctly. On the other hand the Dassault grid exhibits a small kink in the leading edge where the round tip is attached to the wing. However it is not clear how this influences the predictions close to the wing tip.

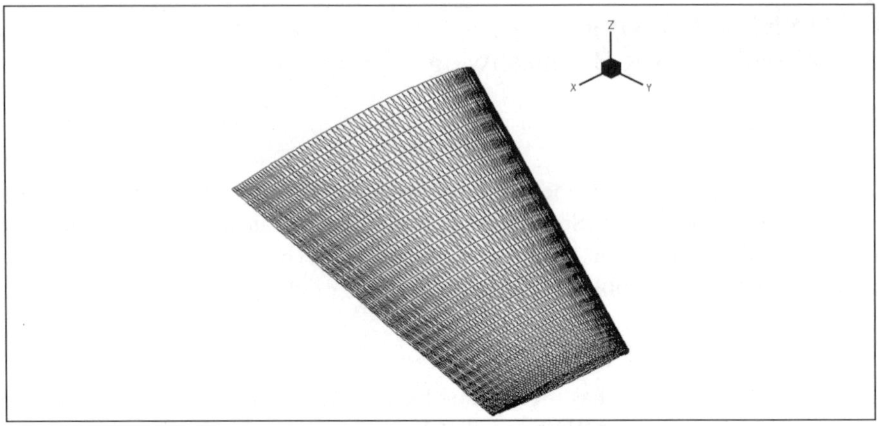

Figure 1 ONERA M6 wing. View of unstructured Dassault grid

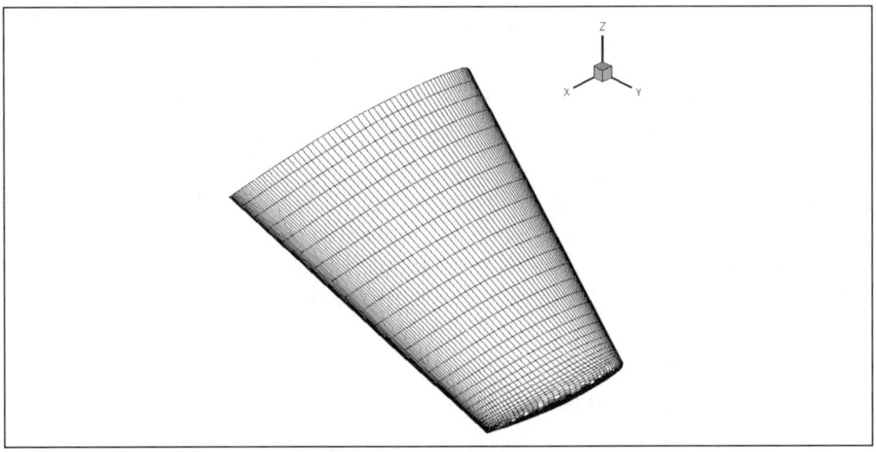

Figure 2 ONERA M6 wing. View of structured DLR grid

4.2.2 Computational details

All computations on this test case were run until the forces, more specific the drag, were converged to a constant value. Additionally DLR aimed at a convergence of the density residual by 4 orders of magnitude. All computations yielded a steady state solution.

Dassault applied the Chen-Patel (19988) k-ε model in linear mode or coupled to the EARSM of Wallin and Johansson (2000). NUMECA used the Spalart-Allmaras (1994) model, the k-ε model by Yang and Shih (1992) and the Menter SST k-ω (1994) model. DLR carried out its computations with the Wilcox k-ω (1988), the LEA k-ω models (Rung et al, 1999), the Wilcox stress-ω (1993), and SSG/LRR-ω Reynolds stress models (Speziale , 1991) with generalised diffusion.

Dassault used a fully implicit time integration whereas NUMECA's method is explicit and embedded into a full multigrid algorithm. DLR combines the explicit approach based on multigrid for the mean flow equations with an implicit scheme for the turbulence equations. Consequently, the CFL numbers used differ between all methods.

Table 1 summarises the details of the different computations. As one can see, the computational effort in terms of iterations or multigrid cycles seems to depend mainly on the complexity of the turbulence model, comparing the one- and two-equation model computations of NUMECA or the two-equation and Reynolds stress model computations of DLR. The EARSM approach applied by Dassault does not degrade the convergence with respect to the linear two-equation model.

Table 1 Details of computations

Participant	Model	Acronym	CFL no.	Iteration no.
Dassault	Chen-Patel k-ε	CP	10	2000
	Chen-Patel k-ε + Wallin-Johannson EARSM	WJ	10	2000
DLR	Wilcox k-ω	Wi88	8	1500
	LEA k-ω	LEA	8	1500
	Wilcox-stress-ω + GGDH	DSM-Wi	8	2500
	SSG/LRR-ω + GGDH	SSG/LRR-ω	8	2500
NUMECA	Spalart-Allmaras	SA	2	1000
	Yang-Shih k-ε	YS	2	1500
	Menter SST k-ω	SST	2	1500

4.2.3 Results

The computational results are compared with experimental pressure distributions at four selected spanwise sections. Figure 3 shows the results obtained with the eddy viscosity models. As one can see, the agreement with the measurements is generally good for all models. The deviations at $\eta = 0.20$ are probably due to the differences at the symmetry plane between the experiment and the computations.

Some differences between the numerical results occur with respect to the shock resolution, which might be partly due to the grid resolution and the numerical dissipation of the respective method. Note the very good shock prediction in the Dassault simulations with the Chen-Patel k-ε model.

For $\eta = 0.99$ all simulations with eddy viscosity models fail to predict the correct pressure level behind the shock. The occurring differences between the results near the trailing edge again seem to indicate the influence of the grid resolution and the numerical dissipation.

Figure 4 shows the results obtained with the EARSM type models, where the LEA k-ω model has been included because it is the linear truncation of an explicit algebraic stress model. As one can see, the EARSM approach leads only to small differences in the results compared to the respective baseline eddy viscosity models. The major influence seems to be an upstream shift of shock positions which is clearly seen in Dassault's results and less pronounced in the DLR results.

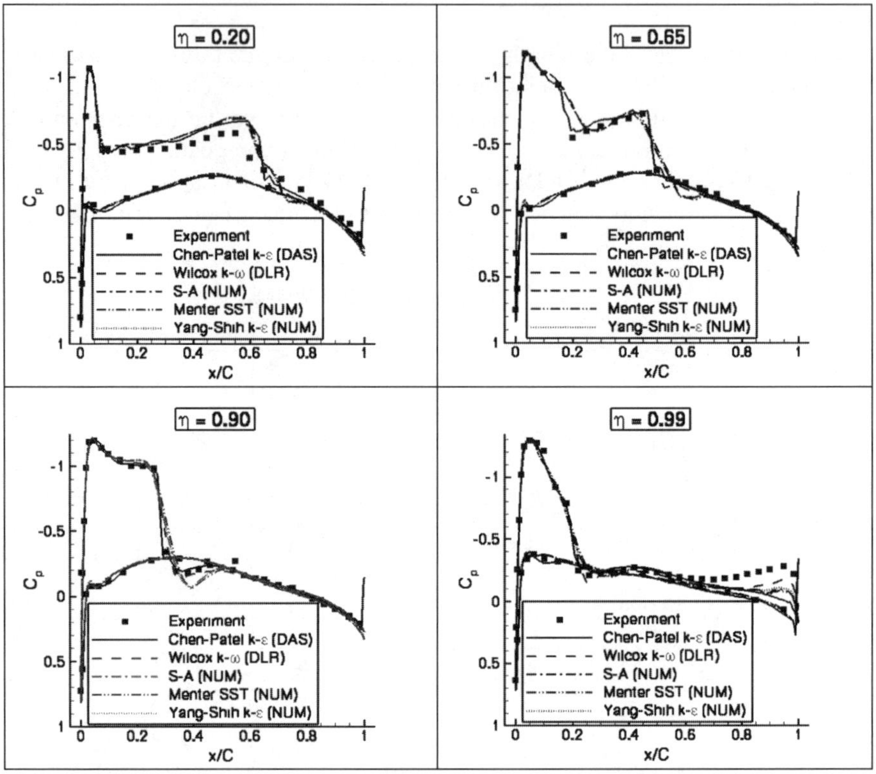

Figure 3 ONERA M6 wing. Pressure distributions for eddy viscosity models

Finally, Figure 5 shows the DLR results for the Reynolds stress models in comparison with the Wilcox k-ω model. As one can see, there is virtually no difference between the computed C_p distributions, except at η=0.99. This probably confirms the findings for the eddy viscosity models that most of the differences in the results are mainly due to different grids and different numerical methods.

Nevertheless, at the wing tip there is a definite influence of the turbulence model on the result. As one can see, at η=0.99 the Reynolds stress model predictions are clearly in better agreement with the measurements than the prediction with the Wilcox k-ω model. The reason for that improvement is the improved resolution of the tip vortex due by the Reynolds stress models due to their exact production term.

Figure 6 depicts the Iso-C_p contours in a plane x/b=0.90 (i.e. normal to the flow at zero incidence) for the Wilcox k-ω and the SSG/LRR-ω model. As one can clearly see, the tip vortex is predicted stronger and more concentrated with the Reynolds stress model than with the eddy viscosity model.

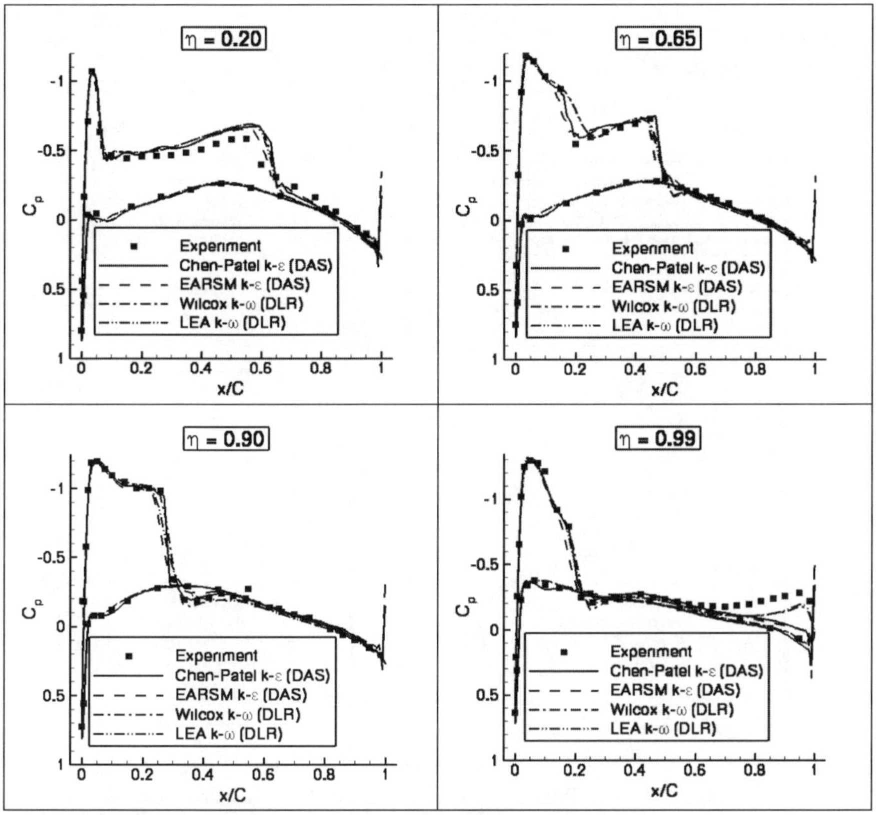

Figure 4 ONERA M6 wing. Pressure distributions for EARSM type models. Comparison with baseline eddy viscosity models

4.3 Conclusion

The flow around the ONERA M6 wing has been computed with various turbulence models, using three different codes on two different grids. Generally good agreement with the measured C_p-distributions has been achieved in all simulations. Most of the differences between the results seem to be due to the different grids and different numerical methods used.

An influence of the turbulence modelling on the result has been observed for the EARSM approach, which shifts the predicted shock position upstream with respect to the respective baseline eddy viscosity model. Reynolds stress models improve the prediction of the pressure distribution close to the wing tip due to an improved resolution of the wing tip vortex.

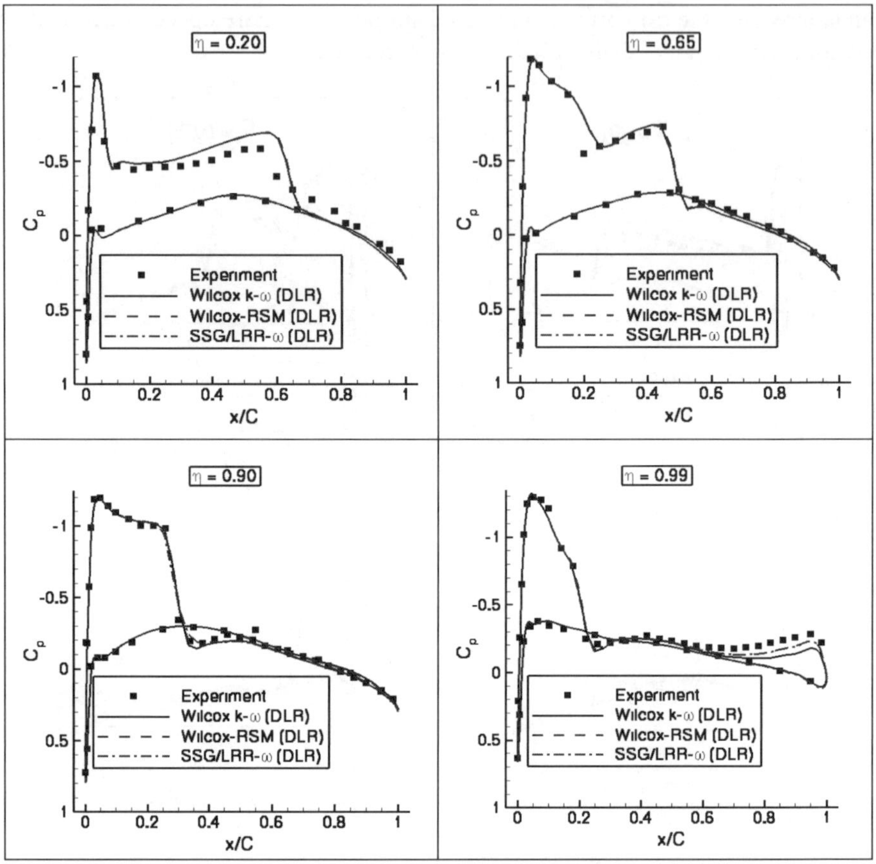

Figure 5 ONERA M6 wing. Pressure distributions for Reynolds stress models. Comparison with Wilcox k-ω model.

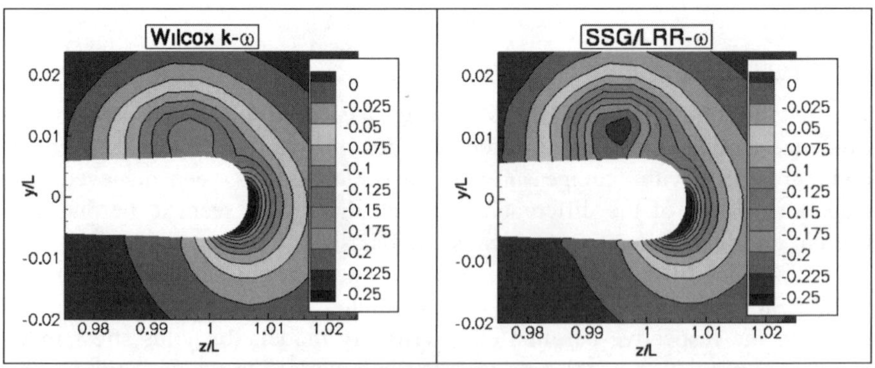

Figure 6 ONERA M6 wing. Iso-C_p lines in plane x/b = 0.90 at the wing tip. Comparison of Wilcox k-ω model with SSG/LRR-ω Reynolds stress model

5 OAT15A airfoil in wind tunnel (Alenia, Dassault)

N. Ceresola, Alenia

5.1 Experimental arrangement and flow conditions

The geometrical and experimental data have been provided by ONERA [1]. The testcase geometry consists of a straight wing profile in the wind tunnel, with both extremities fixed on tunnel walls.

The arrangement corresponds to the 370 mm x 390 mm test section of ONERA T2 tunnel. Shapes of the upper and lower wall of the test section are adapted during the experiment in order to match the streamline curvature so minimizing the wall interference. The airfoil chord length is 150 mm. Transition is imposed at 7% of the chord.

Experimental data are available from static pressure probes and laser velocimetry, and are referred to an upstream Mach number of 0.7536, a Reynolds number based on chord of 3.53 10^6 and an angle of incidence α=10°. The flow conditions considered here give rise to a shock wave on the suction side of the airfoil, with no shock induced separation.

A sketch of the test case geomrtry is shown in Figure 1.

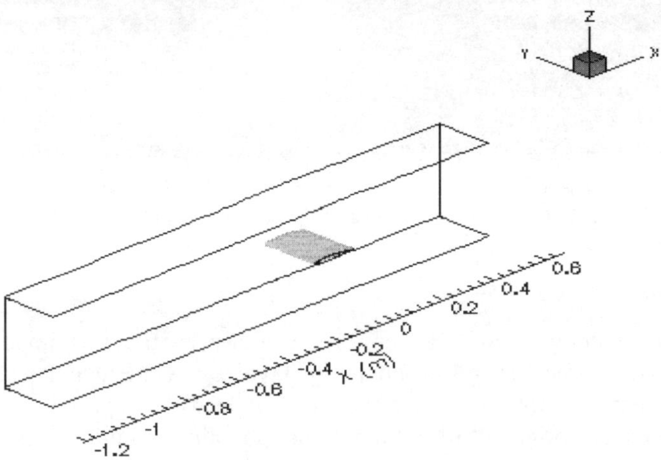

Figure 1 Testcase geometry

5.2 Computations

5.2.1 Grids

Both grids used for the computations reproduced the geometry of the wing in the wind tunnel test section. The deformation of upper and lower walls due to wall adaptation were properly taken in account.

Dassault generated an unstructured grid made by 800 k points. A very fine mesh was used for the description of the flow around the airfoil (y^+=1). A coarser resolution (y^+=5) was used for describing the walls, in order to minimize the global size of the mesh.

Alenia made its computation on a multiblock structured grid that have been provided by ONERA. The grid is made of 1.2 million points, with a resolution of airfoil surface and wind tunnel walls similar to that featured by Dassault grid.

A comparison between the structured (ALA) and unstructured (DA) grids is shown in Figure 2. The latter seem to offer a better resolution in the region of the shock wave on the suction side, to obtain, as it will be shown in the following, a better resolution of the shock wave.

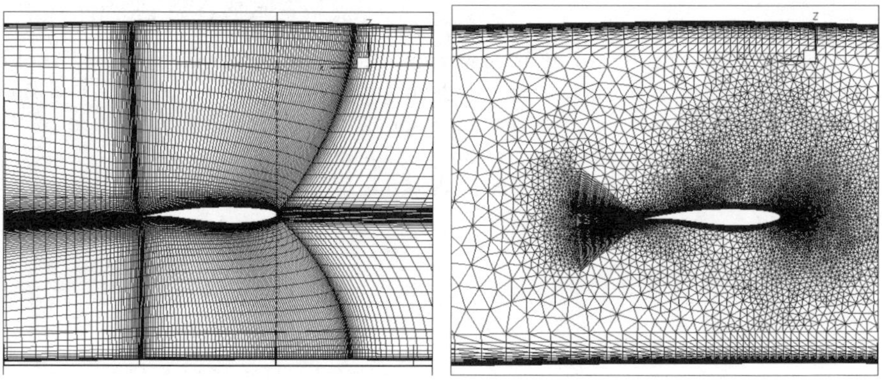

Figure 2 ONERA grid (left) and DA grid (right) at symmetry plane.

5.2.2 *Boundary condition*

In the Alenia computations, an uniform velocity profile was imposed at inlet, neglecting the effect of inflow boundary layer on wind tunnel walls. Constant freestream entropy, total enthalpy and flow angle were imposed at inflow, while the pressure was computed from the internal domain. At outlet, pressure was kept fixed, while computed values of the other quantities were taken.

Dassault managed to impose the pressure level at the exit which induces the correct mass flow rate upstream of the profile, as it results from LDV measurements. A ratio of P/Pi=0.6797 was found to satisfy this condition.
In addition, an inlet boundary layer profile was prescribed at the upper and lower tunnel walls.

5.2.3 Turbulence models

A k-ω turbulence model (WJ-H) was used in the computation by Alenia. It includes the Wallin-Johansson Explicit Algebraic Stress Model as the constitutive model.

Dassault used in its computations two versions of the k-l turbulence model (Sm95 and Sm97), respectively without and with the non equilibrium correction.

5.3 Results

The Cp distributions along the profile at symmetry plane are shown in Figure 3, in comparison with experimental data. The shock wave predicted by Alenia (WJ-H) is positioned slightly upstream with respect to the experiment, while DA-Sm95 is too upstream and a good agreement with respect to shock position is obtained with DA-Sm97. As previously noted, the shock wave in DA computation is better resolved due to the greater resolution of the unstructured grid in this region. At the intersection between the tunnel wall and the profile, the Cp distributions computed by Sm97 and by WJ-H are equivalent (Figure 4).

The Cp values on the upper wall, at symmetry and at the wind tunnel side, respectively in Figure 5 and Figure 6, are reasonably well predicted in ALA WJ-H and DA Sm97 computations. It is worthwile to note however, that different outflow pressures were imposed in the two computations, so introducing an element of uncertainty in the comparison of results. It is demonstrated anyway that the effect of the presence of wind tunnel walls was properly taken in account, as both computations modelled the incoming boundary layer (Alenia computation) had only a small influence on results. Similar considerations can be drawn looking at the Cp distributions on the lower wall, in Figure 7 and Figure 8, respectively.

Looking at the Cp contours on the upper wing surface in Figure 9, the two-dimensionality of the flow from about a quarter span to midspan is evidenced. A stronger effect due to shock-boundary layer interaction on the side wall is present in the DA result.

From wall friction lines in Figure 10 no shock-induced separation is evidenced over the whole wing span, while a small effect of the interaction of wing and wall boundary layers is evidenced in the trailing edge region. Both computations predict a separation bubble to be present at the trailing edge, in the interaction region between wing and wall boundary layers.

Finally, in Figure 11 a comparison is made between the computed longitudinal velocity and the LDA measurements on a traverse at x=40 mm after trailing edge, y=165 mm from the symmetry plane.

A shift in the wake position seem to indicate that the experimental and numerical effective incidences are slightly different. To match the LDA measurements in wake requires probably a more careful reproduction of experimental inflow-outflow boundary conditions.

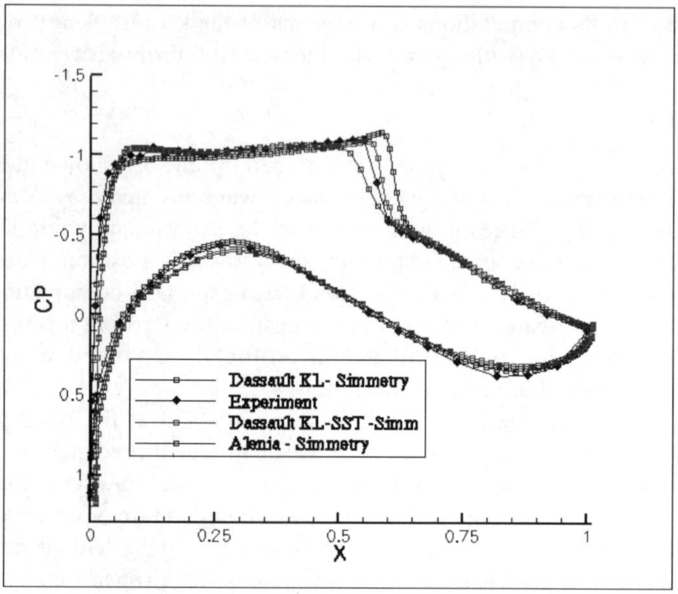

Figure 3 Cp values on profile at symmetry

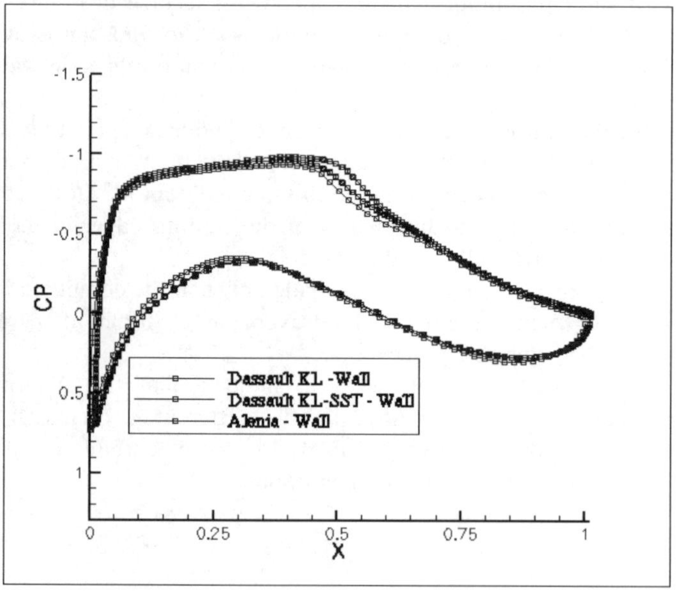

Figure 4 Cp on profile section at intersection with wind tunnel wall

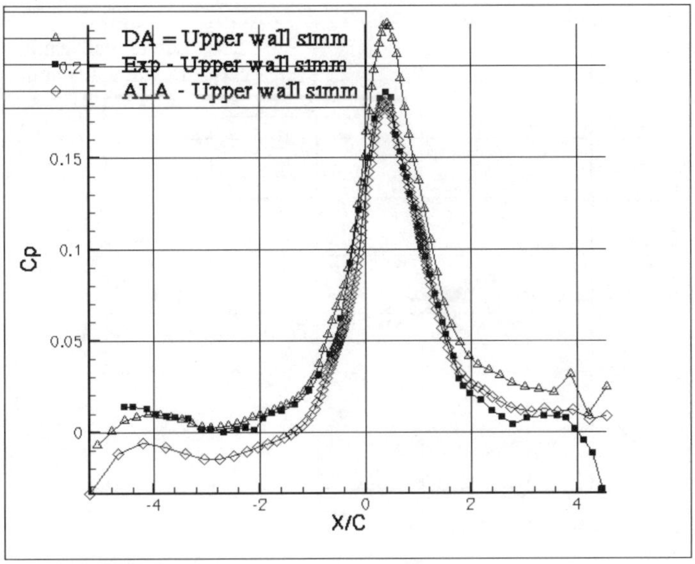

Figure 5 Cp on upper wall at symmetry plane

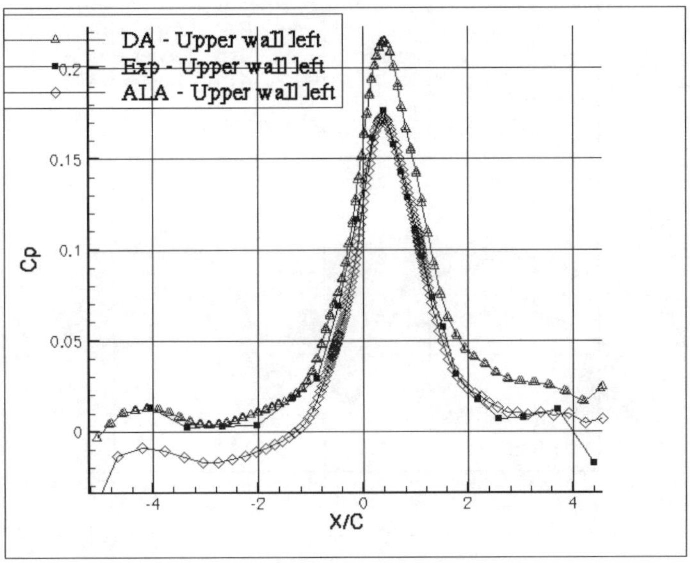

Figure 6 Cp on upper wall at wall intersection.

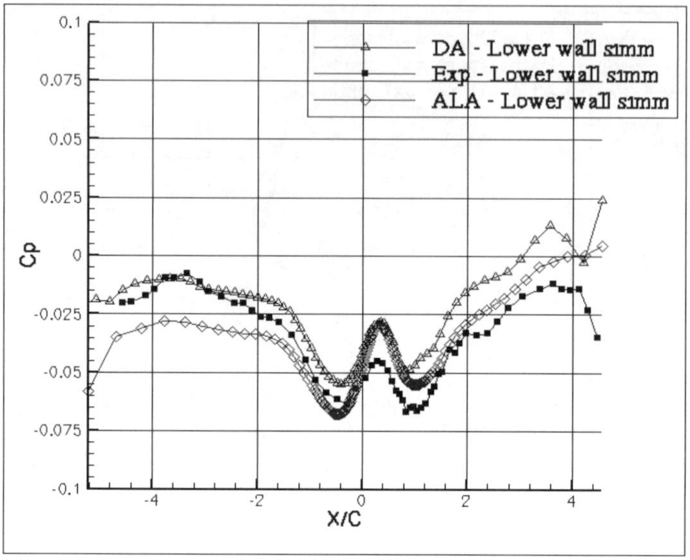

Figure 7 Cp on lower wall, at symmetry plane

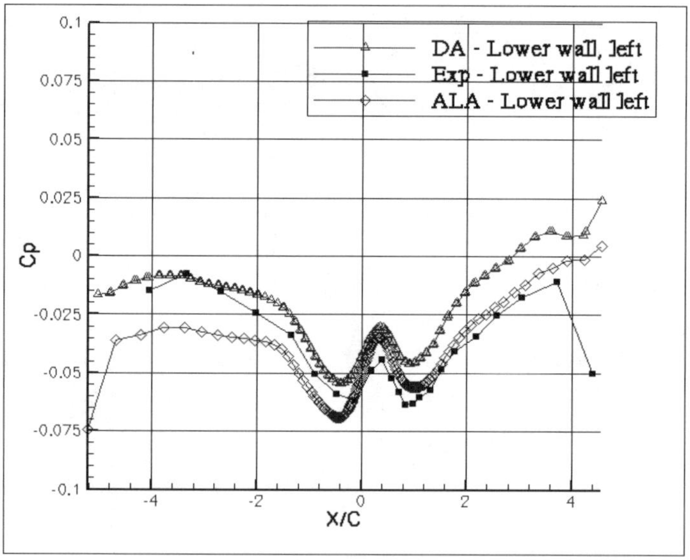

Figure 8 Cp on lower wall, at wall intersection

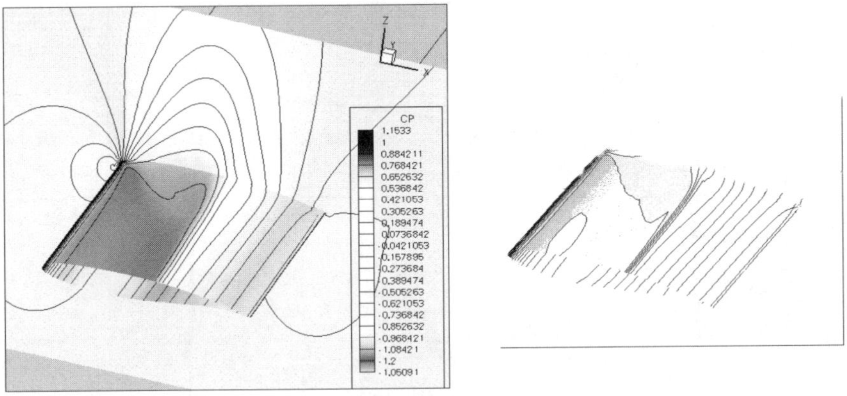

Figure 9 Wall Cp contours

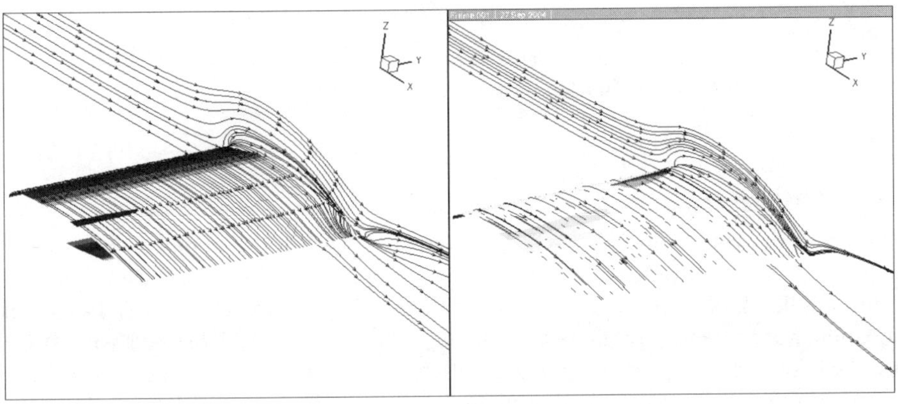

Figure 10 Wall friction lines

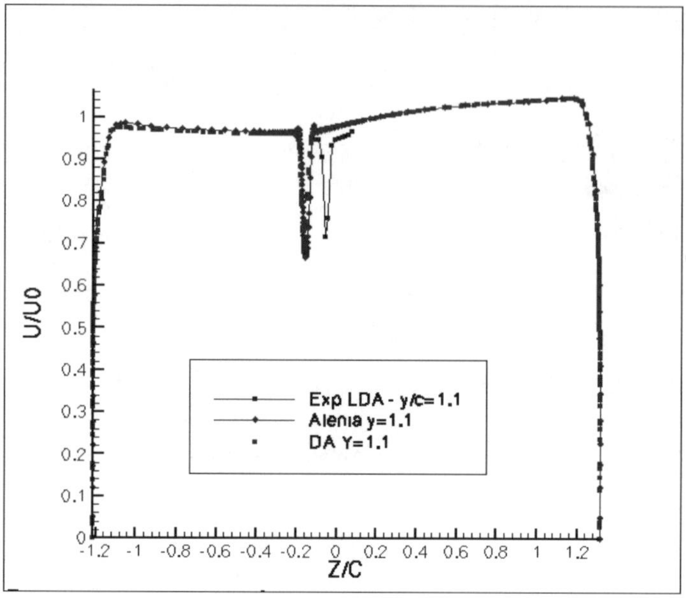

Figure 11 U-velocity component at x=190mm, y=165 mm line across the wake, comparison with LDA data.

5.4 Conclusions

The flow in wind tunnel past OAT15A profile has been computed with WJ-H, Sm95 and Sm97 turbulence models. In the computations made with Sm95 model, a too aft shock position was obtained on the profile at symmetry plane. The shock location was correctly predicted by Sm97, while the results obtained with WJ-H showed a slightly upstream position of the shock wave. Equivalent results, in terms of Cp distributions, were obtained with the two latter models on the upper and lower wind tunnel walls, according with the experimental data. In evaluating the results, it should be taken in account that inflow-outflow boundary conditions were not exactly equivalent in DA and ALA computations, as ALA imposed a constant inflow profile and a different outflow pressure. In both cases, however, a correct boundary layer thickness at the entrance of the test section was obtained, so that the effect of wind tunnel blockage was properly taken in account.

However, for a careful simulation of velocity components in the airfoil wake, and hence a meaningful comparison with the experiment, the experimental inflow velocity and turbulence profiles and outflow pressure distribution should be carefully matched.

6 NACA0012 beyond Stall

Keith Weinman, DLR, Institute for Aerodynamics and Flow Technology

Abstract

Flows with massive separation are challenging for conventional URANS turbulence models. In this study the flow about a NACA0012 airfoil was computed over a range of angles of attack (α) varying from zero to ninety degrees for both URANS and DES models. The detailed analysis at α = 60 degrees is presented in this section. The DES methods are compared against their parent URANS models and it is seen that for this separated flow, DES methods outperform their URANS counterparts.

6.1 Test case description

Flow about a symmetric NACA0012 aerofoil is considered, with the angle of attack (α) being varied while the on-flow Mach number and Reynolds number are held constant. At low to intermediate α, the flow remains attached to the aerofoil. However, as α is increased the flow separates. Traditional RANS and URANS models do not return good engineering estimates in flows with strong separation. By examining variations of model performance at α=60 degrees it is possible to compare the traditional RANS/URANS solutions against those obtained using DES. Note that the airfoil has a chord length c equal to unity, and that the span of the airfoil is equal to the chord length. For this problem the following boundary conditions have been set by all partners:

(1) Far-field boundary at outer extents of grid,
(2) Viscous wall boundary (no laminar region) at airfoil surface,
(3) Periodic boundaries imposed at (x,y,0) and (x,y,c).

All DES computations have been restarted from an initial RANS calculation using an on-flow Mach number M_∞ equal to 0.3 with a Reynolds number (Re) of 100000.

6.2 The list of partner contributions

Table 1 provides a short description of the contributions provided by each partner within this test case. Inspection of the table shows that a range of turbulence models and differing meshes have been used. As noted in the abstract, we consider only the mandatory DES test case at α equal to 60 degrees.

Table 1 Partner contributions to the test case

Partner	DES	URANS	Models	Grid			α (degs.)
				C	M	F	
DLR	X	X	SA,SST,XLES	X	X	X	45,60
IMFT	X	X	SA,KW,KW/OES, KE/OES	X			60
SPTU	X	X	SA, SST	X	X	X	20,40,60
TUB	X	X	SALSA,LLR, CEASM	X	X	X	20,45, 60,90
ICST		X	NLEVM, LEVM	Own mesh			< 20

6.3 Grids

SPTU has provided partners with a set of two-dimensional meshes which were then adapted by the respective partners for their code input formats. Initial meshes were generated with a prescribed span width which could be then later adapted as required by individual partners. The mesh topology that has been used for all DES calculations presented within this section is shown in Figure 1.

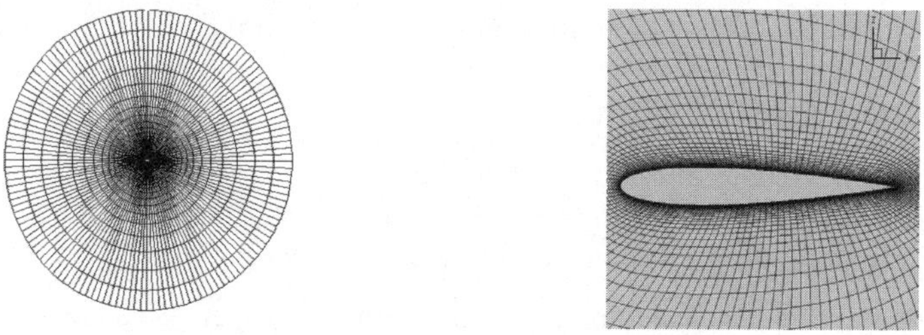

Figure 1 This figure illustrates the grid topology used in the NACA0012 calculations. The grid is structured and some care has been taken to ensure that the grid is sufficiently fine in the separated regions of the flow so that the LES mode of the DES is active in these regions.

Guidelines suggested by (Spalart, 2001) for the construction of DES grids suggest that additional grid resolution may be required in the region of the wake so that an acceptable level of unresolved wake turbulent kinetic energy is found. Using the definitions provided by (Pope, 2001), where VLES is defined as the situation where the filter and grid are too coarse to resolve 80% of the turbulent energy, it is possible that the LES mode of the DES may behave as an unresolved VLES model in some regions of the aerofoil wake. The filter is not explicitly defined in the DES models implemented in this work, so an alternative point of view is that the LES mode of the DES behaves as a RANS model with an eddy viscosity whose eddy viscosity is too low to satisfy the correct scaling of the RANS model. Mesh details are noted in Table 2. From (Spalart, 2001) the number of points required to perform a wall-resolved 3D LES at Re=100,000 is of the order of 10^{14} grid points. In comparison, the number of points used in the DES method is several orders of magnitude smaller. No attempt was made to refine the meshes in the span-wise direction and grid convergence cannot be expected. Proper refinement in the span-wise direction should result in approximately O(1-10 million) points per grid, still considerably less than that required for full DNS or a wall-resolved LES.

Table 2 Details of the meshes used in this study. Note that in following tables no distinction is made between fine meshes used (fine meshes are designated as F) since no consistent refinement strategy has been adopted

Grid	NX	NY	NZ	Num. Pts
C	141	61	24	206424
M	141	81	24	274104
F1	209	101	24	487224
F2	141	101	32	455712
F3	164	82	31	416888
F4	141	81	30	342630

6.4 Discussion of results at α=60 degrees

Detailed discussion of results at this condition is presented in this section. Computed lift and drag coefficients are examined, and some analysis on the frequency content embedded within the lift and drag signal is performed. In order to compare lift and drag, computed values are compared against experimental data obtained from (McCroskey, 1987). The percentage error in the computed results is defined by the following equation

$$E(x,m) = 100\frac{|x-m|}{m}, \tag{1}$$

thus writing E(x,m) = 20 states that the error in the estimation of x in comparison to the experimental estimate m is estimated to be 20 percent. Upper and lower error bounds are not presented (the above measure of error is sufficient for the purposes of this section).

6.4.1 Quantities of interest, definitions, and analysis methods

The principal engineering parameters of interest in this section are the integral values of lift and drag. All flows being studied are unsteady and a statistical analysis is required in order to compare data. In order that statistically significant data is generated, the length of data must be sufficient to ensure meaningful post-processing of data. Measurements may have a long term drift and it is useful to remove any linear trend in the data. To this end, the least-squares fit of a straight line (or composite line for piecewise linear trends) is computed and subtracted from the data. A common procedure to avoid aliasing errors due to an inadequate sample length is to use low-pass filtering techniques prior to the sampling process. The frequency cut-off has been set to the Nyquist frequency (which should guard against aliasing problems). The Nyquist frequency is defined as the inverse of twice the physical time step used in the calculations. The majority of partners have used a time step of 0.025 T^*, where T^* is the convective time scale based on free-stream velocity and airfoil chord length.

6.4.2 Integral values of lift and drag

Comparison is presented on the basis of the grids used in the calculations. In Table 3 the mean values of the integral lift and drag coefficients at 60 degrees α are presented for the coarse (C), medium (M), and fine grids (F). Following the nomenclature selected in Chapter 1, the mean of a time series x(t) is denoted as $\mu(x)$ (here the functional dependence x(t) will not always explicitly stated). Similarly, the standard deviation of the series x(t) is denoted as $\sigma(x)$. The mathematical definitions for the mean and standard deviation can be found in texts such as (Papoulis, 1984). All computations return mean lift and drag coefficients in agreement with experimental data. All errors are smaller than ten percent in magnitude but no single model/implementation combination returns the best agreement in both lift and drag. Series with lower variances tend to result in poorer predictions since the turbulent viscosity is probably too high.

Apart from specific model implementations, there is no evidence of a general grid convergence trend. This is expected as no consistent grid refinement study in the span-wise direction was undertaken. Some specific model/implementation combinations have shown improvement. For example the TUB implementation of the SALSA model showed some improvement on the medium mesh but such improvements are considered circumstantial within the present work.

It is of interest to examine how the base URANS models compare against their DES variants, as seen in Table 4. In all cases, the DES enhancements to the base turbulence models improve the predicted values of mean lift and drag. While there are subtle differences in the implementation of each model, it is clear that the predicted integral values obtained using the DES all lie within 10 percent of the experimental measurement. Standard URANS models give results which lie outside the ten percent error margin for both lift and drag. No overlap is apparent between the URANS models and their derived DES variants.

Table 4 gives additional information regarding the frequency content embedded in the solutions. The Strouhal number is defined as $St = fL/U_\infty$, where L is the chord dimension normal to the on-flow direction ($L = C \sin(\alpha)$), f is the frequency and U_∞ is the free-stream velocity. (Schewe, 1983) noted that the

Strouhal number was approximately 0.2 over a wide range of Reynolds numbers for airfoils with significant vortex shedding.

Table 3 A comparison of integral lift and drag coefficients returned by different DES implementations using different base meshes

Group	Model	Grid	C_{DES}	$\mu(C_l)$	$\sigma(C_l)$	$E(C_l,m)$	$\mu(C_d)$	$\sigma(C_d)$	$E(C_d,m)$
SPTU	SA	C	0.65	0.967	0.25	5.11	1.6	0.42	3.03
TUB	SALSA	C	0.65	0.994	0.32	8.05	1.664	0.50	0.85
TUB	LLR	C	0.78	0.932	0.20	1.30	1.571	0.31	4.79
TUB	CEASM	C	0.65	0.965	0.19	4.89	1.628	0.30	1.33
BD	SA	C	0.65	0.99	-	7.61	1.65	0.00	-
BD	CP	C	0.65	0.99	-	7.61	1.68	1.82	-
DLR	SA	C	0.45	0.982		6.73	1.65		0.00
DLR	SA	C	0.65	0.875	0.16	4.89	1.453	0.25	11.49
TUB	SALSA	M	0.65	0.927	0.24	0.76	1.65	0.37	0.00
TUB	LLR	M	0.78	0.938	0.21	1.95	1.57	0.34	4.85
TUB	CEASM	M	0.65	0.977	0.19	6.15	1.64	0.31	0.61
IMFT	SA	M	0.65	1.2	-	30.43	1.8	-	9.09
TUB	SALSA	F	0.65	0.987	0.39	7.28	1.66	0.66	0.61
TUB	LLR	F	0.78	0.986	0.43	7.17	1.654	0.73	0.0
TUB	CEASM	F	0.65	0.924	0.19	0.4	1.55	0.30	6.06

Table 4 Comparison of DES and URANS predictions at $\alpha = 60$ degrees

Model	Grid	URANS		DES		Strouhal Number	
		$E(C_l,m)$	$E(C_d,m)$	$E(C_l,m)$	$E(C_d,m)$	URANS	DES
SALSA	C	42.7	35.4	8.04	0.84	0.12	0.14
SALSA	M	31.2	25.2	5.69	0.121	-	0.14
LLR	C	34.2	30.5	1.30	4.79	0.11	0.16
LLR	M	30.3	22.5	1.25	4.82	0.13	0.16
LLR	F	16.8	10.7	7.19	0.24	0.13	0.16
CEASM	C	34.9	31.1	4.93	1.29	0.16	0.14
CEASM	M	16.8	10.7	6.23	0.65	0.13	0.15
CEASM	F	19.0	31.2	0.39	5.75	0.11	0.16
SA	C	48.9	27.2	6.73	0.0	0.09	0.15
SA	C	50.0	40.61	7.61	0.0	0.13	0.15
CP	C	47.83	41.82	7.61	1.82	-	0.14

Measurements made by (Sohankar et al., 1996) have shown that the shedding frequency of vortices about a rectangular cylinder corresponds to a Strouhal number of about 0.14. It is then expected that vortex shedding from an airfoil at high α should lie in the range 0.2-0.14.

Examination of Table 4 shows that the Strouhal numbers in the DES series lie within the range expected, while those predicted by the URANS calculations are, in general, lower. The calculation of embedded frequency information is discussed in the following section.

6.4.3 Real time characteristics of lift and drag

Figure 2 illustrates a typical time series obtained for lift for the SA model in URANS and DES modes. The lift coefficient returned via the RANS calculation appears deterministic in nature and by inspection several frequency components are present. Phase differences between these components appear to be relatively constant. In comparison, the DES calculation does not (appear to) support any simple deterministic signal.

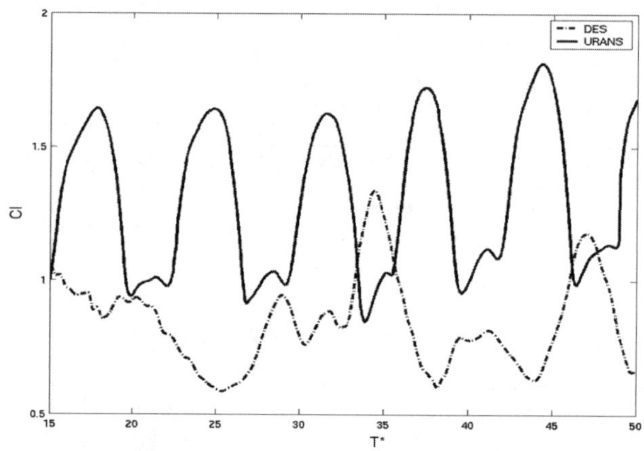

Figure 2 A comparison of the lift coefficient C_l computed using the URANS and DES branches of the SA turbulence model

(Swalwell, 2004) suggested rotation of the aerodynamic co-ordinate system by α into normal and tangential components as follows:

$$C_t = C_l \sin(\alpha) - C_d \cos(\alpha), \tag{2}$$

$$C_n = C_l \cos(\alpha) + C_d \sin(\alpha). \tag{3}$$

A better estimation of the frequencies associated with vortex roll-up and shedding is then obtained as is shown in Figure 3, where the normal (Cn) and tangential (C_t) force components are compared against the original data. Evidence of a shedding process in the C_t component is evident, with a frequency beat typical of vortex shedding present. The original time series shows a more chaotic signal with no well-defined shedding processes being evident.

It is interesting to investigate the frequencies which are embedded within the computed lift signals. In Section 6.6.2, it was noted that a Strouhal number in the range 0.2-0.14 should be expected. Examination of the spectral power density for the computed time series provides additional information about the signal frequency content. It is useful to examine the spectral density for both the original and rotated force systems. Figure 4 illustrates the spectral density for the original lift component.

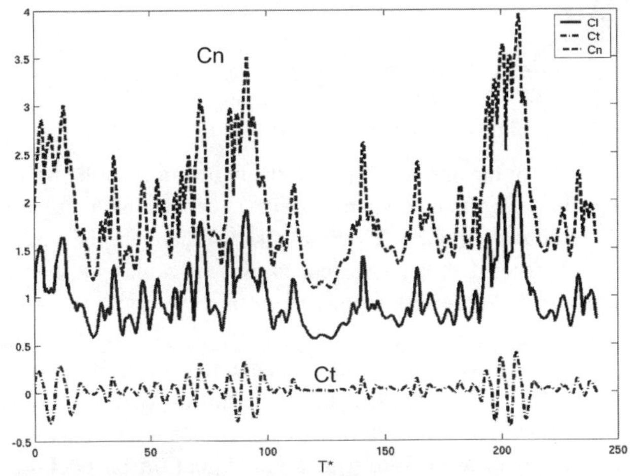

Figure 3 Rotation of the lift and drag forces into a force system orthogonal to the chord of the airfoil can help in the identification of any shedding processes which may be present. Here a clear frequency beat is observed in the force component tangential to the airfoil. Note that T* is time normalised on the convective time scale of the flow

It is reasonable to assume that higher frequency events have more energy per unit mass of fluid, however a significant proportion of the spectral energy is found at frequencies much lower than those associated with vortex shedding processes. The picture is then that vortex shedding is intermittent, since the more energetic vortex shedding events appear to have only a small contribution to the total energy spectrum. The spectral power density for the transformed system is shown in Figure 5. While examination of the tangential force component in the rotated system clearly indicates the presence of frequencies that may be attributed to vortex shedding, the additional information regarding the intermittency of the vortex shedding process cannot be easily deduced in the power spectrum of the rotated co-ordinate system. It is apparent that the frequencies predicted by the Strouhal relationship are present with very little energy being seen at lower frequencies.

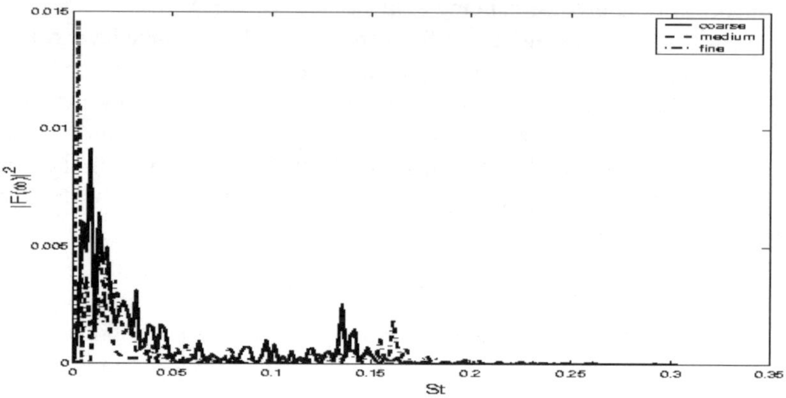

Figure 4 Examination of the Spectral Power Density function for the lift component in the aerodynamic coordinate system exhibits a considerable noise at frequencies less that the Strouhal number. The results shown here are for a SA model solution. Some influence of grid refinement is evident as can be seen from the higher Strouhal number observed on the fine mesh

Examination of individual results suggests subtle differences in the time series returned for the different model and implementation combinations which are not fully understood at present. For example, the persistence of a frequency beat in the time series appear to be more likely in a two-equation based DES or a NLEVM based DES than in a DES-SA. This observation suggests that the sub-grid scale model may inherit some of the weaknesses of the base RANS model. This is a point for further study.

6.4.4 The influence of vortex shedding on the integral flow characteristics

It can be expected that the detachment of vortices from the leading and trailing edges will influence measured values of integral values of lift and drag. By observing the development of the unsteady solutions it is possible to correlate vortex shedding processes with the integral values of lift and drag. Examination of Figure 6 shows that the detachment of a leading edge vortex from the airfoil corresponds to a condition of maximum lift. Similarly, the detachment of the trailing edge vortex appears to be correlated with a point of minimum lift.

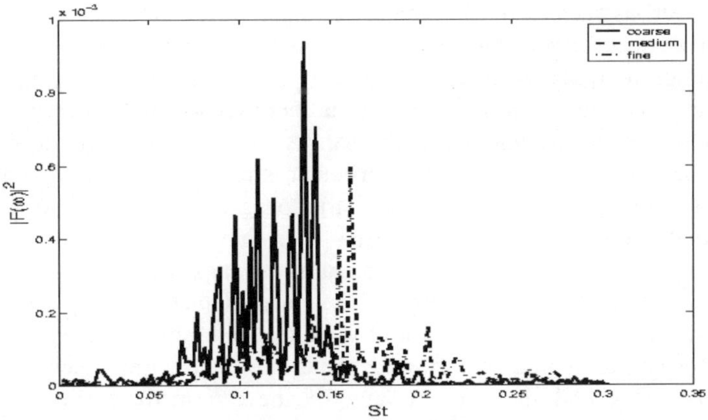

Figure 5 Low frequency noise contamination is reduced when the tangential force component in the rotated system is examined. Frequency resolution is improved with grid refinement however the span-wise resolution between the three calculations shown has remained unchanged. Note that intermittency of the vortex shedding process cannot be deduced from this figure

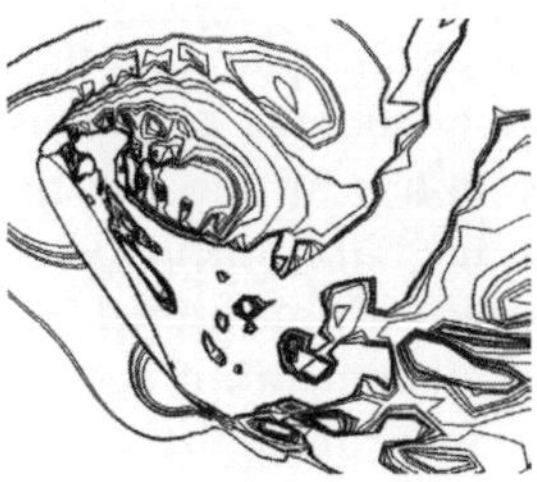

a) Roll-up and separation of leading edge vortex (Maximum C_l)

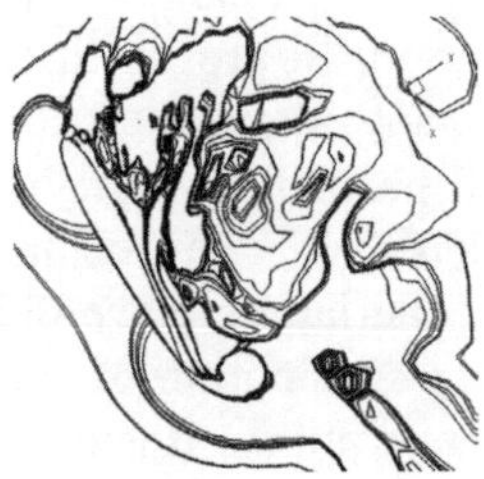

(b) Intermediate structures persist for some significant time

(c) Roll-up and separation at the trailing edge (Minimum C_l)

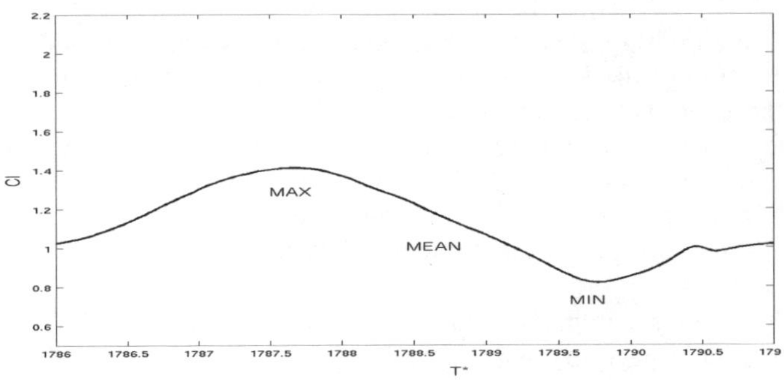

Figure 6: This figure shows the correlation between the integral lift values and the detachment of vortices from the leading and trailing edges of the airfoil

However, as seen from the time series data, vortex detachment events are not the most frequent events observed, and the flow state remains in an intermediate condition representative of the state shown in Figure 6(b) for a significant fraction of time. It is important to note that the topology of the lift curve shown in Figure 6(d) is relatively simple – the lift changes as a monotonically decreasing function of time until the minimum lift coefficient is reached, at which point the lift curve changes the sign of its slope and starts to again increase monotonically from its minimum value. The lift time series also shows non-monotonic behaviour (not shown in figure) between the minimum and maximum lift values. During these periods of the flow it is not clear if the vortex detachment processes correlate well with the maximum and minimum lift values, and initial investigation suggests that the correlation is poorer during these periods. The vortex interactions in the flow are complex and the role, if any, of the intermediate state in the shedding processes is not understood.

6.4.5 Velocity structure investigations

While spectral plots can show frequencies present (in a signal) some information is lost, since the Fourier transform removes the local dependence on a coordinate by integration over the full coordinate space. Additional information about the unsteady processes occurring can be obtained from the probability distribution functions for the series. Definitions of probability distribution functions and their derivation can be found in many standard textbooks, for example (Papoulis, 1984). A study of the resolved probability distribution functions for the velocity components can illustrate additional information about the DES method. At present only the single point Eulerian velocity field statistics are measured and used to compute the probability distribution of each velocity component and the pressure at a number of predefined points. It is more useful to examine the second order velocity structure function $\langle u_i' u_j' \rangle$, however this is costly in terms of time and resources and so, at present, only single point statistics are considered. In order to facilitate examination of the Eulerian velocity density function, several points are chosen in the field and the velocity and pressure at these points are then sampled.

In order to simplify statistic comparisons, velocity data are normalised by the standard deviation of the U velocity component at the leading edge of the section. Note that the span-wise locations at 25 %, 50% and 75% span are chosen for convenience. No attempt has been made in this work to study the influence of span-wise distance (distance between periodic boundaries) and/or span-wise grid resolution on computed solutions. Comparison at points P2 and P3 will be presented in this section. In all respects the data traces at the other span-wise locations are similar.

Figure 8 shows a comparison of the U velocity components obtained for the DES-LLR model (fine grid) and two different implementations of the DES-SA model computed by SPTU(1) and the DLR(2) (both using coarse grids) at position P2. Note that the velocity time-series derived from both SA-DES models are similar and that both differ from the DES-LLR trace, which appears to resolve some higher frequency component.

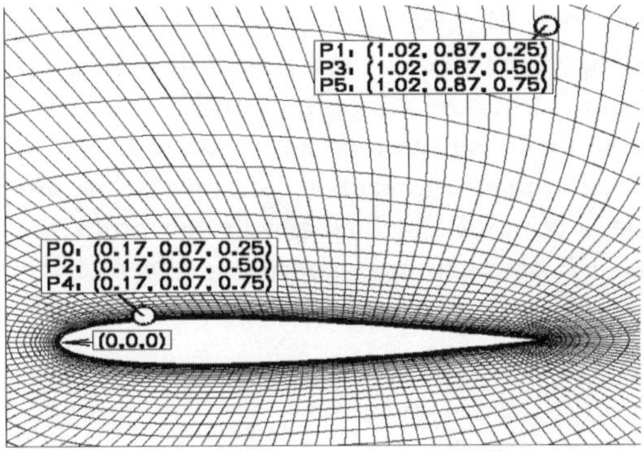

Figure 7 The figure shows the sample point locations chosen for further statistical analysis. Note that coordinates given in the figure [Point id.: (x,y,z)]are scaled on the airfoil chord length and that the span of the airfoil (z coordinate) is equal to its chord length

The SPTU code is a cell-centred structured method, while the DLR method uses a dual grid unstructured method (Schwamborn et al., 1999). Despite significant numeric differences, the resolved time series are similar. However the DES-LLR calculation has been computed on a finer mesh and this is likely to be partially responsible for the higher frequencies embedded in the DES-LLR result.

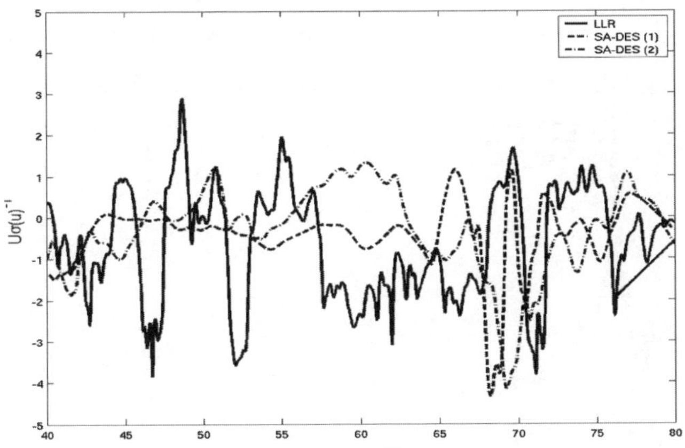

Figure 8 Comparison of the resolved U velocity component obtained with different DES model and implementations

The DES methodologies used in this work apply a LES scheme to regions which are sufficiently far away from walls where the local turbulent Reynolds number is high. Since the low-order statistics of LES are usually insensitive to the choice of sub-grid scale (SGS) model (Wilcox, 1998), the choice of sub-grid scale model should not significantly influence the low-order statistics returned by the LES part of DES. The probability distribution functions for different DES models should be similar at the same point in the LES portion of the flow. Using data sampled at the probe points, probability distributions are obtained by normalising the histogram of events by the total number of events (ensuring that the integral of the histogram over the event space is unity and thus conserving probability). Since the sample size for the DES-LLR calculation is approximately three times the length of the DES-SA results, the DES-LLR statistics are cleaner and contain less noise.

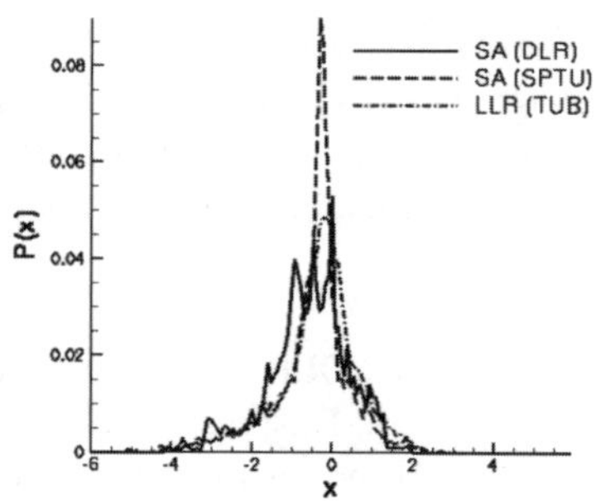

(a) U component: Drift velocity is negative.

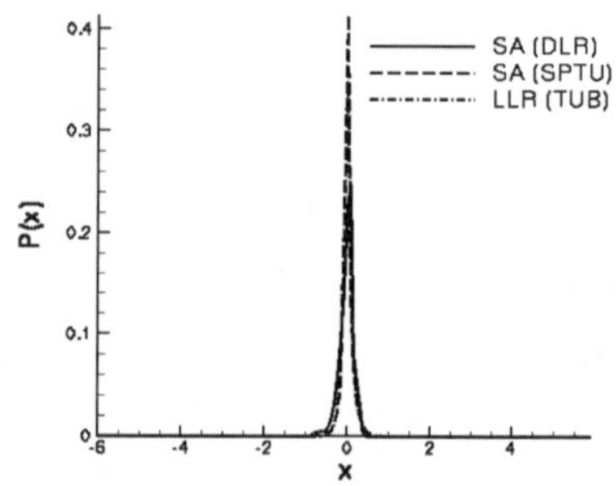

(b) V component: Small deviations in velocity space due to wall influence. The distribution is similar to a delta function.

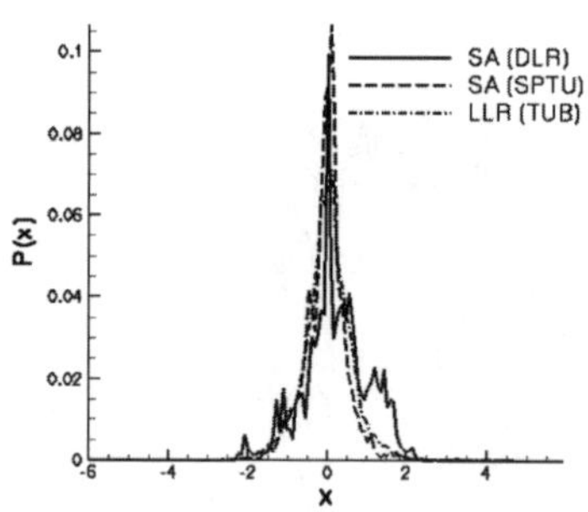

(c) W component: This velocity component appears to have a normal distribution.

Figure 9: Probability distribution functions of the velocity field components at point P2

Figure 9 illustrates the computed probability density functions for the U, V and W velocity components at position P2. Examination of this figure shows some interesting features. Firstly it is seen that the magnitude of fluctuations in the velocity space are very similar for all calculations, implying that the flow dynamics obtained with the different model/grid combinations are equivalent. The mean of the U component (shown in Figure 9(a) is negative. This must be expected for a point lying inside the region of influence of the suction peak near the leading edge of the airfoil. Here a particle with no inertia will exhibit a mean drift velocity towards the position of the suction peak. Figure 9(b) shows that the probability distribution of the V component (the wall-normal velocity) is normally distributed and that the influence of the wall in constraining the magnitude of the velocity fluctuations is significant. The W velocity component (span-wise component), shown in Figure 9(c), may be normally distributed, however more data points are required to confirm this. Since at least one of the velocity components has a non-normal distribution, the turbulence field cannot be isotropic.

Figures 10(a) to 10(c) provides a comparison of the probability distribution functions at position P3. It is apparent that since the range of velocity fluctuations is much greater in the wake region, a significantly larger number of points are required for the statistical analysis in comparison to the point P2. Due to statistical noise it is difficult to draw any solid conclusions about the probability distribution for the U component of velocity, however a clear negative drift velocity is observed for V which can be justified by physical argument. The span-wise component (W) demonstrates a very similar distribution in velocity space as seen in Figure 9(c). Thus the dynamics of the span-wise velocity component at P2 and P3 do not appear to differ greatly, which is an unexpected result, and might suggest insufficient span-wise resolution. As at point P2, the flow is anisotropic.

It is interesting to note that a close correspondence between the DES-LLR and the DES-SA statistics is seen which suggests that the influence of the LES part of the DES model is independent of the underlying sub-grid scale model to first order. Note that since the probability distribution functions for the velocity components are not Gaussian, the turbulence fluctuations cannot be isotropic in nature so that the implications of calibrating DES against an isotropic turbulence must be examined in greater detail.

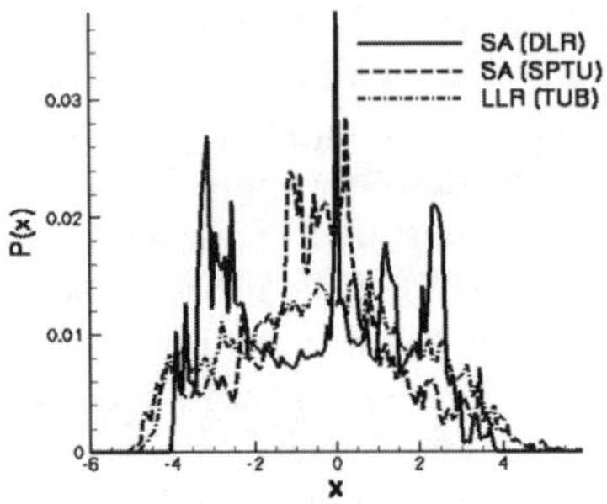

(a) U component: There is insufficient data to obtain a clean distribution function.

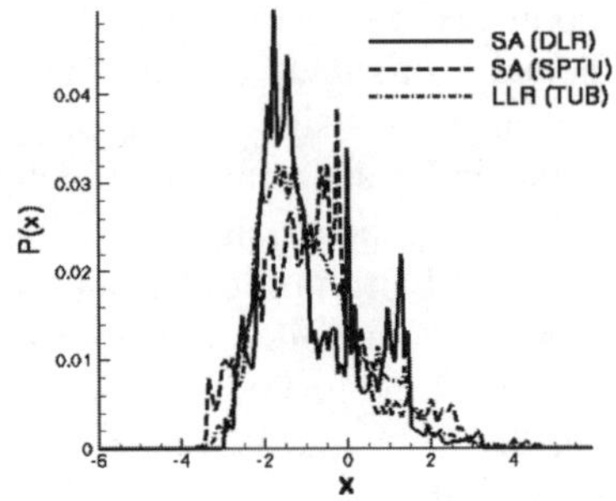

(b) V component: This component demonstrates a negative drift velocity.

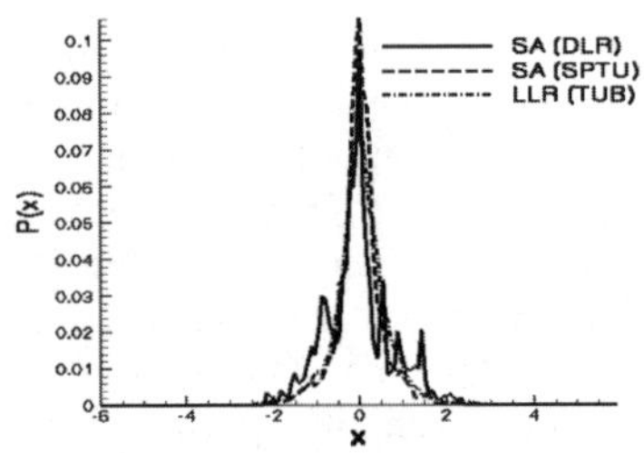

(c) W component: The statistical distribution is similar to that of W at P2.

Figure 10: Probability distributions of velocity components at point P3

6.5 Conclusions from the test case

(Shur et al., 1999) have shown that at low α, where the boundary layer remains fully attached, a DES method combined with a properly chosen grid yields results which are no worse than those returned by a RANS method. It is when the boundary layer is fully separated that RANS methods generally fail. The results shown in this test case show that the DES methodology can be adopted successfully to complement RANS models for the successful computation of flows with significant separation. Significant improvements in the prediction of integral quantities against measurement are obtained, and comparison of the first order statistics computed using DES and URANS suggest that the DES results are superior and at least qualitatively correct in their unsteady dynamics.

7 MRTT

J.J. Guerra, EADS-CASA

Abstract

This chapter summarises the MRTT test case, including the description of the case itself and the model, as well as computations and results obtained by EADS-CASA. The test case comprises three different test conditions in transonic regime. The computations for each one of the three conditions have been made using two different turbulence models, the SALSA and the k-g model.

7.1 Introduction

In the context of the FLOMANIA Project, one must look on this test case from an industrial point of view. It represents a complex test case with complex flow and complex geometry in a real context of an industrial company. Firstly, the used wind tunnel tests are involved in a wider test campaign for other purposes; then, the results available hardly will serve for exhaustive validation of turbulence models. Secondly, and complementarily, the interest as industrial company is to explore and to calibrate the capabilities and reliability of the methods being used intensively at this moment in time, some of them implemented during the FLOMANIA Project.

7.2 Test case description

MRTT is an acronym of Multi-Role Transport Tanker. It is a civil aircraft that has been modified for air-to-air refuelling purposes. The main aerodynamic modification is the inclusion of a Pod-Pylon geometry in the lower side of the wing, near the tip, which induced flow separations in some combinations of pylon designs and flow conditions. The model and the wind tunnel tests presented here, correspond to an early status of the geometry in the design.

7.2.1 Configuration

The geometry model used for computations is displayed in Figure 1. It represents a clean wing with the added pylon and pod and the outboard flap track fairing.

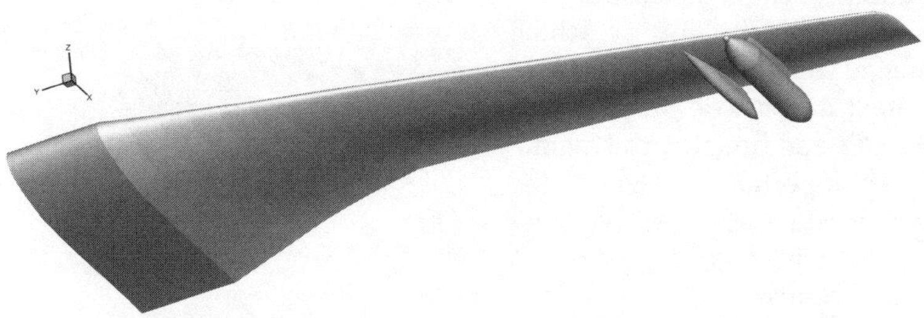

Figure 1 Geometry model (wing+flap track fairing+pylon+pod)

This configuration provides complete geometric detail in the zone of interest (the pylon) and in the vicinity. At the same time the rest of geometry is cleaned, the engine and fuselage are removed and no other flap track fairings are included. The effect of fuselage on lift is provided replacing it by straight wing. This simplification is made in order to avoid an unnecessary large mesh.

7.2.2 Flow Conditions

Typical flow conditions for commercial aircrafts in cruise regime are characterised by transonic Mach number, high Reynolds number and low angle of incidence. For these flight conditions, the flow is attached, but the changes of geometry for this configuration produced detached flows as it was shown by the wind tunnel test visualizations.

For this test case, three flow conditions were selected and they are reflected in Table 1. The three settings showed detached flow close to the wing/pylon junction

Table 1 Flow conditions

Mach	0.78	0.80	0.84
AoA	1º	1º	0.2º

The reason for using the three cases is that the flow separation may be induced by different causes. On one hand, the shock due to the presence of the pylon is stronger for higher Mach number. On the other hand, the angle of attack (AoA) affects also to the shock intensity, but in addition, higher AoA increases the cross flow in the lower side of the wing, particularly in outer part, inducing separations in the outboard side of the pylon. In the inverse sense, lower AoA worsens the offloading effect (the concavity of the profile at the rear part).

7.2.3 Experimental data

The wind tunnel tests for aircrafts take shape in very extensive campaigns covering the flight envelope for different configuration settings. Usually, the results are provided in terms of global forces, surface pressure and flow visualizations. That is the case for the present model, of which experimental data were obtained in such context and whose purpose was to obtain the additional aerodynamic loads It is important to remark the lack of measurements of velocity profiles, which would help to validate the results.

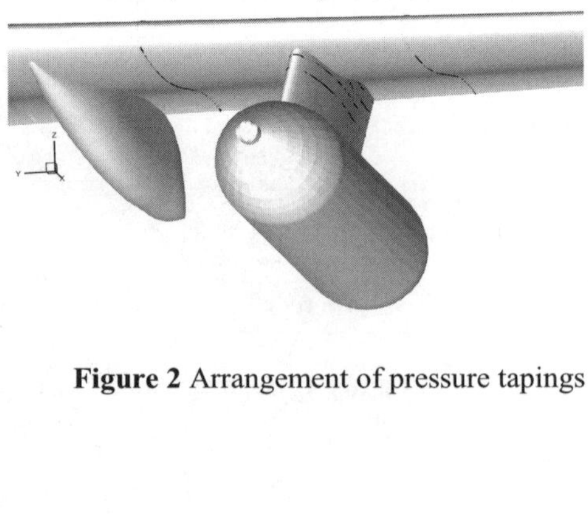

Figure 2 Arrangement of pressure tapings

The wind tunnel model is a complete model of an A310 with a scale of 1:9.5. The tests were performed at the Onera S1 wind tunnel.

Fig. 2 shows the arrangement of pressure tapings in different sections. There are four sections at the pylon, close to the wing. The sections with tapings on the wing were pre-existing to the inclusion of the pod/pylon. Unhappily, the two sections nearest to the pylon are out of the detached flow zones.

Fig. 3 includes pictures showing the detached flow. They cover two of the three cases computed. The flow patterns for M=0.78 and M=0.80 are very similar.

Figure 3 Oil flow visualizations at the wind tunnel for M=0.84 (upper) and M=0.78(lower). Left pictures for the inner side of the pylon and the right ones for the outer side

7.3 Computations

7.3.1 Mesh

The computational mesh is structured multiblock. Only one side of the model is meshed and computed, assuming symmetry. A layer of blocks in C-C topology is disposed in a glove fashion around the wall. The statistics of the mesh are in table 2. Fig. 4 shows detail of the surface mesh in the region surrounding the pod and pylon.

Table 2 Statistics of the Mesh

Blocks	88
Nodes	$2.69\text{x}10^6$
Wall Nodes	37256

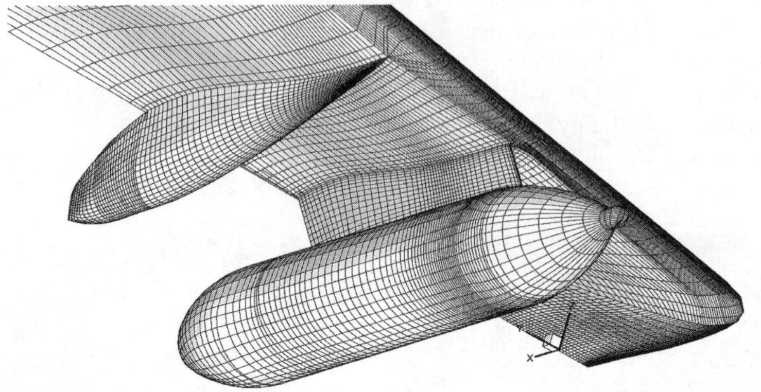

Figure 4 Surface mesh: Detail of .the outer wing with the pod-pylon and flap track fairing

Preliminary computations were made for the case of Mach=0.84 in order to extract information of Y^+ at the first cell layer over the wall. After that, the mesh was adapted in order to obtain values of Y^+ under 1.0.

7.3.2 Computational Parameters

The results have been obtained using the EMENS code and using two different turbulence models, k-g and SALSA. Computations have been carried out using four multigrid levels. Final iterations in single grid were performed but they did not affect the solution (only the convergence level was improved slightly).

The computations try to reproduce the wind tunnel test conditions. The Reynolds number for computation is set to 6.8×10^6 (based in the mean aerodynamic chord of the wing). The transition is prescribed through the input file to the code at the same position than the strips of the wind tunnel model: For the wing, the transition is fixed at 15% of local chord in both sides (upper and lower) and for the pylon the transition is forced at 8% of their chord (slightly upstream of that of the wing). The Mach numbers and angles of attack are those reflected in table 1.

7.4 Results

In the following sections, the results are presented in terms of pressure plots. For each case (M and AoA) there is a set of four plots representing the C_P for the respective four sections on the pylon where tapings are available (see Fig 2). The plots start in the section away from the wing and end in the closest one (note that the two sections near the wing do not intersect the whole pylon).

7.4.1 Mach 0.78

The computations using the SALSA and the k-g produce similar results, and both turbulence models reproduce attached flows. At the same time, the computational results contrast with the experimental ones because of the appreciable differences in terms of pressure, as it can be seen in Fig. 5. The experimental results present higher levels of negative values; the suction peaks at both sides of the pylon are higher for the experiment and in addition, at the inboard side, that peak is forward. In general this describes a more adverse situation of the flow around the pylon for the experiment than for computations.

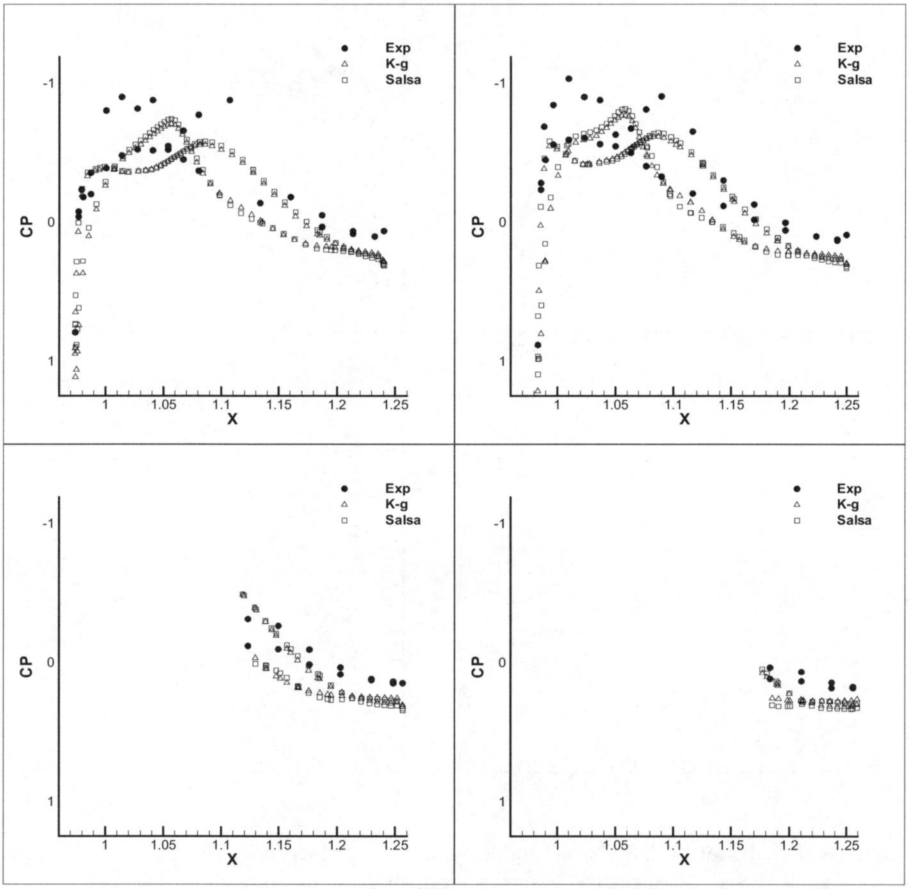

Figure 5 Results for M=0.78, α=1° and Re=6.8×10^6 C_P plots comparing results of k-g and SALSA turbulence models vs experiment at the pylon sections

7.4.2 Mach 0.80

As it can be seen in fig 6, the results for this test are very similar to those of M=0.78. The more adverse conditions at the pylon exhibited by the experiment can be, partly, the cause of having detached flow in experiments but not in computations in either tests (M=0.78 and M=0.80). Although the detachments may induce a different pressure distribution around the pylon and wing in this region, the exhibited differences between pressures seem to be excessive.

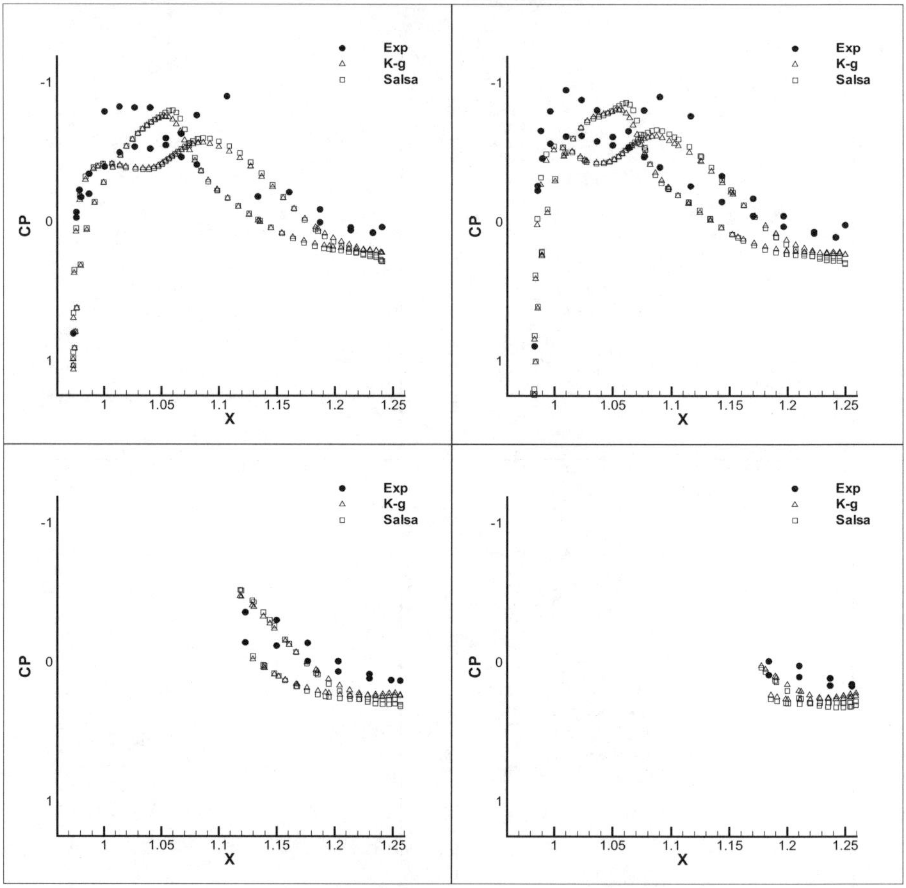

Figure 6 Results for M=0.80, $\alpha=1°$ and $Re=6.8x10^6$ C_P plots comparing results of k-g and SALSA turbulence models vs experiment at the pylon sections

7.4.3 Mach 0.84

The results for this test (see fig 7) compare in a radically different manner to those of the other two tests. Firstly, the computations using the SALSA model do provide detached flows while the k-g model do not. Despite the differences between the three sets of results for this test at the inboard side of the pylon, the pressure values compare fairly well at the outboard side for all them, at least in the 75% of chord forward, where the flow is attached. At the inboard side, the SALSA model overpredicts the detachment with a forward location of the separation point, and hence a large region of detachment. As in the previous tests, the suction at the inboard side is further forward in the experimental results.

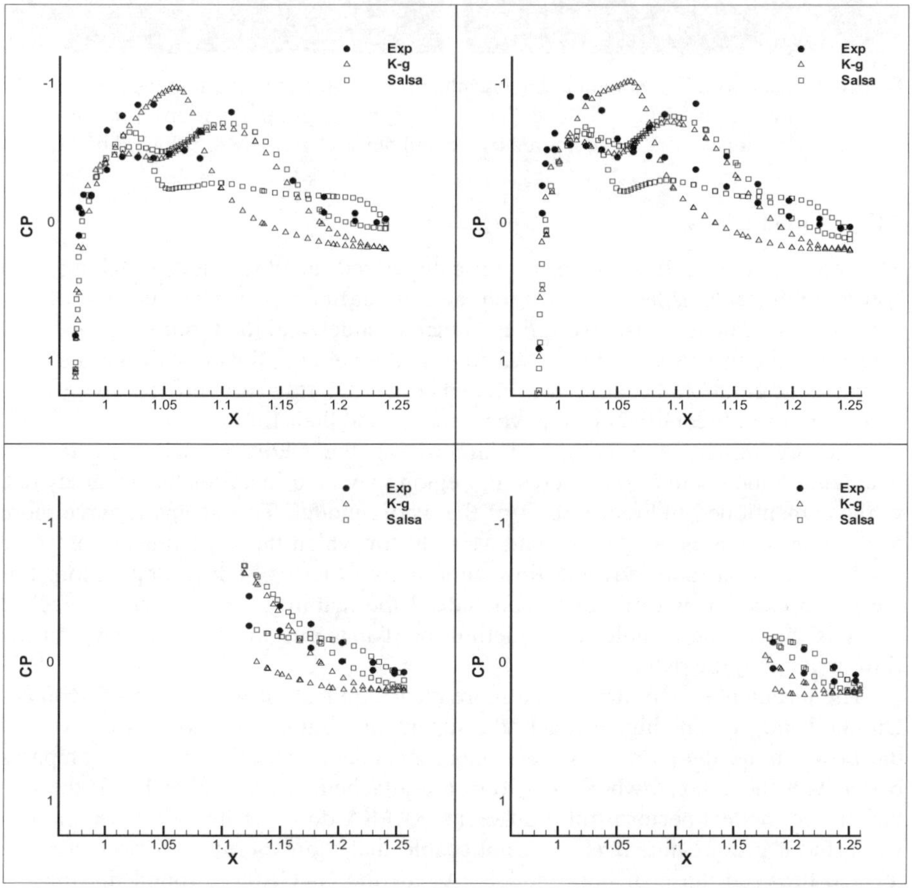

Figure 7 Results for M=0.84, α=0.2° and Re=6.8x10^6 C_P plots comparing results of k-g and SALSA turbulence models vs experiment at the pylon sections

Fig 8 shows a 3D representation of the solution provided by the SALSA. The detached flows are involved in a big bubble at the inner side of the pylon.

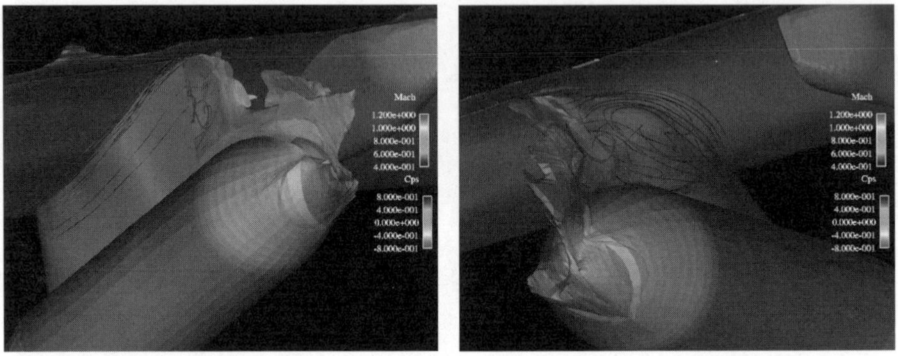

Figure 8 Visualization of computational results for M=0.84, α=0.2° using the SALSA TM. showing the detached flows. View of both sides of the pylon from the rear part. The model surface is coloured by Cp and the stream lines are coloured by Mach

7.5 Conclusions

The conclusions extracted must be categorized at three levels taking into consideration the aspect focused on and its influence in the success of the comparisons: The test case itself, the geometry model, and the turbulence models. Regarding the test case, it is very sensitive to the nominal flow conditions and also to the configuration because the incidence of the flow around the pylon is a secondary or indirect effect of the whole model and their lift distribution.

The preliminary assumption of simplifying the whole model geometry but maintaining the detail at the interesting region may be critical for the accuracy due to those mentioned indirect effects of the whole model. This seems to have more to do with the cases at M=0.78 and M=0.80 for which the experimental pressure levels describe a more adverse flow conditions in terms of flow separation. For these two cases, it is difficult to conclude if the inability for producing detached flows is due to the turbulence modelling or if it is due to the differences in the flow arriving to the pylon.

The turbulence modelling has a more clear implication in the case of M=0.84. On one hand, for this higher mach, the supersonic regions are more extensive and the flow around the pylon has more local dependency and the pressure compares better with those areas where the flows are attached. On the other hand, the two turbulence models perform fully different: SALSA does produce flow separations while the k-g does not. It is even noticeable that the experimental pressures are located between those of both models. As for the NACA0012 round tip, the k-g model predicts higher levels of eddy viscosity than SALSA. In general, the present case compares similarly as for the Obi test case, although for the present case it is very difficult to compare, with experiment, the flow pattern of the detached flow.

8 Prediction of Cross-Wind Stability for a Generic Train

T. Rung and D. Wu, Bombardier Transportation (BT)

Abstract

External aerodynamics has not been widely established in the rail-vehicle industry until recent years. Nonetheless, the subject is of fundamental importance in some respects, e.g. aerodynamic loads due to the head-wave or the slip-stream of a train, running resistance and cross-wind stability. The latter is the dominating safety issue when attention is drawn to high cruising speeds. The objective of this study is to scrutinise the predictive prospects of unsteady, scale resolving *Detached Eddy Simulations* (DES) for cross-wind train aerodynamics. Attention is restricted to a mirrored pair of generic end cars of the German ICE2 high-speed train. The example included refers to a yaw angle of 30° and a Reynolds number - based on the length of the first car - of $Re \approx 10^7$. Computational results are reported for DES, supported by companion steady and unsteady RANS simulations and wind-tunnel measurements. More comprehensive consequences on the stability of the vehicle are briefly addressed by means of a quasi-static mechanical analysis. The aerodynamic loads obtained from the DES approach are in fair agreement with experimental data and outperform RANS predictions slightly. Results indicate that – in terms of the maximum allowable cross-wind speed – the predictive failure returned by DES corresponds roughly to only 1% of the actual value. Moreover, DES provides some insight into potential risks for an excitation of natural frequencies due to cross winds, which might detract the riding comfort.

8.1 Objectives and Motivation

The cross-wind stability of a train is a primary safety issue, particularly when attention is given to higher cruising speeds. About 30 wind-induced accidents are reported until today, see Fig. 1. The most significant aerodynamic loads are the side force, lift force and roll moment at the centre of gravity, followed by the pitching moment. These loads are the principal contributors to the crucial roll moment located at the foot point of the vehicle's leeward wheels. The latter has to be balanced by the weight-induced restoring forces and moments. The endeavour for optimising the passenger capacity in line with an increased 'usable length' has put non-loco-hauled solutions with distributed traction and light steering cars into the commercial focus of modern train concepts. The related consequence is a substantial tare-weight reduction of the end cars, which are usually most critical with respect to cross-wind stability issues.

The flow field and the aerodynamic loads are governed by the leeward vortex which evolves downstream of the nose in proximity to the upper corner of the vehicle's leeward side wall. The process is mainly controlled by the resultant yaw angle. Figure 2 illustrates the cone shaped vortices occurring on the leeward side of the vehicle when exposed to a yaw angle above 10°. The strong suction force associated with the upper vortex core supplements the destabilising windward pressure force. Moreover, the vertical position of the vortex is rather high, which explains its pronounced influence on the roll moment. The restoring gravitational force is partially neutralised by the aerodynamic lift.

Figure 1 Examples of railway overturns due to cross wind occurring in France (28.01.1994, left) and Austria (17.11.2002, right).

As regards flow-physics modelling, the primary challenge is aligned to the dominating influence of the lee-vortex, i.e. the predictive capabilities with respect to strong curvature influences and open separation. The structural equilibrium assumption inherent to any eddy-viscosity model of industrial importance is known to be the major contributor to the misrepresentation of curvature. Moreover, the flow is likely to reveal a significant level of unsteady mean motion. Accordingly, either full second-moment closure models – which are able to mimic curvature influences – or hybrid RANS/LES methods – which aim to numerically resolve the detached vortex motion – appear to be a viable approach. The study aims to assess the predictive prospects of DES for cross-wind train aerodynamics. Attention is restricted to a generic ICE2 configuration cruising at roughly 200 *km/h*, where critical side-wind speeds around 30 *m/s* are associated with 30° yaw angles. Results are reported in comparison to U/RANS, while both methods employ the same mesh, identical background turbulence models and their respective best-performing numerical framework.

Figure 2 Illustration of the leeward vortices of a train exposed to cross winds (left); Flow visualisation from a wind-tunnel experiment (middle); Predicted pressure contours half a car downstream of the nose for 30° yawing (right).

8.2 Description of Test Case

The investigations are concerned with the flow over a generic ICE2 end coach. The test case was experimentally studied in the T-103 wind tunnel at the Russian TsAGI institute. As illustrated by Figure 3, a 1:10 scale leading car model of length $L_{tr} \approx 2.65\ m$, supplemented by a shorter end-car dummy, was the focal point of the experiments. The train was mounted with a ground clearance of $0.02\ L_{tr}$ on a thin elliptic plate. Yaw angles between $0° < \Psi < 60°$ were measured for wind speeds between *30 m/s* and *70m/s*, which corresponds to Reynolds numbers of $Re_{Ltr} = 5.1 - 11.8 \cdot 10^6$. The height of the vehicles is $h = L_{tr}/6.5$. The blockage ratio of the nozzle varied between *4-8%* for the investigated yaw angles of 10°, 20° and 30°. An external six-component gauge was used to measure the aerodynamic forces and moments on the leading car. Data averaging was performed over 4 seconds for each angle with a sampling frequency of 100 *Hz*. A small amount of static pressure measurements on the train surface is available.

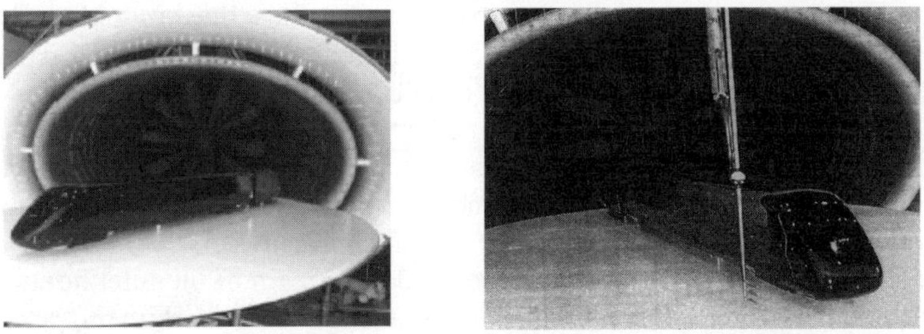

Figure 3 Illustration of the employed 1:10 scale ICE2 model in the TsAGI T103 open wind tunnel measuring section with splitter plate, turning table

The coefficients for the aerodynamic loads are usually calculated as follows

$$C_i = \left. \frac{F_i}{0.5 \cdot \rho \cdot U^2 \cdot A} \right|_{i=x,y,z} \qquad C_{Mi} = \left. \frac{M_i}{0.5 \cdot \rho \cdot U^2 \cdot A \cdot l} \right|_{i=x,y,z}$$

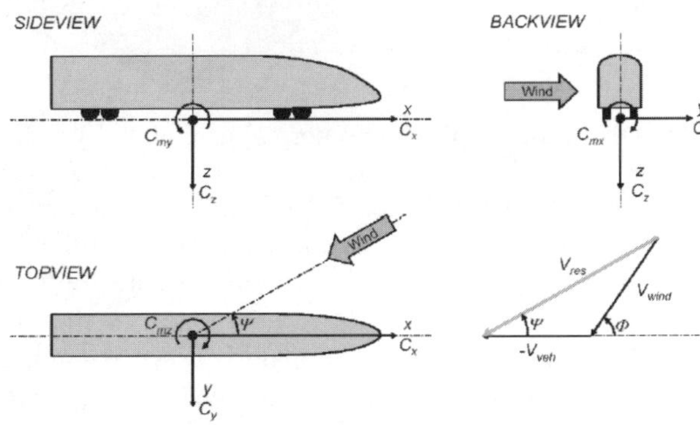

Figure 4 Utilised coordinate system and employed aerodynamic coefficients

where F_i is the force and M_i is the moment in coordinate direction i. The air density is depicted by ρ, U denotes the approaching air speed. In the present case of a 1:10 scale model A and l represent a fixed reference area and length of $0.1m^2$ and $0.3m$. Accordingly, a typical time scale is obtained from $T_0 = l/U = 0.3/U$. The height of the investigated model corresponds to approximately $h = 1.35\ l$. The respective length of the first car reads $L_{tr} = 8.85\ l$. Separate experiments were conducted with and without bogies and spoiler. Figure 4 depicts the above described the utilised coordinate system.

8.3 Computational Modelling Framework

The numerical procedure employed by BT is based on the framework outlined in chapter II.3. The focal point of the BT contribution is the comparison of DES and U/RANS using the k-ε background turbulence model as described in chapter III.2. All turbulence models were operated in low-*Re* number mode supplemented by a universal, all-resolution boundary condition (cf. chapter III.3). No significant *Re*-number effect was observed in the experiments, thus the computed Reynolds number was assigned to Re_{Ltr}=$1.06 \cdot 10^7$, which is close to the highest experiment-tally studied value. Note that no attempt has been made to mimic transition.

As displayed in Figure 5, the computed 1:10 scale model lacks all details like bogies, spoiler or plough. A mirrored end car was used instead of the shorter end-car dummy utilised in the wind tunnel. Computations were performed in a hexahedral physical domain, where the ground measures $5.4L_{tr} \times 4.2L_{tr}$ and the height reads $0.7L_{tr}$. The train was located L_{tr} downstream of the inlet boundary and arranged under an incidence of 30º to the lateral edges of the domain. The ground clearance of the train was in accord with the experimental data, i.e. $0.02\ L_{tr}$.

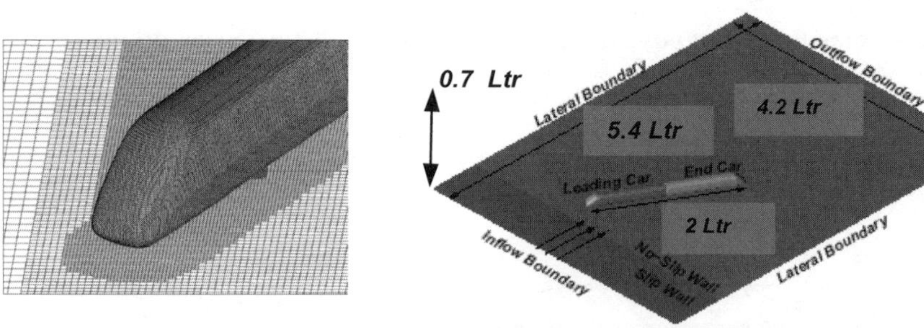

Figure 5 Details of the employed numerical grid and physical domain.

The employed unstructured grid features $5.7 \cdot 10^6$ cells. As depicted by Figure 5, local grid refinement is introduced in the near surrounding of the train and in the wake region. The spatial resolution in the vicinity of the train corresponds to $\Delta X_i/l \approx 0.02$ or $\Delta X_i^+ \approx 1000$, which is of similar size of the boundary layer thickness δ. Due to the prohibitive computational expenses associated with low-*Re* practices, the viscous sublayer has been mostly bridged ($Y^+ \approx 10$ - 250). The inflow turbulent intensity is assigned to $Tu = 0.5\%$ and the inlet turbulent viscosity is prescribed by $\nu_t/\nu \approx 1$. An outlet condition is employed for the pertinent boundary, while the upper and the lateral borders are defined as slip walls. The initial section of the ground was defined as a slip wall, in order to achieve the highest possible agreement with the experimental setup. At $0.3L_{tr}$ upstream of the train head, the floor boundary condition switched to a no-slip wall and the build-up of a turbulent boundary layer was initiated. Mind that the flow around the front of the train remains attached due to the high (nose) radius based Reynolds number ($> 2{\cdot}10^5$; Cooper , 1985).

Time integration is imposed in a strictly implicit manner, using a 1st-order scheme. The time step was prescribed as $\Delta t = 2{\cdot}10^{-1}{\cdot}T_0 = 10^{-3}s$. The corresponding CFL number based on the typical spatial step on the refined leeside of the vehicle is $CFL \approx 10$. An improved temporal resolution is considered not tenable for large industrial applications. The numerically addressed frequencies are estimated by $f \leq 1/(5\Delta t) = 200Hz$, with a reliable resolution up to $f \approx$ 80-100Hz ($Str \approx$ 0.4-0.5). Some 3500 iterations (700 T_0) were performed for the initial computation in order to remove initial transients. Another 3000 time steps – corresponding to 3s or 600 T_0 – were performed to resolve the transients of the flow. Accordingly, the lower bound for comparisons is estimated at $f \approx 1.6Hz$ ($Str \approx 0.008$). U/RANS simulations were based on third-order upwind biased TVD scheme for the approximation of advective fluxes, while the DES applies a mixed scheme with 95% CDS and 5% UDS everywhere.

8.4 Results

Figure 6 depicts the evolution of the principal loads, i.e. the roll moment (C_{Mx}) next to the lateral (C_y) and the lift force (C_z), returned by the steady-state RANS and the transient URANS computation. The variations of the side force and roll moment are rigidly aligned and hardly distinguishable in Figure 6.

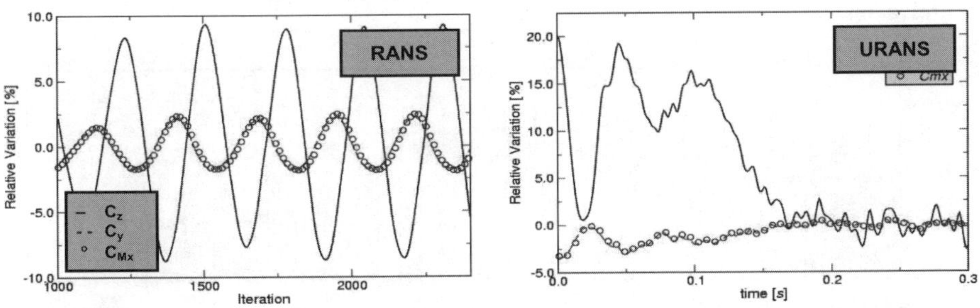

Figure 6 Evolution of RANS (left) and URANS (right) load predictions

Figure 6 reveals, that the mean forces were well converged for the RANS simulation but all loads still hold an amount of oscillation owing to the transient nature of the flow. The variations of the side force and roll moment are limited to roughly ±3%. On the contrary, the lift coefficient is oscillating by ±10%. The actual flow involves massive separation and features a significant amount of transients. As perhaps expected, the level of force convergence remains insufficient for RANS. The latter motivates the assessment of unsteady approaches in terms of both URANS and DES. Analogous time evolutions obtained from the URANS computation are also displayed in Figure 6 (right). Apparently, all oscillations rapidly ebb away and the URANS simulation approaches a steady state. As confirmed by Figure 7, the frequencies modelled by the *k-ε* turbulence model in the lee-vortex region drop down to roughly 80*Hz* while operating in the URANS mode. The maximum, numerically addressed frequency was estimated around 200*Hz*. Therefore, URANS fails to separate modelled and resolved scales and returns excessive levels of eddy viscosity for the upper regime of the resolved scales.

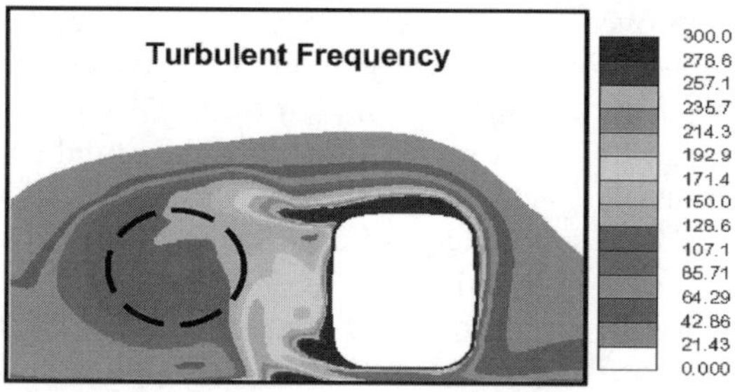

Figure 7 Modelled frequencies obtained from the URANS computation

In contrast to the URANS results, DES returns pronounced transient motion. Figure 8 provides an overview of the predicted unsteady force variations as compared to the respective mean values obtained from the DES. Again, it is seen that the variations of C_{Mx} and C_y are rigidly aligned and limited. When attention is drawn to the transient behaviour of the (wheel unloading) lift-force C_z, a significant variation up to ±50% is seen.

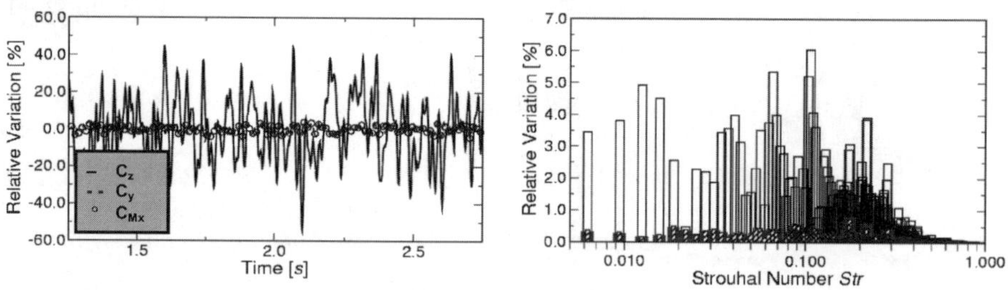

Figure 8 Transient loads as seen from the DES computation (left: time space; right: frequency space)

As indicated by Figure 8, a number of unsteady force contributions are found in the range of $0.01 \leq Str = f{\cdot}l/U \leq 0.1$. For cruising speeds between 200*km/h and* 250*km/h* the latter corresponds to $0.2 \leq f \leq 2.5Hz$. By reference to multi-body simulations, these frequencies are close to a natural body-bounce frequency of

0.75*Hz* which is damped by only 4.7%. Consequently, cross-winds might impair the riding comfort.

When attention is directed towards the evolution of the roll-moment, it is seen that the aerodynamic loads are primarily generated in the nose region. Figure 9 outlines that roughly *55%* of the roll-moment is generated over the initial *35%* of the first car. Moreover, a significant, wheel-unloading suction force is identified around the leeward A-pillar of the nose, which further contributes to the destabilisation of the vehicle. In conclusion, about half of the leeward roll moment is generated over the first 25% of the vehicle. As indicated by the build up of the roll moment illustrated in Figure 9, the two approaches (RANS and DES) display no substantial differences as regards the evolution of aerodynamic loads.

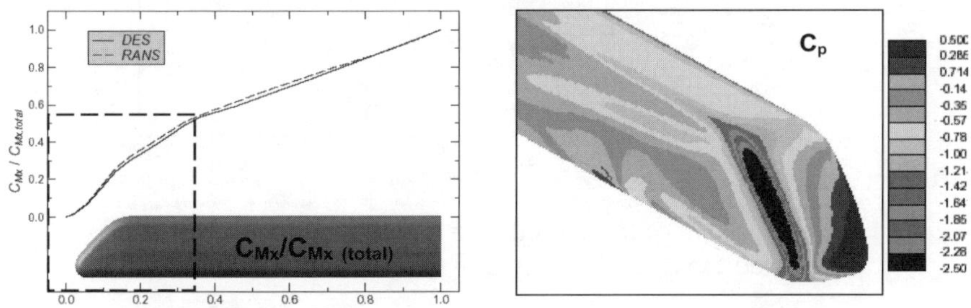

Figure 9 Evolution of the roll-moment coefficient (left) and time-mean distribution of the normalised pressure obtained from DES (right)

Figure 10 provides an overview of the downstream development of the lee-vortex by means of the computed contours of the time-mean pressure obtained from DES in five selected cross sections of the first car. The figure also indicates the

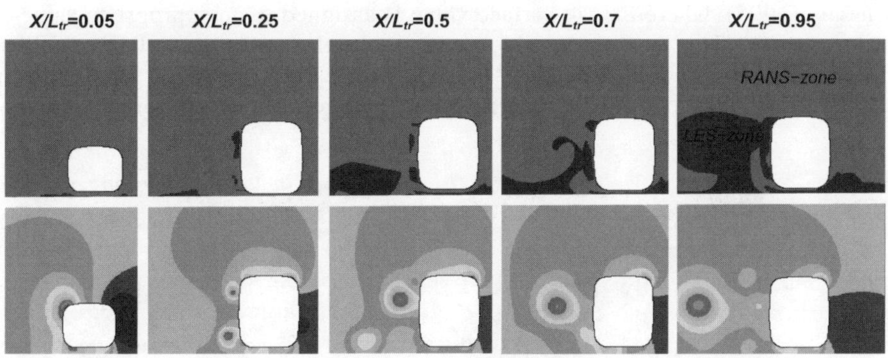

Figure 10 Distribution of RANS and LES mode (top) and contours of time-mean pressure (bottom) for five selected cross sections of the first car as returned by DES

respective split between the LES and RANS mode at a typical time instant. It is seen, that the onset of the LES mode occurs far downstream of the nose region, i.e. at 50% of the vehicle length.

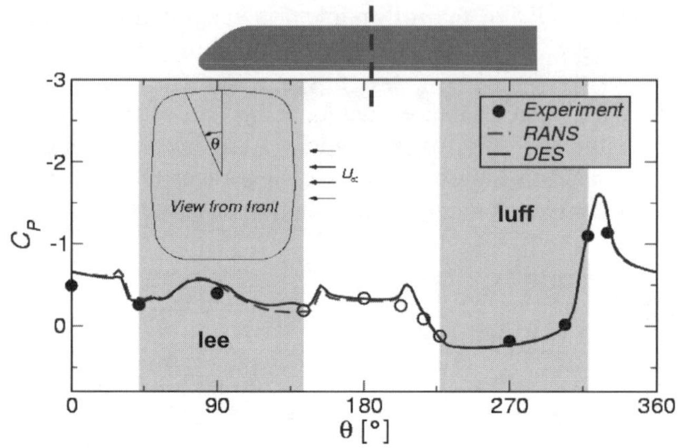

Figure 11 Comparison of measured and predicted time-mean pressures for a selected cross section at $X/L_{tr} = 0.5$.

Figure 11 assess the predictive agreement of RANS and DES by comparing the predicted circumferential pressure distribution at $X/L_{tr} = 0.5$ with the experimental data. In line with the above mentioned findings, both methods offer a fairly reasonable predictive agreement with the measured pressure distribution and no substantial difference is found between RANS and DES. Similar behaviour is obtained for other cross sections. Figure 12 depicts the predicted lee-vortex structure for two different times obtained by DES. The plotted iso-surfaces refer to the medial eigenvalue λ_2 of the tensor $S^2_{ij} + \Omega^2_{ij}$ as suggested by Jeong and Hussain (1995), where the iso-surface value is assigned to an appropriate value of $\lambda_2 = -1.5 \cdot 10^5$. The lee-vortex, which emerges at the upper end of the A-pillar, departs shortly after its onset and extends far beyond the small inter-car gap. Moreover, it reveals an almost time-independent behaviour except for its far end.

Figure 12 Visualisation of the lee-vortex structures by means of λ_2 iso-surfaces featuring $\lambda_2 = -1.5 \cdot 10^5$ at two different times.

On the contrary, the vortex rolling up at the lower edge of the nose decays within a short distance. Its motion is highly unsteady, presumably representing the main source of the load variations.

Table 1 summarises the predictive accuracies with regards to the force predictions. Both computational approaches return a satisfactory level of predictive accuracy. Improvements from DES are seen for the predicted lift force and pitch moment.

Table 1: Comparison of predicted and measured cross-wind loads.

Load Component	Symbol	Exp.	DES	RANS
Drag Force	C_x	-0.23	-0.24	-0.20
Side Force	C_y	3.72	3.76	3.62
Lift Force	C_z	-1.26	-1.15	-1.00
Roll Moment	C_{Mx}	2.26	2.30	2.18
Pitch Moment	C_{My}	-0.48	-0.07	0.22
Yaw Moment	C_{Mz}	5.41	5.27	5.13

It is instructive to assess the predictive accuracy with regards to the related consequences on the prognosis of the sustainable cross-winds. A quasi-static approach is applied for the mechanical modelling of the vehicle's stability in response to predicted or measured aerodynamic loads. The mechanical model of the driving trailer depicted by Figure 13 employs eleven degrees of freedom. The major elements of the mechanical model are five rigid bodies, i.e. two bogies and the car body, represented by their respective suspended and unsuspended masses and centres of gravity. These principal bodies are connected by a set of primary and secondary springs aiming to model the bogie suspension. Moreover, an anti-roll bar is applied between each bogie and the car body. Mind that the two axles of each bogie are merged in the model. The procedure can account for lateral acceleration due to curved track and cant. Track irregularities are, however, excluded due to their dynamic nature (Diedrichs et al., 2003). The wheel-unloading

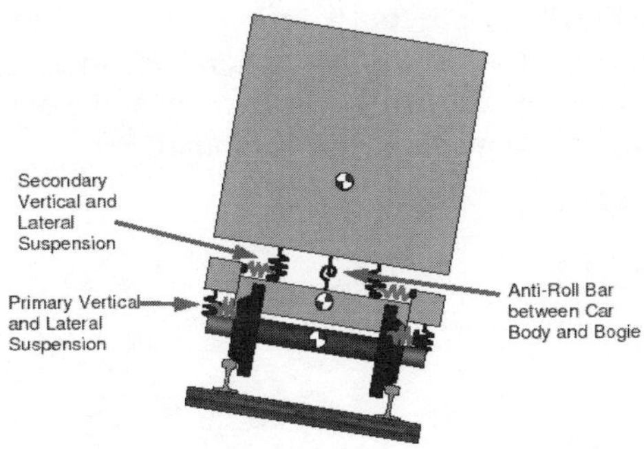

Figure 13 Illustration of the employed mechanical model.

criterion, separately applied to the front or the rear bogie, is employed to assess the stability of the first vehicle. In the present study, the maximum allowable wheel unloading is defined to $0.9Q_o$ where Q_o is the nominal load at rest. As indicated by Table 2, the underestimated lift coefficient, followed by inaccuracies of the predicted roll and pitch moments, are the primary sources of error with respect to the prognosis of the sustainable cross-wind speed. Moreover, the individual errors neutralise each other for the DES simulation, which finally returns an almost perfect agreement. The predictive deviation of RANS is below 3% (1 *m/s*).

Table 2: Comparison of predicted and measured cross-wind stability

Load Component and Symbol		Δ Side Wind at 100% [m/s]	Δ Side Wind DES [m/s]	Δ Side Wind RANS [m/s]
Side Force	C_y	2.00	-0.022	0.054
Lift Force	C_z	4.00	0.349	0.825
Roll Moment	C_{Mx}	-13.00	-0.230	0.460
Pitch Moment	C_{My}	-0.20	-0.171	-0.292
Sustainable Side Wind		***Exp. [m/s]***	***DES [m/s]***	***RANS [m/s]***
Example: 210 km/h Cruising Speed		33.20	33.13	34.25

8.5 Concluding Remarks

The study conveys that DES successfully extends the predictive realm of RANS-based industrial CFD for investigations aiming at an assessment of unsteady aerodynamic loads in conjunction with safety and comfort issues. Due to the dominance of the nose regime, which is governed by steady phenomena, the DES method provides only a small amount of improvements with respect to the predictions of aerodynamic forces due to cross-winds. The overall accuracy of the force-predictions is satisfactory for both RANS and DES. It should be noted that RANS revealed a significant amount of force oscillations at 'convergence' and URANS was afflicted with a spectral overlap between modelled and resolved motion in the lee-vortex regime. The computational effort associated with the DES of a train exposed to cross wind is still considerable and more than two orders-of-magnitude higher than the companion RANS simulation.

9 Analysis of Sound Sources for a Generic Car Mirror

Th. Rung[1], D. Wu[1], U. Bunge[2], C. Mockett[2], and F. Thiele[2]
[1] Bombardier Transportation (BT), [2]TU Berlin

Abstract

The study seeks to scrutinise the predictive prospects of scale-resolving *Detached-Eddy Simulations* (DES) in comparison to unsteady RANS (URANS) for the prognosis of aero-acoustics dipole sources. This engineering realm has recently gained importance for the transportation industry due to the enhanced aero-acoustic noise contributions associated with higher travelling speeds and the increasing legal restrictions for noise emission. As noise-reduction techniques have improved significantly in the area of vibro-acoustics, flow-induced noise is shifted into the focus of vehicle and aircraft manufacturers by means of the development of reliable source-prediction methods. In the present effort, attention is confined to the flow over generic car mirror mounted on a flat-plate. Results are reported for time-averaged and transient pressure signals in comparison with experimental data provided by DaimlerChrysler Research and Technology (Höld et al. 1999; Siegert et al. 1999). Employing the same computational environment, the total sound pressure level predictions obtained from DES outperform URANS by approximately 20*dB*.

9.1 Objectives and Motivation

The application of conventional statistical (RANS) closure models for massively transient flows featuring open vortex separation requires a spectral decoupling between the transient motion and the modelled turbulence. This requirement is often violated when performing unsteady RANS for aero-acoustic investigations, where small time steps and locally refined meshes are employed and resolved motion penetrates into the regime of modelled scales. Direct Numerical Simulation (DNS) and Large Eddy Simulation (LES) seem more suitable for this purpose but the associated computational expenses make them hardly feasible. Accordingly, hybrid RANS/LES methods and DES approaches (Spalart et al., 1997) appear to be a viable compromise between feasibility and computational accuracy. The study aims to outline the predictive benefits and drawbacks of DES. Results are reported in comparison to URANS, whereby both methods employ the same mesh, identical background turbulence models and their respective best-performing numerical framework. Moreover, the impact of the turbulence model and the employed advection scheme in the framework of DES are briefly addressed.

9.2 Description of Test Case

The investigations are concerned with the flow over a generic car side mirror. The test case was measured by Siegert et al. (1999) in the FCTS wind tunnel at the University of Stuttgart. As illustrated by Figure 1, the mirror, which is essentially a half-cylinder supplemented by a quarter-sphere, is centrally mounted on a flat plate which spans $3m$ in length and $1.6m$ in width. The diameter and the height of the half-cylinder are both $D = L = 0.2m$. The investigated free-stream velocity is $U_0 = 39m/s$, thus $Re_D = 520\ 000$ based on the diameter of the mirror. Due to the low Mach number ($Ma = 0.11$), the flow is considered to be incompressible in the simulations.

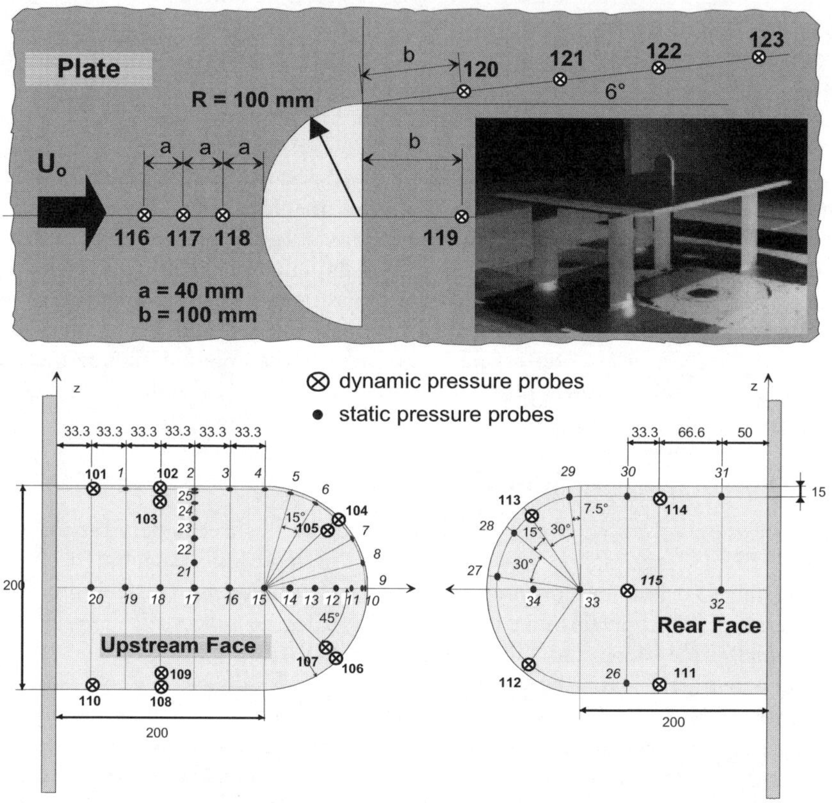

Figure 1 Experimental setup and location of pressure probes reported by Siegert et al. (1999)

Experimental data by means of pressure probes at various locations on the mirror and plate surface were provided DaimlerChrysler Research and Technology (Siegert et al. 1999). As outlined by Figure 1, two groups of pressure probes were installed, the first group consisting of 34 pressure sensors (*P1 – P34*) for time-averaged measurements and the second one with 23 piezo-sensors (*P101 – P123*) returning instantaneous pressure signals. Unsteady measurements were performed for 164 travel times T_0 at a sampling rate of $20kHz$. One travel time corresponds to

$T_0 = D/U_0 \approx 0.005s$. The data was filtered with a $2kHz$ low-pass filter, thus comparisons with experimental data are restricted to frequencies below $2kHz$.

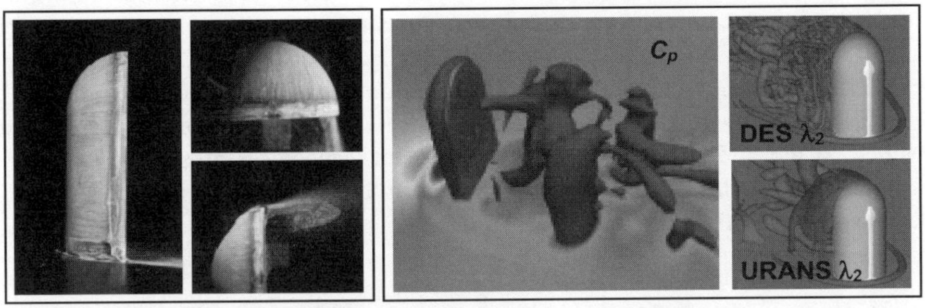

Figure 2 Flow visualisation reported by Siegert et al. (1999) (left) and visualisations obtained from DES and URANS (iso-surfaces of C_P & λ_2, right)

Measured data was only available in terms of the respective Fourier transformed signals with a bin width of $10Hz$, i.e. no time evolution of pressure signals was provided. Since the experiment was set up for aero-acoustic purposes, no artificial transition was implemented in order to avoid additional noise contributions. The location of the transition can be estimated from the oil-flow visualisations displayed in Figure 2, which indicate a small laminar separation at approximately $0.15D$ upstream of the rear-face edge. The flow is considered to be turbulent henceforth. Figure 2 also depicts the shedding of the primary vortex from the lateral edges. A primary Strouhal number of $Sr = f\,D/U_o \approx 0.15 - 0.20$ is expected for the shedding mode according to pertinent measurements. The corresponding frequency yields $f \approx 30 - 40Hz$.

9.3 Computational Modelling Framework

The numerical procedures employed by BT and TUB are based on the framework outlined in the respective parts of chapter II.3 and II.16. The focal point of the BT contribution is the comparison of DES and URANS using the SA background turbulence model. As opposed to this, TUB investigated the predictive differences in response to the background turbulence model, i.e. LLR k-ω and SALSA, when using DES. All turbulence models were operated in low-Re number mode supplemented by a universal, all-resolution boundary condition (cf. chapter III.3). It should be borne in mind, that none of the employed computational frameworks were able to account for laminar regions.

The computed physical domain utilises the same dimensions as the experiments. The stream-wise and lateral edge lengths of the supporting plate are $15D$ and $8D$ respectively. The height of the domain is $7.2D$. The rear-face foot of the mirror is located at $5.25D$ downstream of the leading edge. Two numerical grids, a mandatory structured grid (M1) provided TUB and an unstructured grid (M2), generated by BT have been investigated. The TUB grid consists of $2.6 \cdot 10^6$ cells, the locally refined BT grid involves $1.6 \cdot 10^6$ cells. With the exception of the wall-adjacent region, the grid aft of the mirror's rear face is isotropic for the M2 mesh, featuring cell spacings of $D/\Delta x_i = 100$. Both grids were carefully adjusted to match the required near-wall resolution with $Y^+ = 0.5 - 3$ in the area of primary interest. Figure 3 provides an overview of the surface meshes supplemented by an instantaneous Y^+ distribution for the mandatory mesh. The lateral and the upper boundaries of the domain are assigned to an Euler boundary condition. The adequate streamwise end surfaces of the domain are allocated as the inlet and the outlet. At the inlet, a non-dimensional turbulent viscosity is prescribed as $\nu_t/\nu = 1$. Unlike the experiment, the flow is turbulent in the entire domain and no transition can be expected from the numerical results.

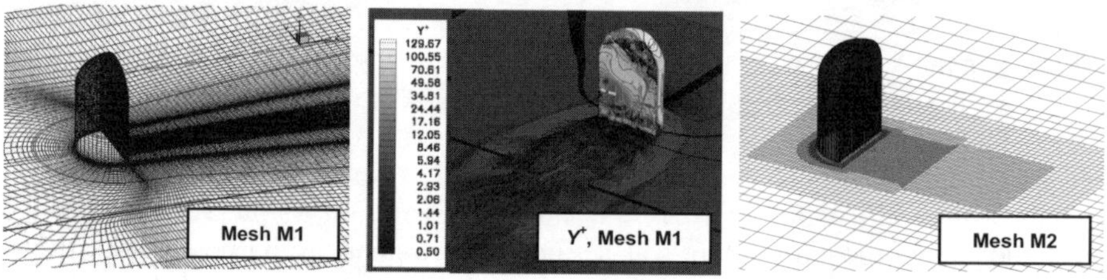

Figure 3 Employed computational meshes (left TUB-M1; right BT-M2) and typical Y^+ distribution (M1 mesh)

Time integration was imposed in a strictly implicit manner, using either a 1st-order scheme (BT) or a 2nd-order scheme (TUB). The time step is prescribed as $\Delta t \approx 2 \cdot 10^{-3} \cdot T_0 = 10^{-5}s$ (CFL ≈ 0.05) for the simulations performed by BT and $\Delta t \approx 2 \cdot 10^{-2} \cdot T_0 = 10^{-4}s$ (CFL ≈ 0.5) with regards the TUB investigations. The time steps are grossly in line with the low-pass filter employed in the experiment and best-practice findings obtained from the investigations of simpler cases throughout the project (i.e. DHIT and NACA0012 beyond stall). The small time step of BT was found necessary due to the low-order scheme. Accordingly, the estimated highest resolvable frequencies correspond to Strouhal numbers of $Sr \approx 5$ ($f = 1/(10 \cdot \Delta t) \approx 1000 Hz$) for the TUB simulations and $Sr \approx 17$ ($f = 1/(30 \cdot \Delta t) \approx 3300 Hz$) for the BT investigations. Initial transients have been removed by 70 time units T_0. All simulations were performed for at least another $70T_0$ in order to assure reliable averages in the energy containing low-frequency regime ($Sr \approx 0.1$) by means of roughly 7 period lengths. The URANS computations were carried out with a third-order upwind biased TVD advection scheme which is the conventional approach to be employed in industrial CFD. As opposed to this, DES simulations of BT employed a flux-blending scheme everywhere (95% CDS, 5% UDS) whereas TUB employed Strelet's (2001) hybrid scheme.

9.4 Results

Prior to the comparison of transient pressure signals, Figure 4 addresses the predictive performance for a selection of time-mean pressures on the surface of the mirror. The agreement between computations and experiments is generally fair with the exception of locations *P10/P11* and *P24/P25* in the separated flow regime on the upstream surface. The respective predictive deficiency can be attributed to a misrepresentation of the laminar separation by the fully-turbulent computational framework. In conclusion, the simulated boundary layer is artificially stabilised and remains attached. The latter is supported by an improvement of the predictive accuracy when compared with data pertaining to a higher (supercritical) Re-number of $Re_D = 740\ 000$ (i.e. 200*km/h*). Here, the experimental flow is already turbulent before the suction peak and does not separate. The choice of the lower Reynolds number for the present computational exercise was due to the larger amount of available transient data.

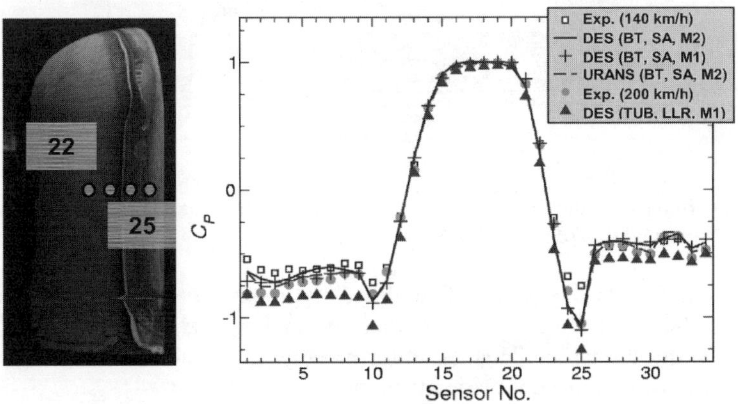

Figure 4 Comparison of measured and predicted time-mean pressures at on the surface of the mirror

Transient Data Analysis

Before taking a closer look at the pressure transients, some formal remarks regarding the data-processing methods shall be made in order to describe the utilised formats of presentation and the respective limitations of the comparisons. For this purpose, the pressure signal recorded at *P123* shall serve as an illustrative example. The recorded transient pressure evolution at the 23 piezo-sensor positions are converted into the frequency domain by means of Discrete Fourier Transforms (DFT) and subsequently analysed in terms of sound pressure levels (*SPL*), viz.

$$SPL = 20\log_{10}\left(\frac{p'}{p_0}\right) dB$$

Here p' denotes the pressure fluctuation, i.e. $p - p_{mean}$ and the reference pressure is $p_0 = 2 \cdot 10^{-5} Pa$. The *Hanning* window-function is applied for the DFT in order to avoid an artificial augmentation of the *SPL* at higher frequencies when transforming non-periodic, finite length data (Fahy, 2001). The impact of the Hanning-window in comparison with the standard rectangular window is depicted in Figure 5. Superficially, the result returned by the rectangular window function – which transforms the raw CFD data – implies a better overall agreement with the measurements at higher frequencies. Due to the severe leakage of none resolved frequencies into resolved scales (aliasing), this agreement should be handled with scepticism.

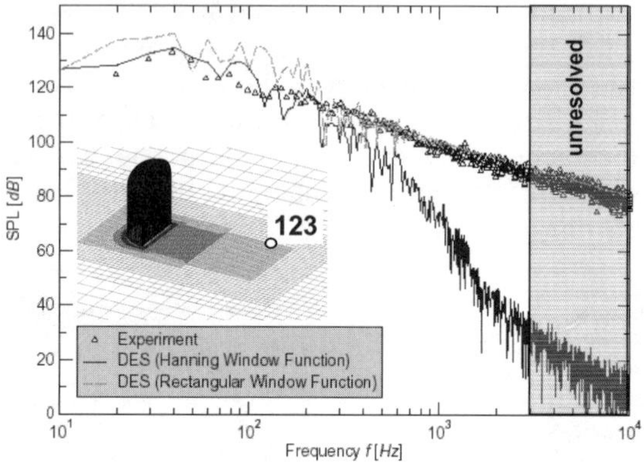

Figure 5 Impact of the Hanning window function on the evaluated *SPL* at *P123* as returned by DES (BT-SA-M2)

Furthermore, Figure 5 suggests that the experimental data was transformed by means of the rectangular window thereby making its correctness at higher frequencies questionable. This is unfortunately impossible to confirm, as the original time-traces were not made available.

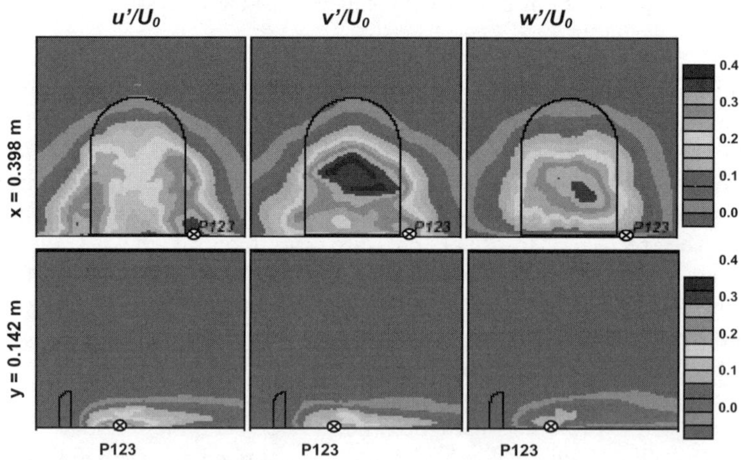

Figure 6 Resolved time-averaged turbulent intensity obtained from DES after $20T_0$ (BT – SA – M2) in two cross-sections through *P123*. Left column: u'/U_0; Middle column: v'/U_0; Right column: w'/U_0. Upper row: x cross-section at $x = 0.398m$; Lower row: y cross-section at $y = 0.142m$

The significant down-drop of the SPL returned by DES with the Hanning window at approximately 1000*Hz* justifies further attention. The reason for this seemingly unfavourable behaviour can be addressed as the grid-dictated limitation of the resolvable spectrum. The maximum spatial step $\Delta = max(\Delta x, \Delta y, \Delta z)$ at *P123* is $\Delta \approx 8mm$, which is close to $\Delta \approx \delta$. The local spatial resolution appears thus sufficient for the detection of the large structures at the pertinent position. Accordingly, the smallest resolvable wave length at *P123* is ideally $L = 2\Delta$. The transient velocity fluctuations yield $u_i' = Tu_i \cdot U_0$ and the maximum turbulent intensity at the location of *P123* can be premised as $Tu_{max} \approx 40\%$. The latter is supported by resolved fluctuating motion in *P123* displayed in Figure 6, which was obtained from the SA-based DES of BT on mesh M2 by averaging 10000 time steps ($20T_0$). An estimate of the maximum resolvable frequency hence reads:

$$f_{\max} = \frac{u'}{L} = \frac{Tu \cdot U_0}{2\Delta} \approx 1000Hz$$

Moreover, the remaining non-symmetries displayed in Figure 6 reveal the insufficient averaging duration of $20T_0$ for a reliable statistics. Analogous maximum frequencies varying within the range between 1000*Hz* to 4000*Hz* can be found for all other dynamic pressure sensor positions on both grids. Bearing in mind that the experimental data refers to a 2*kHz* low-pass filter, the employed time steps and grids generally seems to be adequate for a comparison up to 1*kHz*.

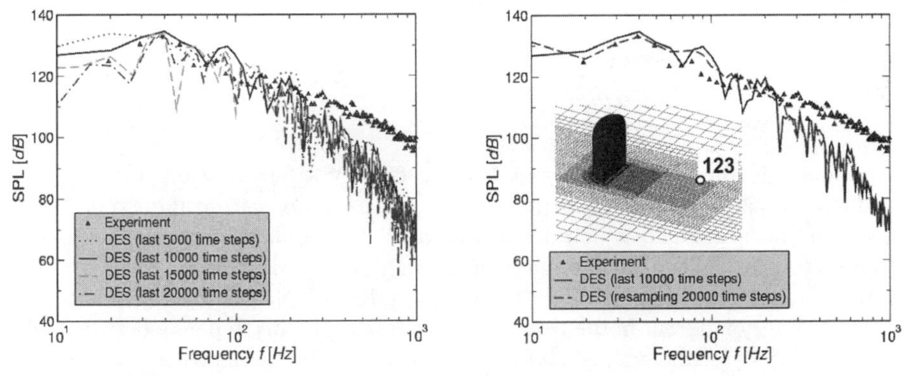

Figure 7 SPL at *P123* as seen in the experiment and returned by DES. Left: Study of the impact of the number of utilised samples; Right: Study of the impact of re-sampling.

A further validation of the numerical results shall be provided in terms of sufficient simulation time and appropriate Fourier transformation. Figure 7 displays four different *SPL's* from the BT-SA-M2 DES simulation, featuring different numbers of time steps for *P123*. The enhanced solution accuracy with increasing number of samples is obvious in the lower end of the spectrum, where the smallest frequencies are statistically only resolvable by means of longer simulation duration. The 5000 time-step *SPL* reveals poor results for frequencies below 40*Hz*, whereas the 10000 time-step *SPL* passably mimics the downward slope if the *SPL* towards the large scale eddies. The remaining two *SPL's* both reveal acceptable performance. As the experimental data are provided with a bin width of approximately 10*Hz*, numerical results shall employ the same bin width as well. This task is achievable for a transient signal of any length and any sampling rate by means of re-sampling Figure 7 (right). The result of the re-sampling exercise suggests that the use of 10000 samples and simple DFT without re-sampling returns sufficiently accurate SPL's in conjunction with BT's results. TUB employed 12500 samples and sliding averages of data windows.

The convergence of the computations can be ascertained by means of the time evolutions of the *total pressure level*, which typically displayed a minor variation in the range of 1*dB* – corresponding to less than 1% (cf. Table 1) – after removing initial transients. The spectral analysis of the obtained pressure transients can thus be performed on a reasonable basis.

Transient Pressure Signals

Tables 1 and 2 assemble the RMS-values of the transient pressure signals at selected locations. The RMS-values are calculated by means of the Fourier Transforms of the experimental and computational results in order to ensure comparability, while attention was confined to the spectral energy inherent in the range from 20*Hz* to 1000*Hz* and a bin width is $\Delta f = 10Hz$. In detail, the RMS-values of are obtained from

$$p_{rms} = 10\log\left[\frac{1}{W}\sum_{N=2}^{100}\left(\frac{p(f_N)}{p_0}\right)^2\right] \quad with \;\; f_N = N \cdot 10Hz \;\; and \;\; W_{\exp} = 1, W_{CFD} = 1.5$$

The values denoted in Table 1 and Table 2 convey a fair agreement of the DES predictions with the measurements. Significant deviations from the experimental values are seen at *P116* which demands a more thorough investigation. The DES results are particularly encouraging in the region that is best resolved by both computational grids (e.g. *P119*). In contrast, URANS clearly fails to portray the spectral energy inherent in the flow at all locations. The discrepancy between DES and URANS typically yields 20*dB*. In the most active area, where the vortices impact on the plate (e.g. *P123*), URANS simulations still underestimate the total level by 10*dB*.

Table 1 Total sound pressure levels for selected pressure sensors as seen in the experiment and in the BT simulations (SA-M2). Locations relative to the mirror.

Sensor	Location	Exp. [dB]	DES[dB]	URANS[dB]
P102	front face – rear edge – lateral	129.3	125.6	103.0
P113	rear face – upper lateral	118.3	115.6	94.0
P114	rear face – lower lateral	122.8	119.6	104.4
P116	plate – upstream –central	131.0	116.5	90.9
P119	plate – downstream – central	122.9	122.0	100.7
P123	plate – downstream – lateral	138.1	138.6	127.6

DES Results depicted by Tables 1 and 2 varied only marginally with *(2dB)* depending on the utilised mesh and background closure. In particular, the two one-equation model predictions reported by BT and TUB on different meshes are in good agreement.

The selected *SPL's* to be investigated closer are those recorded at *P116* and *P119*, presented in Figure 8 and the top of Figure 9 respectively. The former is the centrally located upstream most sensor, while the latter is *D*/2 downstream in the near wake of the mirror's rear face. Typical streamwise and lateral spatial resolutions at *P116* and *P119* can be taken from Figure 9 for the example of mesh M2. Neither URANS nor DES could represent the *SPL* at *P116* correctly. This shortcoming could be attributed to the inadequate grid density upstream of the mirror and the associated insufficient resolution capabilities of both methodologies, although an alternative explanation is feasible.

Table 2 Total sound pressure levels for selected pressure sensors as seen in the experiment and the TUB simulations (M1). Locations relative to the mirror.

Sensor	Location	Exp. [dB]	DES[dB] SALSA	URANS[dB] LLR k-ω
P102	front face – rear edge – lateral	129.3	122.0	120.3
P113	rear face – upper lateral	118.3	115.4	117.2
P114	rear face – lower lateral	122.8	119.6	120.3
P116	plate – upstream –central	131.0	-	-
P119	plate – downstream – central	122.9	120.0	122.0
P123	plate – downstream – lateral	138.1	135.5	135.3

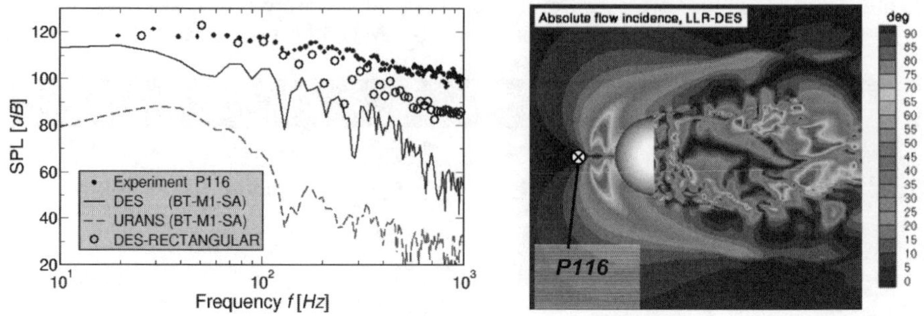

Figure 8 *SPL* at *P116* (BT-SA-M1) and instantaneous absolute flow incidence in the cross-section $z = 0.005m$ returned by DES (TUB-LLR-M1) seen in the experiment and returned by DES

As indicated by the right part of Figure 8, the flow reveals no substantial transients in terms of varying flow incidence and the approaching flow is almost symmetrical. This observation was made in conjunction with both grids and all background models. That such a high experimental SPL is reported for this point despite the low level of activity here is therefore hard to explain, and constitutes a source of doubt as to the validity of the experimental SPL for *P116*. Particular concern refers to the significant leakage from higher to lower frequencies that occurs when employing a rectangular window function at this position.

Turning now to *P119*, Figure 9 reveals a much finer spatial resolution here. Accordingly, the *SPL* returned by all DES simulations is in almost perfect agreement with the experimental data (cf. top of Figure 9). However, the URANS solution still evidently fails to meet the target *SPL*. This is in accordance with the fundamental principles of the approach, i.e. that past a certain point, RANS results cannot be physically enhanced by a refinement of the mesh.

A numerical oddity is presented by the bottom left part of Figure 9, which displays the *SPL* at sensor *P113* as observed in BT's DES computation on the locally refined mesh M2 and the mandatory mesh M1. The striking feature is the dramatic increase of the DES signal featuring two local maxima at approximately $2500Hz$ and $5000Hz$ in conjunction with mesh M2. The origin of this artefact, which was only found for the significantly refined isotropic M2 mesh, is evidently non-physical since the intermediate frequencies are almost entirely absent. Bearing in mind that the M2 mesh features very small numerical dissipation close to the mirror, a closer observation reveals that both the resolved and the turbulent motion are negligible in the region of *P113*. The wiggles are thus caused by a numerical instability which can be attributed to the lack of an adaptive hybrid upwind-biased/central advection-scheme in areas dominated by RANS.

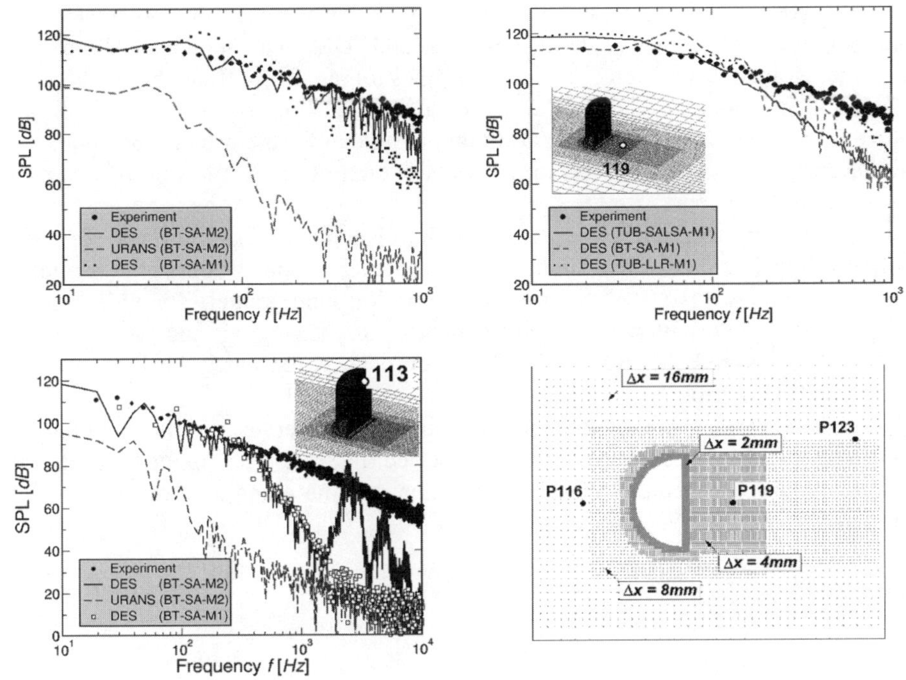

Figure 9: Predicted *SPL* at *P119* reported by BT (top-left) and TUB (top-right); *SPL* at *P113* returned by BT's DES in conjunction with 95% CDS (bottom-left) and the respective positions of selected sensors in the mesh M2 (bottom-right).

9.5 Concluding Remarks

The results convey that DES successfully extends the predictive realm of RANS-based industrial CFD for investigations towards open separation with an aero-acoustic background. Despite some minor deficiencies, the methodology returns noticeably improvements for the prediction of dipole sources. The computed total sound pressure levels are in good agreement with the experimental data and outperform URANS significantly by $20dB$ at most investigated locations.

Due to the acoustic nature of the study, the computational expenses – by means of the required spatial and temporal resolution – are considerable for both RANS and DES. Accordingly, resolution aspects are, of course, crucial to the predictive accuracy but not different for the investigated approaches.

In conjunction with the background closure model of the DES, no significant influence was observed for the investigated flow, as might have been expected. Similarly, the predictive success was not much afflicted by the observed weaknesses, since their influence was mostly confined to the upstream and near wall RANS regions and were too subtle to drastically change the vortex motion.

The SPL returned from a point upstream of the mirror was significantly lower than the experimental value in both URANS and DES calculations. Alternative hypotheses are feasible, on the one hand questioning the sufficiency of the grid resolution in this area, and on the other hand suggesting a possible error in the processing of the experimental data. In any case, due to its upstream position, the influence that this deviation has on the wake region of interest has been shown to be minimal.

The inability to predict the pressure-induced laminar separation is a common deficiency of almost any industrial turbulence-closure approach. Counter measures refer to the RANS part of the background model. By means of a supplementary intermittency-transport equation, laminar regions and transition can grossly be captured (Menter, 2002). The approach could easily be incorporated into the DES concept.

The numerical setup of the DES methodology deserves careful attention as indicated by the spectral oscillations reported by BT. The applied numerical system should thus feature a hybrid advection scheme along a route outlined by Strelets (2001).

10 The TUM Forward-Swept Wing Generic Aircraft

W. Haase, H. Åslund, EADS-M,
M. Kuntz, F. Menter, ANSYS

Abstract

The forward-swept wing, a canard-wing configuration for low Mach numbers measured by the Technische Universität München (Breitsamter&Laschka, 2001) serves as a validation test case for testing turbulence models in the presence of strong vortices. Although the aspect ratio is small and holds for military applications rather than for civil aircrafts, the physical background is very well comparable with similar flow features on commercial aircraft. For both military and civil applications new and improved turbulence models, as they were tested in the FLOMANIA project, are now providing more reliable and robust results for vortex-dominated flows.

10.1 Case description

The current investigation deals with a wing-body-canard geometry according to the geometry description provided by Breitsamter and Laschka (2001). While the wing sections are based on the NACA64A010 airfoil geometry, the fuselage is formed as an elliptic cylinder with a canopy.

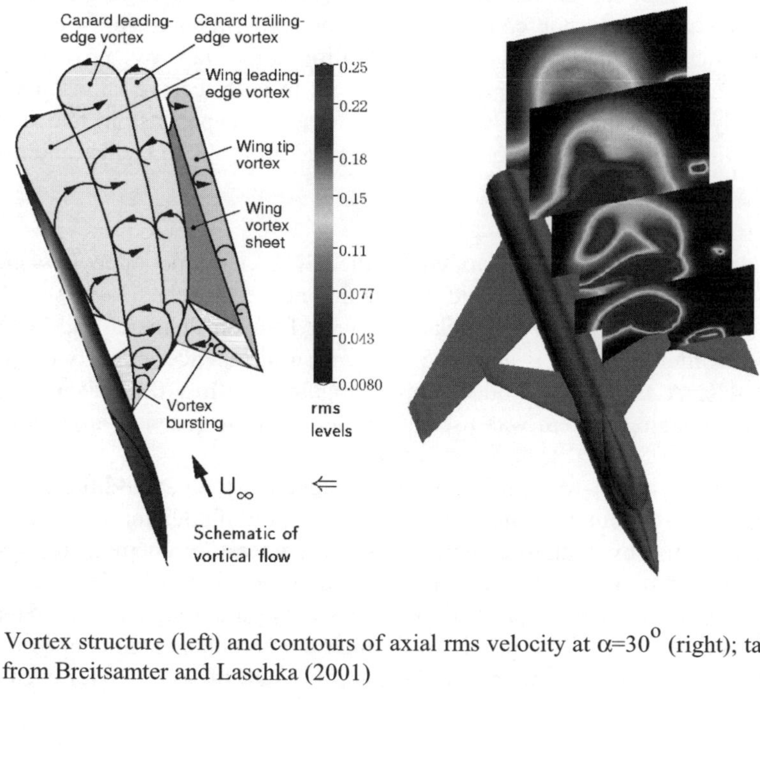

Figure 1 Vortex structure (left) and contours of axial rms velocity at α=30° (right); taken from Breitsamter and Laschka (2001)

A first glance on the generic forward-swept-wing (FSW) geometry model and experimental results can be taken from Fig. 1, presenting the complex vortex structure and contours of axial rms velocities measured at four different cross sections.

Figure 2 X29 – Dryden test Flight center EC87 0182-14

The whole flow case was initiated by in-flight measurements for an X-29A, see Fig. 2 above, at sub-critical (M=0.6) and critical (M=0.9) flight speeds. As flight-test data are often not comprehensive, the experimental study carried out by Breitsamter et al (1997, 1999) is filling this gap by providing detailed measurements on a similar geometry in order to both receive knowledge about the unsteady behaviour of the flow field and to get a deeper insight into flow physics – with the clear view on a comprehensive validation of CFD codes and methods.

10.2 Wind tunnel set-up

The experiments were performed in the Göttingen-type open low-speed wind tunnel of the Institute for Fluid Mechanics at Technische Universität München. The open test section is a circular one with 1.5m in diameter and 3m in length. The maximum velocity is around 55m/s with a turbulence intensity range between 0.3-0.4%. A three-axis model support together with a three-degree-of-freedom probe traversing system was used with the ability to provide an incidence range from 0° to 31.5^{o}.

Dual-sensor hot-wire probes were used together with an additional temperature probe to correct output voltage in case of variation of ambient temperature. Probes were operated by a multi-channel constant-temperature anemometer system and, by means of signal conditioner modules, output voltages were low-pass filtered at 1000 Hz before digitisation and amplified for optimal signal level. Signals were then digitised with 12-bit precision through the 16-channel simultaneous-sampling A/D converter, and together with the sampling rate of 3000 Hz for each channel a

Nyquist frequency of 1500 Hz was obtained. With a sampling time of 26.24 s, each sampling block contained 78,720 points. All in all, random error calculations provided accuracies of 0.2, 1, and 3% for the mean, standard deviation and spectral density estimation, respectively (Breitsamter & Laschka, 2001).

10.3 Test description

With a constant reference velocity of 40 m/s, a Reynolds number of $Re=0.46 \times 10^6$ and a Mach number of M=0.118 was obtained, based on a wing mean aerodynamic chord of 0.167 m. Static pressure and temperature were according to local atmospheric conditions and transition was not fixed. Open test section corrections - together with incidence corrections - have been applied for evaluating force and moment measurements.

Flow field measurements have been conducted in four cross planes, perpendicular to the model axis, x_W/l_W=0.2, 0.5, 0.8 and 1.1, thus the last one already in the wake. Survey points in these cross planes were equally spaced, resulting in a grid resolution of 0.016 in both span-wise and vertical directions (based on the wing span) and a lateral extension of 1.1 wing span.

With the hot-wire probes used in conjunction with their stiff probe holders, no significant effects on vortex breakdown locations and vertical flow structures were detected. This was supported by laser sheet and liquid crystal surface flow measurements at x_W/l_W=0.2, were the largest interference should occur.

10.4 Computational results: mesh generation

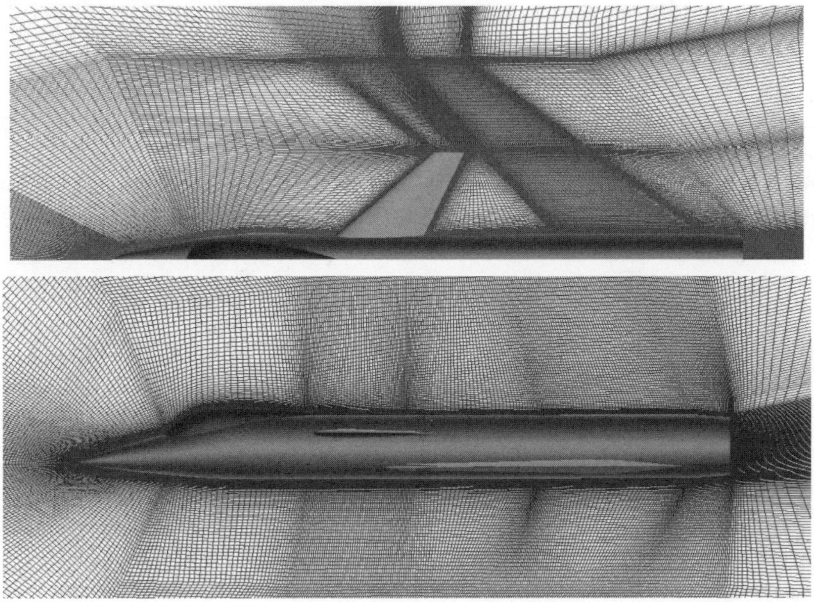

Figure 3 Mesh structure in symmetry and top-view plane (section between canard and wing)

A structured mesh was generated with a fine grid level of 9.91 million points, a medium grid of 1.24 million and a coarse mesh with 0.15 million points. Using ICEM CFD Hexa, the mesh supported full multi-grid on three levels, consisted of 57 blocks, and was generated as a C-type mesh in axial, O-type mesh in lateral and O-type mesh for all boundary layer blocks along the surface. The grid in the vicinity of the aircraft is presented in Fig.3 in symmetry (bottom) and top view (top).

10.5 Comparison of results

Two partners have computed the forward-swept-wing case, ANSYS and EADS-M using the turbulence models specified in Table 1. Highlighted cases (bold) are those that are going to be presented in more detail below.

Table 1 Methods by partners

Partner	**Turbulence model(s)**	**Incidenced computed**
ANSYS/CFX	**SST** k-ε k-ω	10°, 20°, 30°, 40°, 45°
EADS-M	**Wilcox-k-ω without vortex correction** **Wilcox-k-ω with vortex correction**	10°, 20°, 30°

Both partners have thoroughly investigated the complex flow features of this test case. The main effort was put on a proper investigation of the vortices emanating from canard and wing, similar to what is sketched in Fig. 1. It turned out that global forces and moments obtained are only slightly grid sensitive as it can be seen following results shown in Table 2 for 20° incidence.

Table 2 Summary of results for an incidence of 20°

Partner	**Turbulence model(s)**	**Mesh level**	C_L	C_D	C_M
ANSYS	**SST**	**Coarse**	1.0865	0.40000	0.68875
ANSYS	**SST**	**Fine**	1.0913	0.39007	0.72574
EADS-M	**Wilcox-k-ω noCorr**	**Coarse**	1.0201	0.37726	0.66577
	Wilcox-k-ω Corr	**Coarse**	1.0159	0.37423	0.66786
EADS-M	**Wilcox-k-ω Corr**	**Fine**	1.0731	0.40465	0.70963
Measurement			**1.1**	**0.4**	**0.7**

Comparison with measurements seems to be in reasonable agreement when based on fine mesh solutions. Complete lift, moment and drag polars are provided in Figs. 4, 5, and 6, respectively. Results are displayed for the mandatory medium mesh and fine meshes.

In accordance with Table 2, it should be noted that the use of a vortex correction (EADS-M results) did not result in severe changes of the force coefficients. Very likely, the reason might be that – according to the high angles of attack used - the vortices emanating from canard and wing do not really penetrate the wall boundary layers, because they are formed (to a high degree) outside of the boundary layers and thus do not alter the near wall region by induced velocities.

In Fig. 4, lift versus angle-of-attack is provided for the EADS-M results (Wilcox k-ω with vortex correction, Wi88) and the ANSYS results using the SST model in medium and fine mesh. Apart from EADS-M's result at 20°, results seem to be relatively mesh independent as pointed out above. At 40° and 45° ANSYS' results indicate some larger differences between the two meshes, but even here they remain small. In general the comparison between measurements and computations is reasonable, even in the post-stall area beyond 35°. Although difficult to see from the measurements, the slope $dc_L/d\alpha$ seems to be somewhat underpredicted by the computations. This holds for all computations using either k-ε, k-ω or the SST turbulence models.

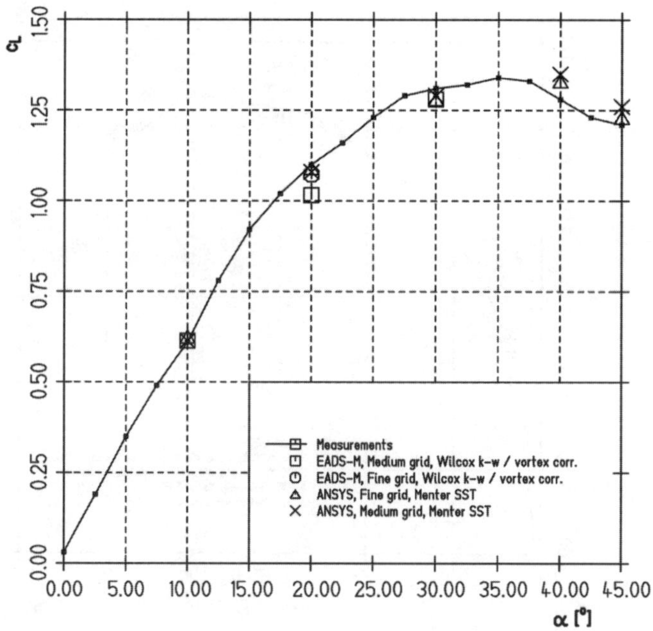

Figure 4 Lift coefficient versus incidence and mesh level

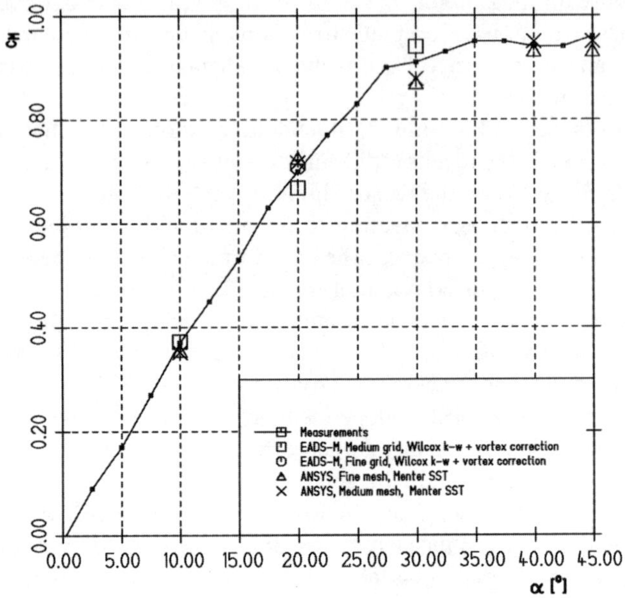

Figure 5 Moment coefficient versus incidence and mesh level

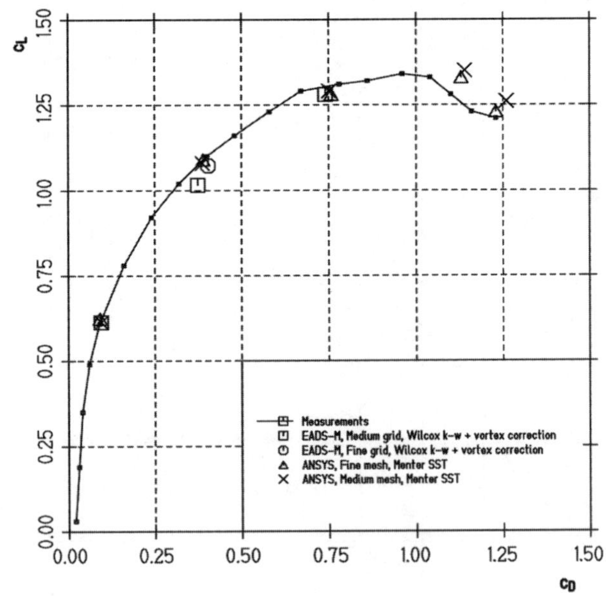

Figure 6 Lift polar and mesh dependence

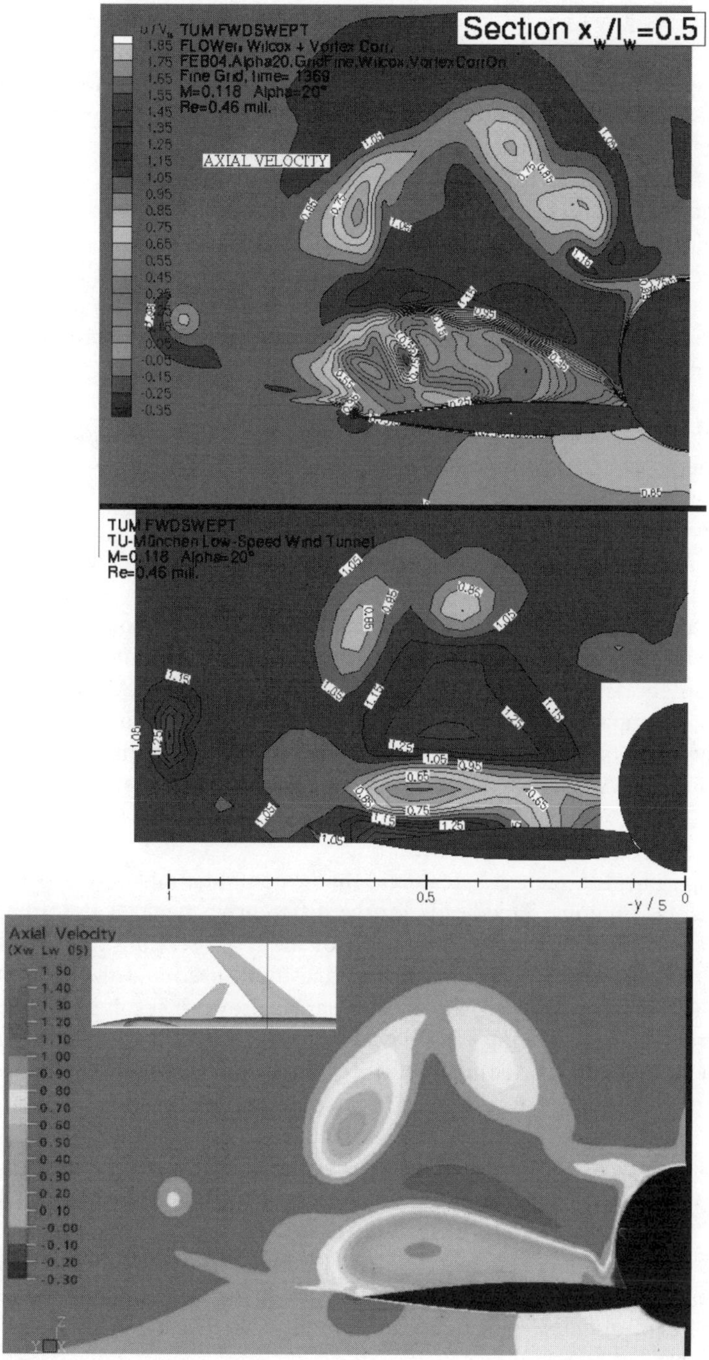

Figure 7 u/U_∞ at x-c=0.5, EADS-M (top), Experiment (middle), ANSYS (bottom)

The comments given above for Fig. 4 also hold – in general – for Fig. 5 exhibiting the pitching moment. Although pitching moment is a quite sensitive quantity in numerical simulations, the agreement with measurements is good, even beyond stall. At 20°, ANSYS' results appear mesh independent while EADS-M's computations show a somewhat larger discrepancy between the two meshes.

Fig. 6 presents the lift polar. It can be seen that ANSYS's computations very well represent the experimental values and that grid dependence is more or less negligible.

When mentioning above that the vortices emanating from canard and wing do not penetrate the wall boundary layers, flow investigation of cross-sections should provide important means of arguing on the accuracy of computations compared to measurements. As it can be detected from Fig. 1, several cross sections have been measured.

In Figures 7 and 8, for the cross-section $x_W/l_W=0.5$, the station close to the wing-leading-edge-body junction, see geometry inserts in the lower part of the two figures, main, Fig. 7, and lateral, Fig. 8, velocity contours are presented and compared to the experimental findings.

Fig. 7 exhibits velocity contours, u/U_∞, for $\alpha=20^o$ incidence. The top figure denotes EADS-M's results, the middle part represents the measurements and the bottom figure shows ANSYS's computations. All simulations have been carried out on the fine mesh, with the Wilcox model (Wi88) plus the Kok rotation correction for EADS-M and the Menter SST model for ANSYS. It should be noted that comparing the level of grey shading in the domain farer away from the wing shows some difference between computations and measurements (middle part of Fig. 7). Analysing the data one can find that there is really a difference in the u/U_∞ data with u/U_∞ about 1.0 in the computations and u/U_∞ around 1.1 (average values) in the measurements.

Comparing the general features, both computations agree well with the experimental findings. This holds for the different vortices, as described in Fig. 1, i.e. the canard vortices in the upper part of the cross-sections and the wing vortex on the left hand side. According to the fine mesh used and the vortex correction applied (EADS-M), the gradient in these areas are well resolved. Flow prediction in the area near to the wing is somewhat off in the EADS-M results. The surface-dominated area is larger compared to the measurements and the structure is much more pronounced. In that area even reverse flow in the main velocity direction has been detected, which holds for both computations, although ANSYS's results are only close to vortex bursting, indicated by the near-to-zero main velocities in the wing area, but show separation in the domain of the wing-body junction - which is not the case for EADS-M's results. It is known however, that for (any) RANS computations, vortices emanating from corners like wing-body junctions and forced by the body (or side-wall) boundary layer(s) are often overpredicted in strength. Unfortunately, measurements in that area are not available (hot wire experiment).

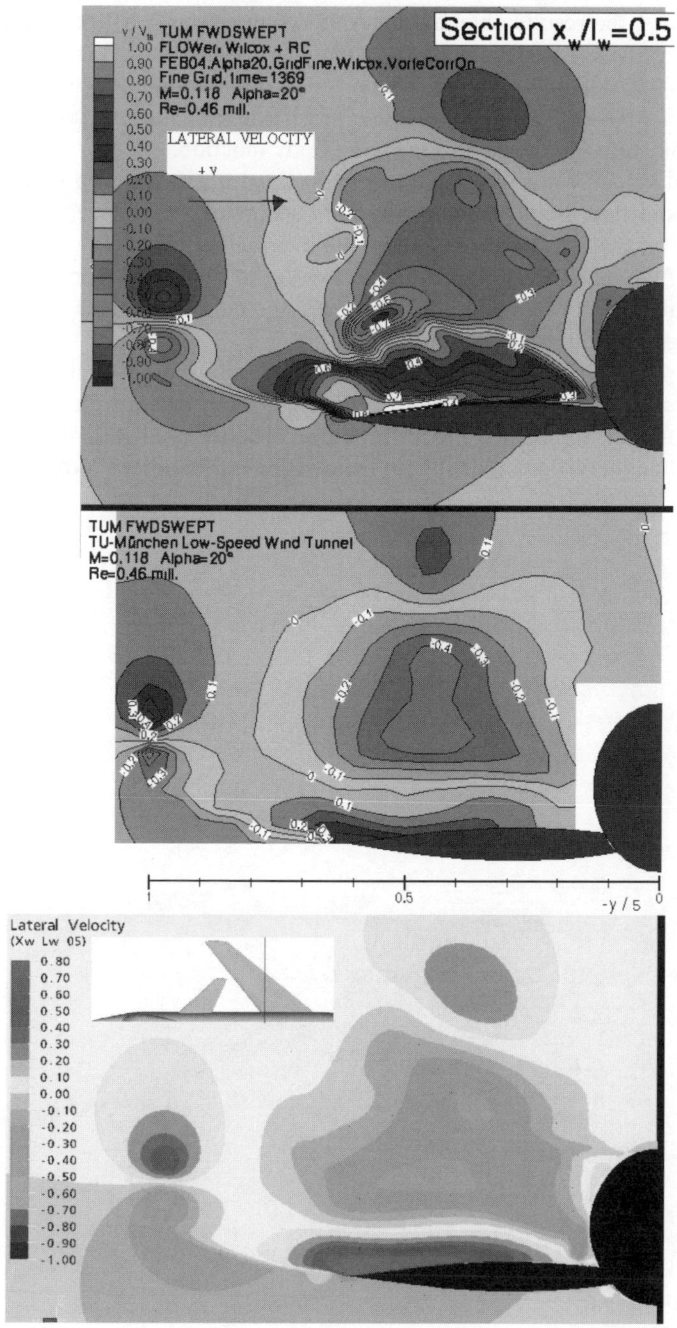

Figure 8 v/U_∞ at x-c=0.5, EADS-M (top), Experiment (middle), ANSYS (bottom)

Nevertheless, the results obtained by ANSYS agree much better in the near-wing area where EADS-M's results show (wing-)vortex bursting. This is, of course, the reason why EADS-M computations do not show a wing-body-area separation as seen in the ANSYS results because the large extent of the wing vortex introduces more momentum - in the area of the wing-body junction.

Lateral velocities are provided in Fig. 8. Canard and wing-tip vortices are well captured and flow field gradients are quite accurately reproduced. Again, EADS-M results for the lateral velocity show the overprediction of the vortex field in the vicinity of the main wing. According to Fig 7, the small differences in the level of grey shading are not related to differences in the data themselves, due to a slightly different graphical representation.

When considering that both numerical methods (the unstructured CFX-5 and the structured DLR-FLOWer code) are of the same or equivalent accuracy then the Menter SST turbulence model provided a better predictive accuracy for this test case than the Wilcox model with the rotation correction by Kok included.

11 TUM Delta wing

A. Davroux, Dassault Aviation

Abstract

A well-known category of flows where Boussinesq's hypothesis is expected to fail is vortex-dominated flows. Unfortunately, many of the flows of industrial interest involve vortices, such as those emanating from the roof of a car, or from the wing tips of airplanes. The delta wing is another example of a flow where the accurate resolution of the physics of the vortex is a key parameter in the computation. This is why this test case was under consideration within the FLOMANIA project.

Several turbulence models were computed by the partners involved in this project, the complexity of which extends from improved two equations models with ad hoc corrections such as the SAERC, to a full RSM computation. EARSM models are in-between, and were shown to improve significantly the results. Comparison with the experiments is globally fairly good for both 25° and 35° angles of attack; a few discrepancies between the models are analyzed thereafter.

11.1 Presentation of the experimental set up

The geometrical parameters of the delta wing under consideration are presented in Fig. 1. It is a sharp leading edge delta wing with an aspect ratio of 1. Measurements were carried out by Breitsamter et al. at the Technische Universität München, in a closed-return wind tunnel with a 1.5m diameter nozzle. For more information about the measurements, please refer to (Breitsamter and Laschka, 1993) and (Breitsamter, 1997).

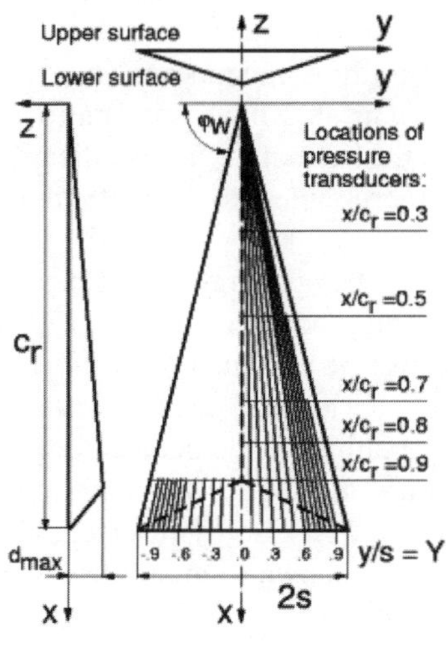

Cr = 0.670 m. (chord length of wing)
Dmax = 0.057 m. (max.thickness at 90% chord)
lμ = 0.447 m. (mean aerodynamic chord)
2s = 0.335 m. (full span)
φ_w = 76° (leading edge sweep angle)
Λ = 1. (aspect ratio)

Figure 1 Geometry of the delta wing

The delta-wing tests were conducted for four angles of attack, namely 12.5°, 25°, 30° and 35°. As mentioned before, two of those will be under consideration in the present study, 25° and 35°. The experimental results show that there is no bursting at 25°, with an increased size of the secondary vortices compared to the 12.5° case. At 35°, however, vortex breakdown (an abrupt disorganization of the vortex) occurs. Its location was estimated to be at mid span of the delta wing, but as noted by the authors, the intrusive technique used for the measurements may have had a significant influence on these results. This point will be analysed and discussed hereafter, with the enlightening experience of computational EARSM results.

Carried-out measurements include pressure distribution at different streamwise stations x/C=30%, 50%, 70%, 80% and 90%. Hot wire measurements systems were used to get detailed insight in the structure of the flow, providing velocity fields in the same sections. In addition, flow visualizations using the laser light sheet technique were performed. The liquid crystal technique was also used; it uses a shear stress sensitive/temperature insensitive liquid crystal coating, which enables a deeper insight into the physics of the flow.

The free-stream velocity is constant and equal to 37m/s. Experimental conditions were ambient static pressure and room temperature (no additional information is available). The experimental Reynolds number, based on the mean aerodynamic chord, is approximately 10^6. No specific treatment for the transition was applied.

11.2 Parameters of the computations

11.2.1 Meshes

Three different eight-block structured meshes were provided by EADS-M, with a number of nodes of 3.10^6, 5.10^6 and 10^7 respectively. The 3 million node mesh was used as a baseline by Alenia, who generated a coarser mesh by taking out every other point in the i-direction, resulting in a 1.5 million-node mesh. The mesh used by EADS-M derived from the 5 million-node mesh which was improved after several computations. Both meshes have a conical structure, a visualization of which is available in Fig. 2. These two meshes are suited for low Reynolds number computations.

A one million-node fully unstructured tetrahedral mesh generated by Dassault Aviation is also available (see Fig. 2). Since the leading edges of the delta wing are sharp, it was decided to take advantage of the use of wall functions. Special care was taken to refine the mesh in the vicinity of the leading edge in order to capture the strong velocity gradients in this area, resulting of the acceleration of the flow from the windward side to the leeward side. A two million-node mesh was derived from this mesh for the computation of the 35° case by resorting to a local refinement in the vortex area. This refinement was shown to greatly enhance the predictions.

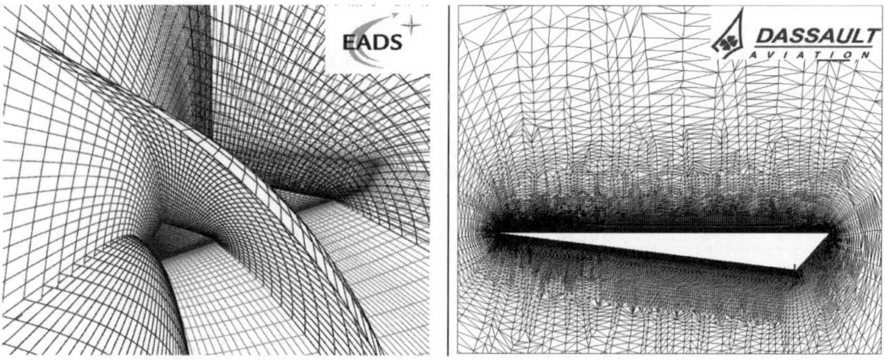

Figure 2 View of the meshes used - left: representation of the conical structure of the structured meshes right: view of the symmetry plane of the unstructured mesh

11.2.2 Inflow conditions

The available flow parameters are the free stream velocity : U_∞=37 ms^{-1}, and the Reynolds number (based on the mean aerodynamic chord) $Re_{l\mu}=10^6$. Stagnation pressure and temperature are unknown. The Reynolds number used for computations was the same for all the partners involved, corresponding to the experimental value. On the opposite, the Mach number (which is not available from the experiment) is different: EADS-M and Alenia used an inflow value corresponding to Ma=0.15, while Dassault Aviation used 0.10. Nevertheless, since the Mach number is very low, the difference is expected to be negligible. This conclusion is enforced by the comparison of the results presented hereafter.

11.2.3 Turbulence models

Alenia performed computations on the 25° and 35° angles of attack. The turbulence model under consideration is the k-Rt, coupled with EARSM of Wallin and Johansson, with the rotation curvature WJ-G.

Dassault Aviation also performed computations on both 25° and 35° angles of attack. The turbulence models used for the computations are the k-ε LS R and the EARSM extension of Wallin and Johansson based on this model.

EADS-M focussed on the 25° angle of attack. The SAE model with and without rotation curvature (SAERC, SAE), and the SST-k□ model, with and without rotation curvature (SST-R) were under consideration. In addition, EADS-M computed this angle of attack with the SSG-ω model, which is of utmost interest in order to evaluate the performance of the improved modelization of turbulence on this relevant test case.

11.2.4 Summary

To summarize the contributions of the partners, one could refer to Table 1. As a result of the various contributions of the three partners, a lot of interesting comparisons can be performed, which deal with the efficiency of the rotation curvature, the improvements of the EARSM extension, and the RSM modelling. Those comparisons are presented in the next section.

Table 1 Summary of the contributions of the partners

	Alenia	Dassault Aviation	EADS-M
Meshes	$1.5.10^6$ nodes	10^6 and 2.10^6nodes	$5.1.10^6$ nodes
	structured	unstructured	structured
Inflow condition	Ma=0.15, aoa=25° and 35°	Ma=0.1, aoa=25° and 35°	Ma=0.15, aoa=25°
Turbulence model	WJ-Rt	k-ε LS R	SAE, SAERC, SAD, SST, SST+R, SSG-ω
	Low-Re model	Wall functions	Low-Re model

11.3 Analysis of the results

Results with several turbulence models are available. They can be classified into three categories: These based on "usual" two-equation models related to the Boussinesq hypothesis (SA, k-ω SST and derivatives based on rotation corrections), these based on an Explicit Algebraïc Reynolds Stress modelling (k-Rt+EARSM used by Alenia, and k-ε +EARSM used by Dassault Aviation), and one with a fully resolved Reynolds stress tensor (RSM used by EADS-M).

11.3.1 Influence of the rotation correction

Two baseline models results are available: the SA model and the TNT model, whose results are compared hereafter.

SAE results

The model shows a much lower eddy viscosity level than that obtained with the k-ω model, with a factor of about 50 to 100. The results of this model are available in Fig. 4. As a result, there is nearly no eddy viscosity production around the vortex. Therefore, the two tested rotation corrections, namely the Dacles-Mariani (SAD) and the Spalart-Shur (SAERC), show no difference at all when compared to the SAE computations, and therefore are not presented here. This can be easily understood, since these corrections are based on a limitation of the production term.

k-ω results

Results obtained with the k-ω SST with and without the rotation correction are presented in Fig. 4, and compared with the SAE results. The first thing to be noted is that the results are in a good agreement with the experiments. The pressure over the first half of the wing is slightly underestimated, as can be seen at location x/C=30% and 50%. This is especially true for the SAE model, whereas the k-ω model improves the predictions. The rotation correction, albeit small, seems to be active only in this area. Unfortunately, the effect is wrong: it tends to decrease the pressure coefficient level whereas it is underestimated with respect to the experimental value. Downstream, the predictions are almost identical.

11.3.2 EARSM influence

Alenia and Dassault Aviation performed computations with EARSM extensions, for both 25° and 35° angles of attack. Alenia used the k-Rt model with the EARSM extension of Wallin and Johansson with rotation curvature, whereas Dassault Aviation used a k-ε LS based EARSM(WJ) without this rotation correction. Computations with the k-ε LS are also available to evaluate the improvements linked to the use of the EARSM formulation.

25°

Comparison of the Cp distribution predicted by the computations and the experimental data is presented in Fig. 5.

Firstly, one can note that the computed Cp distributions are quite identical on the windward side. Although there are no experimental data, this tends to indicate that the slight difference between the inflow conditions (Ma=0.15 for Alenia and Ma=0.1 for Dassault Aviation), does not have a significant influence on the Cp distribution. Note that the discrepancy at x/C=0.9 is related to the base region and the resulting geometrical singularity. Similarly, the modelization does not affect the Cp distribution on the windward side, since the k-ε and the k-ε +EARSM results are identical.

On the leeward side, the results are very similar as well for all the models. The location of the vortex, which corresponds to a peak of Cp, is in good agreement with the experimental data. A tiny difference appears in the neighbourhood of the trailing edge (x/C=80% and x/C=90%) : the WJ-Rt and the k-ε LS present a slight underestimation of the Cp distribution, the k-ε +EARSM being in better agreement with the experimental values.

35°

This angle of attack is high enough for the breakdown to occur. Bursting location was predicted at x/C=62% for the WJ model, and at x/C=60 % for the WJ-Rt, The agreement of the computed location with both EARSM models is fairly good, around 60% of the chord, whereas the bursting occurs nearer to the leading edge at x/C=50% with the k-ε model. These values are to be compared to the experimental location at mid chord. But as explained hereafter, this value may have been influenced by the intrusive technique used for its measurement.

Comparison of the Cp distribution predicted by the computations and the experimental data is presented in Figure 6. The Cp distributions on the windward side predicted by all the computations are almost identical, which is in line with the previous conclusions on the 25° case. On the leeward side, the EARSM results are in very good agreement with the experimental data, though this angle of attack is more challenging than the previous one. In particular, at x/C=50% where bursting was measured experimentally, the comparison is still fairly good, whereas it is not the case for the k-ε model, which already predicts a breakdown of the vortex. Since the agreement in Cp distribution is better with the EARSM models, whereas they would seem to delay the bursting location, the warning of the authors concerning the value measured for the location of the breakdown seems to be justified. If the bursting effectively occurred at x/C=50%, EARSM results would not be in such a good agreement with the experimental Cp distribution, and conversely, k-ε results would be in a better agreement with the experiments. Therefore, this would imply that the experimental location of the bursting was effectively influenced by the way it has been measured. If so, this value would be expected to be underestimated (i.e. the bursting would indeed be located downstream x/C=50%), that would be in favour of a better agreement of the computational results with the experimental data. The only thing that can be taken for granted is that the bursting occurs between 50% and 70% of the chord.

As a general comment it must be added, that the measurements were not (really) intrusive, the upstream influence in terms of vortex bursting is less than 2% chord. This is supported by the fact that the vortex burst above the slender delta wing and not near to the trailing edge recompression area. Moreover, because of the high turbulence intensity and axial velocity in the vortex core, it is the core area that is deformed first, without a (larger) influence on the surface pressure below the vortex.

11.3.3 RSM computations

One of the goals of FLOMANIA was to have an RSM model available in industrial CFD codes. The FLOWer code indeed was used by EADS-M to perform a computation on this test case with the SSG-ω model, thus providing a high added-value contribution. Since this case seems especially well suited for an RSM computation, a more detailed description of this computation is presented.

Computational setup

Computations were run on two different grids of respectively 5M nodes and 10M nodes, the last one being derived from the previous one by doubling each plane in the y-direction (normal to the surface of the wing). Convergence is achieved by resorting to local time stepping, and to a multigrid technique.

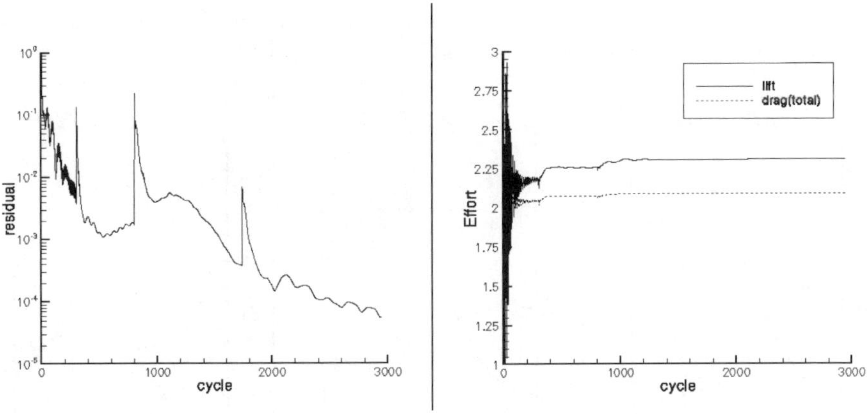

Figure 3 Convergence of the RSM computation on the basic grid left : residual, right : effort (=forces)

As can be seen in Fig 3, the convergence of the RSM computation on the basic grid is quite good, with a residual reduction of three orders of magnitude, and the efforts (lift and drag) converging towards a steady value. Unfortunately, convergence of the turbulence equations (not shown here) is not as good, with a smaller rate of convergence. This would imply that more iterations will be necessary in order to achieve the convergence of the turbulence equations. The same difficulty in convergence was reported, even with two equation models. Nevertheless, since the purpose is to compare with the experimental Cp distribution, the convergences of the residual and of the global efforts are a good indicator of the convergence of the Cp distribution.

The additional cost of a full RSM computation compared with a similar computation (with thc same number of iterations) with a k-ω model is around 1.5 for the CPU time, and about 2 for the memory.

Results obtained with the RSM model

Comparison of the Cp distributions for 25° is presented in Fig. 7 for both computations on the basic and on the fine mesh. The overall agreement with the experiments is good. It can be noticed that the peaks of Cp is higher than those predicted with two equation models (see Fig. 4), resulting in a better agreement with the experiments. As a result, the secondary vortex characterized by a second maximum in -Cp near the leading edge (y/b around 1) is smaller than those predicted by two-equation models. The same behaviour was observed when comparing k-ε and EARSM k-ε based results (see Fig. 5).

Quite surprisingly, at x/C=50%, 70% and 80%, the levels of -Cp are overpredicted. This may be related to the non dissipative nature of the RSM equations, whereas the two-equations modelling, dissipative by nature, tends to underestimate the levels of -Cp.

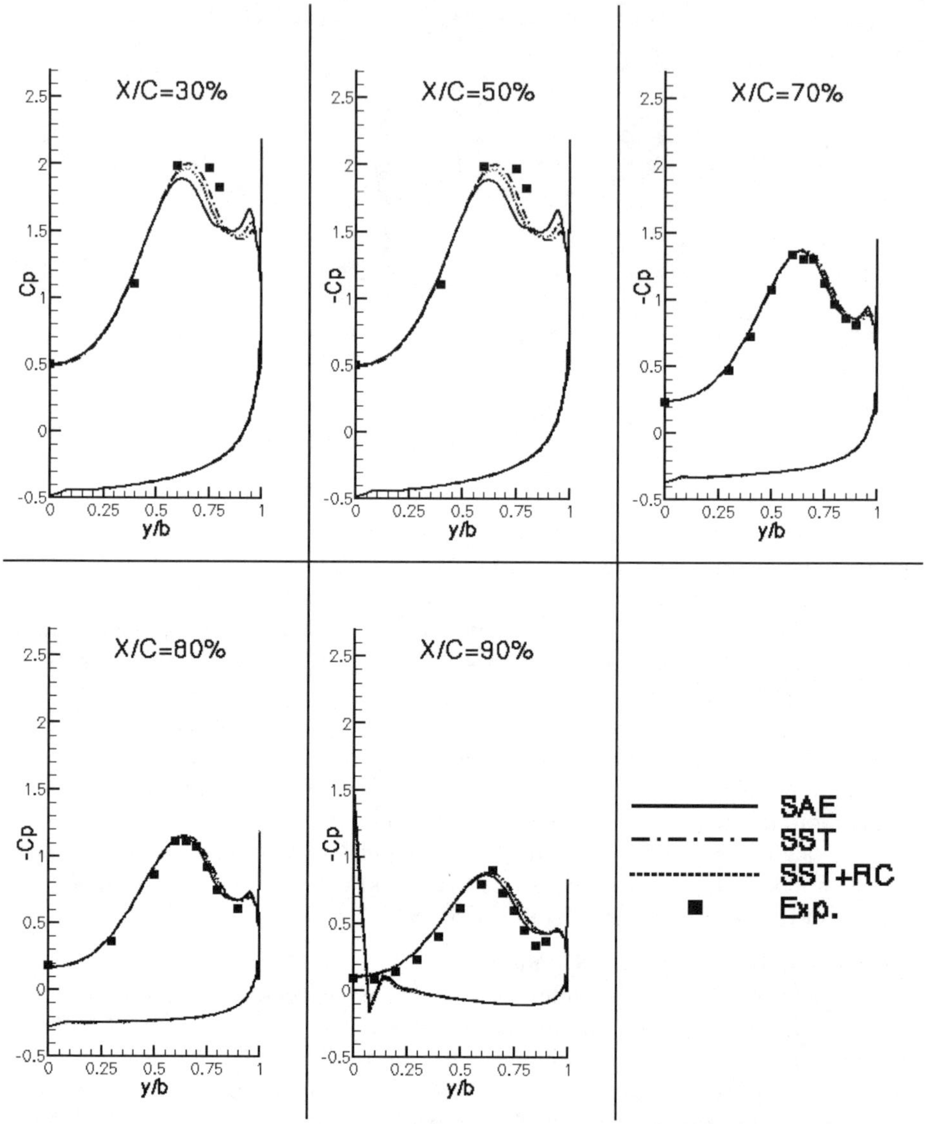

Figure 4 25° - SAE and k-ω computations

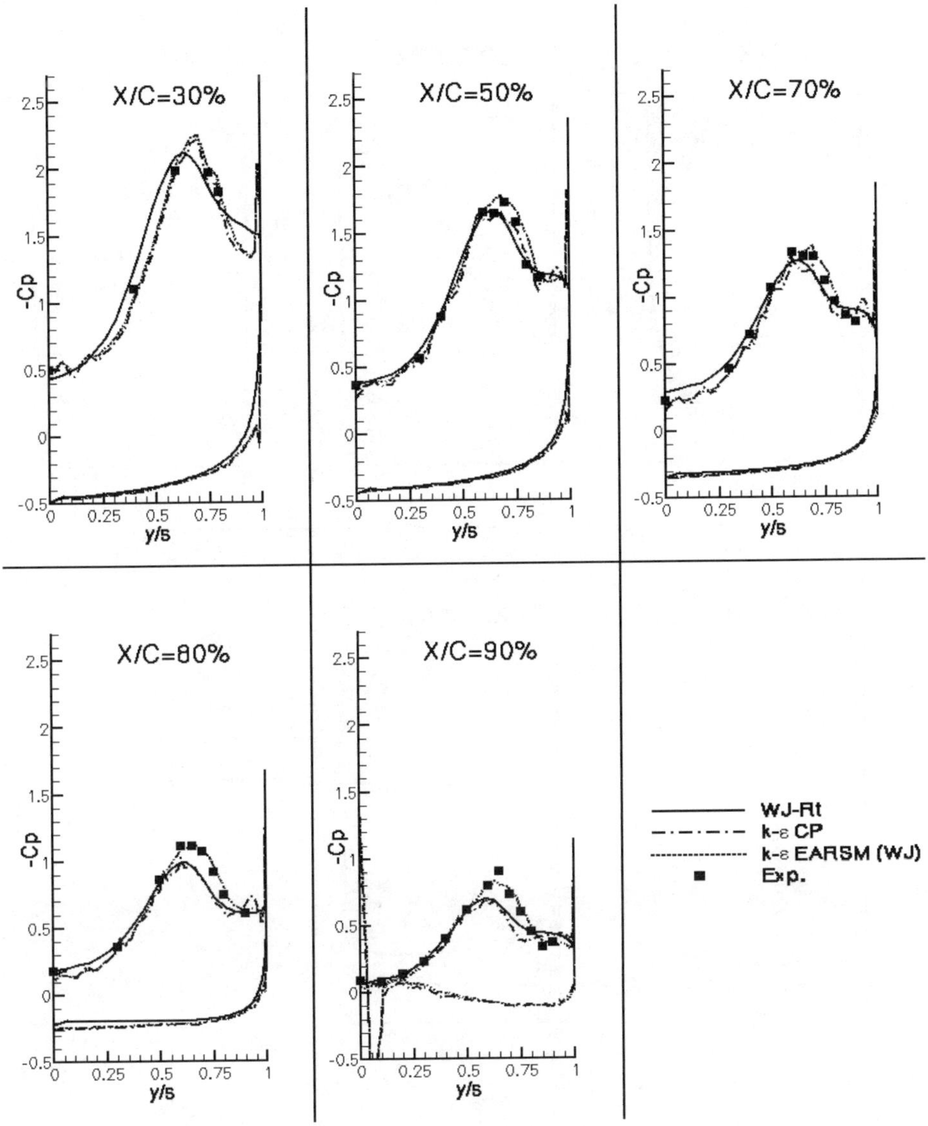

Figure 5 25° - EARSM computations

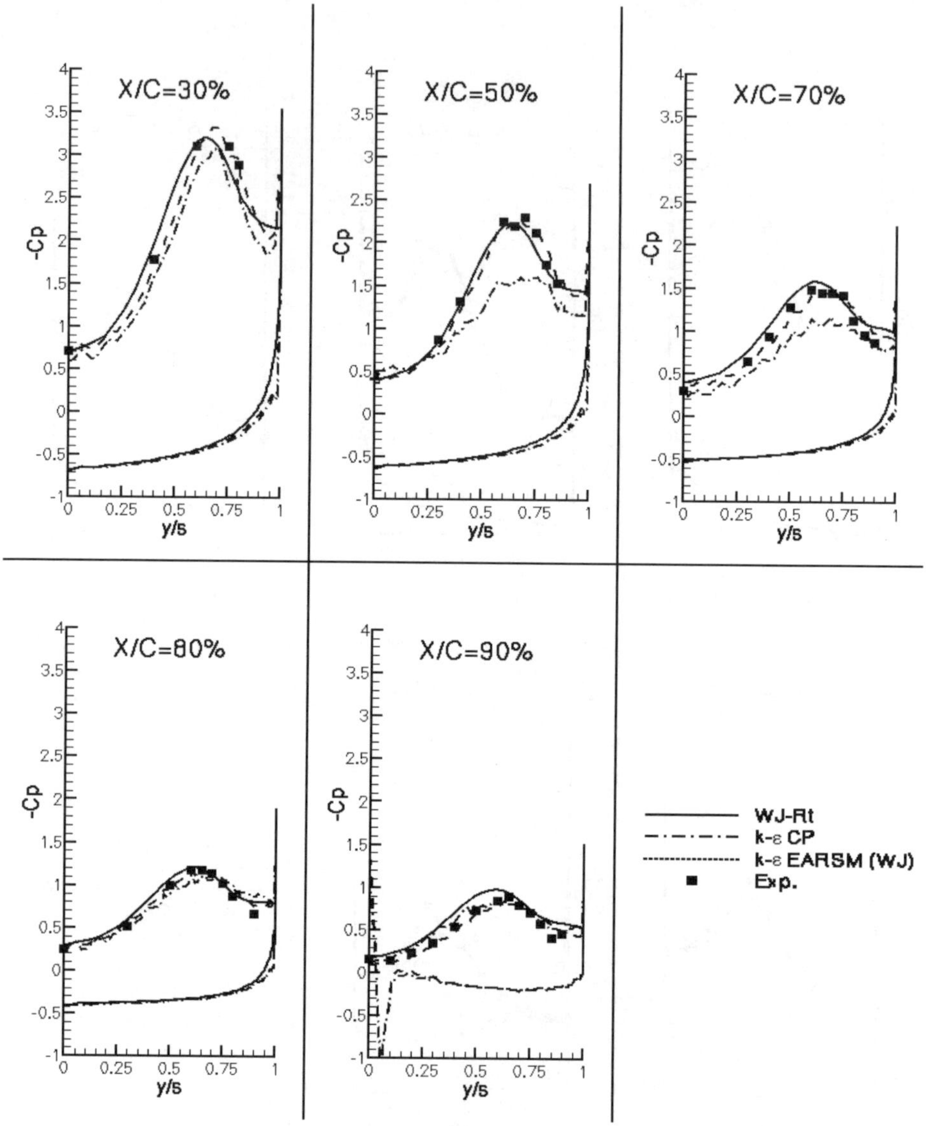

Figure 6 35° - EARSM computations

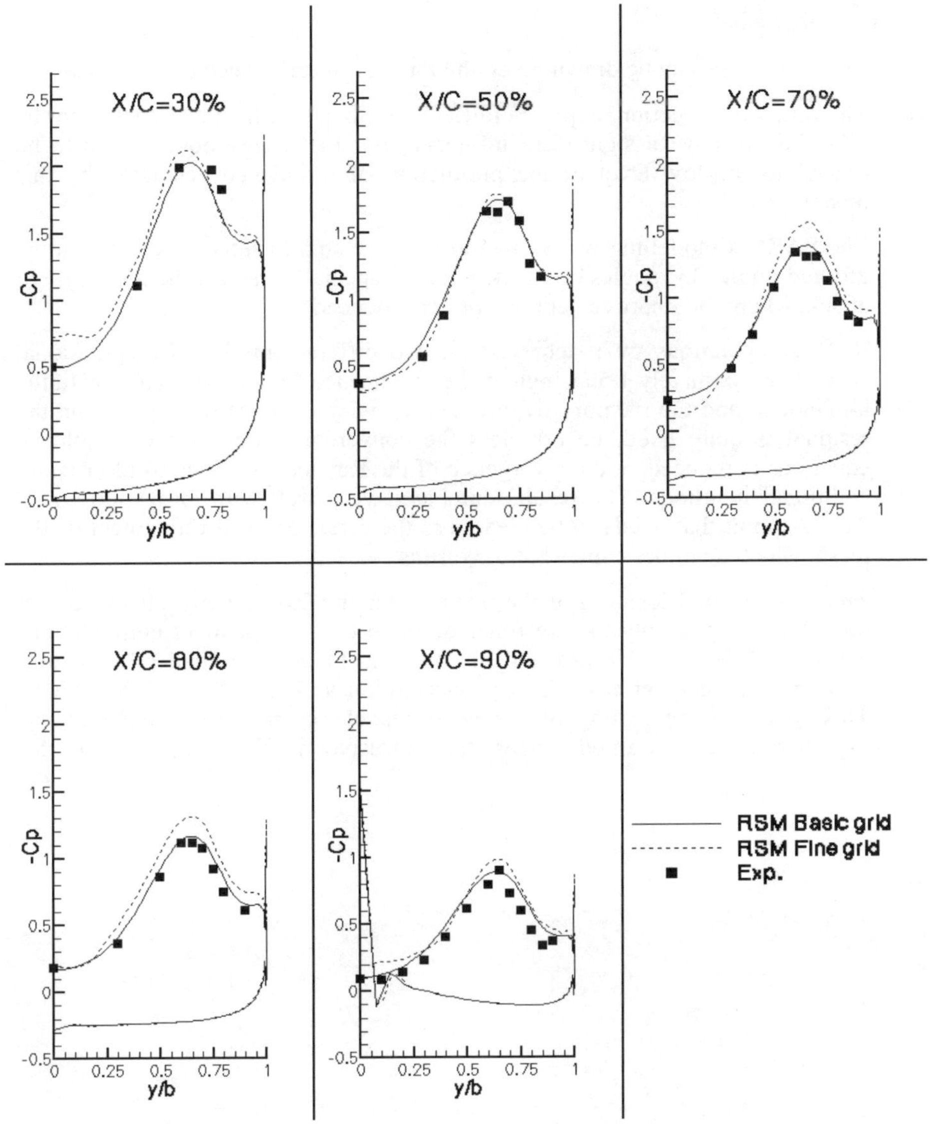

Figure 7 25° - RSM computations on a 5.10^6 nodes and 10^7 nodes grids

11.4 Conclusions

Several conclusions can be drawn given the number of turbulence models used:

1. The rotation correction, applied either to the Spalart-Allmaras model or to the SST did not have a significant influence. For the SAE model, this is to be related to the low level of the production of turbulence generated by this model.

2. The EARSM modelling was shown to improve significantly the results on the 25° and on the 35° angles of attack. This is especially true for the 35° angle of attack, where the improvement is more pronounced.

3. RSM computations were achieved on two different meshes. Computational cost is approximately 1.5 as high as the one needed for a computation with the k-ω model, and the memory requirement is twice as big. Convergence on the residual is quite good, nevertheless the convergence rate of the turbulence equations is reduced, and convergence of the Reynolds stresses would require additional iterations. The results compared very well with the experimental data. A point that needs to be cleared is the effect of local refinement of the mesh, which degrades somewhat the results.

4. The experimental location of the bursting for the 35° angle of attack case at mid-chord is probably underestimated, since two independent computations using EARSM model estimated this bursting to occur around 60%, with a very good agreement with the Cp distribution at X/C=50% and X/C=70%. This was also the feeling of the wind tunnel experts (who conducted the experiments and warned about the intrusive technique used for this measurement).

12 3D circular cylinder

R. Perrin[1], M. Braza[1], E. Cid[1], S. Cazin[1], A. Sevrain[1], M. Strelets[2], M. Shur[2], Y. Hoarau[1] A. Barthet[1], G. Harran[1], and F. Moradei[1]

[1]Institut de Mécanique des Fluides de Toulouse, Groupe EMT2
[2]Federal Scientific Center "Applied Chemistry", St.-Petersburg

Abstract

The main objective of the present study is to analyse the turbulence properties in unsteady, strongly detached flows and to provide a database for improvement and validation of turbulence models, concerning the present class of non-equilibrium flows. The flow around a circular cylinder with a low aspect ratio (L/D=4.8) and a high blockage coefficient (D/H=0.208) is investigated. This confined environment is used in order to allow direct comparisons with realisable 3D Navier-Stokes computations avoiding 'infinite' conditions. The flow is investigated in the critical regime at Reynolds number 140,000. A cartography of the velocity fields in the near wake of the cylinder is obtained by the 2D Particle Image Velocimetry. Statistical mean (Reynolds averaged) and phase-averaged quantities are determined. Furthermore, POD analysis is performed on the data set in order to extract coherent structures of the flow and to compare the results with those obtained by the conditional sampling technique. The Reynolds stresses, the strain-rate and vorticity fields as well as the turbulence production terms are determined. These physical quantities where not measured in this detail up to now, to our knowledge and they are very useful for the development of advanced turbulence modelling techniques for unsteady flows. Based on the IMFT's circular cylinder test-case, computations are carried out by advanced statistical turbulence modelling and by DES. The partners involved are IMFT and STPU.

12.1 Test-case description and experimental set-up

12.1.1 Configuration

The experiment has been carried out in the wind tunnel S1 of IMFT. The channel has a 670x670 mm^2 square cross section. The cylinder spans the width of the channel without endplates and has a diameter of 140 mm, giving an aspect ratio L/D=4.8 and a blockage coefficient D/H=0.208. The upstream velocity U_{inf} at the centre of the channel is 15m/s, so the Reynolds number based on the upstream velocity and the cylinder diameter D is 140,000. The free stream turbulence intensity, measured by hot wire technique in the inlet was found 1.5%. All the quantities have been dimensionless by the U_{inf} and D.

12.1.2 Measurements

Wall pressure measurements have been carried out with a pressure transducer Validyne DP15-20. The frequency response is flat up to about 500 Hz. The acquisitions have been achieved during 2 to 5mn at a sampling frequency of 1 kHz. The uncertainty on the Cp coefficient is estimated to 3.5%.

For the PIV measurements, a double-pulsed Nd-YAG laser Quantel (2x200mJ) and PCO-sensicam cameras (1280x1024 pixels) have been used. The seeding particles used have been DEHS. A typical size of the particles is 1µm. The system, both camera and laser, has operated at a frequency of 4Hz. The measurements have been carried out in the near wake of the cylinder in the (x,y) plane located at the middle span position $z=0$ (Figure 1a).

2C-PIV

The camera was equipped with a 35mm objective lens at a diaphragm aperture of 11. The size of the measurement area has been 238 x 188 mm (1.34D x 1.7D). The software used to analyse images is a product of IMFT ("service signaux-images"). The algorithm is based on a 2D FFT cross-correlation function implemented in an iterative scheme with a sub-pixel image deformation, according to Lecordier & al. (2003). The flow has been analysed by cross-correlating 50% overlapping windows of 32×32 pixels. This yielded fields of 77 x 61 vectors with a spatial resolution of 3.13mm (0.0224D). This resolution is proven to be sufficient for the evaluation for the evaluation of the major part of the turbulent stresses, according to tests that have been carried out by using smaller PIV planes and a calculus of vector with an interrogation window of 16x16 pixels. Nevertheless, the smallest-scale turbulence beyond the above resolution cannot be provided in the energy spectrum, as is generally the case for any PIV experiment.

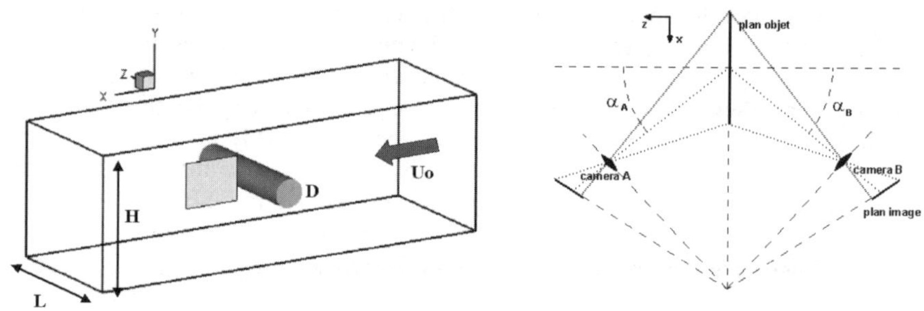

Figure 1 Left: flow configuration, right: Scheimpflug configuration

Reynolds averaging

About 3000 pairs of images were analysed to generate converged turbulence statistics with the 2C-PIV. The uncertainties are estimated, using a 95% confidence interval. They are 0.02 for U, 0.03 for V mean components, and 0.015 for $\overline{u^2}$, 0.02 for $\overline{v^2}$ and 0.01 for $\overline{uv}$ correlations. For the 3C-PIV, 2570 instantaneous flow fields were acquired. The differences between the 2C-PIV and 3C-PIV results are lower than 0.03 for the mean components and lower than 0.04 for the correlations.

Phase averaging
The nearly periodic nature of the flow, due to the von Kármán vortices, allows the definition of a phase and the calculation of phase averaged quantities. The flow is classically decomposed into a mean component, a periodic fluctuation and a random fluctuation as $U_i = \overline{U_i} + \tilde{U}_i + u_i'$ (Hussain and Reynolds (1975)). The phase average quantity is then $\langle U_i \rangle = \overline{U_i} + \tilde{U}_i$. The phase averaged quantities have been measured with the 2D-PIV technique. The trigger signal used as an indicator of the vortex shedding is the pressure on the cylinder at an angle $\theta=70°$ with the forward stagnation point. This location is near the separation and upstream the transition. Therefore, the signal has a strong quasi periodic component at the Strouhal frequency, and it is not very affected by turbulent fluctuations. A typical pressure signal and its power spectrum are shown on Figure. 2.

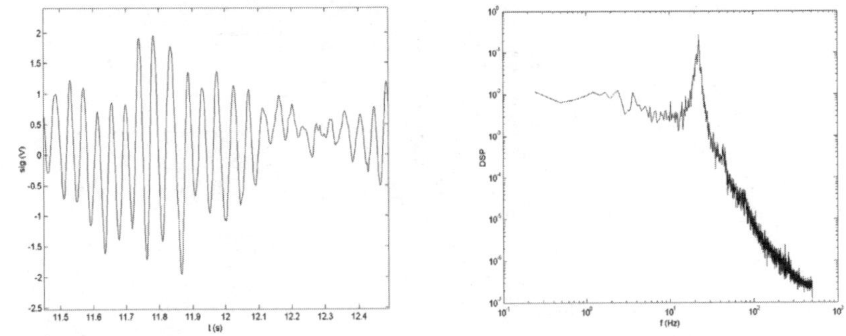

Figure 2 Wall pressure signal at θ=70° (left); power spectrum (right)

Both, PIV images and pressure signal are acquired and stored to obtained phase averaged quantities. By post-processing the pressure signal, the phase is determined at each instant of acquisition of PIV. The flow fields are then ranged in 16 classes corresponding to phase angles dividing a period ; each class is in fact a window of width 2π/128. Statistics are then performed in these classes. From the trigger signal, the phase of the flow is determined by using the Hilbert transform technique, as in Wlezien & al (1979). The Hilbert transform allows calculation of the instantaneous envelope and phase from a band limited signal. Before applying the Hilbert techniques, a bandpass filter is applied to the signal. The independence of the results from these parameters has been checked. Finally, about 170 images are collected per classes and averaged. The estimated uncertainties are 0.07 for $<U>$, 0.1 for $<V>$, 0.05 for $<u^2>$, 0.08 for $<v^2>$, 0.04 for $<uv>$.

12.2 Experimental results

12.2.1 Flow regime

Measurements of mean wall pressure coefficient have been carried out in the median section z/D=0, around the cylinder, every 10° for different Re numbers from 65,333 to 191,333 (Figure 3a). The mean pressure drag coefficient is evaluated by integration of the pressure (Figure 3b). The base-pressure coefficient, ($-C_{pb}$) is found higher than in non-confined flow conditions, because of the blockage ratio. This yields a drag coefficient higher than in a non-confined case. The drag decrease shows that the flow is at *the beginning of the critical regime.* This regime occurs at lower Reynolds number than reported in Roshko (1961), because of the free-stream turbulence intensity (1.5%).

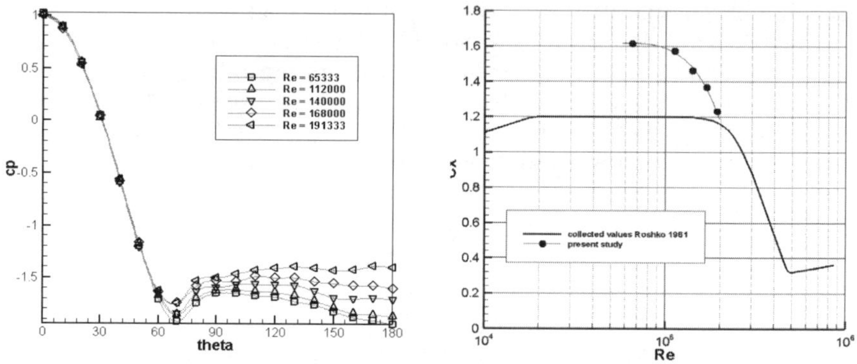

Figure 3 Mean wall pressure coefficient around the cylinder (left); mean pressure drag coefficient versus Reynolds number

12.2.2 Reynolds averaged fields

The topology of the mean flow at Re=140,000 is studied in this section according to Reynolds averaging decomposition. Streamlines, iso-U and iso-V contours measured by 2C-PIV are shown in Figure 4. Capital letters indicate Reynolds averaged quantities. As expected, a two symmetric eddies pattern is obtained, due to the averaging of the passage of the alternating vortices, resulting in a symmetric pattern for U and in an antisymetric one for V. As expected, the mean spanwise component W, measured by 3C-PIV is found to be null. The dimensionless recirculation length l_c is found 1.28 ±0.03 with 2C-PIV and 1.23 ±0.03 with 3C-PIV. Values between 1.1 and 1.4 are found by several authors (Cantwell & Coles (1983), Norberg (1998),...) in the same Re number range. Given the difference in the boundary conditions (blockage, aspect ratio and inlet turbulence intensity) and experimental details having an important influence especially in the critical regime, where the global parameters vary rapidly with *Re* number, the present results appear to be reasonable.

The mean velocity gradients, then the strain rate tensor S and the rotation rate tensor ω are calculated with a central difference scheme. Figure 5 shows the iso-contours of $S_{12} = \frac{1}{2}\left(\frac{\partial u}{\partial y} + \frac{\partial v}{\partial x}\right)$ and $\omega_{21} = \frac{1}{2}\left(\frac{\partial v}{\partial x} - \frac{\partial u}{\partial y}\right)$. As expected, a two-lobe antisymmetric configuration is reached. The mean strain rate and the mean rotation rate are of the same order of magnitude. The vorticity maximum is found at the location x/D=0.6 and y/D=0.55.

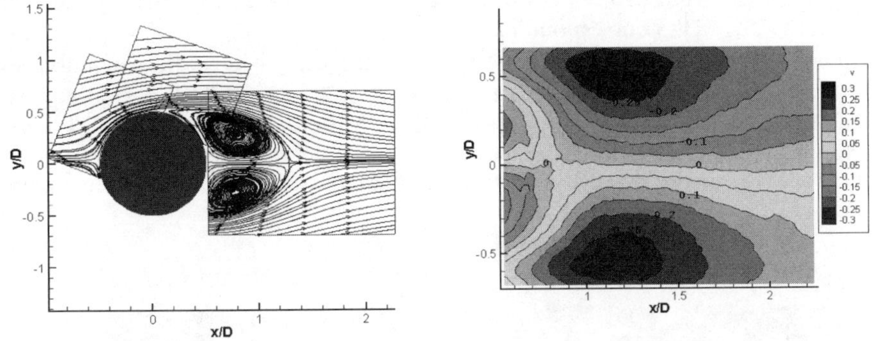

Figure 4 Streamlines showing in detail the near-wall separation region (up); mean iso-*U* velocity field

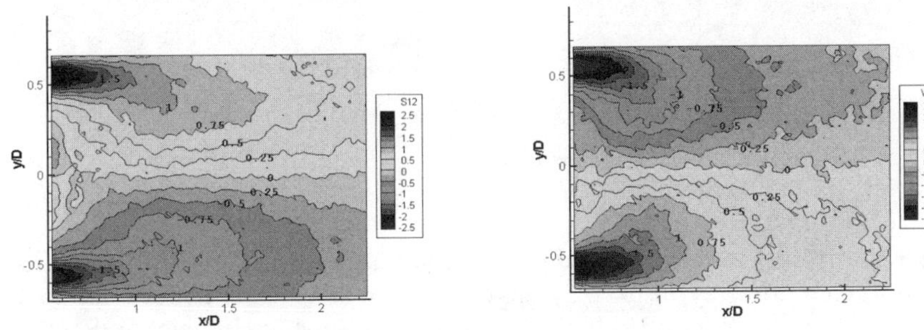

Figure 5 Left: mean strain rate S12 ; right: mean vorticity rate ω12

Figure 6 represents the iso-contours of the Reynolds stress tensor components. The $\overline{u^2}$ component has a two-lobe structure with maximum values located near *x/D*=1 and *y/D*=±0.5. On the rear axis, the maximum value is found to be at *x/D*=1.3, that is close to the recirculation length. $\overline{v^2}$ has a one- lobe structure with the maximum value 0.6 on the rear axis at *x/D*=1.4. Concerning the shear stress field $\overline{uv}$, the maximum values (±0.2) are located on either side of the wake centre line at *x/D*=1.4 and *y/D*=±0.3. As observed by many authors, it is noticeable that all the components have their maximum value near the vortex formation region. The turbulent production term $P = -\overline{u_i u_j}\frac{\partial \overline{U_i}}{\partial x_j}$ is evaluated. Significant values coincide with the turbulent kinetic energy, but the maximum values are located near the shear layer at *x/D*=0.6 and *y/D*=±0.55.

12.2.3 Phase-averaged fields

The power spectrum of the wall pressure signal (Figure 2) exhibits a peak at the frequency 22.5 Hz, corresponding to the vortex shedding at a dimensionless frequency (Strouhal number) $St=fD/U_\infty=0.21$. Figure 7 shows the streamlines of the phase-averaged flow at phase angles 0, $\pi/2$ and π. The periodic vortex shedding is clearly shown. The dimensionless vorticity is represented on Figure 8 at four phase angles in the mean period. It is shown that the vorticity peak at the centre of a vortex decrease from 3 to 1 (in absolute value) when the vortex moves

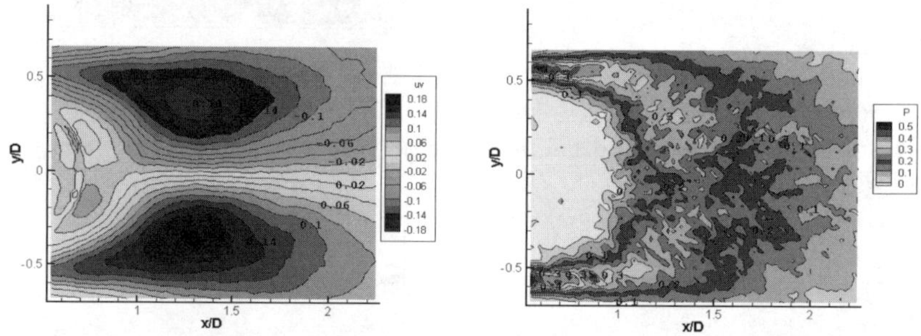

Figure 6 Mean shear stress (left); mean turbulence production

downstream from $x/D=0.6$ to $x/D=2$. Furthermore, the region of significant vorticity (taken arbitrary to $\omega>0.5$) increases from a width of about $0.8D$ to $1D$. The vortices centres (identified with the Q criterion, Jeon & Hussain, (1995)) have been marked for each phase angle and the mean trajectory is shown on Figure 10. At $x/D=2$, the trajectories seem to be nearly parallel to the rear axis at a distance $y/D = 0.25$, as found by Cantwell & Coles too. The longitudinal mean celerity of the vortices is evaluated by taking the derivative of the trajectories, to reach a value of 0.7 U_{inf} at $x/D=2$.

The Reynolds stresses at constant phase are evaluated. Figure 10 shows the normal components $\langle u^2 \rangle$, $\langle v^2 \rangle$, and the shear stress $\langle uv \rangle$ at the phase angle $\varphi=\pi$. The general topology is found comparable to the result of Cantwell & Coles. This indicates that the normal stresses have high values near the centre of the vortices, while the maximum of the shear stress are located around the vortices. This is observed downstream of the formation region. In the formation region, significant values of $\langle u^2 \rangle$ and $\langle uv \rangle$ are found in the shear layers. This topology is also found comparable to the results of Leder (1991) who measured phase averaged quantities in the near wake of a flat plate by LDA technique.

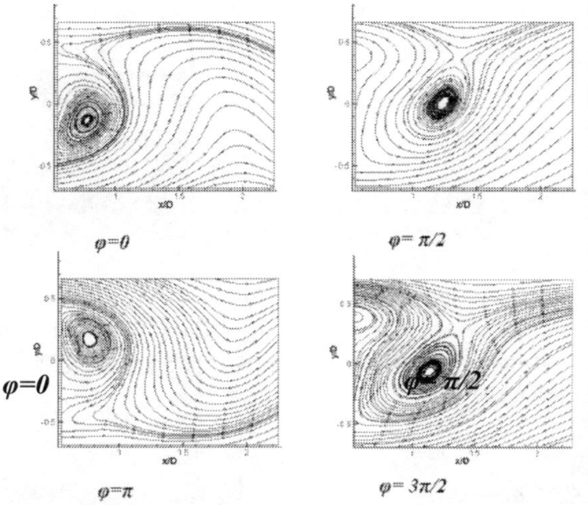

Figure 7 streamlines of the mean velocity fields at constant phase angles.

The global Reynolds stress (in statistical sense) can be decomposed in two contributions: the periodic motion and the random motion. The definition of this decomposition ensuring that the two contributions are uncorrelated, the Reynolds stresses can be expressed as:

$$\overline{u_i u_j} = \overline{\tilde{u}_i \tilde{u}_j} + \overline{\langle u_i' u_j' \rangle}$$

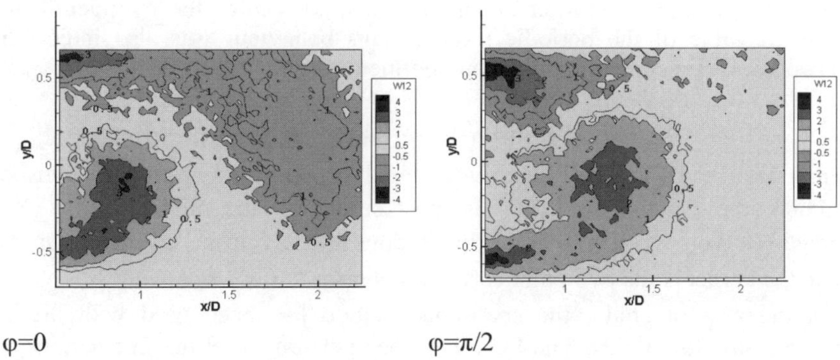

Figure 8 Iso-contours of the mean vorticity at constant phase.

Figure 11 shows the two contributions of the shear stress according to the Hussain & Reynolds (1975) decomposition.

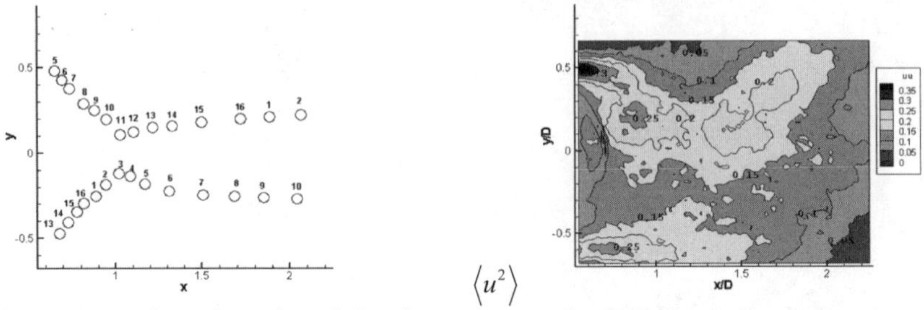

$\langle u^2 \rangle$

Figure 9a Left: trajectories of the alternating vortices. Circles indicate the centres of the vortices and number indicates the phase (1:φ=0 … 16: φ=30π/16). Right: $<u^2>$ at phase, □=180°.

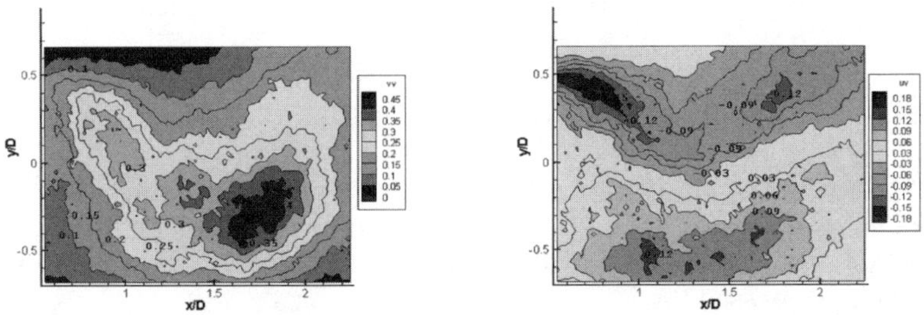

$\langle v^2 \rangle$ $\langle uv \rangle$

Figure 9b Reynolds stresses $\langle v^2 \rangle$ and $\langle uv \rangle$ at constant phase, ϕ=*180°*.

The topology and the level of these two terms are comparable, the maximum values of the random motion being located quite nearer the cylinder than the maximum value of the periodic motion. This behaviour was also indicated by Cantwell & Coles concerning the non-confined cylinder experiment.

12.2.4 Coherent structure identification by means of the POD

From a data set U(X), the Proper Orthogonal Decomposition analysis consists in searching the function $\phi(x)$ that is most similar to the members of U(X) on average (Berkooz & al. (1993)). This is done by solving an eigenvalue problem where the kernel is the two point correlation tensor $R(x,x') = \overline{u(x)u(x')}$.

In the present study, the snapshots method has been used with the 3000 instantaneous flow fields. The POD has been performed on the fluctuations away from the mean field.

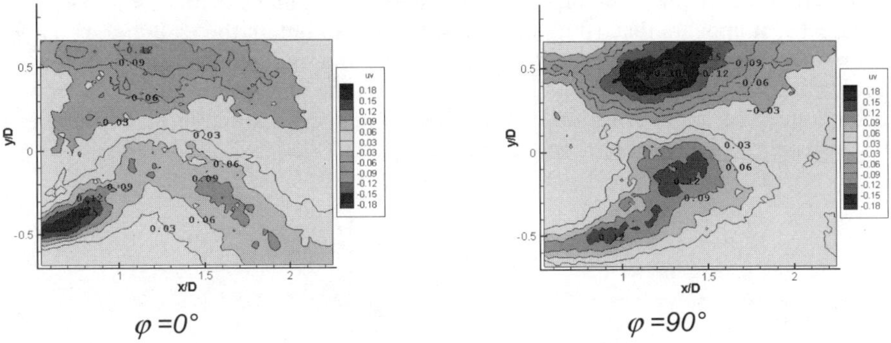

Figure 10 Shear stress iso-contours at two phase angles in a period

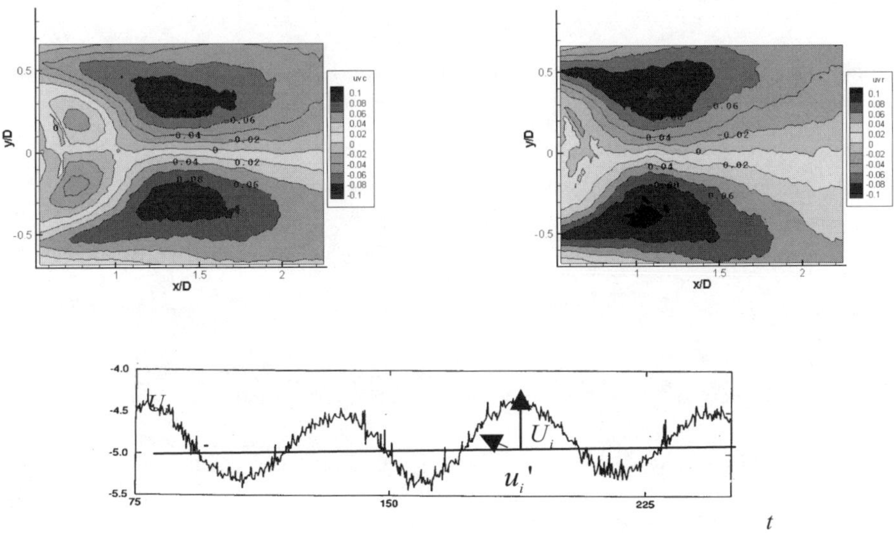

Figure 11 Shear stress contributions of the periodic motion and of the random motion according to the decomposition $U_i = \overline{U}_i + U_i + u_i'$

The average operator used is the statistical mean. Figure 12 shows the energy of the 100 first modes. Beyond the 10th mode the energy drops under 1% and then diminishes quite slowly. This is due to the highly turbulent nature of the flows. Furthermore, the energy diagram indicates a discontinuity in the slope evolution beyond the 10th mode towards a less rapid drop, due to the raise up of the incoherent random turbulence. Therefore, it can be reasonably conjectured that the first decade of the modes corresponds mainly to the coherent motion. This criterion will be adopted for the coherent pattern reconstruction. The reconstruction of an instantaneous field with 5, 10 and 20 modes is shown on Figure 13. It appears that 10 modes are sufficient to obtain the essential of the von Kármán vortices. Furthermore, a log-log modes diagram shows a reduction of the decreasing slope concerning the energy of the modes, towards a chaotic behaviour. The slope criterion can be used to suggest a suitable mode number to reconstruct the coherent part with a sufficient detail.

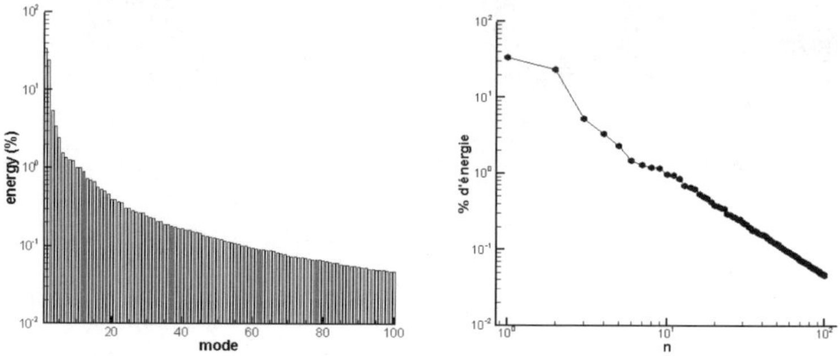

Figure 12 Energy of the POD modes (in % of the total) in decimal scale (left) and in log-log scale (right).

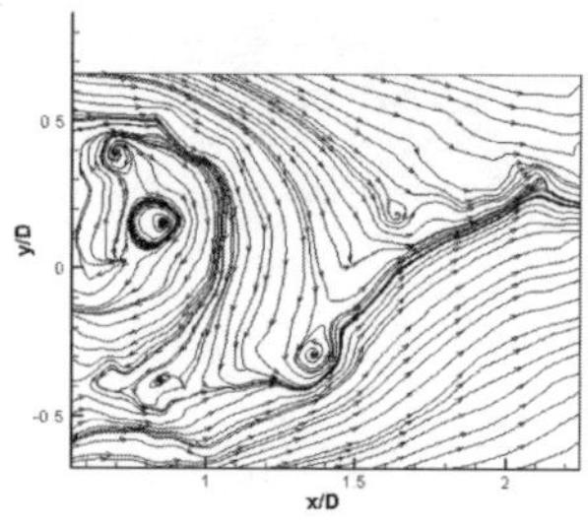

Instantaneous field

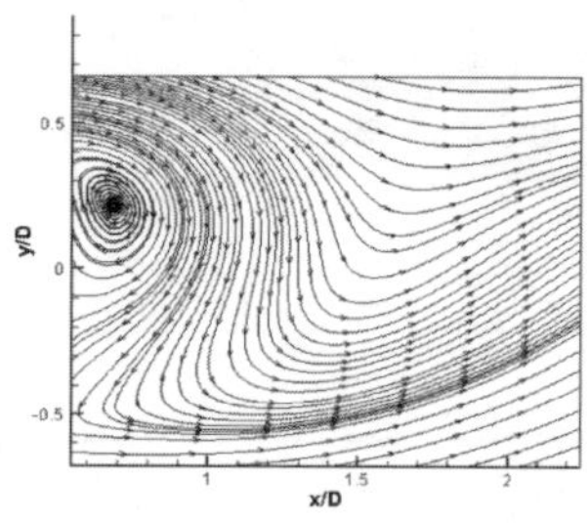

4-mode reconstruction

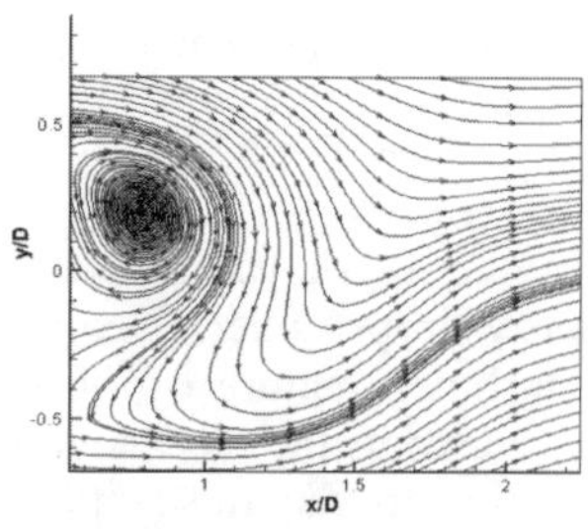

10-mode reconstruction

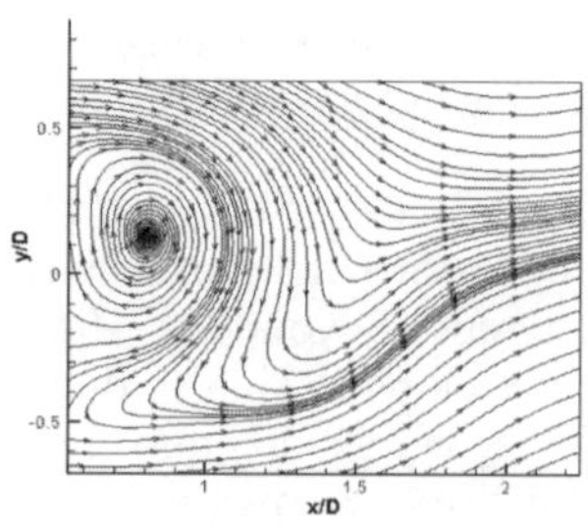

20-mode reconstruction

Figure 13 Reconstruction of an instantaneous PIV field with *N* modes

Figure 14 shows the energy spectra at different downstream positions in the near wake, according to the "in-house" LDV collaborative study of IRENAV and IMFT (Djeridi et al, 2003). The data acquisition is performed with 4kHz. These spectra show that the turbulence spectrum undergoes a slope modification in the inertial range. By using a band-pass filtering, the fluctuation spectrum is made distinct (blue line) in the inertial range and in the dissipative scales one. The slope modification comparing to the (-5/3) one of the dissipative range is clearly pointed out. It is due to the non-linear interaction between the coherent process and the random turbulence in the inertial range and it translates a non-equilibrium behaviour. By using a two-component spectral splitting, it can be derived that the turbulence length scale undergoes a significant reduction in the non-equilibrium regions characterised by this slope modification.

12.2.5 Spectral analysis and OES turbulence modelling

The power spectrum of the turbulence fluctuation is shown on fig.3. An analysis of the scales in the spectral domain, based firstly on the LDV data, (Djeridi, Braza *et al*, 2003)) yields the same order of magnitude for the eddy-diffusion coefficient as achieved by the DSM. Indeed, it has been shown that the spectrum slope changes from the equilibrium value (-5/3) in the vicinity of the organised motion peaks, fig. 4. The same fact was reported after the DNS data by (Braza *et al*, 2001). It is recalled that present spectrum is not self-similar as in case of homogeneous isotropic turbulence. Taking into account these modifications, the nonlinear transfer of energy has been quantified in an analogous way as adopted from the Kovasznay hypothesis (Hinze, 1975). The turbulence kinetic energy spectrum, in case of equilibrium takes the form: $E(\kappa(n))=(\gamma_\kappa\ \alpha_n)^{-2/3}\ [1-\kappa(n-1)/\kappa(n)](5/3)e^{(n)}\ /[\kappa(n)-\kappa(n-1)]$ and in the general case of non-equilibrium by: $E(k(n))=(gk\ an)-(2/5)p(\ [1-k\ (n-1)/k(n)](-p)e(n)\ /[k\ (n)-k(n-1)]$, where $(-p)$ is the slope value in the inertial range, n being the number of multi-component spectrum splitting (Hoarau *et al*, 2002, 2004). Additional energy spectra by the time-resolved PIV are shown in the 'Circular cylinder' chapter and a detailed analysis is carried out in the DESIDER European research program, under way. Whenever only one organised mode arises, n=1 and κ is the wavenumber. It has been obtained that the velocity scale $k^{0.5}$ (k is the turbulence kinetic energy) is reduced by a factor of 4.5. This corresponds to a reduction of the length scale by 0.22 in the non-equilibrium regions, comparing with the equilibrium turbulence length scale $l_{RANS}=k^{3/2}/\varepsilon$. According to the relation $\nu_t= C\mu\ k^{0.5}\ l$, the equivalent eddy-diffusion coefficient $C\mu$ takes the value of 0.02.

12.2.6 Numerical simulation of the IMFT Cylinder test-case

The numerical simulation of the IMFT cylinder test case has been carried out by two partners, SPTU and IMFT, by using the DES and the OES approaches respectively. For the description of them, the reader refers to section "*Task4.4:DES and zonal approach*", to the ref. (Dervieux, Braza and Dussauge,

1998). The spectrum splitting and a schematic representation of this approach is given in figure 16.

OES simulation

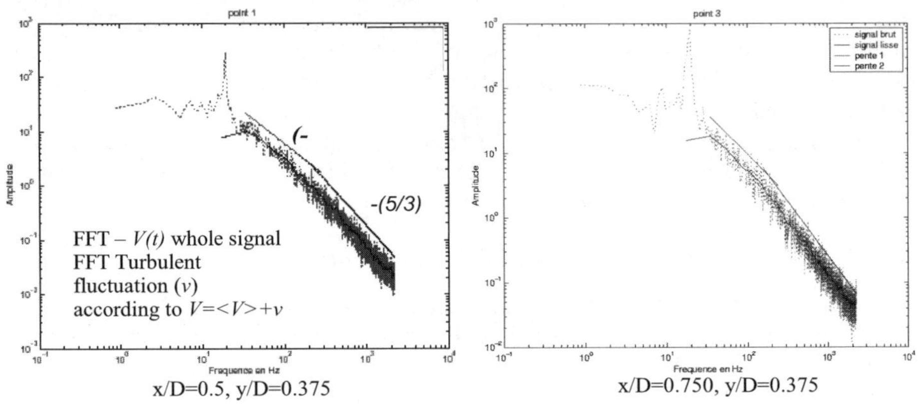

Figure 14 LDV energy spectra (Djeridi *et al*, 2003) at different positions in the near wake showing the spectrum shape and slope modification in the inertial range due to the coherent structures detachment, comparing to the Kolmogorov equilibrium value of (-5/3) that stands for the dissipative, high wave-number part of the spectrum.

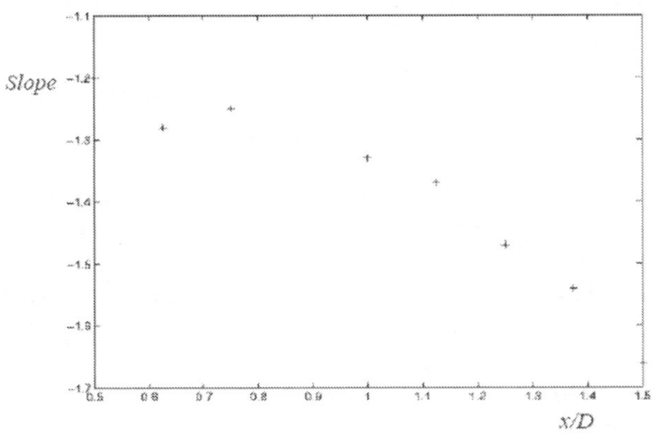

Figure 15 Inertial range slope variation in the energy spectrum versus *x/D*

IMFT used the OES approach in 2D in comparison with a two-equation modelling in URANS, the k-ω-SST one, (Menter, 1994). A comparison of the global parameters is : C_d=1.45 with OES and 1.35 with the k-ω-SST.

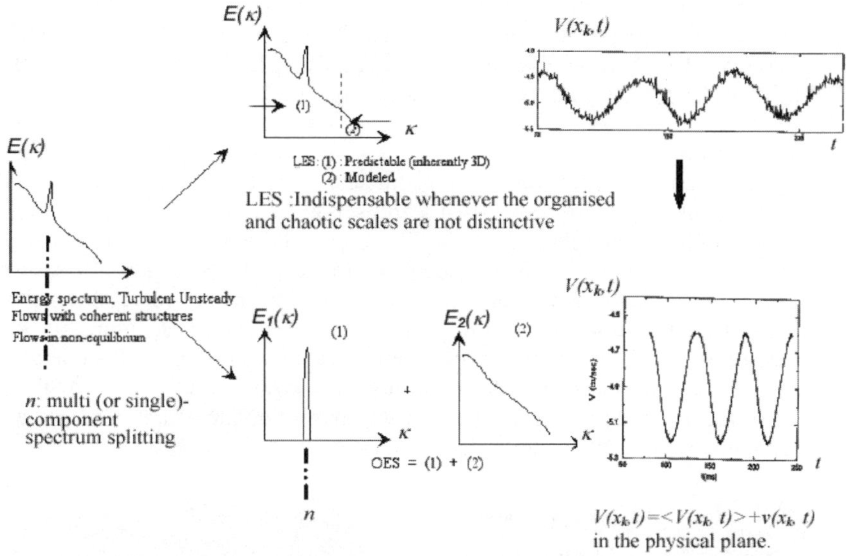

Figure 16 The OES (Organised Eddy Simulation) approach: Due to the property of distinction of the organised scales from the chaotic ones, a suitable macrosimulation approach is applicable, yielding solution of the time-dependent phase-averaged Navier-Stokes equations involving an adapted statistical turbulence modelling.

The experimental drag coefficient, $C_{d\text{-}exp}$=1.44. As expected, the OES approach ensures a lower level of turbulence production.

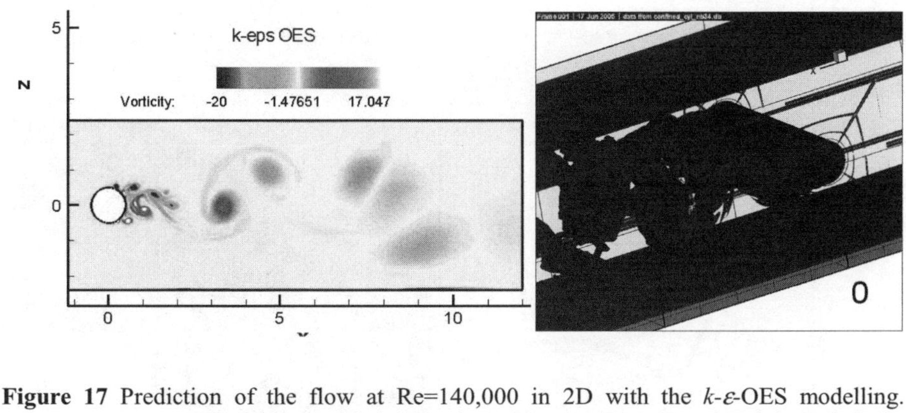

Figure 17 Prediction of the flow at Re=140,000 in 2D with the *k*-*ε*-OES modelling. Instantaneous iso-vorticity; formation of the von Kármán and Kelvin-Helmholtz eddies. Left: median plane; right: instantaneous iso-ω_z vorticity.

In fig. 17 the vortex shedding formation and especially of the Kelvin-Helmholtz vortices past the separation point are shown. This is rarely achieved by URANS computations at high Reynolds number. In fig. 17 (right), the different classes of 3D eddies are shown as well as the secondary instability.

DES simulation

SPTU performed DES (3D) concerning this test case, first by using slip boundary conditions in the upper and lower walls and secondly by using no-slip conditions. A 3-block computational mesh of 830,000 nodes total is used and the SA (Spalart & Allmaras) turbulence model has been employed in DES. Figure 19 shows the iso-U and iso-V Reynolds averaged fields that compare very well with the experiment (see figure 4). The same stands for the streamlines, figure 20. A very good comparison with the IMFT experiments is achieved by the DES comutations (fig. 20,). Figure 20 shows the 3D complex flow structure by means of the DES computations. The interaction of the von Karman ondulated main vortices and of the streamwise and vertical smaller-scale ones is clearly achieved.

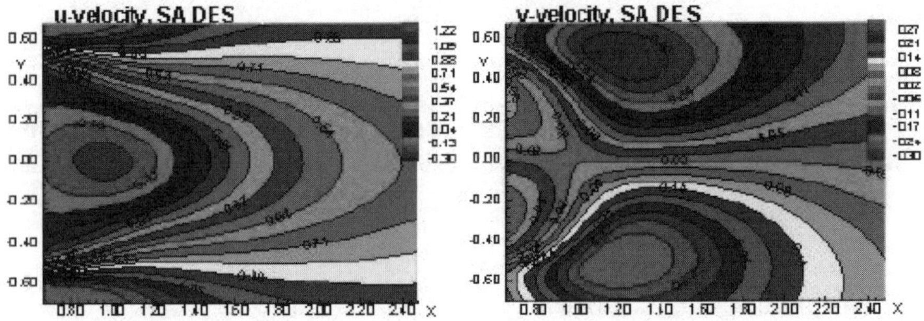

Figure 19 DES prediction of the flow at Re=140,000 in 3D with the *SA* modelling: Reynolds-averaged U and V velocity fields (SPTU, M. Strelets & M. Shur).

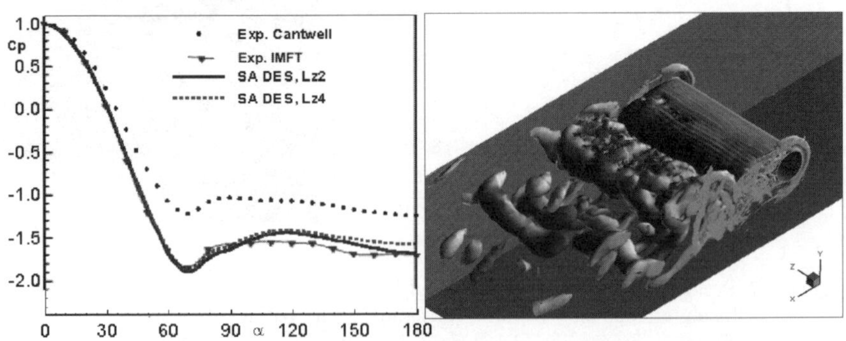

Figure 20 Streamlines in the median plane (up-left) and pressure coefficient (up-right), flow visualisation by swirl-isosurfaces according to DES approach (SPTU).

13 AS28 wing-body configuration (w/o nacelles)

C. Gleyzes ONERA

13.1 Configuration

The configuration that is tested corresponds to a half wing-body model, without nacelles, at scale 1/7.5 (configuration AS28). A view of this configuration is given in figure 1. Two grids were provided by AIRBUS/F, according to the wall boundary condition (WBC) used. The grids are structured multi-block grids (39 coincident blocks), one for low Re computations with the standard no-slip condition (referred to as MF, with 3.00M points) and one adapted to wall function computations (referred to as WF, with 2.57M points). These two grids only differ in the distribution of the points normal to the surfaces where a no slip condition is applied. Note, that the grids are provided at 1/1 scale. Consequently, legends concerning the span sections are given either in metres at aeroplane scale or in percentage of the half-span (half-span b=27.8m).

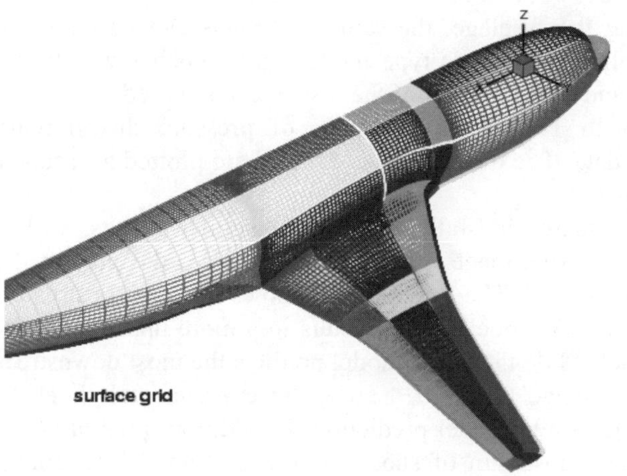

Figure 1 Surface grid for AS28

Unfortunately, this test case comes with only few experimental data: Cp distributions are available for some span sections, as well as global forces (lift and drag).

13.2 Computational conditions

Computations are performed for $\alpha=2.2^o$, Mach=0.8 and Re=10.2 10^6. Four turbulence models have been tested: Wi88, Sm93, TNT and the one equation model SA, for two types of WBC: MF and WF. All the calculations have been performed using the same integration scheme (LU-SSOR) and a maximum of identical parameters (artificial dissipation, Harten parameter...). Some special procedures had however to be applied to get convergence for the Sm93 model. Calculations have been performed with a multi-grid method.

Global force evolutions are used to check the convergence of the computation. Usually, 1000 to 1500 iterations were enough to get the solution.

13.3 Results using low Re WBC

Figures 2 to 7 present the contours of the skin friction coefficient on suction and pressure side of the wing, for 3 of the 4 models tested (As Wi88 and TNT results are similar, only the last one has been plotted). This representation is very global and allows a qualitative study of the results.

Globally, C_f distributions on the wing are very similar, on one hand for the two k-ω type models and on the other hand for the two other models. The two groups have in fact different C_f levels, which are significantly lower for the two k-ω type models. Differences are mainly visible on the inner wing (Y/b < 30%) where friction coefficients in the region between 30 and 60% of the chord differ by around 30 to 40% both on lower and upper surface.

Concerning the fuselage, the same comments can be made, with even some kind of "amplification"; k-ω type models give much lower friction, both in the cockpit area, and in the aft part of the fuselage, compared to other models.

Figures 8 to 11 present examples of pressure distributions compared to experimental data. The distributions of "-Cp" are plotted as a function of the local reduced chord.

For the pressure side, all models give similar results, with generally good agreement with experiments. On the suction side, the predicted shock location depends on the model. The Sm93 model shows the most upstream shock position, followed by the SA model, which results in a more upstream shock position than the TNT model, while the Wi88 model predicts the most downstream position. For sections in the range 0 < Y/b < 50%, the experimental shock location is even upstream of the Sm93 model prediction. In addition, pressure levels are not well predicted, either upstream of shock or downstream. When going tip-wards, the experimental location of shock moves downstream, compared to experiments, with better-computed shock intensity.

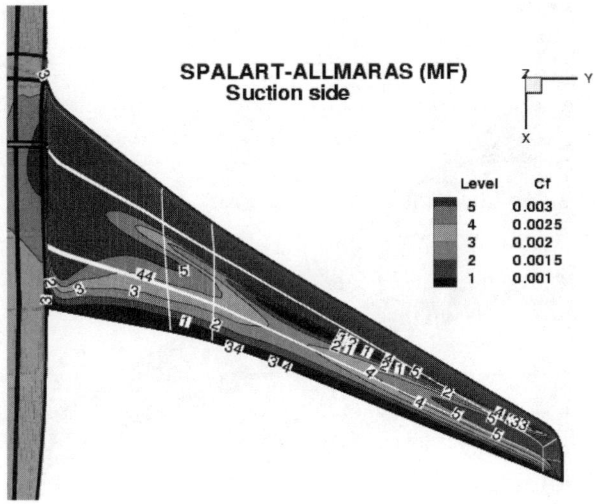

Figure 2 Skin friction coefficient. Suction side. SA model. MF

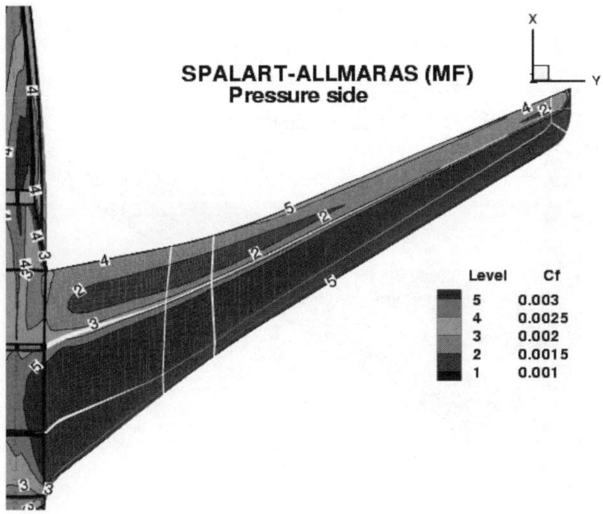

Figure 3 Skin friction coefficient. Pressure side. SA model. MF

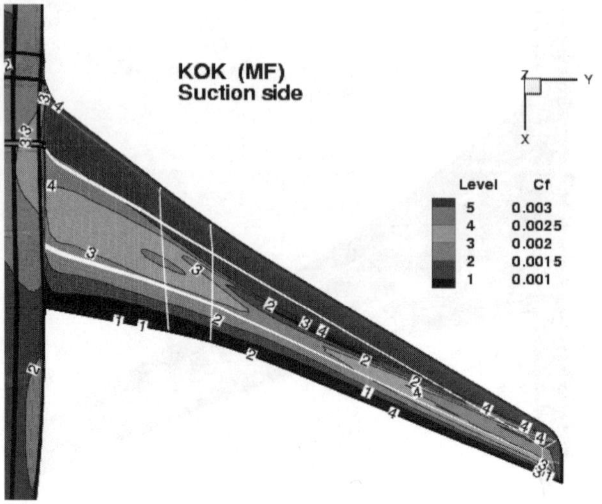

Figure 4 Skin friction coefficient. Suction side. TNT model. MF

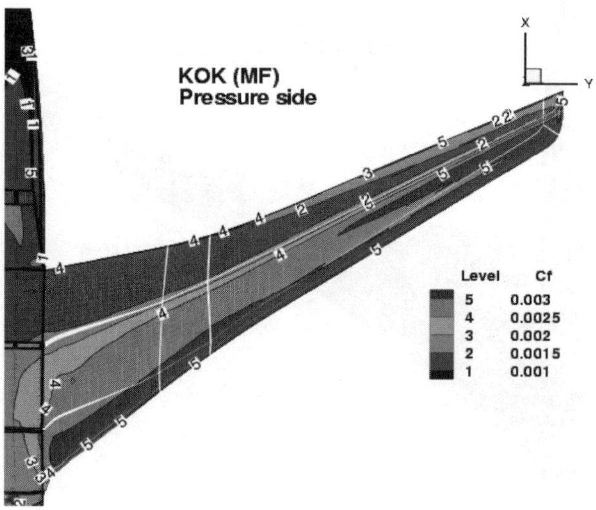

Figure 5 Skin friction coefficient. Pressure side. TNT model. MF

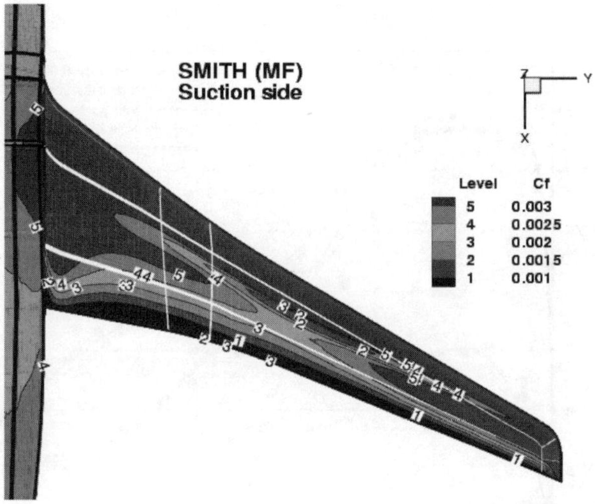

Figure 6 Skin friction coefficient. Suction side. Sm93. MF

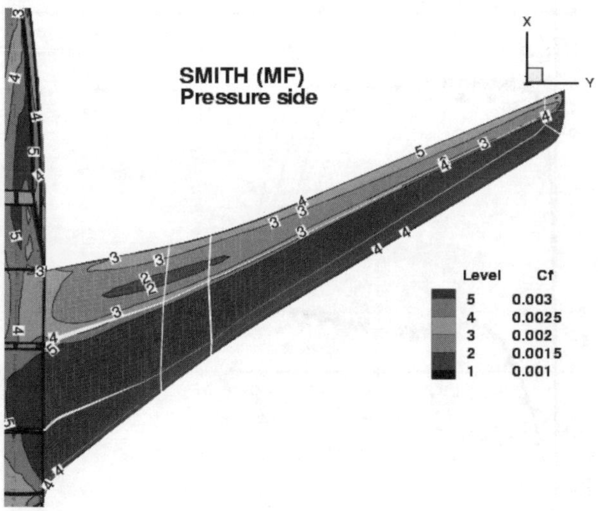

Figure 7 Skin friction coefficient. Pressure side. Sm93. MF

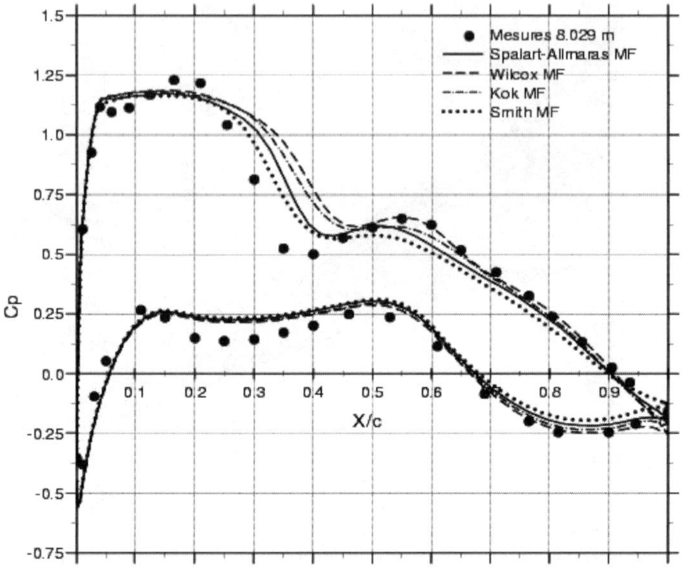

Figure 8 Cp distributions at 29% half-span. Low Re WBC

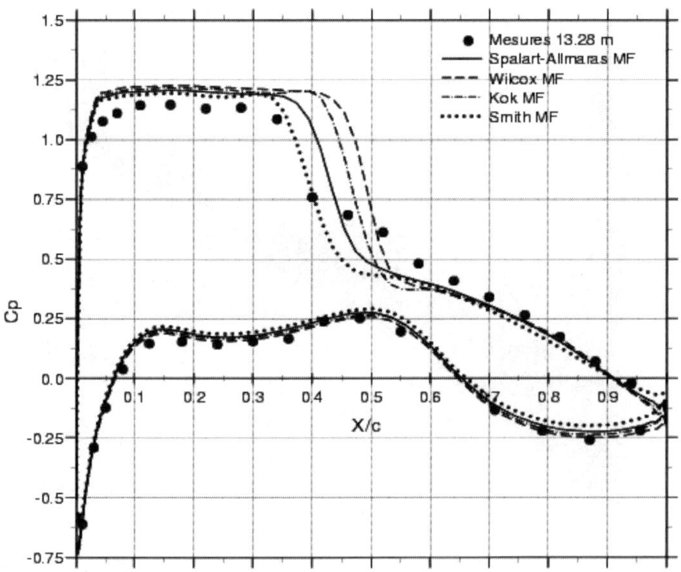

Figure 9 Cp distributions at 48% half-span. Low Re WBC

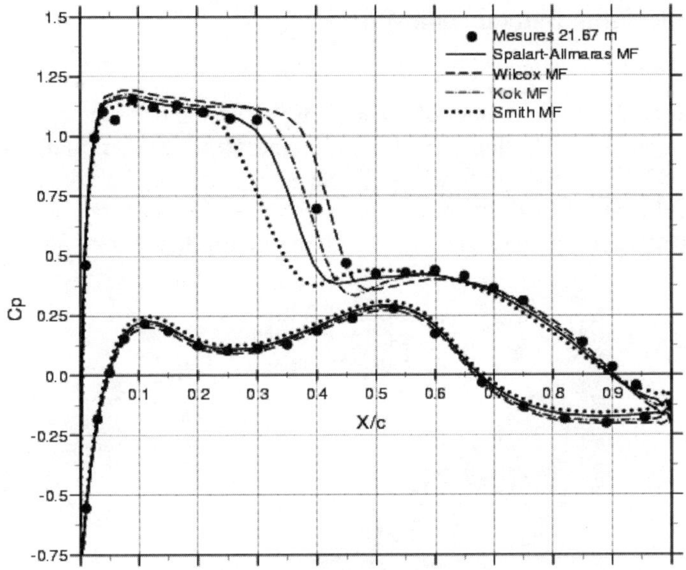

Figure 10 Cp distributions at 78% half-span. Low Re WBC

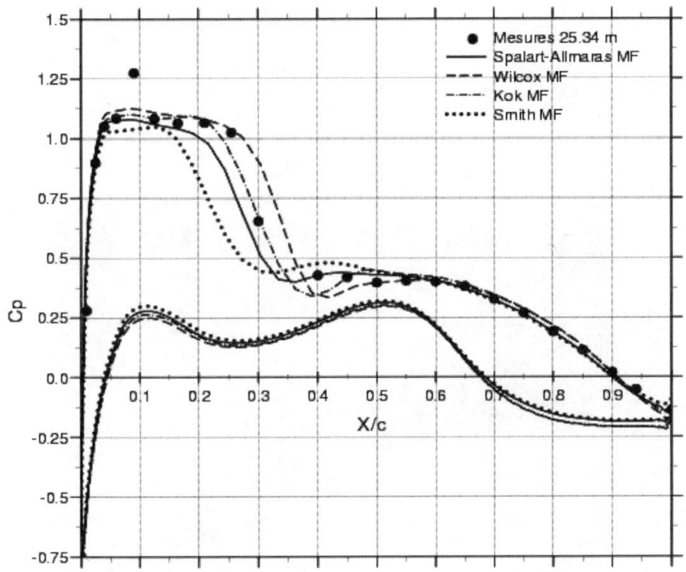

Figure 11 Cp distributions at 91% half-span. Low Re WBC

13.4 Results using wall function WBC

13.4.1 Preliminary comments

Figure 12 and 13 show contours of the size of the first cell, in the case of the wall function computation. On the wing, levels around 50 to 200 are found, with some peaks around 250 in the high C_f areas. These levels may be slightly too high in the last 30% of the chord (mainly on pressure side) where adverse pressure gradient is present. Y^+ levels are perhaps also a bit high on the fuselage (up to 350).

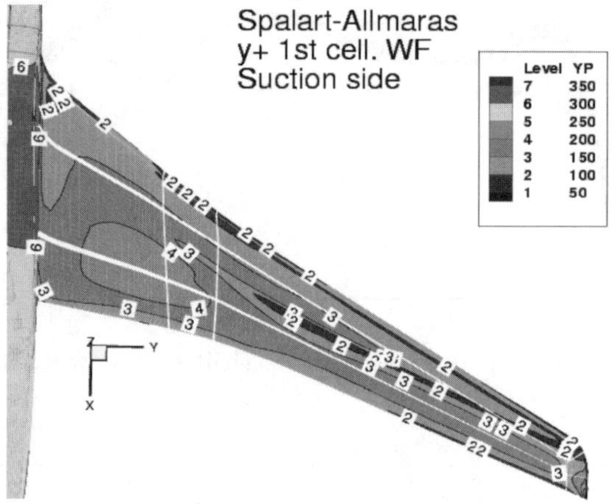

Figure 12 Size of first grid cell. Suction side. SA model. WF

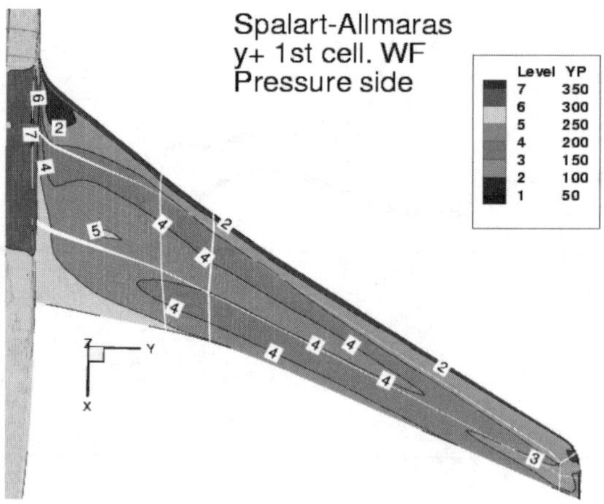

Figure 13 Size of first grid cell. Pressure side. SA model. WF

13.4.2 Results

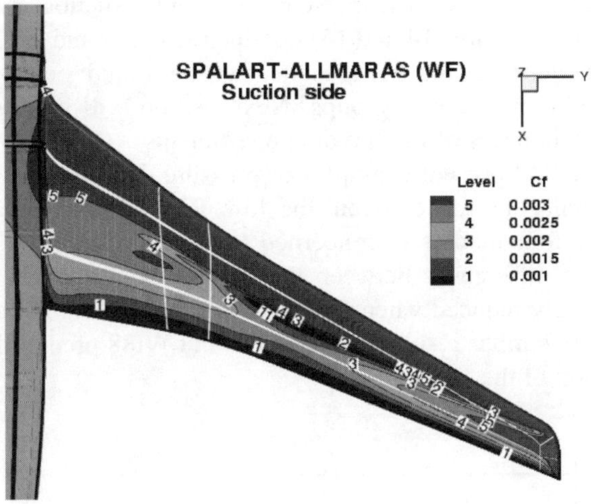

Figure 14 Skin friction coefficient. Suction side. SA model. WF

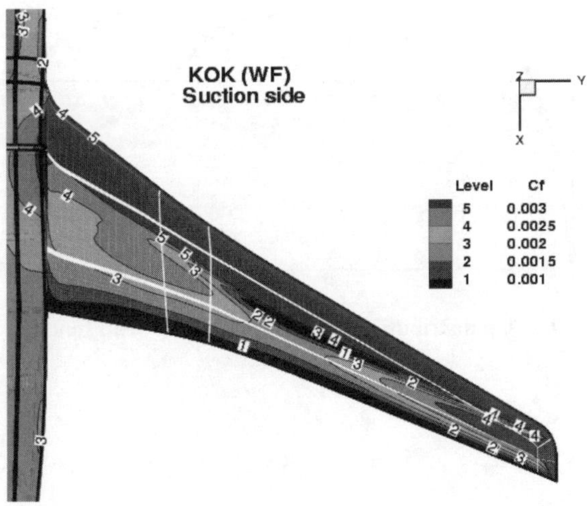

Figure 15 Skin friction coefficient. Suction side. TNT model. WF

In practice, two kinds of behaviour occur: that of the k-ω type models on one hand, and that of the other models (SA, Sm93) on the other hand. Consequently, for these calculations, we will present here only suction side skin friction coefficient contours (figures 14 and 15) obtained with SA and TNT models.

Similar behaviour as for MF results can be found but with much lower discrepancies between the two groups. Next section will give a global view of these differences, in terms of lift and drag coefficients.

Figures 16 and 17 present examples of pressure distributions. Conclusions are qualitatively similar to those given for low RE WBC calculations as far as behaviour of the four models is concerned (to be compared to figure 9 and 10). However, smaller differences between the models exist: the range of location of the shock is strongly reduced when using WF instead of MF. In particular, SA and TNT provide very similar results, while Sm93 and Wi88 predict the minimum and maximum location of the shock.

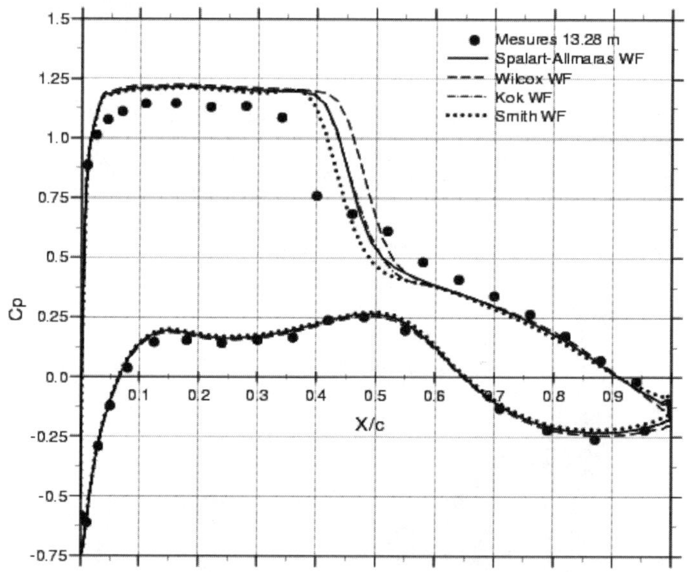

Figure 16 Cp distributions at 48% half-span. Wall function WBC

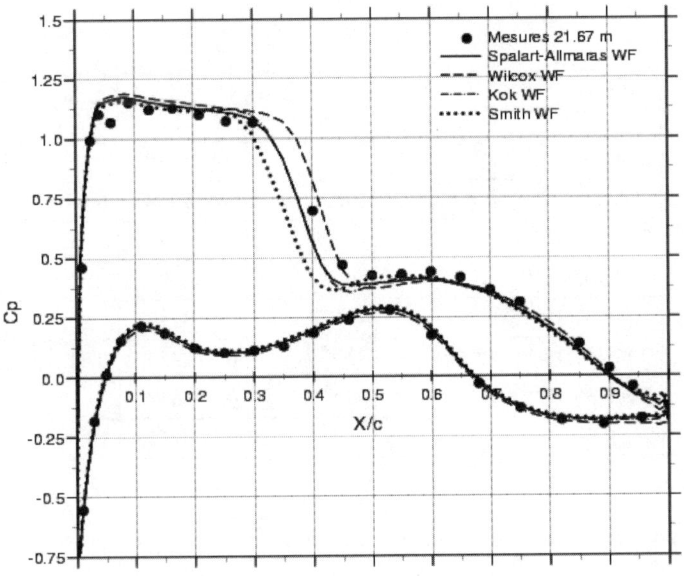

Figure 17 Cp distributions at 78% half-span. Wall function WBC

13.5 Global results

The following Table 1 presents an overview of the results, as total lift and drag coefficients obtained for the eight computations. Experimental drag and lift are 2.91 10^{-2} and 0.5, respectively.

Columns 2 and 5 give the drag and lift obtained from a raw integration of the pressure and viscous forces on the surface. Third column gives, for drag coefficient, difference (at iso-model) between low Re calculation (as reference) and wall function calculation. The same definition holds for column 6 concerning the lift coefficient. Columns 4 and 7 give, finally, the difference to experiment.

Table 1 Overview of results

Model	Cd	Δ % Cd : MF/WF	Δ % Cd : /Exp	Cl	Δ % Cl : MF/WF	Δ % Cl : / Exp
SA MF	$2.80890\ 10^{-2}$		-3.5	0.487059		-2.7
SA WF	$2.79225\ 10^{-2}$	-0.6	-4.2	0.519558	6.65	+3.7
Wi88 MF	$2.72264\ 10^{-2}$		-6.9	0.546344		+8.5
Wi88 WF	$2.88563\ 10^{-2}$	6.0	-0.8	0.544095	-0.4	+8.1
TNT MF	$2.60592\ 10^{-2}$		-11.7	0.518287		+3.5
TNT WF	$2.82961\ 10^{-2}$	8.60	-2.8	0.518353	0.01	+3.5
Sm93 MF	$2.82935\ 10^{-2}$		-2.8	0.452648		-10.5
Sm93 WF	$2.78587\ 10^{-2}$	-1.50	-4.5	0.500261	10.5	+0.1

As a first comment, we can note that all models tend to underestimate the drag, and most of them overestimate the lift. Looking at the basic drag and lift data, discrepancies between the models look smaller when using wall functions. Going further in the analysis, it can be noticed that Sm93 and SA models seem particularly sensitive to boundary conditions as far as lift coefficient is concerned. Depending on the WBC used, we find 6.6 % difference for SA and 10% for Sm93. The drag coefficient is much less affected by the choice of boundary condition (between 0.6 and 1.5% differences depending on the WBC).

This conclusion is totally inverted (less than 0.5% on C_l, but 6 to 8.6% on C_d) when looking to k-ω type models.

13.6 Conclusions

The performance of four turbulence models has been tested on the AS28 configuration with two types of WBC (low Reynolds – MF – and wall functions – WF), for a Mach number of 0.8 and a Reynolds number of $10.2\ 10^6$.

The following conclusions can be drawn:

- Wall function calculations seem to significantly reduce the discrepancies between the models.
- The k-ω type models give C_l results nearly independent of the wall treatment. Opposite, they are very sensitive to the boundary condition as far as C_d is concerned.
- The Spalart-Allmaras model gives globally the best results, with a reasonable computing effort.

14 Generic air intake (ONERA)

C. Gleyzes, ONERA

14.1 Configuration

The configuration that is tested corresponds to a generic air intake, in zero axial velocity condition, with a low velocity transverse wind. AIRBUS/France provided the grid, as well as reference conditions. The grid consists in a structured multi-block topology with 10 coincident blocks, with a total of 852000 grid points. The grid is designed for low Re turbulence models. No grid was supplied for wall function version of the models. Boundary conditions for the far field corresponds to U and W=0, with a Mach number in the negative Y direction of 0.062.

The downstream condition (in the exit plane of the duct) is a constant static pressure field, corresponding to an isentropic Mach number of 0.5.

No-slip conditions are applied to the air intake lips (inner and outer walls) , but the hub is computed with a simple slip condition. It should be pointed out here that the leading edge of the intake lip is not in a vertical plane, but is slightly tilted.

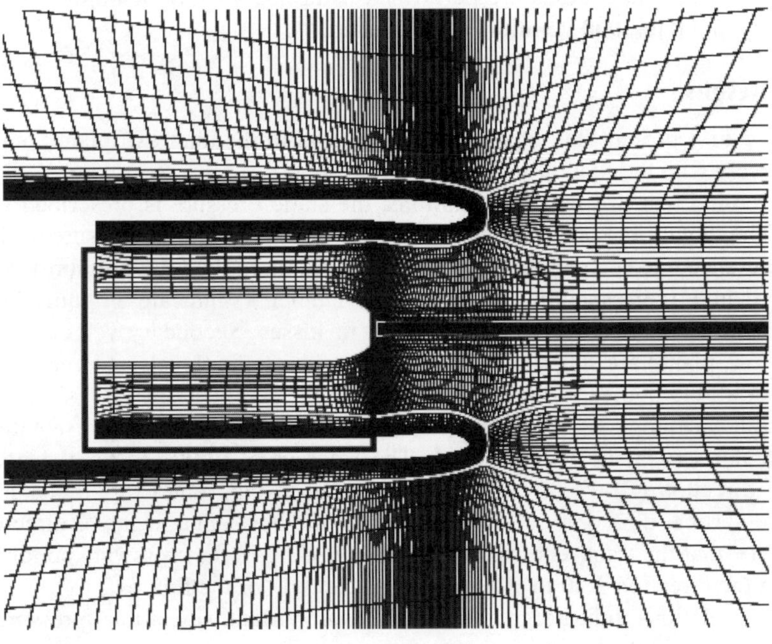

Figure 1 Section of the mesh in a horizontal (X, Y) plane along X-axis

Figure 1 gives a cross section of the grid in a horizontal plane along the air intake axis In this representation, crosswind flow is from bottom to top, and the duct exit section is at left.
This generic air intake comes without any experimental data. The objective was to look at the performances of some turbulence models in this difficult configuration.

14.2 Computation conditions

The computations are performed using only low Re models. Five different basic models have been tested: BL, Wi88, TNT, Sm93, SA and SARC.
SARC model was tested to look at influence of the correction for curvature effects. These effects should be present around the lips plane because of the 90° turn and in the windward part of the duct, because of the 3D separation at windward lips (see figs 9 and 11). The BL model is, according to Airbus, the one giving the best agreements with experiments for configurations where experimental results are available. In our case, as no experiment exists, this model is intended to be used as a reference.

14.3 Convergence

The global force evolutions are used to check the convergence of the computations. Computations are performed using the LU-SSOR implicit scheme. Generally, 3000 to 4000 iterations were needed.

14.4 Results

Figures 2 to 7 present the total pressure contours in the duct exit plane, obtained respectively with the five turbulence models tested.
Let us remind that, for these calculations, the static pressure is prescribed in this plane. The "total pressure" plotted is the ratio of the local total pressure to the far field reference one (i.e. far field boundary condition on the total pressure). Values greater than 1 (corresponding to level line number 4) indicate a "gain" in total pressure, while values below 1 correspond to losses. Secondary velocity vectors (in the Y-Z plane) are superimposed. Scales are the same (levels and vector length scale) for all the figures.

For an unambiguous discussion of these results, a "clock" representation of this circular exit plane will be used. The reference "12:00" is for the positive vertical direction (Z), "3:00" for the positive horizontal one (Y), "6:00" for the negative vertical direction (-Z) and "9:00" for the horizontal negative direction (-Y). Notice that the transverse velocity corresponding to the crosswind condition is going in the direction 3:00 to 9:00 (i.e. from the right of the figures 2 to 6).

All calculations indicate, in a more or less important way, a total pressure loss in the area at 3:00, due to a more or less important separation region on the windward side of the air inlet. We can also notice a total pressure gain (total pressure levels number above 4) in various regions, the location and the size of which depend on the model used. In fact, the way turbulence models predicts the strongly 3D separation at the windward region of the inlet lip is the key point to the development of the flow in the duct itself.

The following table 1 gives the maximum, minimum and mean pressure ratio for the five models.

Table 1 Maximum, minimum and mean pressure ratio for the five models

	Pi/Pi_o min	Pi/Pi_o max	Pi/Pi_o mean
BL	0.849	1.0143	.975
Sm93	0.852	1.014	.967
Wi88	0.845	1.033	.954
TNT	0.843	1.069	.960
SA	0.847	1.048	.960
SARC	0.847	1.067	.953

The level of the minimum of this pressure ratio is nearly constant whatever model used, with local total pressure around 15% below far field total pressure. The size of these low pressure areas is however significantly different according to the model. If we look, for instance, at the area where pressure loss are greater than 10% (below level 2), we can notice that SA, SARC and TNT predict large areas, while for BL and Sm93, these areas are limited to the wall region. Wi88 also predict limited areas, with however an additional small region around "11:00".

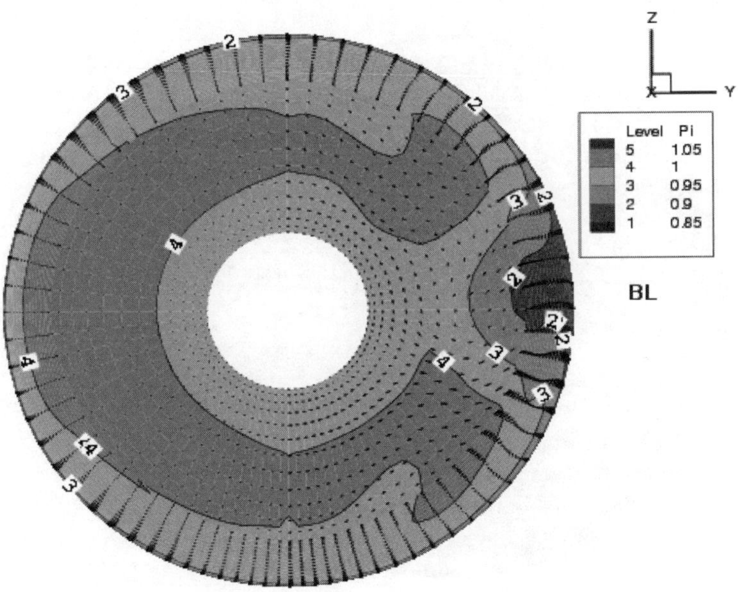

Figure 2 Total pressure in the duct exit plane. BL model

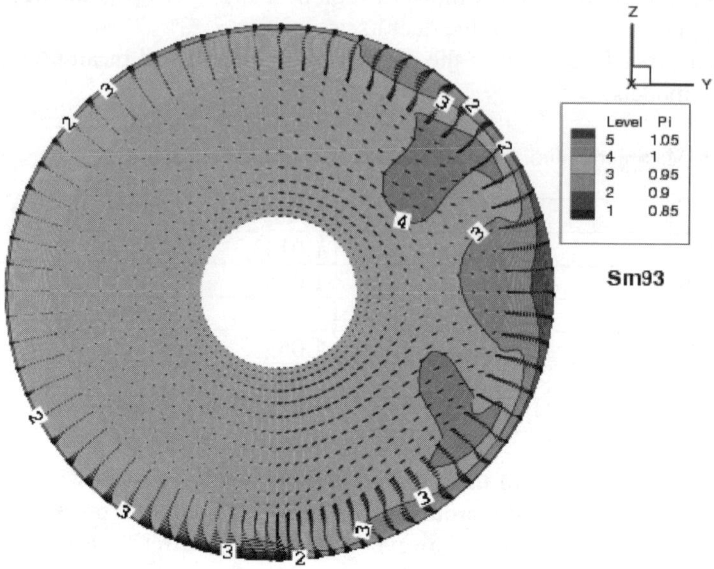

Figure 3 Total pressure in the duct exit plane. Sm93 model

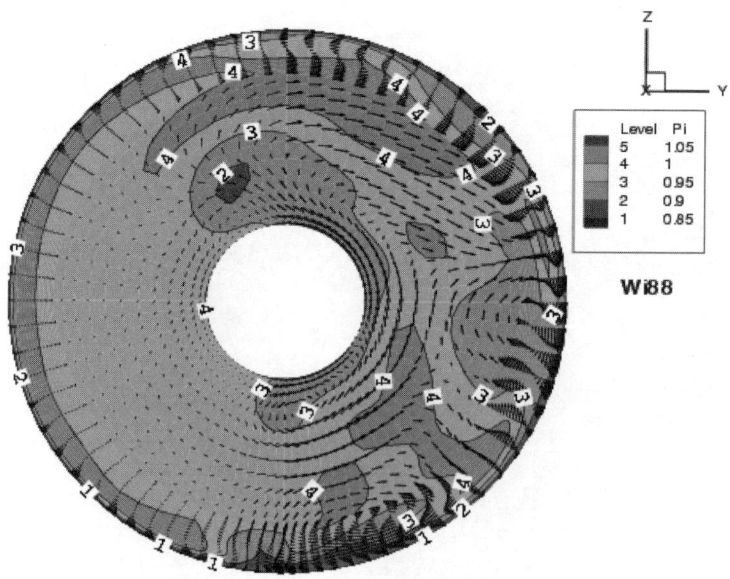

Figure 4 Total pressure in the duct exit plane. Wi88 model

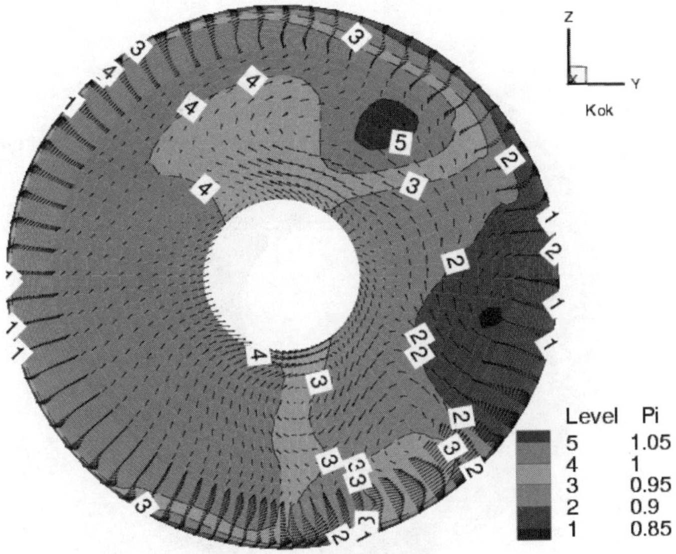

Figure 5 Total pressure in the duct exit plane. TNT model

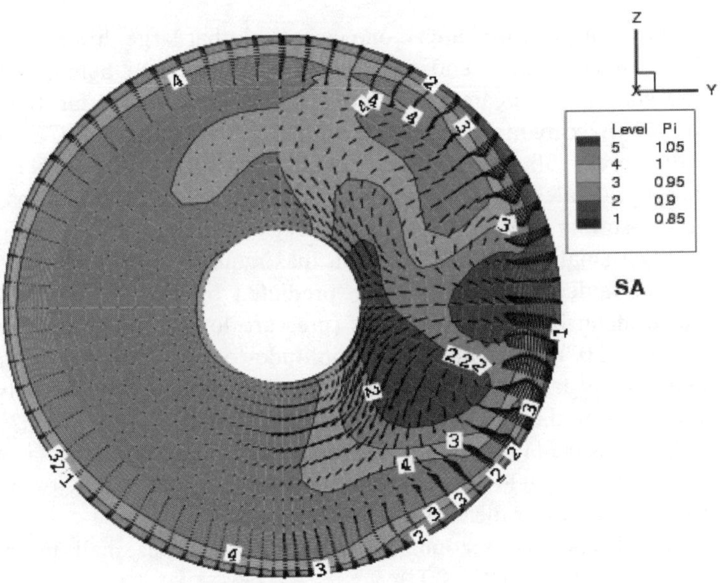

Figure 6 Total pressure in the duct exit plane. SA model

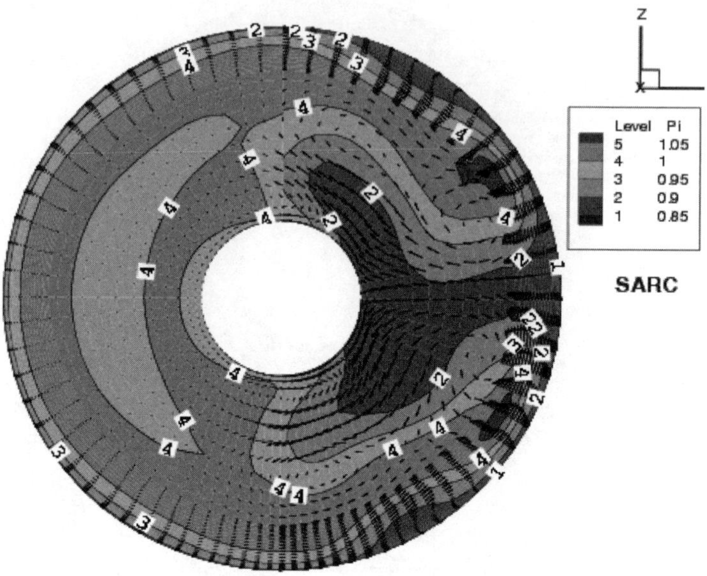

Figure 7 Total pressure in the duct exit plane. SARC model

For the "overpressure" areas, one can check that large differences exist depending on the model. Sm93 and BL models (this last one being, normally, the most representative) indicate moderate differences with the far field reference total pressure (maximum overpressure around 1.5%). In addition, Sm93 predicts much smaller areas of overpressure, compared to BL. Wi88, TNT, SA and SARC models indicate, on the opposite, fairly large pressure gains. The maximum overpressure is 6.9% for TNT, with very large areas of pressure gain. These areas are much smaller for Wi88, with a maximum overpressure of 3.3%. In fact, the areas of under and overpressure predicted by these two k-ω type models are located at fairly similar positions (pressure loss around "3:30" and overpressure around "1:30"), but with larger amplitude for the TNT model. In addition, these models show fairly large secondary vortices around position between "3:00" and "5:00", not predicted by the BL and Sm93 models.

As far as mean pressure is concerned, all models give the same level of around 96% of the far field total pressure within less than 0.5%, except for BL which gives slightly lower mean pressure losses.

For all models, secondary velocities are quite small in the lee-ward area (between "7:00" and "11:00").

Finally, one can notice that the models providing the most visible effects (Wi88, TNT, SA, SARC) indicate a pattern which is non symmetrical (around the X,Y plane), although for the other kind of models (BL, Sm93) the level lines are symmetrical. This is due to the fact that the leading edge is slightly tilted relative to the (Y,Z) plane, and that the three-dimensionality of the separation at the windward lips is significantly greater for the former models.

It should be pointed out, however, that these plots are simple cuts at constant X, and that they only show the trace of a more complex 3D turbulent flow pattern.

The correction in the SARC model slightly changes the distribution of under and over pressure areas, but not in a very drastic way. The slight difference is that for SARC, a small overpressure area appears in the lee-ward area, with low value however, and pressure losses are higher in the wind-ward area. Secondary velocities are not significantly different.

Figure 8 gives an example of velocity vectors predicted by the SA model which provides one of the highest "pressure gain", while figure 9 gives same results for the reference BL model.

The plots correspond to a radial cross section in the duct. The frame plotted in figure 1 gives an image of this area. The right-hand side is close to the hub leading edge, the left-hand side to the exit section (in which is prescribed the static pressure). Flow is exiting the duct in the right hand side section. The cut is in the (X,Y) plane and the crosswind direction is from bottom to top. Lower part is the windward part of the duct, i.e. the section at "3:00" according to the convention described in section 14.4. Let us stress here that this 2D view is only a cut of a strongly 3D flow, but it gives anyway some information on the trace of this flow in this plane.

Figure 8 shows the trace in this (X,Y) plane of a low velocity recirculation region on the windward side (i.e. in the lower part of the plot). Close to the hub (on which a slip condition is applied), we can see stream wise oscillations of the magnitude and direction of the mean velocity, with some local acceleration above the recirculation region. The velocity profile in the exit section (lower left of the figure) shows significant variation in the Y direction (i.e. in the radial direction), with a lower mass flow, compared to leeward side (upper left) which looks much more uniform.

Figure 9 shows a qualitatively similar pattern, with however a more stable flow around the hub with almost no oscillations on the direction of the mean flow. BL model also predicts a more uniform velocity profile and a higher mean velocity at windward duct exit (lower left), compared to SA.

Flow pattern for the lee-ward side (upper part of the figures) are very similar for the two turbulence models.

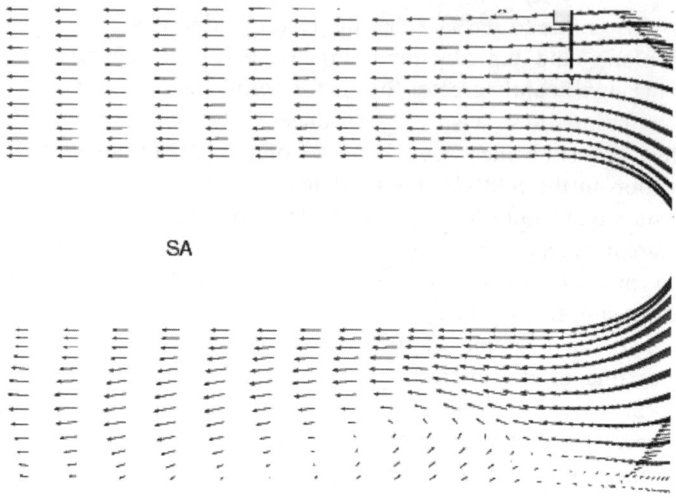

Figure 8 velocity vectors in the X-Y plane. SA model

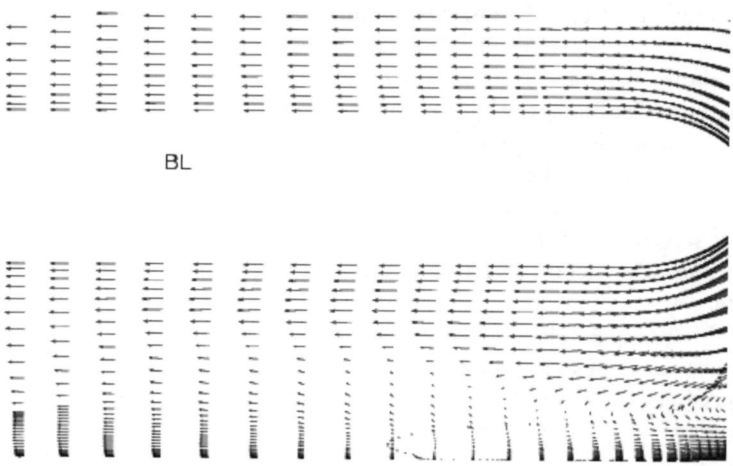

Figure 9 velocity vectors in the X-Y plane. BL model

To try to get a view of the three-dimensionality of the flow, and the corresponding differences between the results obtained respectively with SA and BL models, we have plotted, in figures 10 and 11, a perspective view of 4 cross sections of the half windward part of the air inlet. The most right cut corresponds to a section slightly downstream of the hub nose, while the most left one is the exit of the duct (i.e. the section plotted in figures 6 and 2). Flow in the duct is from upper right to lowe left (i.e. along X axis) and cross-wind direction is in the (-Y) direction (lower right to upper left). Only the secondary velocities are plotted (in the Y-Z plane).

The outer rings correspond to 4 cuts of the flow on the external part of the inlet, with the cross-wind conditions. The disks represent the flow in the duct, with the trace of the vortices induced by the 3D separation at the windward lip of the inlet. The central blank area is the hub.

For both figures, the scale of the arrows representing the secondary velocity vector are the same.

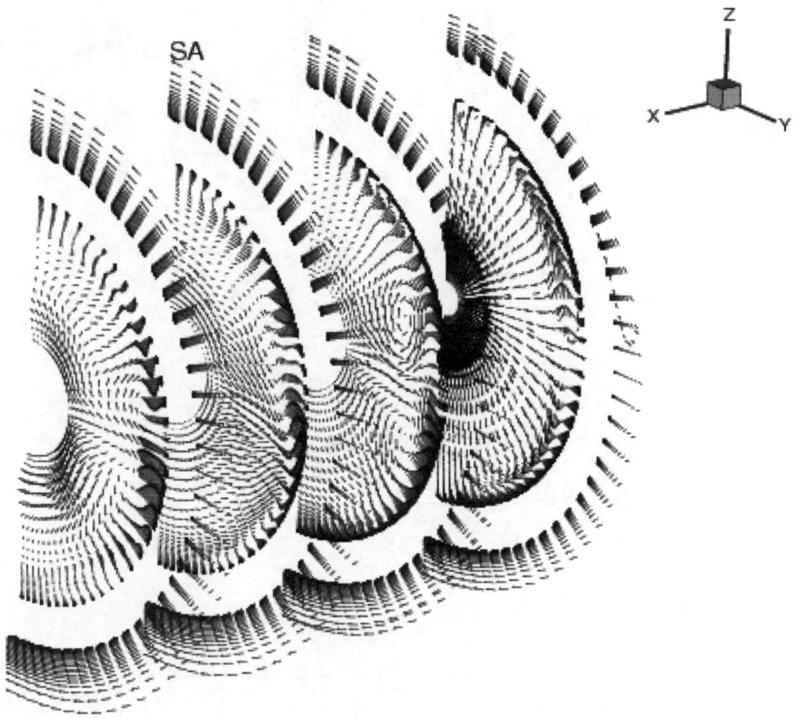

Figure 10 velocity vectors in four Y-Z planes in the air intake. SA model

For the SA model (Fig 10), in the most right section, we can see the trace, on the secondary flow, of the 3D separation at the wind-ward lips induced by the crosswind. This pattern is nearly symmetrical (relative to a X,Y plane). Downstream (i.e. going left-wards), this separation produce longitudinal vortices visible in the second cut plane (still counting from right to left) which remain present in the exit plane. The maximum of the magnitude of the secondary velocities is around 30% of the magnitude of the reference "isentropic" Mach number used as boundary condition in the exit plane of the duct.

For the BL model (fig 11), the trace of the 3D separation is much less extended. Some traces remain visible in the second cross section, but there is no evidence of the presence of longitudinal vortices in the exit plane (left section). In addition, the magnitude of the maximum secondary velocities in this section is three times smaller than for the SA model.

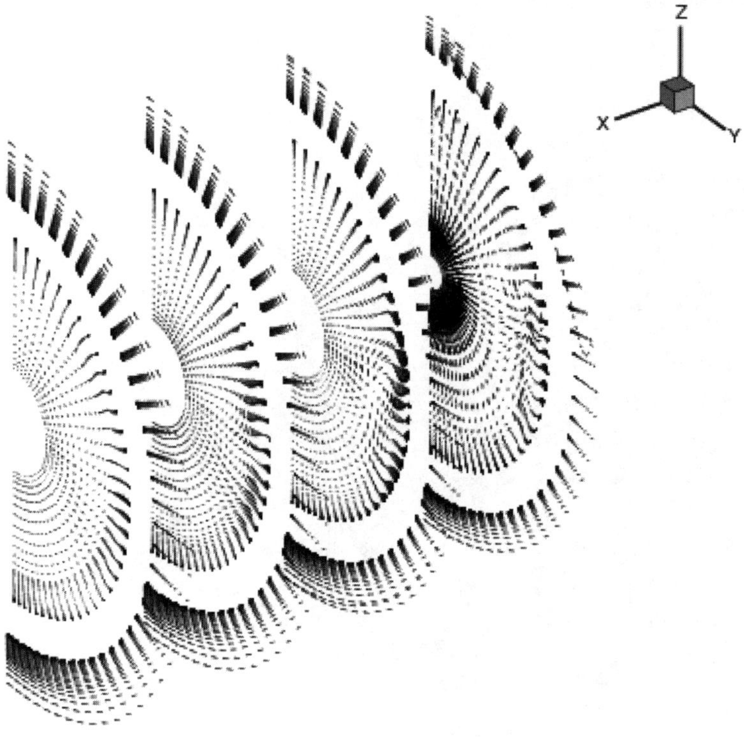

Figure 11 velocity vectors in four Y-Z planes in the air intake. SA model

14.5 Conclusions

Due to the lack of experimental results, it is difficult to draw definite conclusions concerning the validity of the turbulence models tested. One can however say, if we assume that the BL model is giving the right trends, that the SA and k- ω type models are overestimating the importance of the separation. However, in terms of periodic oscillations between high and low total pressure for a rotating fan, SA model seems to provide the worse conditions.

15 Ahmed Car

M. Kuntz, and F.R. Menter, ANSYS Germany GmbH

Abstract

The present chapter describes the computations carried out for the Ahmed Car, a generic car geometry with a slanted back. The challenge of the comparative study is the correct prediction of the flow topology especially at the back of the car. A variety of results are computed by 6 partners with different RANS, DES and LES simulations for the 25° case.

15.1 Introduction

The Ahmed car is a generic car geometry including a slanted back (see Figure 1). Despite its geometric simplicity, it creates a flow-field, similar to the one found for real-life cars. The main features of the flow are the separation and reattachment at the slant and the wake development behind the body. Different slant angles have been studied in the experiments in order to investigate the changes in the flow structure and the global forces due to this variation in the geometry.

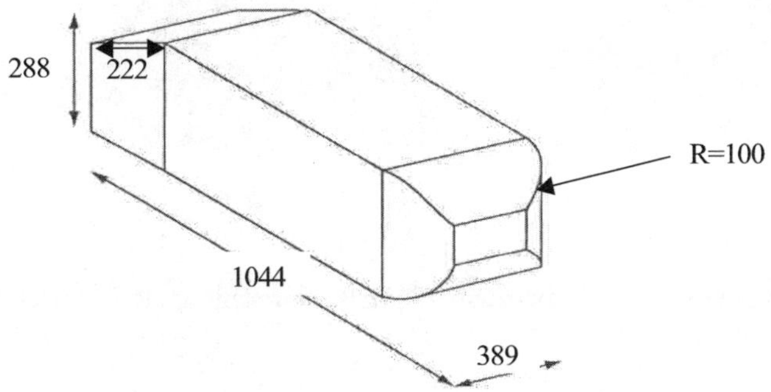

Figure 1 Ahmed car geometry

15.2 Geometry and flow conditions

The first experimental studies on this shape were carried out by Ahmed (1984). The present comparison is mainly based on the data of Lienhart et al. (2000) and Lienhart and Becker (2003) at the LSTM (low speed wind tunnel) of Erlangen. In this experiment, the Reynolds number with respect to the length of the car was $Re_L=2.6\text{x}10^6$. The slant angle was varied between 0° and 40°. The emphasis of the experimental investigations was on the flow structure in the slant region and downstream of the body.

The body is mounted on 4 stilts (diameter 30[mm]), so that it is located 50[mm] above of the ground. The cross-section of the tunnel is 1.87x1.4[m^2]. The studies were conducted in a ¾ open test section (i.e. floor but no sides or ceiling) with a blockage ratio of 4%. The velocity and air temperature were controlled by a computer-feedback system.

Considering the geometrical simplicity of the Ahmed model and the Reynolds number studied, the flow at the rear of the car is the most challenging part for a CFD simulation. The flow structure is strongly dependant on the slant angle. For the 25° case, the flow displays a three-dimensional behavior at the rear of the body. A separation bubble is present on the upper part of the slant. The flow reattaches near the middle of the slant. Two counter-rotating lateral vortices start on the edges of the slant. They are a result of the induced lift due to the turning of the flow. In contrast, the 35° case has a virtually two-dimensional behaviour over the slant. The separation takes place at the top of the rear window and the flow separates over the whole of the rear of the body. The side vortices are much weaker for this case, due to the reduced flow turning (see Figure 2). More details are given by Gilliéron and Chometon (1999).

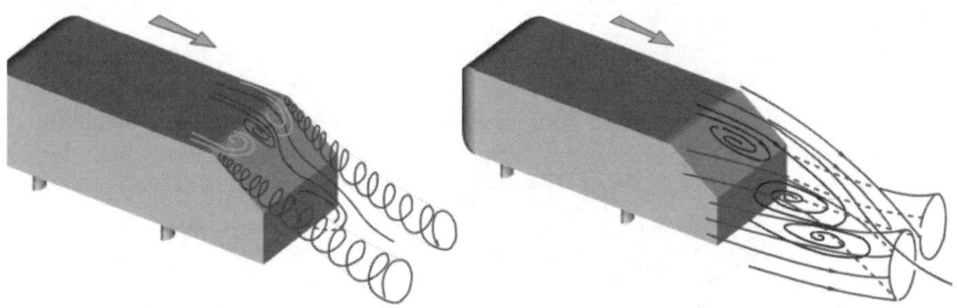

Figure 2 Flow topology for slant angle 25° (left) and slant angle 35° (right). Courtesy Lienhart, LSTM Erlangen

In the frame of the FLOMANIA project, all participating partners focused on the 25° case, because the correct prediction of the flow structure on the slant (separation with reattachment) is a challenging task. Previous comparative studies (Manceau, 2002 and Durand et. al., 2002) demonstrated a strong sensitivity of the turbulence model and the grid density used on the flow structure.

15.3 Computational grid and set-up

Different grids are used by the partners for the computation of the Ahmed body (see Table 1 and Figure 3). A common hex grid used by most of the partners is provided by CFX. Additionally, two partners created hybrid grids for their computations. The grids generated for steady-state RANS computations cover only half of the body using a symmetry condition. Furthermore, special grids for running the DES and LES models are available.

Table 1 List of grids

Type	Grid size (million nodes)	Grid generation software	Used by partner
Hexahedral	1.30 (half model)	ICEM-HEXA	ICSTM, CFX, SPTU, NUMECA
Hybrid	0.82 (half model)	NUMECA	NUMECA
Hybrid	1.98 (half model)	CENTAUR	DLR
DES	1.71 (full model)	ICEM-HEXA	CFX
LES	16.5 (full model)	ICEM-HEXA	CHALMERS

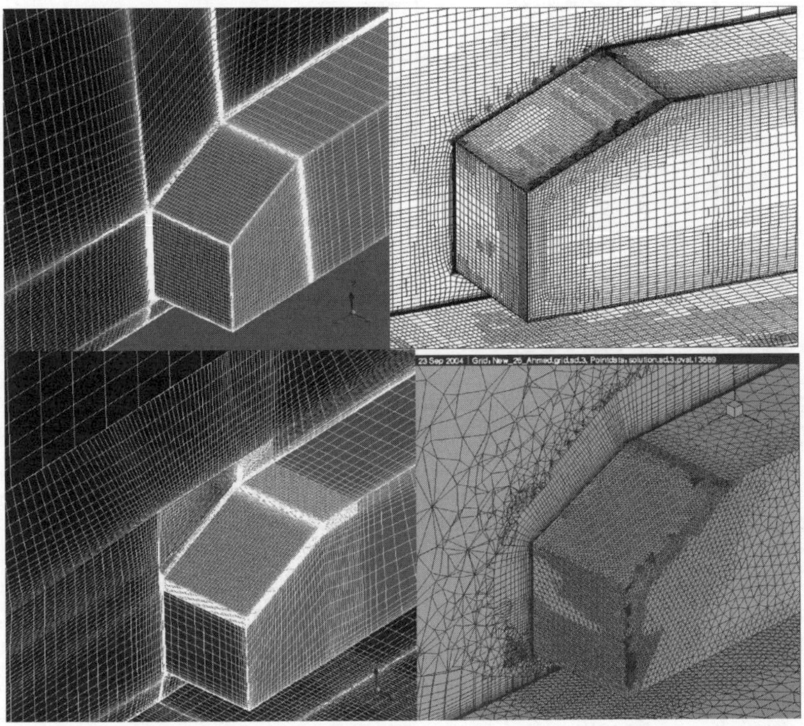

Figure 3 Grid views (top: hexahedral grid, NUMECA hybrid grid, bottom: CFX DES grid, DLR hybrid grid)

The following boundary conditions have been set:

- Inflow condition: The bulk inflow of 40 m/s is specified 2.1 m in front of the body in order to match the experimental velocity profiles at 40 mm in front of the body. The turbulence intensity at the inlet is set to 1 % and the turbulent eddy viscosity equal to the laminar viscosity.
- A relative pressure of 0 Pa is set at the outflow boundary.
- Free slip wall conditions are set at the top and the side walls.
- No slip wall conditions are set for the body surface and the ground

15.4 Contributors and their contributions

Computations of the Ahmed car are carried out by 6 different partners. The list of contributors and their models used for the computation are listed in table 2.

Table 2 List of contributors

	EVM	NLEVM	SSG	DES	LES
ICSTM	LS	AJL-ω, AJL-ε	SSG-Ch		
CFX	LS,SST		SSG	DES-SST	
STPU	SST, SA, SARC			DES-SA, DES-SARC	
NUMCEA	YS, SST, SA, SARC				
DLR	SA, SST				
CHALMERS					LES

15.5 Results

The present chapter summarizes all results of partners computed during the FLOMANIA project. The comparison of results is carried out per turbulence model (or turbulence model group). This kind of grouping allows the comparison of results computed with different codes using the same model (on the same or different grid). The following groups are defined:

- k-ε models: Launder and Sharma (LS) and Yang and Shih (YS)
- SST models
- SA models
- DES/LES models
- NLEVM models (AJL) and RSM models (SSG)

Due to the large number of data sets generated during the FLOMANIA project, the comparison of results has to be limited to a few quantities. The main information about the flow structure can be extracted by analyzing the flow velocity component in the freestream direction. The main interest is in the flow field on the slant and the wake region. Therefore plots are presented for different velocity profiles on the slant at y=0 mm (symmetry plane for steady state runs) at y=180 mm (near side of car) and in the wake (y=0 mm).

15.5.1 Comparison of results for k-ε based model

The present chapter shows the comparison for k-ε based turbulence models, namely the Launder and Sharma model (LS, 1974) used by ICSTM and CFX and the Yang and Shih model (YS, 1992) applied by NUMECA. The results from ICSTM and CFX are computed on the hexahedral grid, whereby the NUMECA results are obtained on unstructured adapted mesh. The LS model implementation of ICSTM includes the solution up to the wall, whereby CFX uses scalable wall functions. NUMECA improved the robustness of the YS model ('YS new' model, details can be found in the deliverable D-4.2-30). The comparison of velocity profiles can be seen in
Figure 4, Figure 5 and Figure. All models predict attached flow on the entire slant. In general, the k-ε based models miss the flow separation at the onset of the slant and do not predict the separation bubble and the reattachment in the middle of the slant. Despite this shortcoming, a good agreement is obtained for the flow recovery in the wake region. This indicates, that the overall flow topology with two strong side vortices is predicted well. This behavior of the k-ε model is however the result of a cancellation of errors. As the model misses the separation, there is no need for an adequate representation of flow reattachment on the slant.

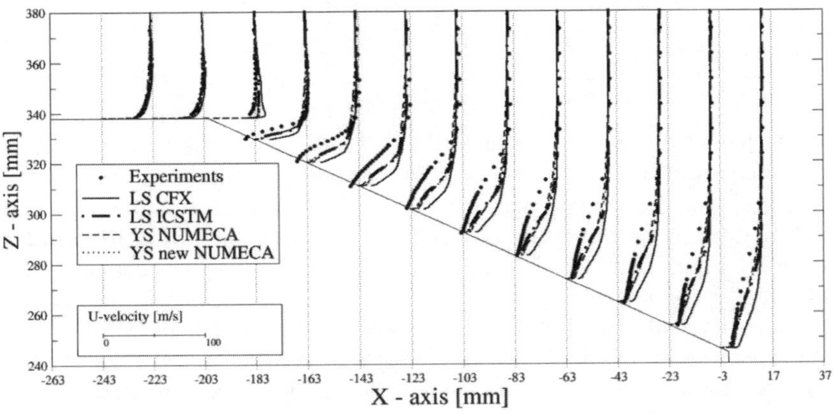

Figure 4 Velocity profiles y=0 mm on slant

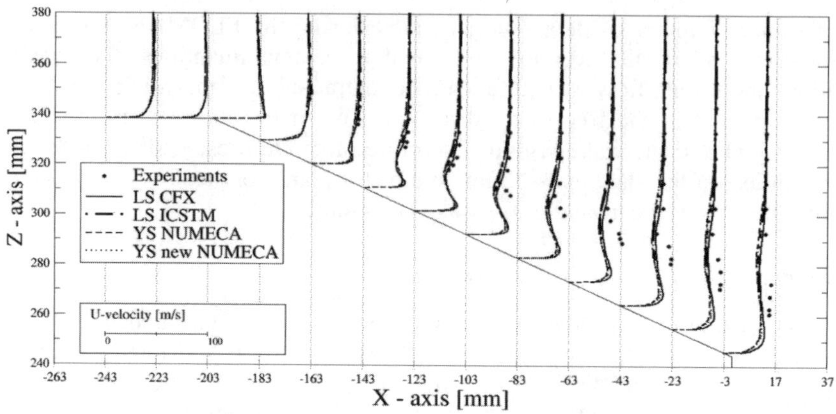

Figure 5 Velocity profiles at y=180 mm on slant

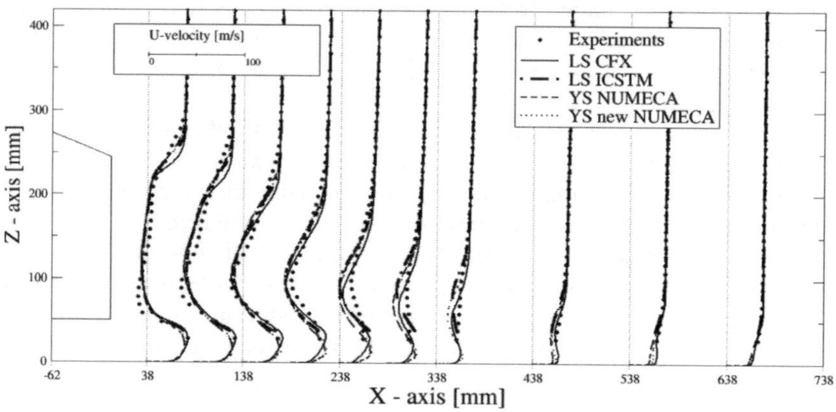

Figure 6 Velocity profiles at y=0 mm in wake

15.5.2 Comparison of results for SST model

The present chapter describes the comparison of codes using the SST turbulence model in the codes of different partners. CFX and SPTU computed results on the hexahedral grid with 1.3 million nodes, DLR on the hybrid grid with 2.0 million nodes and NUMECA on a hybrid grid with 0.8 million nodes. Velocity profiles are presented in Figure 6, Figure 7 and Figure 8. The SST model has demonstrated in a large number of test cases to be capable predicting flows with separation bubbles quite accurately. It is however well known that RANS models with an optimized capability for the prediction of the separation onset tend to underpredict the turbulent stresses in heavily separated flows. This usually leads to an underprediction of the boundary layer recovery downstream of reattachment.

Almost identical velocity profiles are obtained by CFX and SPTU using the same grids. The results of NUMECA show a slightly smaller separation on the slant, whereas the DLR results show a stronger backflow in the mid plane of the slant, but a no separation on the side of the slant (like the CFX and SPTU data). The agreement with the experimental data at the side of the slant and for the flow recovery is better for the results computed on the unstructured grids (Figure 7 and Figure 8). This represents a change in flow topology. The reason for the differences between the structured and the unstructured grid results are not clear.

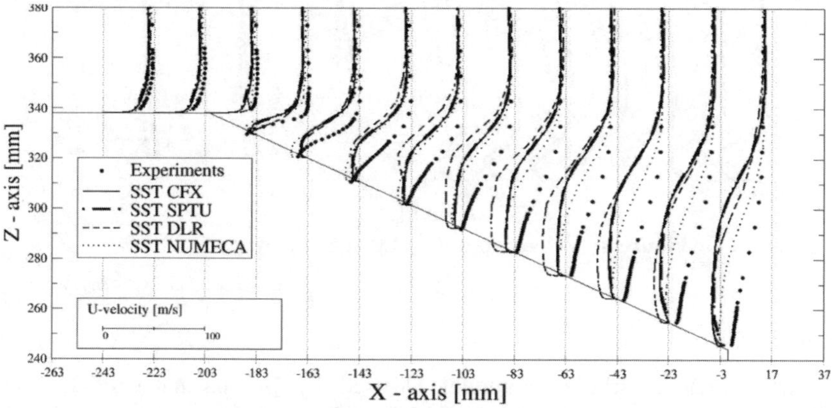

Figure 6 Velocity profiles y=0 mm on slant

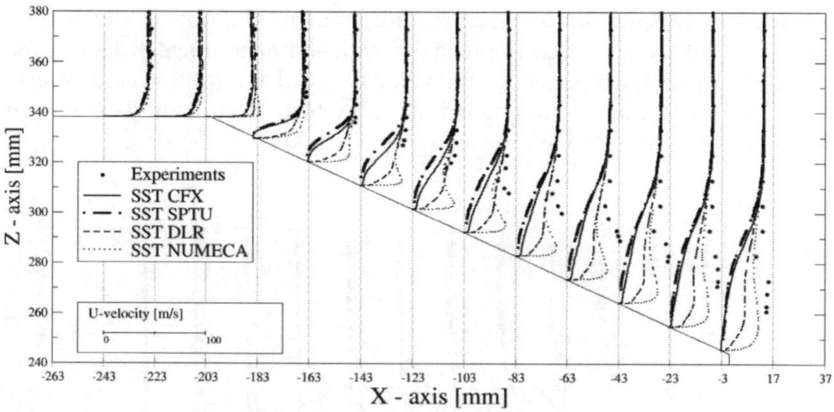

Figure 7 Velocity profiles at y=180 mm on slant

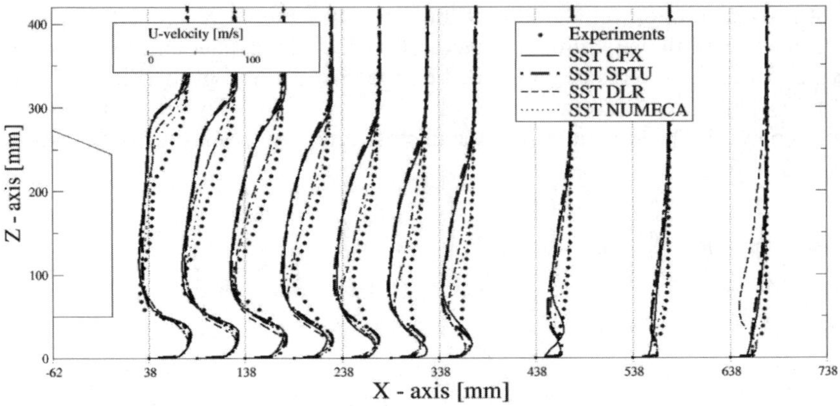

Figure 8 Velocity profiles at y=0 mm in wake

15.5.3 Comparison of results for SA model

The Spalart–Allmaras turbulence model is used by the partners SPTU and NUMECA on the hexahedral grid with 1.3 million nodes and on different unstructured grids by DLR and NUMECA. The results of SPTU and NUMECA in Figure 9, Figure 10 and Figure 11 show a quite similar behavior as the k-ε based turbulence models. Despite the adverse pressure gradient, the flow is attached on the whole slant, but shows a good agreement with the experimental data in the wake region. The overall agreement of the velocity profiles for the structured grid and NUMECA's unstructured grid is good. A different flow structure is predicted by the SA model of DLR showing a separation in the middle of the slant.

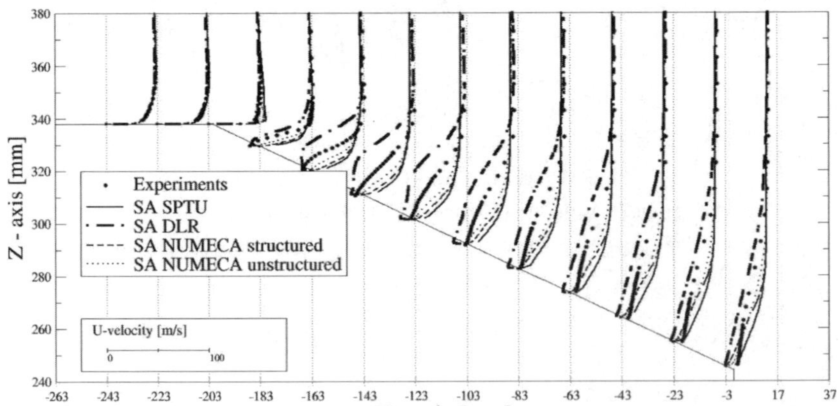

Figure 9 Velocity profiles y=0 mm on slant

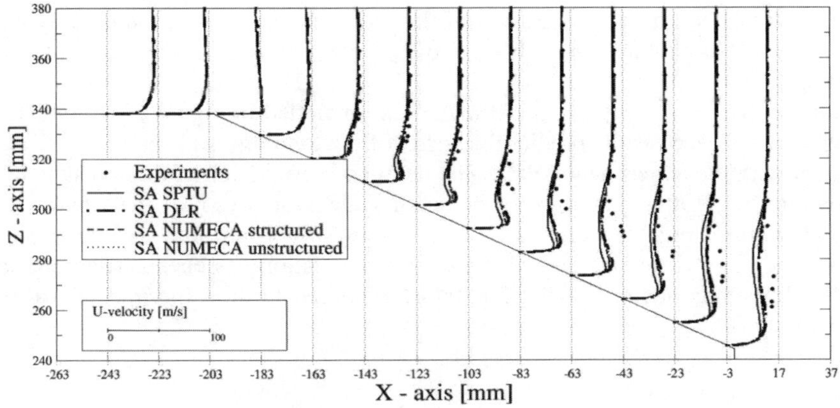

Figure 10 Velocity profiles at y=180 mm on slant

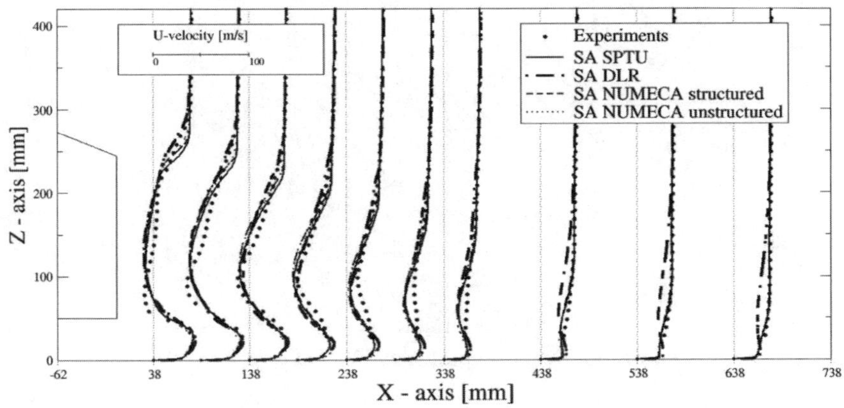

Figure 11 Velocity profiles at y=0 mm in wake

15.5.4 Comparison of DES / LES Models

The present chapter shows the comparison of computational results for different DES methods and the LES method. The DES methods are based on different RANS models: the SST-DES model of CFX and the SA-DES model and SARC-DES model of SPTU. The results are computed on hexahedral grids and presented in Figure 12, Figure 13 and Figure 14. The computed velocity profiles show differences to the profiles of the underlying RANS models. The DES extensions of the RANS models perform typically somewhat better for the prediction of the separation bubble. Improvements can be seen e.g. for the zonal SST-DES model of CFX showing a less strong separation in the middle of the slant and an attached flow on the side of the slant. The SA based DES models show the same tendency as the underlying RANS models: no separation predicted by the SA-DES model and strong separation predicted by the SARC-DES model.

Despite the significant increase in computational costs, all DES models still fail to predict the correct flow structure. A reason for this, also discussed in

Manceau (2002) could be the inability of the DES models (and of the underlying RANS models) to capture the high level of turbulent stresses downstream of the separation line.

Additionally LES results of CHALMERS are included in Figure 12, Figure 13 and Figure 14. These results predict the correct flow topology and show an overall agreement with the experimental data. The detailed presentation of these results by Krajnović (2004, 2005a, 2005b) showed small-scale structures near the separation point indicating a high level of turbulent stresses, which are also visible in the experimental data. Note however, that the LES simulation has been carried out at a smaller Reynolds number of Re=200.000 (compared to the experimental value of Re=768.000).

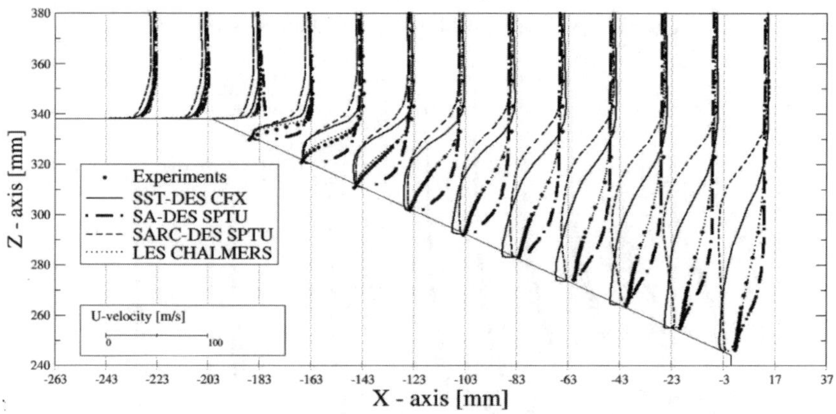

Figure 12 Velocity profiles y=0 mm on slant

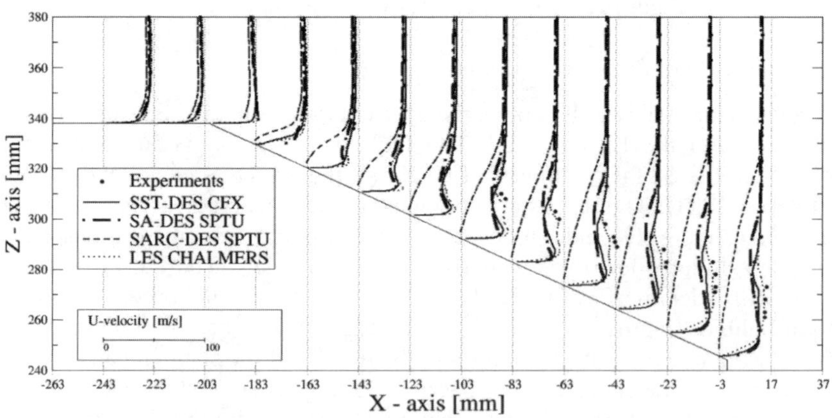

Figure 13 Velocity profiles at y=180 mm on slant

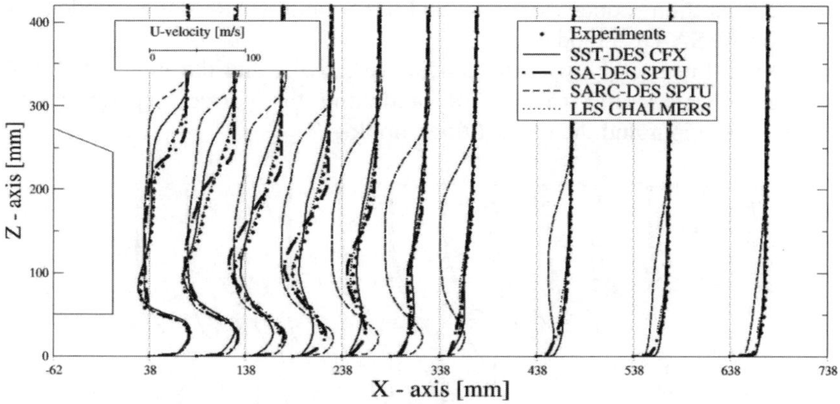

Figure 14 Velocity profiles at y=0 mm in wake

15.5.5 Comparison of NLEVM / RSM Models

Non-linear eddy viscosity models and Reynolds stress models are compared in the present chapters. ICSTM computed the flow field with the quadratic EVM/EARSM model AJL (Abe et al., 2003) based on the ω-equation and the ε-equation. The SSG model was selected as common Reynolds stress model in the FLOMANIA project. CFX uses this model in combination with wall functions, ICSTM solves the equations up to the wall. All results are computed on hexahedral grid, results are given in Figure 15 and Figure 16.

The AJL models predict a separated flow on the whole slant (midplane) comparable to the SST model results. Therefore, the flow recovery in the wake region differs strongly from the experimental data. The SSG results of ICSTM and CFX show a different characteristic on the slant. The ICSTM result (without wall function) reattaches shortly before the end of the slant, while the CFX result (with wall functions) never separates. Both SSG results agree well with the experimental data in the wake region.

15.6 Conclusions

A variety of computational results for the Ahmed car geometry with 25° slant angle have been presented and discussed in the present chapter. This geometry results in a challenging flow with a strong sensitivity of the flow topology to the turbulence model and the computational grid.

Different RANS, DES, Reynolds Stress and non-linear eddy viscosity turbulence modes have been applied. Each RANS model showed the characteristic behaviour already known from other basic test cases. k-ε and SA models show a limited tendency to predict the flow separation, resulting in attached flow. Other models predict separation, but fail to capture the flow reattachment on the slant.

DES extensions based on the SA and the SST model show in principal the same tendency concerning flow separation as their underlying RANS models, e.g.

too strong separation zones for the SST-DES and SARC-DES model, no separation for the SA-DES model.

Also the non-linear eddy viscosity models (e.g. AJL) and the Reynolds Stress models (e.g. SSG) are not capable of predicting the correct separation and reattachment behaviour and the correct flow topology.

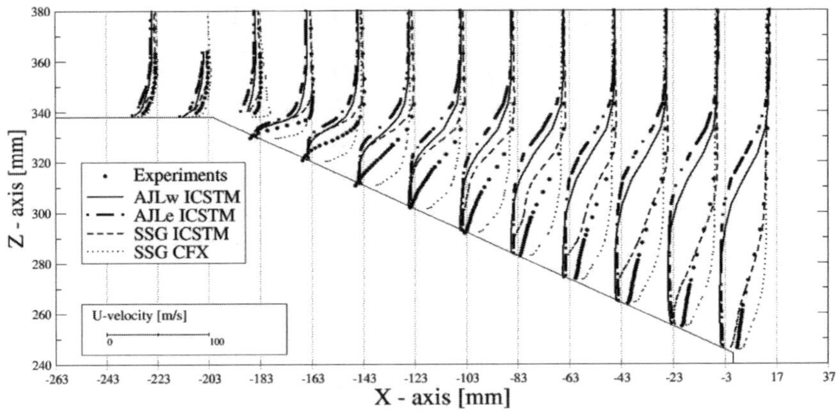

Figure 15 Velocity profiles y=0 mm on slant

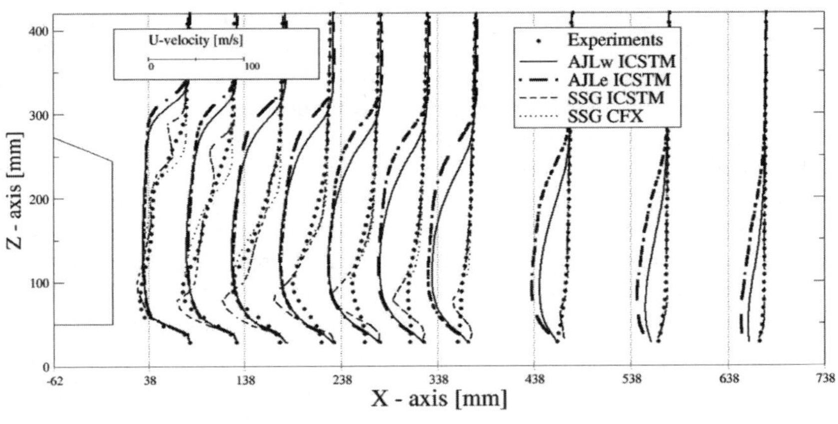

Figure 16 Velocity profiles at y=0 mm in wake

It should be noted that the 25° Ahmed car body is an extremely sensitive test case (see discussion in Manceau 2002). This can be seen by the fact that different codes with the same turbulence model produced different flow topologies. It is also known that small changes to a model can flip the flow topology from "stalled" to "attached". The results from the DES and more so the LES simulations suggest, that some unsteady flow content will have to be taken into account in order to arrive at reliable predictions. The case might therefore not be well suited for the evaluation of steady-state RANS models.

16 3D low hill with complex separation

Lars Davidson, Chalmers University of Technology

16.1 Introduction

Six partners have computed the flow around the axisymmetric 3D hill, namely Chalmers (Davidson, Dahlström, 2005) , EDF, FOI, ICSTM (Wang et al, 2003; Temmermann et al, 2004), SPTU and UMIST. The configuration consists of an axisymmetric hill, see Fig. 1. The creast of the hill is located at x=z=0, and the height of the hill is H. The Reynolds number is $Re=U_{in,\infty}H/\nu=130{,}000$. LDV measurements have been carried out at the plane x=3.69H (Byun et al, 2004) and in the centre plane z=0 (Byun and Simpson, 2005). Measured surface pressure and friction data are also presented in Byun et al, 2004. Profiles of the streamwise velocity and the Reynolds stresses were measured at x=0 without the hill. SPTU and UMIST used these profiles as inlet boundary conditions upstream of the hill at $x\approx-4$. ICSTM and FOI carried out a separate boundary layer simulation to create steady inlet boundary conditions. Those partners who have carried out unsteady simulations have created unsteady inflow conditions from white noise (EDF) or pre-cursor channel DNS fluctuations (Chalmers). The turbulence models and meshes used by the partners are summarized in Table 1 (the turbulence models are presented in detail in Chapter III).

Table 1 Turbulence models and meshes.

Partner	Method	Turbulence model	Mesh
Chalmers	Hybrid LES-RANS	1-eq	130x160x82
EDF	LES	SM*	800 000**
FOI	RANS	WJ-st-w	150x120x81
ICSTM	RANS	Shima98	128x96x96
UMIST	RANS	Wi88	128x160x96
SPTU	RANS	SA	115x65x67

*Smagorinsky SGS model , **unstructured mesh.

16.2 Results

The flow accelerates over the hill. At the lee-side the flow decelerates and a three-dimensional re-circulation takes place. In the centre plane the flow separates at $x\approx H$, see Fig. 2 and reattaches at $x\approx 2H$. The height of the separation region in the centre plane is smaller than 0.1H. The separation line at $x/H\approx 1$ can be seen in Fig. 3. A saddle point is formed at the centre of the separation line (z=0) and from this point the separation line extends out to $z=\pm 0.5H$. For $z>0.5H$ the separation line is deflected downstream and at $x\approx 1.2H$, $z\approx\pm 0.8H$ it ends in two focus points (see Fig. 3).

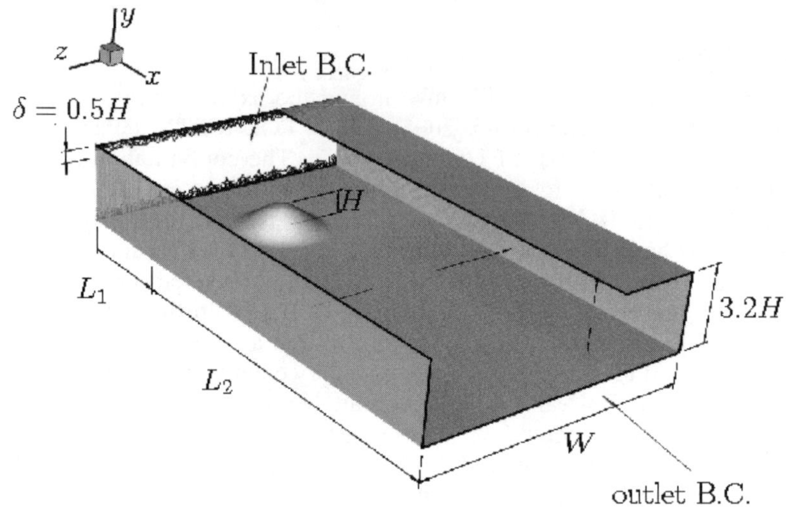

Figure 1 3D hill. Configuration.

The flow in the region of the centre plane approaching the separation line is deflected toward the sides. At the edge of the separation line (z ≈±0.8H), the flow is due the separation region forced slightly further away from the centre plane and after that continues in the downstream direction. Some of this fluid is entrained into the separation region from the sides (1.2<x/H<2, z ≈±0.8H). Fluid from above is also entrained into the separation region. The streamwise flow in the centre plane passing above the recirculation region impinges at the bottom wall downstream of the reattachment of the recirculation region. It is believed that the impinging process downstream of the reattachment creates the two secondary vortices (Ω_x) one of each side of the centre plane. The flow direction of these vortices is such that fluid near the walls is directed away from the centre plane (see Fig. 12).

The predicted vector fields in the centre plane are shown in Figs. 4-6. As can be seen, the recirculation region predicted by all RANS simulations (Fig. 4-5) is much too large. The model that predicts the smallest recirculation bubble is actually the Wi88 k-ω model. The recirculation region predicted with hybrid LES-RANS (Fig. 6a) is in fairly good agreement with the experiments whereas that obtained with LES (Fig. 6b) is slightly too large.

The predicted streamwise velocities at x/H=3.69 are compared with experiments in Fig. 7. The agreement between RANS predictions and experiments is poor, especially near the centre plane (z=0) where the predictions show a much too large velocity deficit. This is of course a consequence of the too large predicted recirculation bubbles at the lee-side of the hill which were observed in

Figs. 4-6. The predicted velocity deficit is smallest for the k-ω model which is consistent with the relatively small recirculation bubble in Fig. 5b.

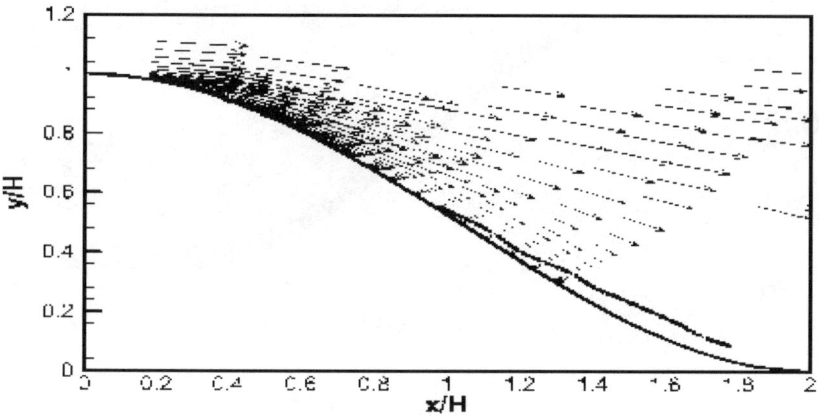

Figure 2 Experimental velocity vectors in the symmetry plane (Byun and Simpson, 2005).; the dotted thick line indicates the position of zero velocity

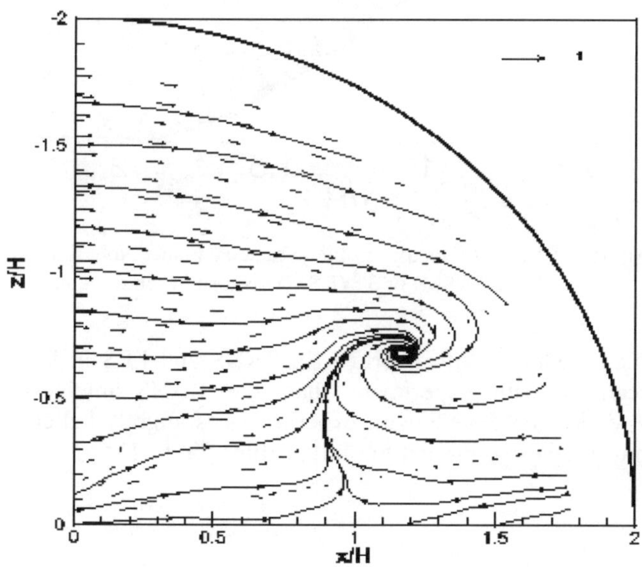

Figure 3 Experimental velocity vectors close to the surface (Byun and Simpson, 2005)

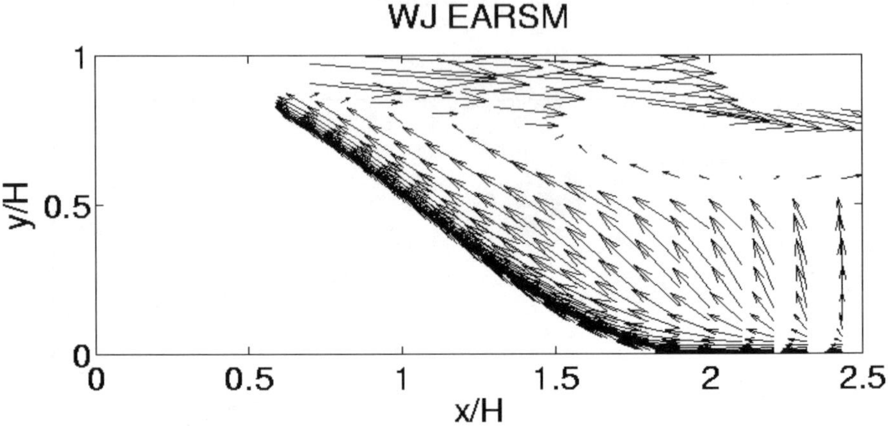

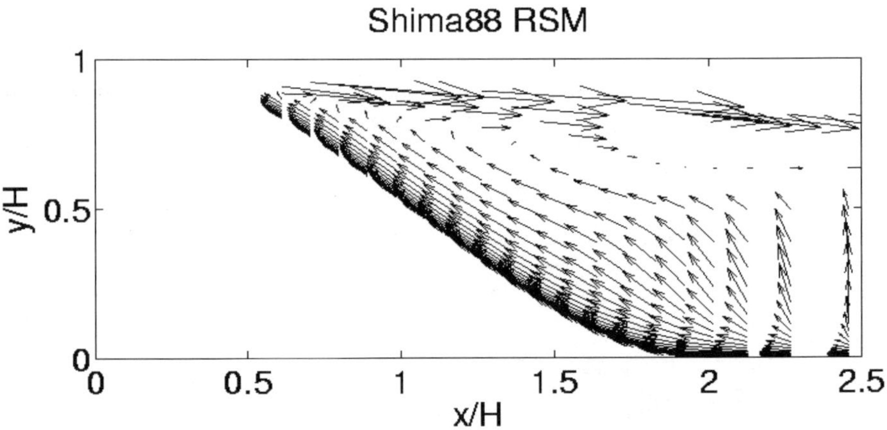

Figure 4a (top), b (bottom) Velocity vectors in the symmetry plane. Note that only every 2nd (EARSM) or 3rd (RSM) vector in each direction is shown

The velocity profiles predicted with hybrid LES-RANS and LES in Fig. 8 is in much better agreement with experiment than the RANS-simulations. The agreement of the hybrid LES-RANS with experiments is slightly better than the LES. However, both simulations predict too full profiles for $|z/H|>1.3$.

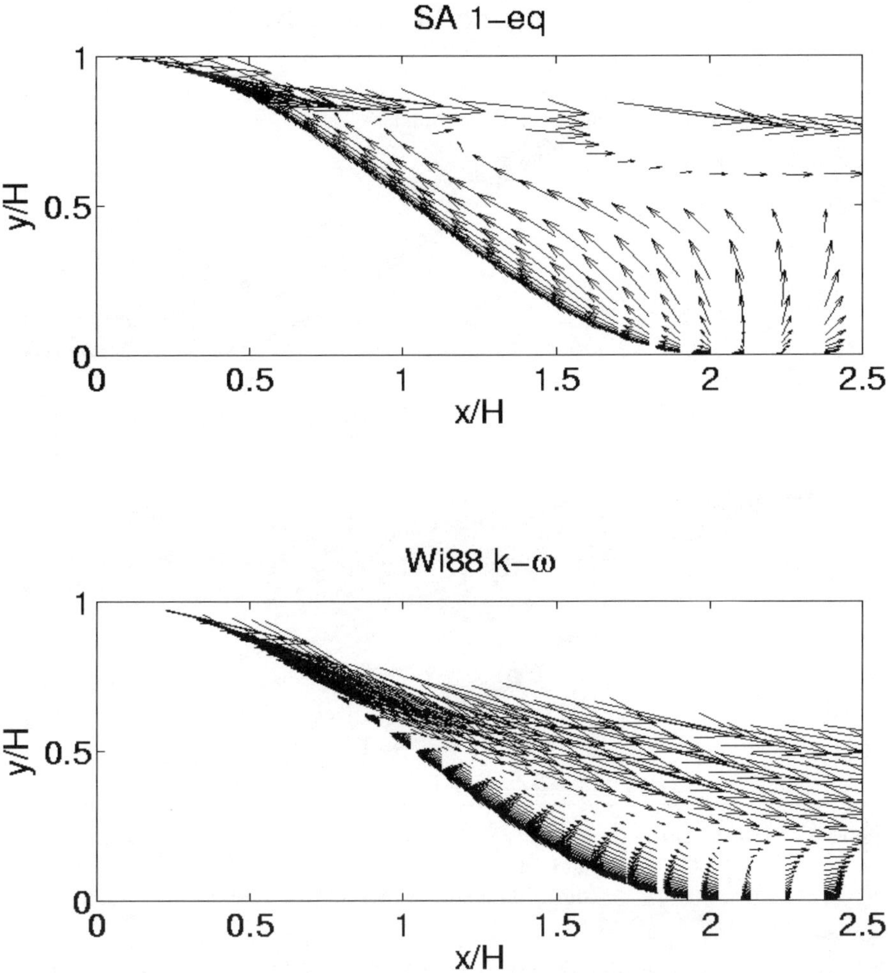

Figure 5a (top), b (bottom) Velocity vectors in the symmetry plane. Note that only every 2nd vector in each direction is shown

Figures 9-11 present the secondary velocity vector fields at x/H=3.69 which can be compared with the experimental vector field in Fig. 12. The experimental velocity field presents a large clock-wise vortex extending from the centre plane out to z/H≈-2 . All RANS models show a completely different flow picture. They all show a *counter* clockwise vortex near the centre plane which is fairly strong and large with all models except with the k-ω model. Further away from the centre, the SA model and the k-ω model do predict a rather large clockwise vortex.

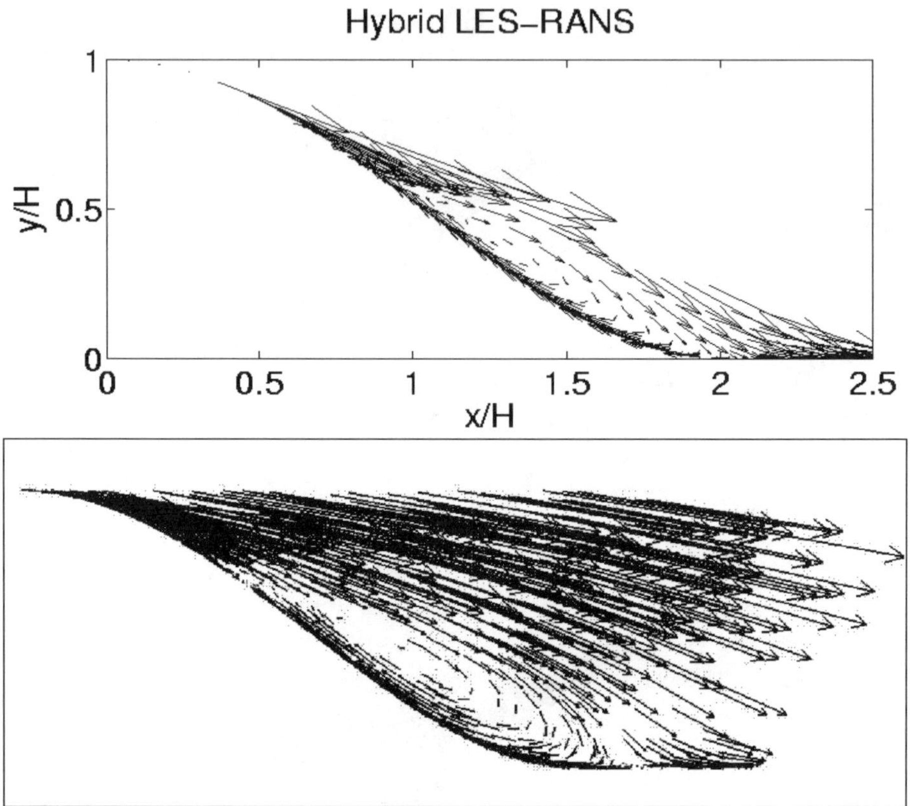

Figure 6a (top), b (bottom) Velocity vectors in the symmetry plane (hybrid LES-RANS and LES). Note that only every 2nd /3rd vector in the x/y direction is shown (hybrid LES-RANS)

The secondary flow predicted with hybrid LES-RANS and LES agree well with experiments although the location of the vortex predicted by LES is located slightly too far away from the centre plane. When studying the vector field in y-z planes further upstream (Davidson, Dahlström, 2005), it is found that the large secondary clockwise vortex (Fig. 11) is formed after the reattachment. The flow in the centre plane impinges on the floor closing the recirculation bubble. This seems to be the mechanism which creates the secondary vortices. In the RANS simulations the vortex near the centre plane rotates in the *opposite* direction. This vortex exists already at x/H=1 and it is probably created on the lee-side of the hill.

The predicted spanwise velocity components are compared with experiments in Figs. 13 and 14. Again, it can be seen that the agreement with experiments for RANS is very poor but for hybrid LES-RANS and LES the agreement with experiments is fairly good.

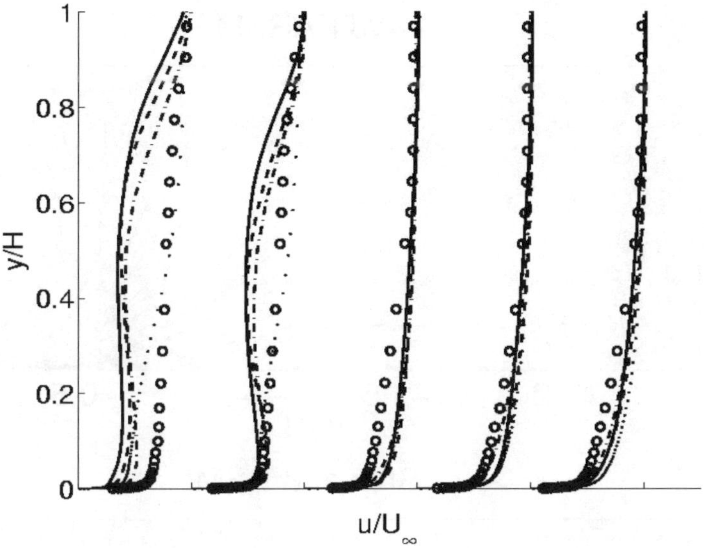

Figure 7 Streamwise velocities <u>/U_{in} at x/H=3.69. RANS models. The profiles are taken at the location (from left to right) z=0, z/H=-0.33, z/H=-0.81, z/H=-1.30, z/H=-1.79. Solid lines: WJ-st-w; dash-dotted lines: SA model; dashed lines: Shima88 RSM; dotted lines: Wi88 k-ω model. Markers: Experiments (Simpson et al 2002}

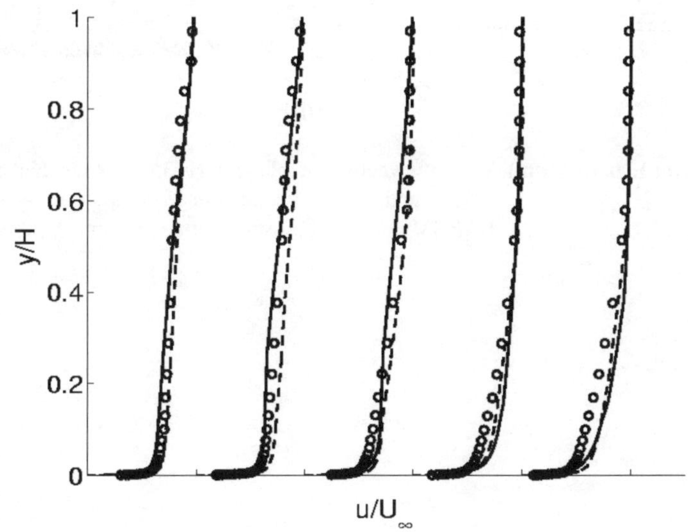

Figure 8 Streamwise velocities (see caption in Fig. 7). Solid lines: LES; dashed lines: hybrid LES-RANS; markers: experiment by Simpson et al (2002).

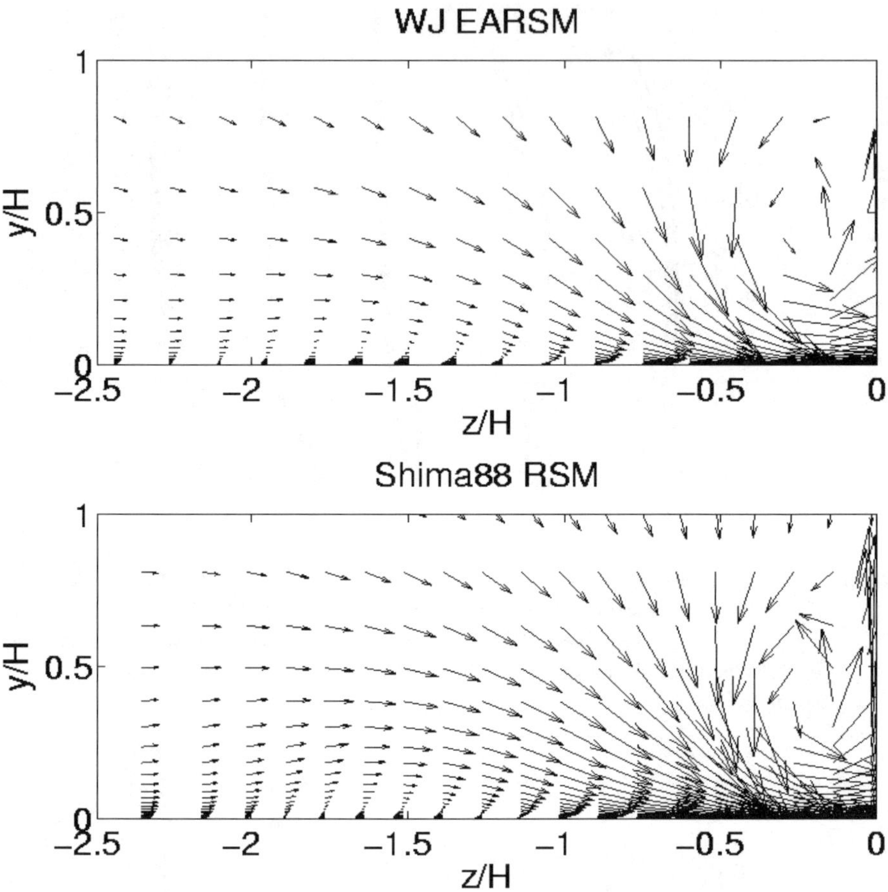

Figure 9a (top), b (bottom) Velocity vectors in plane x/H=3.69. Note that only every 2nd/3rd and 3rd/4th vector in z/y direction is shown for EARSM and RSM , respectively

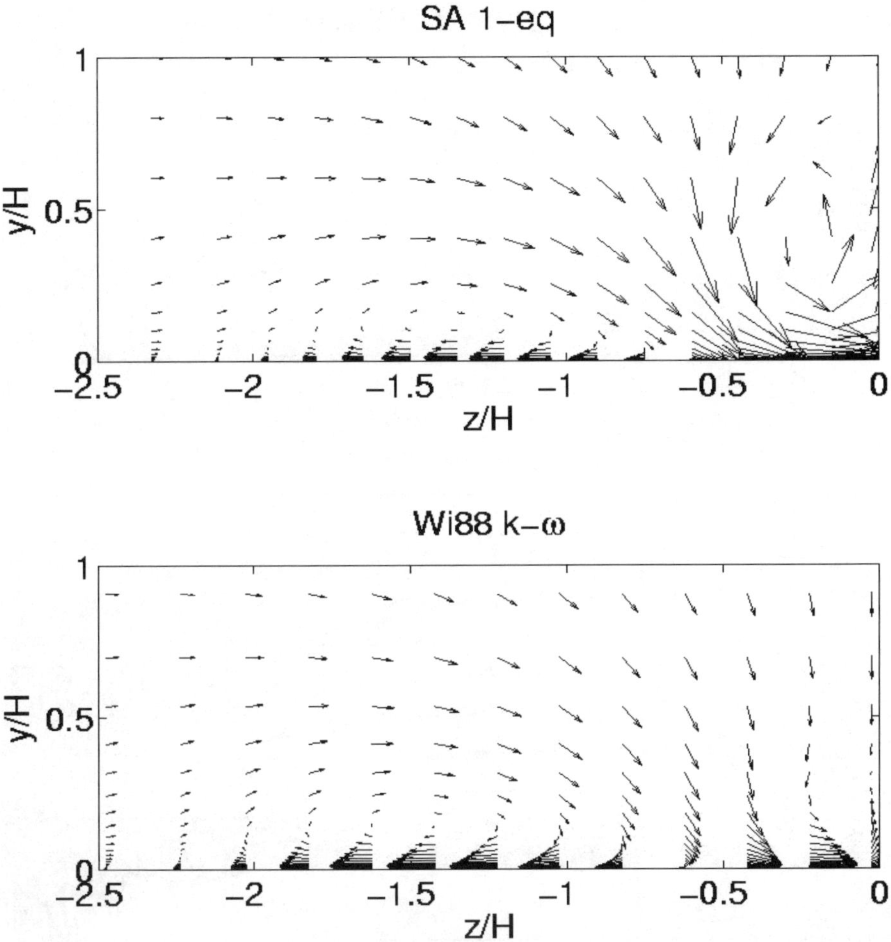

Figure 10a (top), b (bottom) Velocity vectors in plane x/H=3.69. Note that only every 3rd/3rd and 2nd/4th vector in z/y direction is shown for SA and Wi88 , respectively.

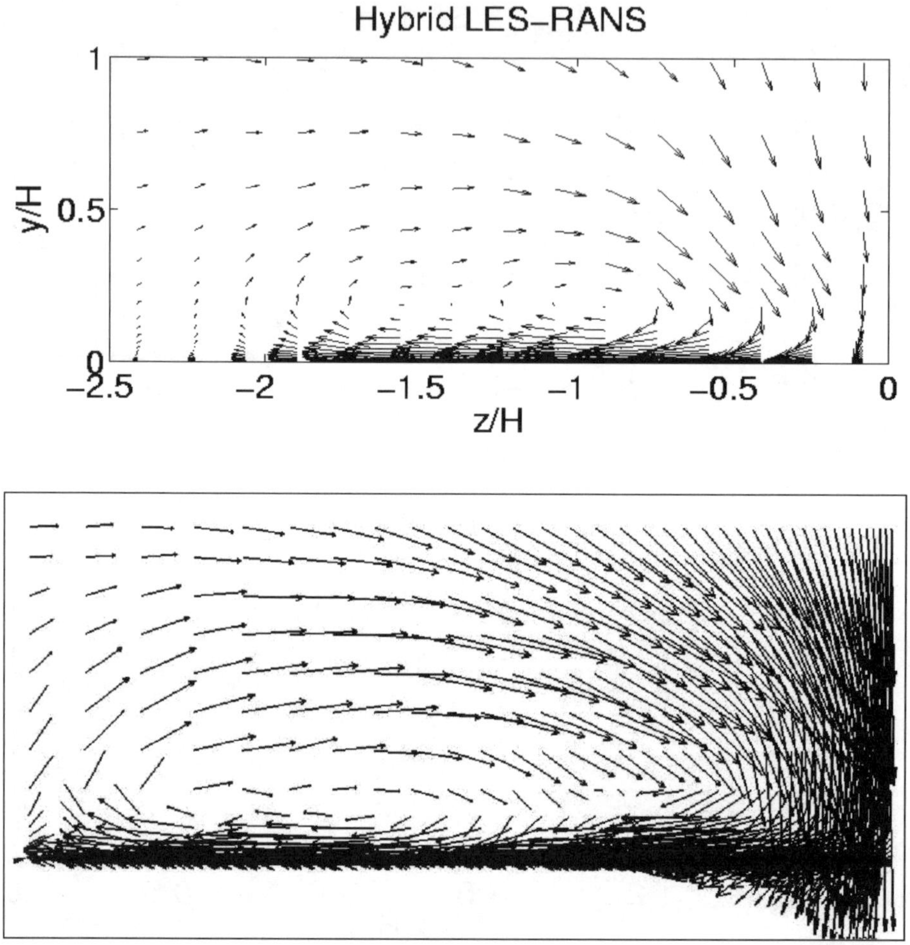

Figure 11a (top), b (bottom) Velocity vectors in plane x/H=3.69 (hybrid LES-RANS and LES). Note that only every 2nd/3rd vector in z/y direction is shown for hybrid LES-RANS.

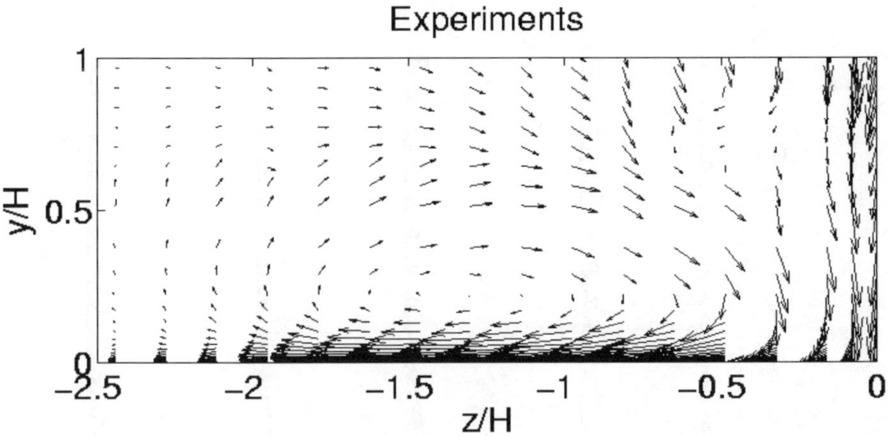

Figure 12 Experimental velocity vectors in plane x/H=3.69 (Byun et al, 2004)

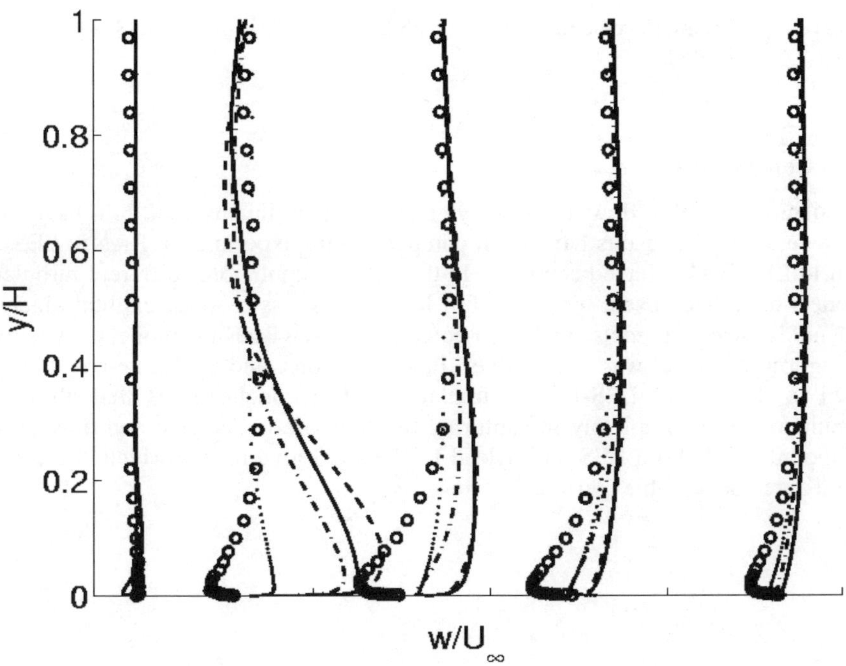

Figure 13: Spanwise velocities obtained with RANS (see caption in Fig. 7).

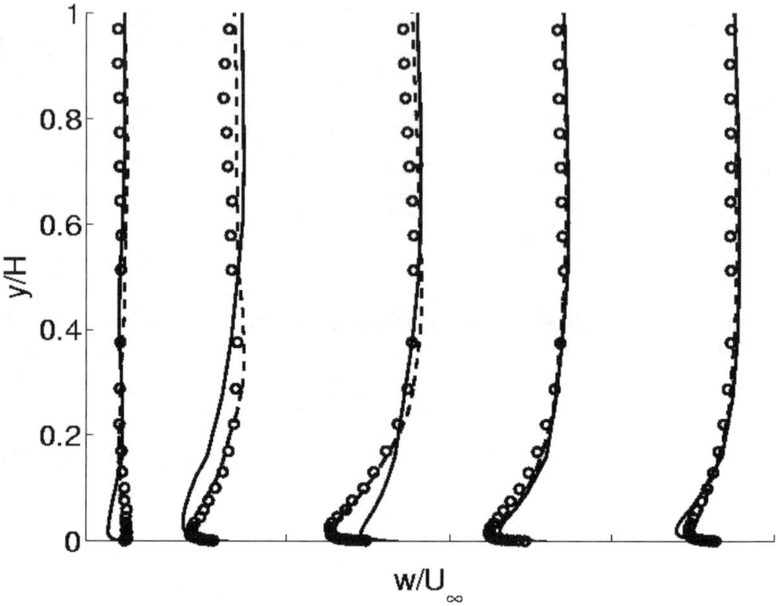

Figure 14 Spanwise velocities obtained with LES and hybrid LES-RANS (see caption in Fig. 8).

16.3 Conclusions

Simulations of the flow over a symmetric, three-dimensional hill have been presented and the results have been compared with experiments. RANS, LES and hybrid LES-RANS have been used. In the RANS simulations, different turbulence models have been used, including full Reynolds stress models, explicit algebraic Reynolds stress models and two-equation eddy-viscosity models. A simple Smagorinsky model was used in the LES-simulations, and a k_{sgs} one-equation was used in the hybrid LES-RANS-simulations. The conclusion is that all RANS simulations fail completely in capturing the flow on the lee-side and downstream of the hill while both LES and hybrid LES-RANS give results which are in fairly good agreement with experiments.

17 Separation behind 2d hills

D. Laurence and J. Uribe, UMIST

Abstract

The flow over a series of two-dimensional hills has been computed by partners ICSTM and UMIST with 8 models. The flow separates smoothly after the first hill and reattaches naturally before the second, making it ideal to test the ability of the different models to predict these difficult features. Two equation models generally fail early separation or are affected by slow post-attachment recovery. The wall resolved SSG and the AJL NLEVM provide better predictions.

17.1 Test case description

The flow over a series of hills is studied, where the hills have a polynomial shape described by Almeida et al. (1992). The Reynolds number based on the hill height and the bulk velocity on the crest of the hill is Re = 10 595. The computational domain is shown in figure 1 where the dimensions are Lx = 9h and Ly = 3.036h. The flow is periodic in the stream-wise direction.

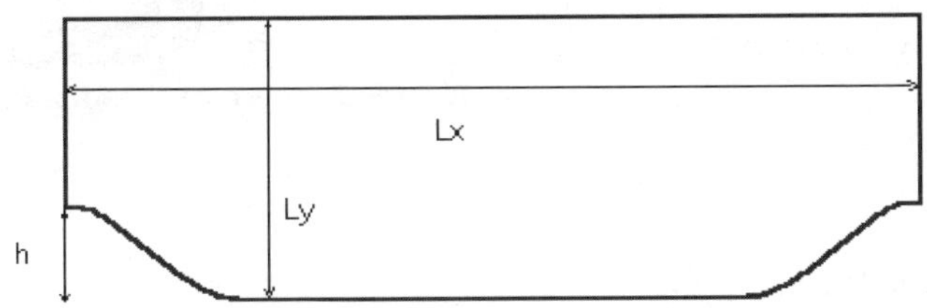

Figure 1. Computational domain

Two highly resolved LES calculations have been carried out by independent groups (Temmerman and Leschziner. (2001) and Mellen et al (2000)) with good agreement between them. The data used for comparison is the one obtained by Temmerman and Leschziner using a mesh with 4.6×10^6 nodes with a maximum value of $y^+=0.5$. The calculation shows separation after the top of the hill at x/h = 0.22 and reattachment at x/h = 4.72.

The flow proved to be very sensitive to the near-wall treatment and the sub grid-scale modelling. The smooth separation near the top of the hill creates a large recirculation bubble and a natural reattachment is obtained before reaching the beginning of the next hill, which is an attractive feature for testing turbulence models. The experimental data is not clearly fully periodic nor 2D, hence the LES data has been used to compare different turbulence models in recent papers (Jang et al. (2002), Wang et al (2004), Laurence et al (2005).

The case was studied in two ERCOFTAC/IAHR workshops (2001, 2002), and in the latter only two hills were computed using the LES data as inlet conditions. The idea was to investigate the influence of the periodic boundary conditions in the stream-wise direction and see whether the errors produced by different models were due to the model itself or to propagation due to the periodicity. Although in many of the contributions, there was a slight difference between the results compared to the periodic case, it could be concluded that the build-up of errors was not dramatic and the periodic case is suitable to estimate models' capabilities to predict separation and reattachment.

17.2 Meshes used

The periodic mesh used can be seen in figure 2, it is block structured and has 20640 elements (120x172) with a maximum y+ = 0.62. It has a geometrical expansion near the wall to resolve correctly the near wall region. After a mesh refinement study, it was found to be fine enough and therefore suitable for use for low-Reynolds models with integration up to the wall.

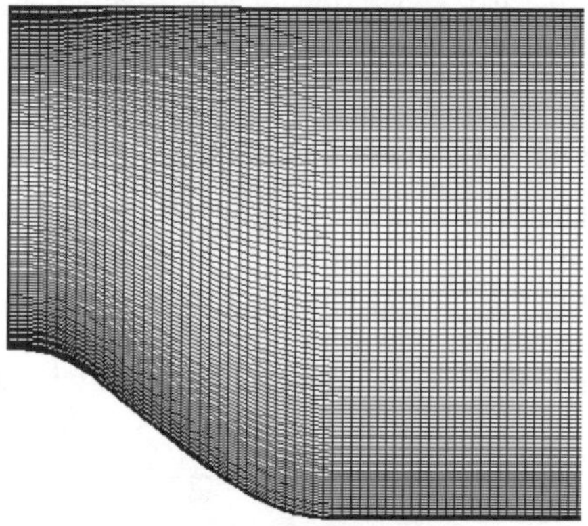

Figure 2 zoom of recommended mesh.

17.3 Results

Two partners contributed to the case, ICSTM and UMIST.

The models used by ICSTM were the Abe et al. NLEVM [AJL], the Wallin-Johansson [WJ] with ω equation, and the SSG second moment closure with the Chen formulation for near the wall integration [SSG-Ch]. UMIST used the Launder-Sharma k-ε [LS], the k-ω [Wi88], the Shear Stress Transport [SST], the φ model [V2F-Lau] and the SSG with scalable wall function.

The resulting streamlines can be seen in figure 3 where they are compared with the LES results.

The separation and reattachment positions can be seen in table 1. The separation point is very sensitive to the model used. The first 4 are the ω based models which tend to give an earlier separation, quite close to the LES, while a too large recirculation is then produced. The last four are ε based models. The SSG-Ch seems acceptable in global terms, while the use of wall functions (SSG-SCWF) significantly reduces the size of the recirculation. The last two models show a late separation, and predictably, as the streamlines at separation point consequently point downward, a shorter recirculation is produced, however less so with the V2F model.

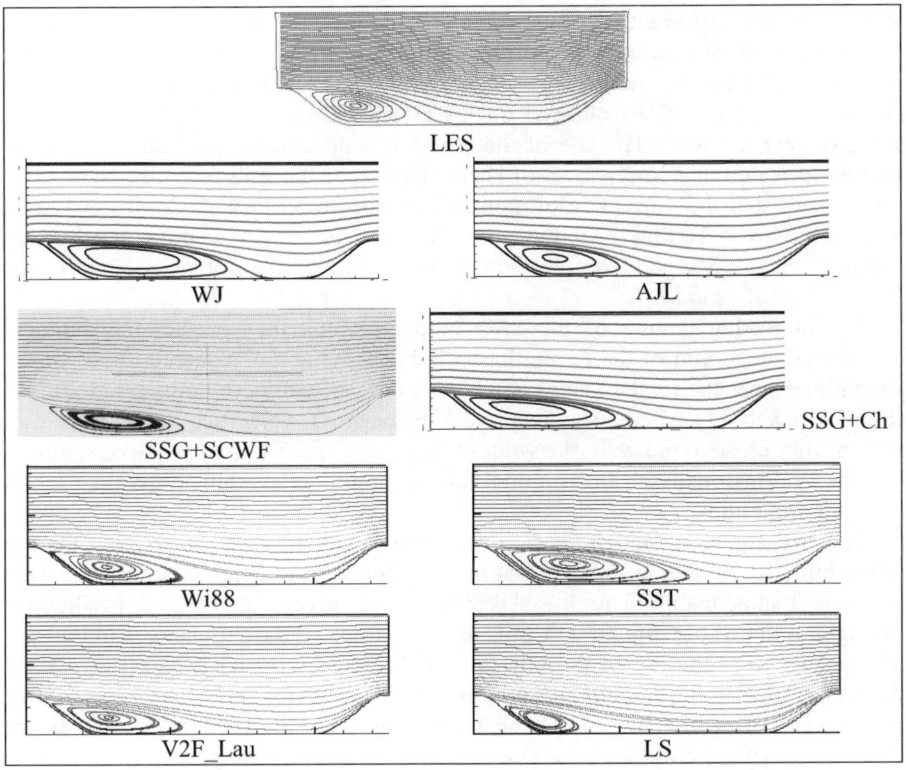

Figure 3 Streamlines for various models

Table 1. Separation and reattachment points

Model	Separation	Reattachment	Length	%
LES	0.22	4.72	4.5	100
AJL	0.26	5.3	5.04	112
WJ	0.22	6.1	5.88	130.66
SST	0.283	7.54	7.257	161.26
Wi88	0.283	5.63	5.34	118.82
SSG-Ch	0.21	4.85	4.64	103.11
SSG-SCWF	0.55	3.67	3.12	69.33
LS	0.38	2.93	2.55	56.8
V2F_Lau	0.38	4.49	4.11	91.33

In figures 4 to 9, profiles are shown for the velocities, the vertical and shear stresses at three different sections. At the top of the hill, only the second moment closures are able to capture the correct sign of the shear stress, and therefore, the inflexion point on the stream-wise velocity profiles. The low-Reynolds version is correct in the bulk of the channel but over-predicts the peak of shear stress and velocity near the wall. The use of the scalable wall function predicts an accurate velocity profile but a lower level of shear stress near the wall (see also figure 10). This shows that the correct resolution of the anisotropy near the wall is of great importance for this flow. All the non-linear models fail to capture the hill-top features, but give a good representation in the rest of the domain, specially the Abe et al. [AJL] model.

An interesting feature of the SSG models is that they predict reattachment before the actual end of the recirculation bubble giving an erroneous shape of the streamlines near the reattachment point. This can be seen in the profiles at x/h = 5 (Figures 8 and 11) where despite predicting negative velocities in the near wall region, very close to the wall, the velocity increases and becomes positive again.

The Wallin-Johansson [WJ] model fails to predict reattachment giving a larger recirculation bubble.

The two-equation models tend to give poor levels of shear stress near the top of the hill (fig. 5), due to the fact that they do not recover quickly enough after the previous reattachment. The k-ω [Wi88] model gives fairly good predictions throughout the whole domain but fails also to capture the right changes in the sign of shear stress at the top of the hill. Although the SST model is based in part on the k-ω, it gives a too large recirculation region and therefore has not enough time to recover before the top of the next hill (fig. 5). This feedback effect is the difficulty in such periodic cases: since the SST predicts low-speed flow at the top of the hill, the separation is, as expected, very large, which yields lower velocities at the top of the next hill etc. ending with a shift of some of the mass-flow towards the upper part of the channel. Conversely, the Launder-Sharma [LS] model gives a correct stream-wise velocity profile at the top of the hill, but only because the reattachment is predicted far too early, giving it more time to recover, by chance, before the next hill. This test case thus highlights the consequences of the slow recovery after reattachment featured by most 2 equation models, as is well known from back-step flow test cases.

The predictions of the φ-model [V2F_Lau] are quite close to those of the k-ω, showing that the elliptic relaxation method improves the k-ε model by damping the normal stress near the wall and therefore predicting the right levels of turbulent viscosity. Note that the φ-model gives results similar to the original version of the V2F model (Durbin 93) which was also tested on this case (Laurence et al. 2005).

All of the two-equation models fail to give accurate levels of shear stress in the whole domain, quite dramatically underestimating the value produced by the LES (fig. 7 & 9).

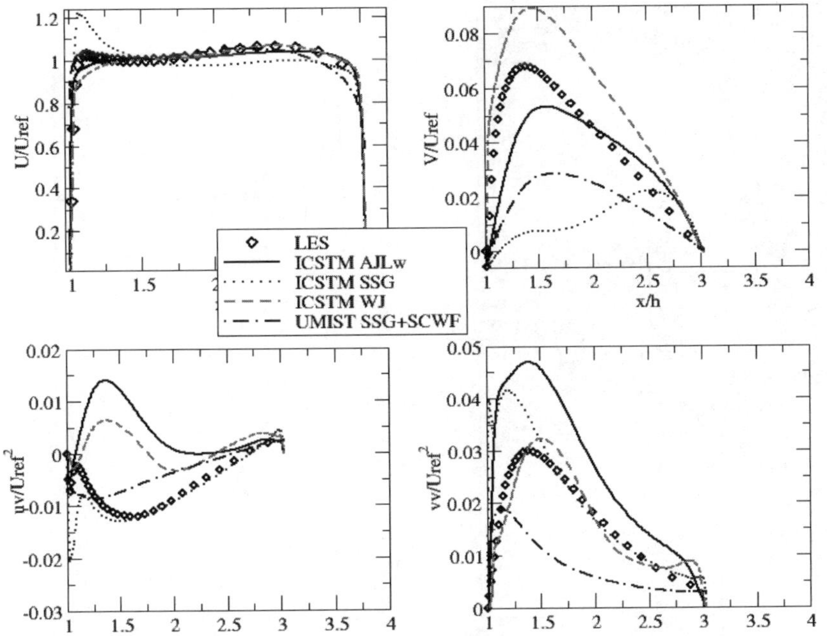

Figure 4 Profiles at x/h = 0.05 Non-linear and Second Moment closures

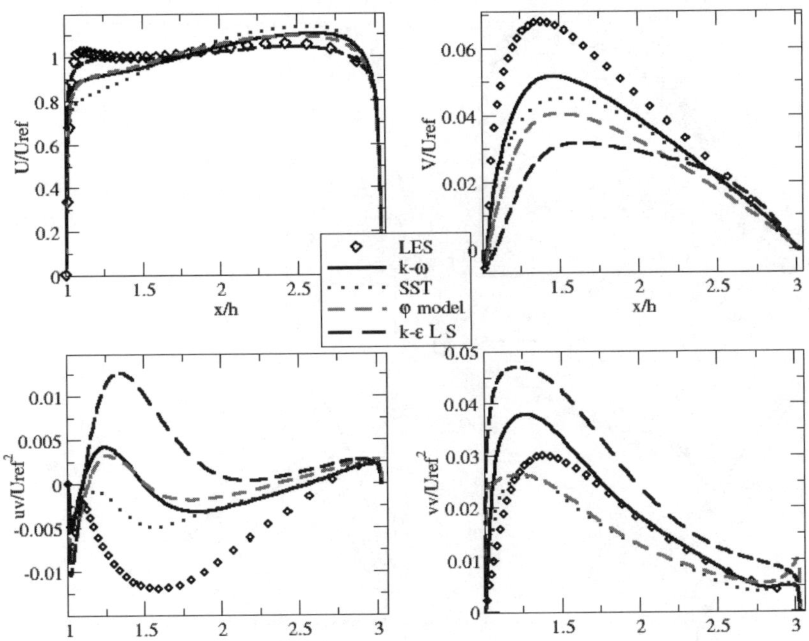

Figure 5 Profiles at x/h = 0.05. Two-equation models

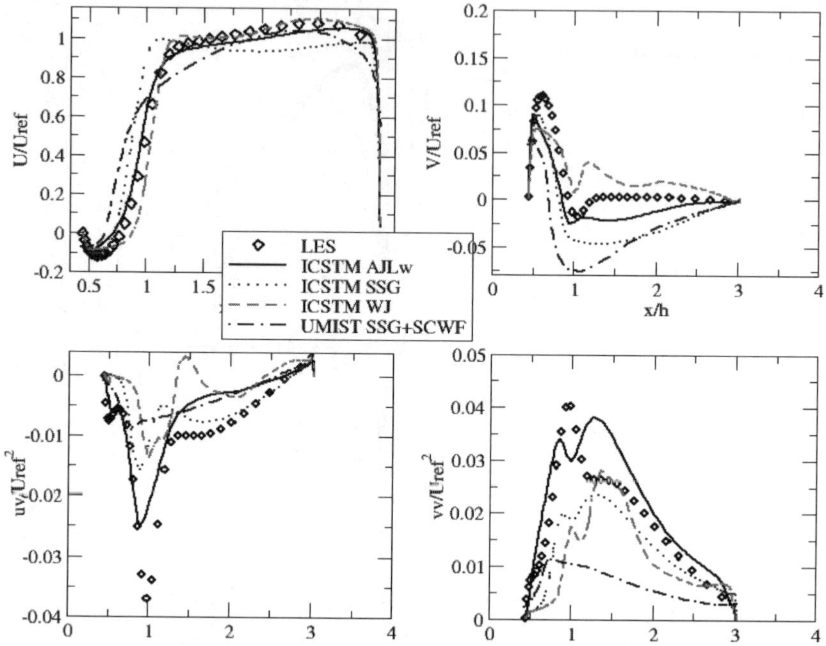

Figure 6 Profiles at x/h = 1 Non-linear and Second Moment closures

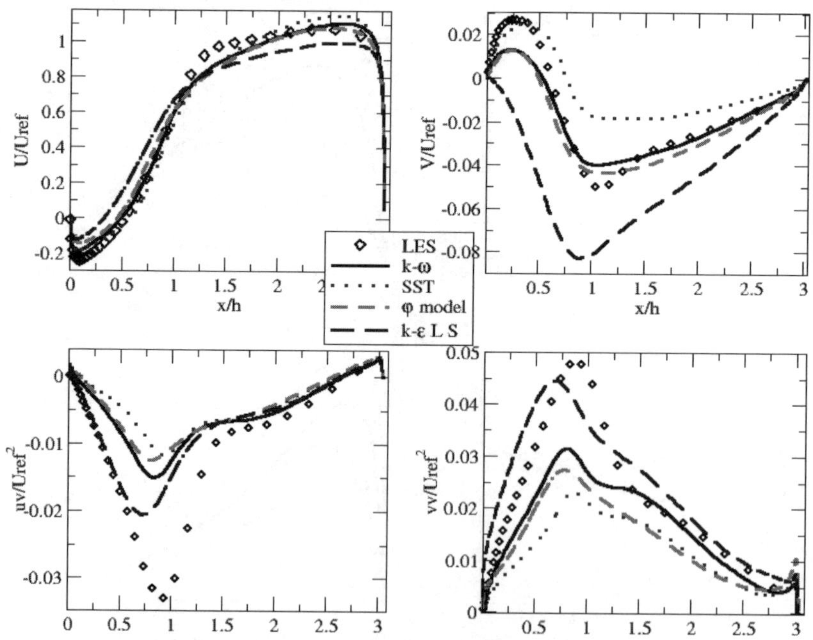

Figure 7 Profiles at x/h = 1. Two-equation models

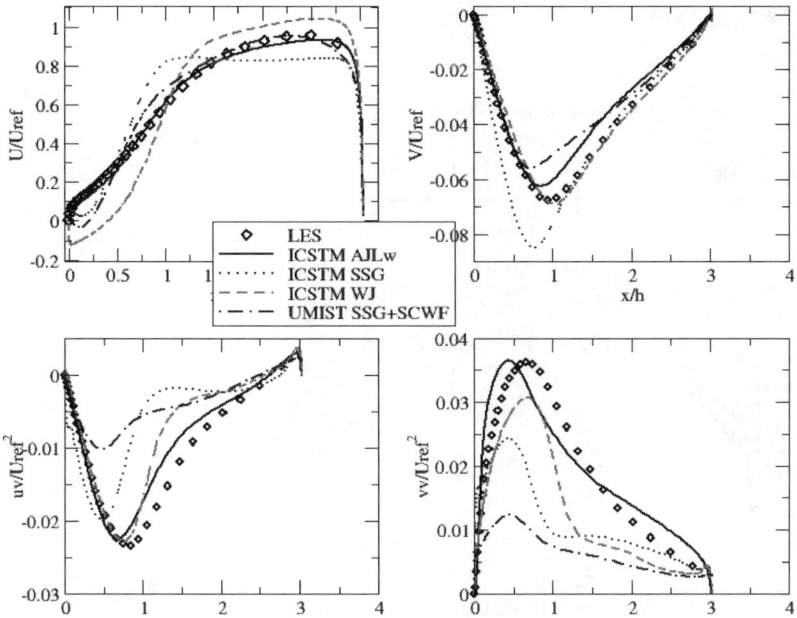

Figure 8 Profiles at x/h = 5. Non-linear and Second moment closures

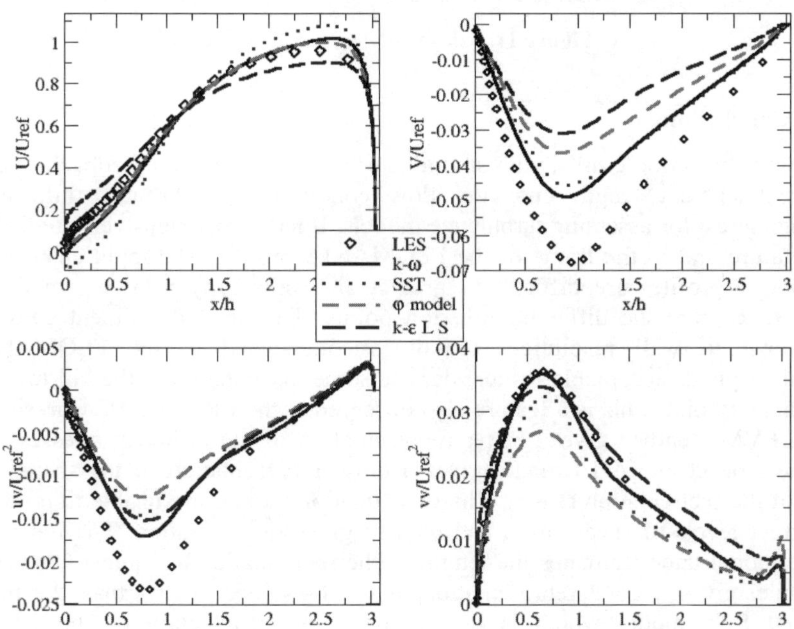

Figure 9 Profiles at x/h = 5. Two-equation models

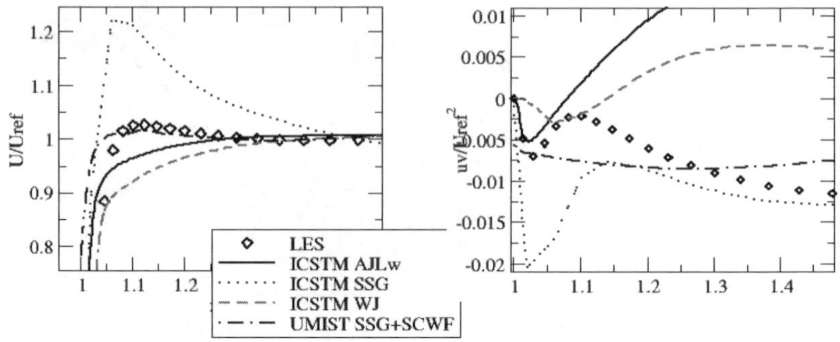

Figure 10 Close up view at x/h = 0.05

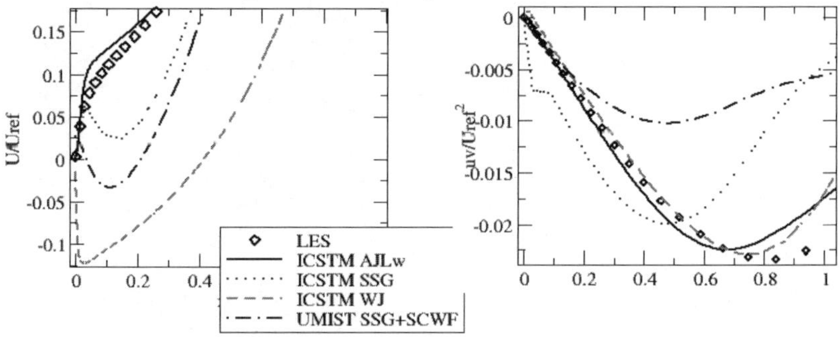

Figure 11 Close up view at x/h=5.

17.4 Conclusions

The flow over periodic hills, with smooth surface separation and feedback effects from reattachment length onto the flow conditions for the next hill, is a challenging test for assessing turbulence models. It has been extensively studied in the literature and in the frame of the FLOMANIA project. The rapid changes in the velocity profiles are difficult to predict and anisotropy resolving model is needed to capture the different inflexion points. The Second Moment Closure, with down to wall resolution, initially recommended to the FLOMANIA consortium yields acceptable results, despite some inaccuracies at the hill-top and reattachment point. This is a feature also observed at the ERCOFTAC workshops. The NLEVMs tend to give a better representation of the turbulent stresses, but only the Abe et al. [AJL] model gives a close representation of the shape and length of the recirculation region. The two-equation models are unable to produce the correct levels of shear stress and tend to give late separation. The Launder-Sharma model underestimates the length of the recirculation by almost 50%. The k-ω model gives a good representation of the flow, even better than the more advanced SST model which is uncommon. The introduction of the elliptic relaxation improves the predictions of the k-ε model, giving results similar to the k-ω.

18 The Aerospatiale A-airfoil

W. Haase, EADS-M

Abstract

The A-airfoil has served as a validation test case in previous EU projects, EUROVAL, ECARP and LESFOIL, using RANS approaches for the first two projects and LES methods in LESFOIL. Nevertheless, and demonstrated below, there is still considerable interest in further investigation of that airfoil case for both turbulence model validation and testing of numerical schemes and methods. In that respect, new RANS computations for the full three-dimensional geometry (wing in wind tunnel) resulted in obvious 3D effects that put the two-dimensional approaches "somewhat" in question, or vice versa.

18.1 Case description

The A-airfoil, a 0.6 m chord length single airfoil was designed at Aerospatiale and tested in the ONERA F1 1.5 x 3.5 m wind tunnel (Rodde, 1987, Houdeville et al., 1987), as well as in the ONERA F2 1.4 x 1.8 m wind tunnel (Gleyzes & Capbern, 2003). For each configuration the results comprise pressure distributions and skin friction values up to about 15 different angles of attack. The drag coefficient was obtained using a total pressure probe within the wake. Wall suction was applied at the wind tunnel's lateral walls – apart for the F1 measurements at $Re=2.1 \cdot 10^6$. Transition was tripped at x/c = 0.3 on the pressure side, while on the suction side a transitional separation bubble was detected with turbulent reattachment at x/c=0.12 - in the case of $\alpha=13^o$. For incidences beyond 16°, flow visualisation studies identified a laminar transitional bubble with turbulent reattachment at about x/c = 0.15.

Experimental results in the F2 tunnel were obtained without wall suction and include pressure and skin friction distributions, boundary layer thicknesses, and LDV measurements for velocity and Reynolds stress profiles in boundary layer and wake. LDV measurements confirmed that the flow was two-dimensional up to about 13°, which can be easily detected from Fig. 1 where oil flow patterns are presented.

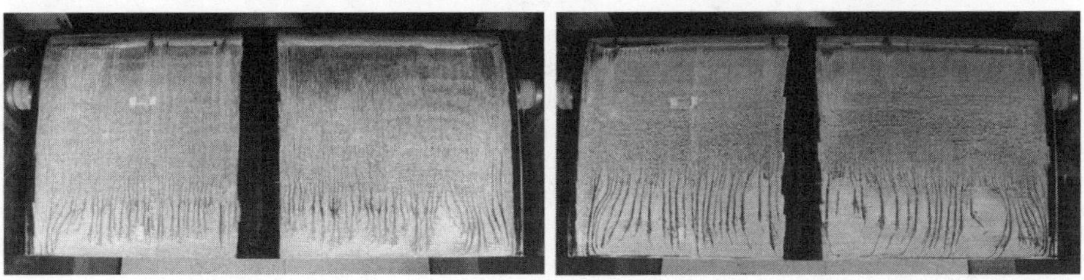

Figure 1 Oil flow patterns for A-airfoil (Gleyzes, 1988) for $\alpha=12^o$ (left) and 13° (right)

Three-dimensional effects appeared in the F1 wind tunnel at even higher incidences due to a larger test section. Under these conditions, the lift coefficient is assumed to be under-estimated in the F2 tunnel for an angle of attack greater than 13° while the drag coefficient – obtained from LDV far wake profiles in the F2 tunnel - was found to be slightly over-estimated with respect to F1 measurements.

The F1 results are judged to be the appropriate reference for the global coefficients C_L, C_D and for pressure and skin friction distributions, while the F2 results are considered to be more appropriate for boundary layer quantities, displacement and momentum thicknesses as well as velocity profiles.

The mandatory flow data for the present studies were: $Ma=0.15$, $Re=2.1 \cdot 10^6$, $\alpha=13.3^o$.

18.2 Mesh generation

To obtain a mandatory grid which was able to produce (nearly) mesh independent solutions, a very fine structured C-type mesh was constructed with 1024 times 256 cells in the wrap-around and wall-normal direction, respectively.

A grid-dependence study using the Johnson-King (JK) model resulted in three consecutive grids exhibiting comparable physical flow structures. In particular, the laminar separation bubble on the upper surface was predicted in the three finest grids. It is seen to be an absolute must, when aiming at grid dependence, to obtain in all these meshes the same flow physics, otherwise a grid-dependence study would render meaningless. However, in order to achieve this goal, roughly 130 points (±15% from leading to trailing edge) have been placed inside the boundary layer to cope with all flow features of interest resulting in a mesh structure for the coarse mesh (256x64) that is presented in Figure 2.

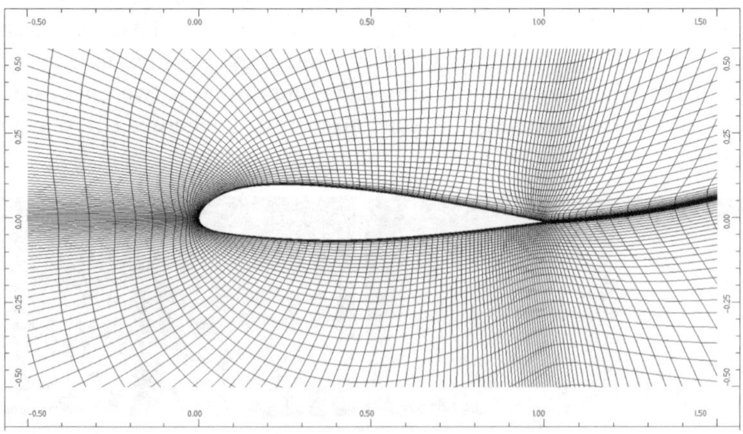

Figure 2 Mesh structure for mandatory mesh in coarse version (256x64 volumes)

The nearly constant number of grid points resolving the boundary layer allowed for accurately predicting the boundary layer history - starting at the stagnation point already. Additionally, quite small y^+ values have been obtained, in order to keep them below one for all the three fine meshes.

Since the A-airfoil features a thick trailing edge, a closure of the geometry contour was necessary, prior to generating the mesh. The approach adopted by EADS-M kept the upper surface unchanged and modified only the lower surface in a way which did not violate the thickness ratio of the airfoil. Moreover, the adjustment was carried out in a way that allowed for a smooth geometry distribution (second derivative) starting on the surface at maximum thickness, which is about 40% of the chord. The geometry changes caused by closing the geometry did not reveal even minor differences in circulation around the airfoil.

Table 1 Grid dependence study

Mesh / **Properties**	**256x64**	**512x128**	**1024x256**	**Measurements**
y+ (upper surface)	0.6-0.9	0.3-0.4	0.1-0.15	
# points in boundary layer.	30-40	55-75	105-145	
Trailing edge separation at	0.7951	0.8556	0.8945	
C_L	1.5269	1.5424	1.5694	1.562
C_D	0.024572	0.020978	0.018794	0.0208
C_M	0.003649	0.006948	0.005718	

Despite all efforts to obtain (as far as possible) accurate solutions for the underlying grid dependence study with solutions converged on all subsequent meshes, results on the three finest meshes still vary slightly as it is demonstrated by Table 1.

The reason for the still varying values is related to effects caused by the laminar separation bubble. None of the CFD methods used by the partners was able to predict accurately the flow inside that bubble where transitional/unsteady/oscillatory effects have been discovered, with increasing unsteadiness on the finer meshes. Fig. 3 is exhibiting skin friction distributions for the mesh dependence study.

The increase in oscillatory behaviour in the laminar separation bubble around x/c=0.11 on the upper surface results in an increase of the skin friction peak downstream, changing the boundary layer history considerably.

It must be noted in addition that the so-called mandatory grid with 1024x256 mesh points has been used by all partners. However, for some computations it was decided to use the next coarser mesh level and for 3D computations, including the sidewalls of the wind tunnel to investigate on 3D effects, grids used were used that were based on the mandatory grid.

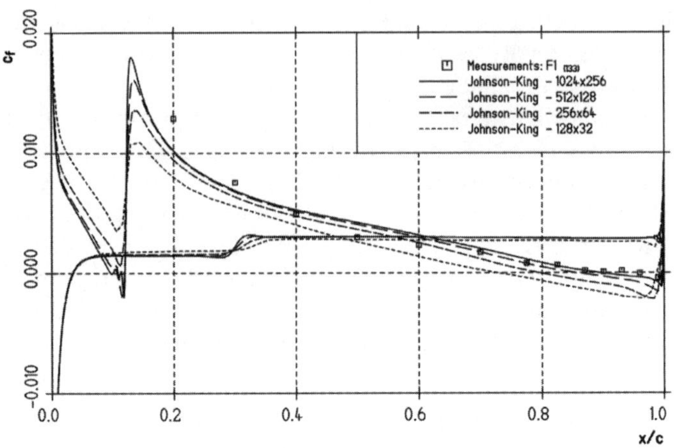

Figure 3 Skin friction distribution for mesh dependence study

18.3 Partner contributions

Table 2 is providing a summary of main results derived by the participating partners together with models and methods in use. EADS-M has been put in parentheses as runs have only been performed for the above mentioned mesh dependence study. Results are presented below for all those turbulence models being written in bold in Table 2.

Table 2 Results and methods by partners

Partner	**Turbulence model(s)**	C_L	C_D
ANSYS/CFX	**SST + transition model**	1.5706	0.01856
DLR	**LEA k-ω** **Wilcox k-ω** Wilcox RSM + SGDH **Wilcox RSM + GGDH** SSG/LLR-ω + SGDH **SSG/LLR-ω + GGDH**	1.588 1.648 1.659 1.643 1.585 1.588	0.0170 0.0163 0.0169 0.0170 0.0184 0.0184
ST. Petersburg University	**SA**	1.6590	0.02019
ONERA	**SA** Smith k-l Wilcox k-ω	1.65186 1.61703 1.5982	0.019167 0.027917 0.019247
(EADS-M)	**JK**	1.5694	0.018794
Measurements		**1.562**	**0.0208**

First of all, Fig. 4 provides a summary of pressure distributions for all those test cases indicated in bold in Table 2. Although the overall agreement seems to be acceptable, the trailing edge pressure indicating up to what extent turbulent separation is predicted is somewhat off when comparing with the measurements. It is well know that the extent of the trailing edge separation zone is an additional measure for circulation, thus the over-predicted suction peak on the upper surface, is a result of this discrepancy in trailing edge pressure. It is interesting to note, compare the left graph of Fig. 5, that the Johnson-King turbulence model is providing the best agreement compared to measurements. This has been detected already in Haase et al (1997), underlining again the fact that the JK model is - for 2D applications - a favourable turbulence model. (It should be noted that the EADS-M results in Fig. 5 are represented by triangles, they appear however like a dotted line as the symbol height is quite tiny).

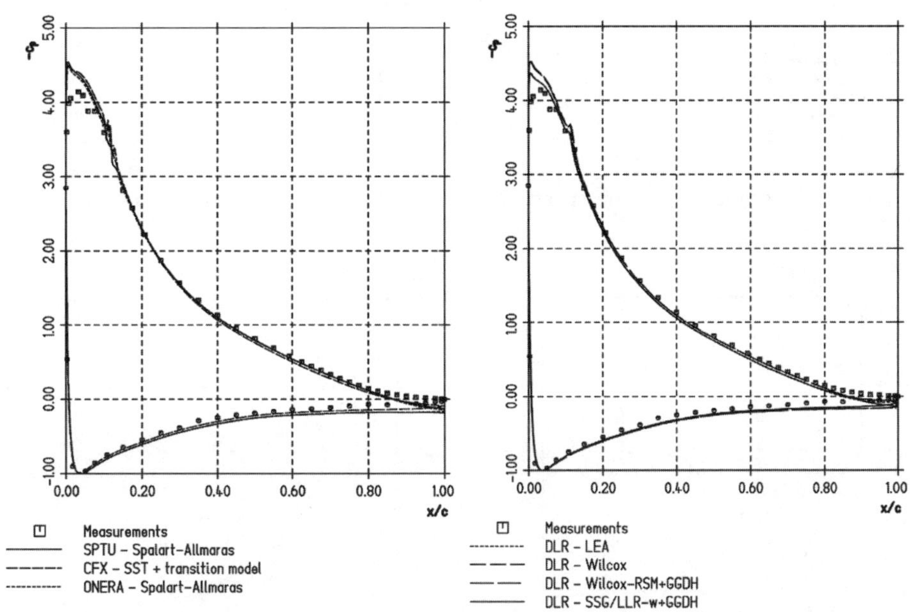

Figure 4 Pressure coefficient distributions

In Fig. 6 cross plots of skin friction results on upper and (partially) lower surface are presented. It can well be detected (as shown in Fig. 4 already for SPTU's, ANSYS' and ONERA's pressure distributions) that the aforementioned comments concerning oscillations in the area of the laminar separation bubble in the nose area and the subsequent friction peak are shown by all partner contributions, compare Fig. 6 (left).

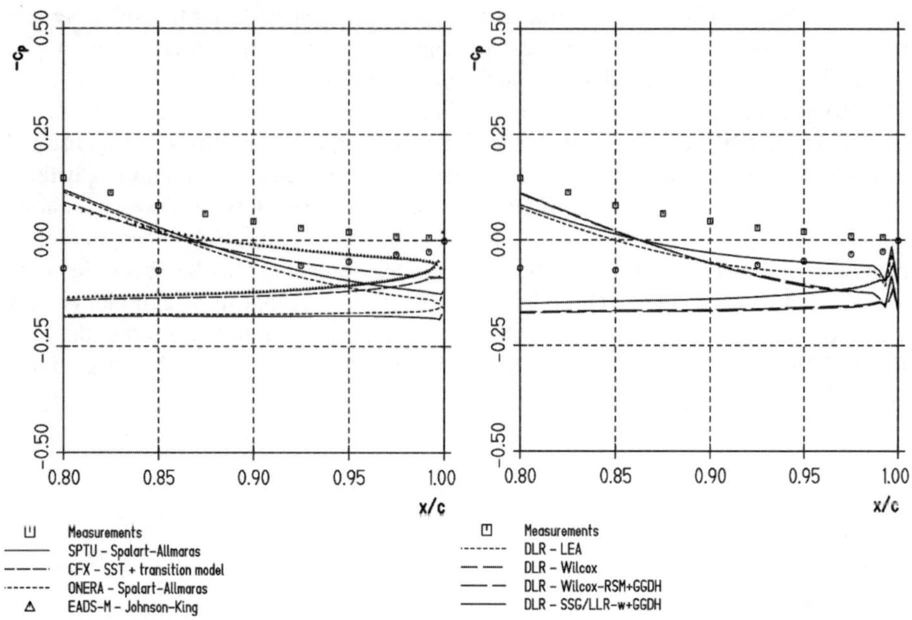

Figure 5 Pressure coefficient distributions in the trailing edge area

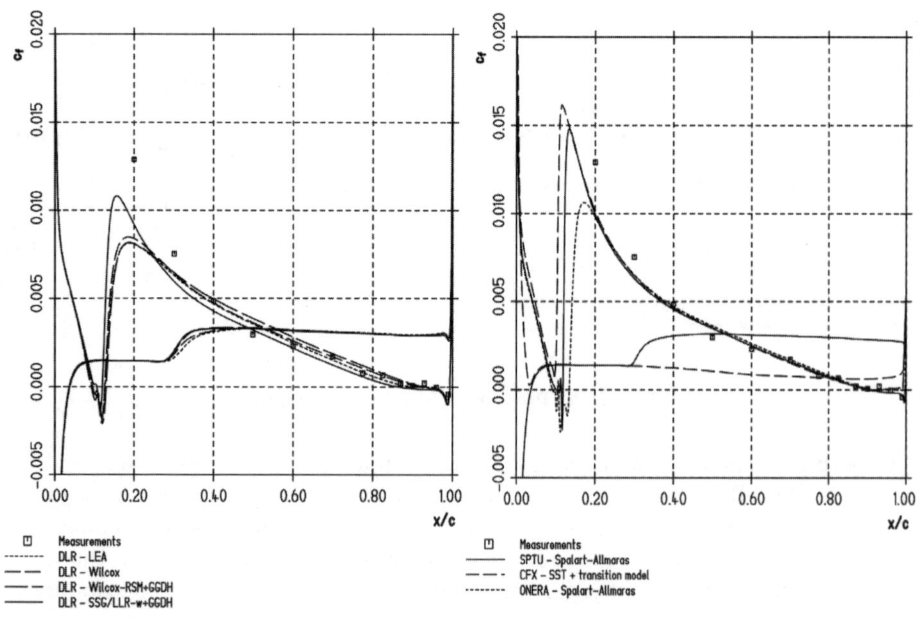

Figure 6 Skin friction distributions

When cross-comparing results by all partners obtained with different turbulence models and different numerical methods, Fig. 6 (right) presents results obtained by only one partner, DLR, which means that results do not rely on different numerical methods but on different turbulence models only, hence differences are less pronounced. It can be seen that the new SSG/LLR-ω model provides the better skin friction behaviour right after turbulent re-attachment. Moreover, it predicts turbulent separation earlier (a bit too early?) which is exactly in line with the pressure distribution at the trailing edge. And oscillations in the laminar region do not show up in all DLR results, which can be seen already when comparing Fig. 4 left (ONERA, ANSYS and SPTU results) with Fig. 4 right.

To better investigate on accuracy of the numerical results, velocity profiles are provided in the Figs 7 and 8 for station x/c=0.3 and x/c=0.96, both measured normal to the upper wall.

It seems not to be clear why the computed momentum of flow is somewhat under-predicted at x/c=0.3 which in turn correlates with the larger boundary layer thickness. Having mentioned that the laminar separation bubble at about 11 to 12% of the upper surface is problematic for achieving steady results in the very fine mesh and transition tripping right at the end of the bubble is questionable, the over-prediction of the main velocity outside of the boundary layer – shown in all numerical results – cannot, nevertheless, be verified. When, however, comparing results for the main velocity component with the normal velocities at the same chord-wise stations, see Figs. 9 and 10, it becomes obvious that at x/c=0.3 also the normal component is massively over-predicted.

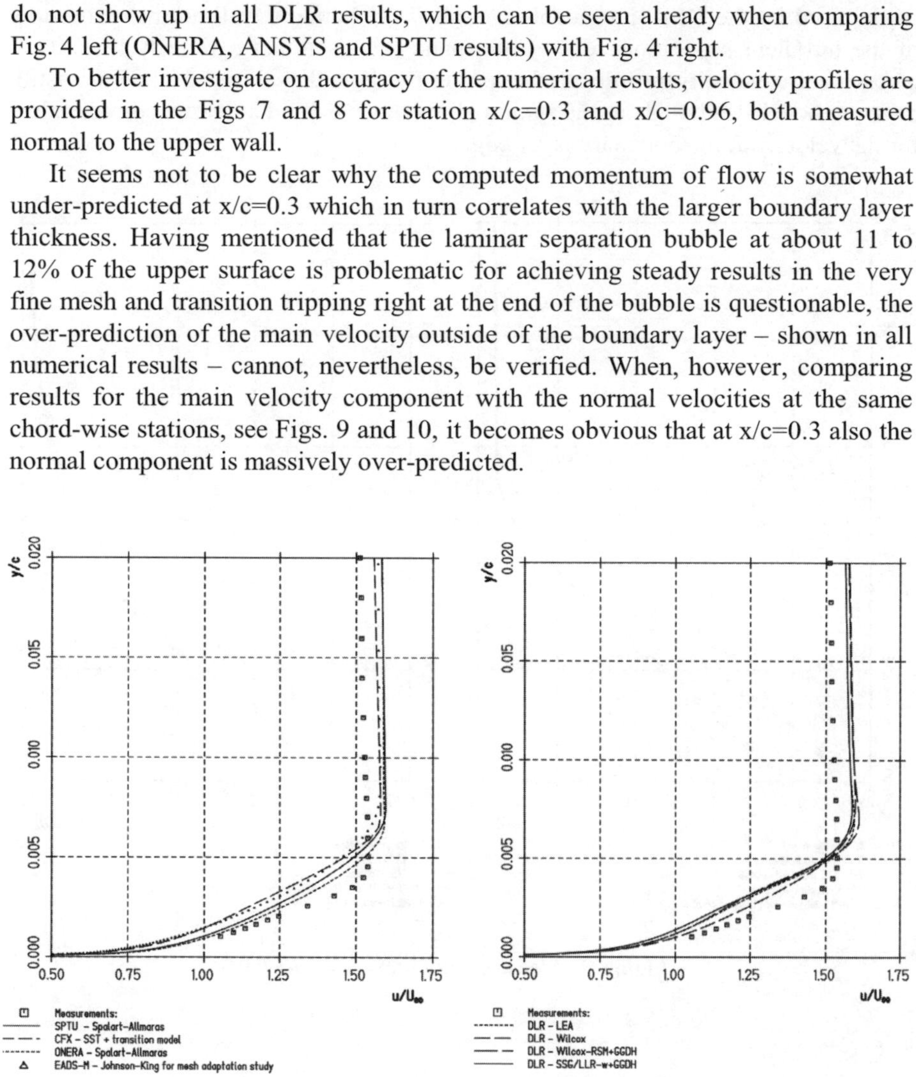

Figure 7 Main velocity profiles at x/c=0.3

Comparing results for x/c=0.96, the DLR results are still over-predicting boundary layer thickness, Fig. 8 right, comparable to the Johnson-King model results. This has been detected already when investigating the momentum loss thickness – carried out for the grid-dependence study – that exhibited a non-decreasing momentum loss thickness in the trailing edge region.

When stating that the extent of the separation bubble has a profound influence on circulation around the airfoil this normal extent of this separation bubble can be taken from with Fig. 8. Apart from the Johnson-King model and the SSG/LLR-ω model used by the DLR, no model is "really" predicting a wall-normal extension of the turbulent separation zone. Unfortunately, this is not a remedy towards a more precise velocity profile prediction as the boundary layer thickness is still over-predicted – although at this station all methods exhibit at least a correct value for the velocity at the boundary layer edge.

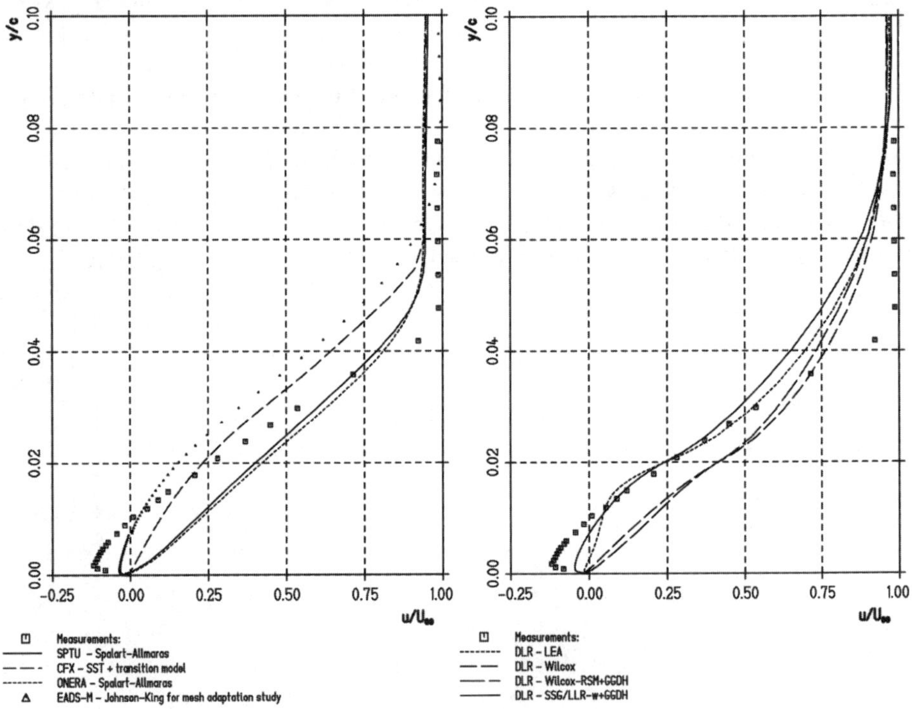

Figure 8 Main velocity profiles at x/c=0.96

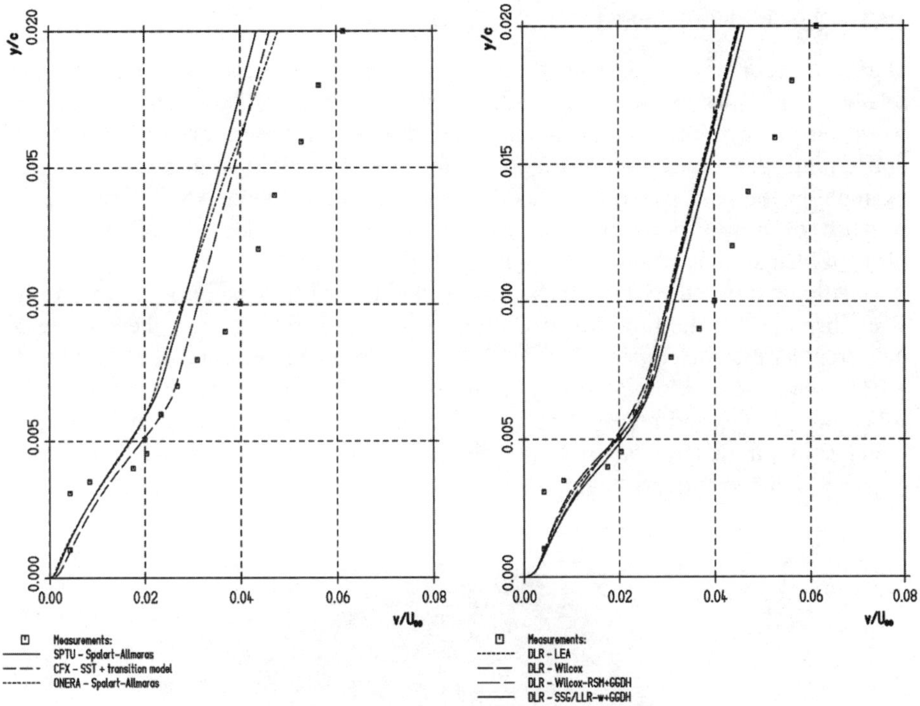

Figure 9 Wall-normal velocity profiles at x/c=0.3

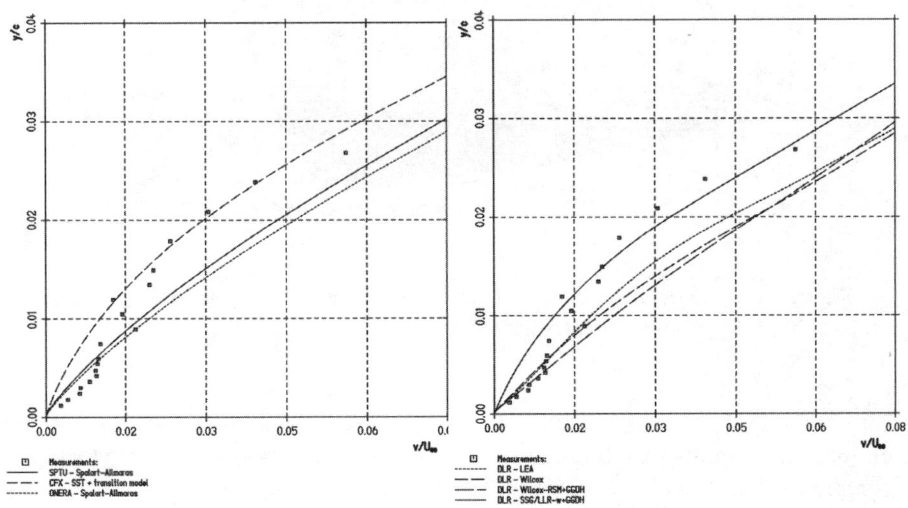

Figure 10 Wall-normal velocity profiles at x/c=0.96

18.4 The A-airfoil at 13.3° – a 3D measurement?

During the context of work in FLOMANIA, several discussions did take place related to the question whether the A-airfoil is a useful candidate for a validation work on a two-dimensional basis. This discussion based on recent debates concerning the transonic RAE2822 airfoil, being 'THE' transonic validation example in the past, but which was discarded from the FLOMANIA test case list because of a too big three-dimensional influence - making a 2D validation obsolete. Do similar arguments also hold for the A-airfoil?

Getting back to the LDV measurements and the oil flow patterns shown in Fig. 1, confirming that the flow was two-dimensional up to roughly 13°, there seems to be a remarkable discrepancy between the 13° measurement and the (corrected) 13.3° computation. More precisely, comparing the 3D computations carried out by SPTU with the experiments – and let consider the 0.3° difference in incidence not being a major driving factor – the differences are crucial and the comparison becomes somewhat questionable.

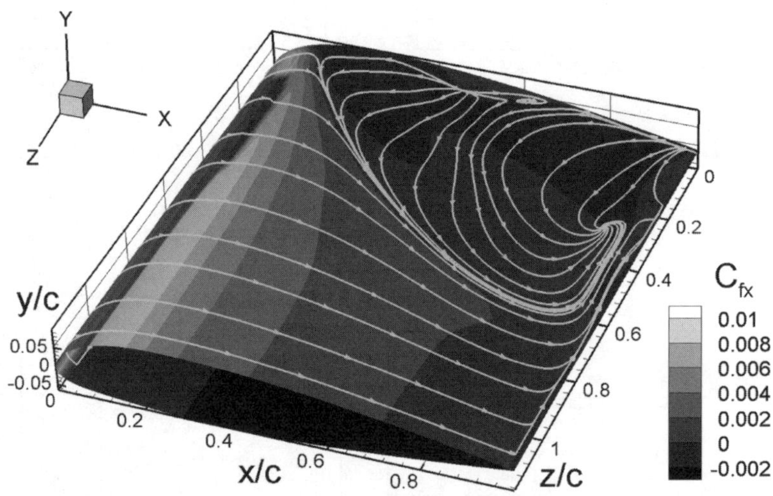

Figure 11 Skin-friction contours and streamlines on the upper surface – a 3D numerical investigation

To underline this previous statement, Fig. 11 presents skin-friction contours and on top of this wall streamlines on the upper airfoil surface computed by taking the complete wind tunnel geometry into concern. Obviously, the results show large tunnel wall influences, not showing up in the corresponding measurements. In the vicinity of the symmetry plane - and this plane is suggested to serve as the basis for 2D computations – the flow becomes nearly two-dimensional.

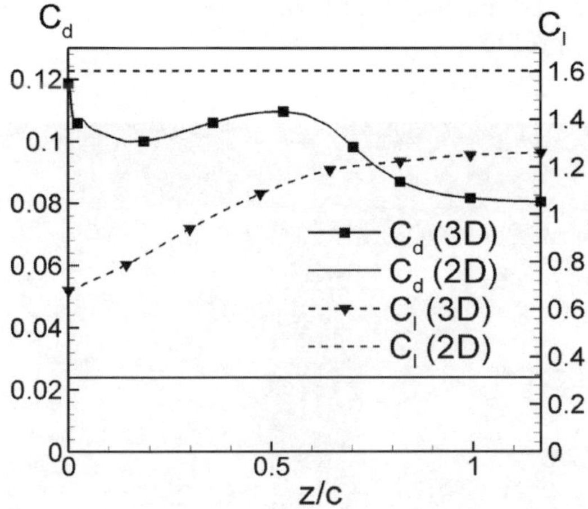

Figure 12 Local drag and lift distributions – a comparison between 2D and 3D approaches

However, one should note that a symmetry boundary condition was applied in the computations – but the 3D influence is definitely not a minor aspect. The latter has to be underlined when comparing Fig. 12, presenting local lift and drag distributions. First of all, the differences between 2D and 3D results near or at the symmetry plane are indicating a dilemma because they merely do not fit, i.e. the 3D results at the symmetry plane are different to the 2D results. But why?

One answer could be that it is often recognised using RANS approaches that they (clearly) over-predict vortical flow structures in corner geometries. This would lead to more pronounced friction in the airfoil-wind tunnel intersection, as it is likely the case in the 3D computations, thus a deeper insight is needed.

Consequently, the final results do not indicate the 13° case as being governed by 3D effects.

Moreover, according to intensive discussions with the experimentalists, no 3D effects have shown up at incidences of 13° or 14 °. The "somewhat random pattern" in Fig.1 is due to the fact that skin friction is almost diminishing. This means that the time scale for the oil pattern has to be extremely large in order to drive the oil flow – a time span that was not applicable in the measurements. Furthermore, the oil was applied with a brush, leading to a non-uniform initial thickness of the oil film and hence to the "slightly 3D, but random" pattern.

The latter is justified by another experiment at 14° incidence where two different runs - both oil-painted - have been carried out. Again, due to the very low skin friction in the aft 20% part of the airfoil, results have been obtained exhibiting "some" 3D patterns. However, these 3D effects are likely related to local disturbances caused by the brush painted initial conditions coupled to the very low skin friction. Thus, *in neither case side wall effects were visible* which is supported by LDV measurements at different span-wise stations exhibiting neither significant differences in the chord-wise velocity profiles nor a significant transverse velocity profile.

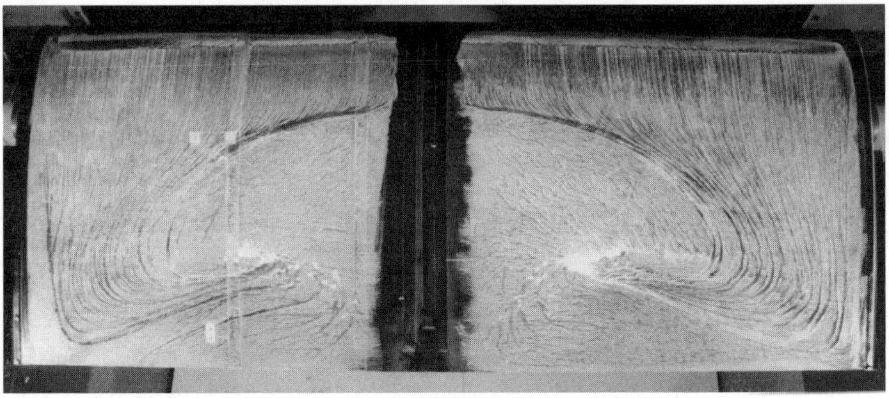

Figure 13 Oil flow picture for 16° incidence

An idea was brought up by C. Gleyzes from ONERA, who was very much in touch with both measurements and computations for the A-airfoil, namely to compare the experiment at 16 °, provided in Fig. 13, with the 13.3° incidence of the computation (see again Fig. 11). Although Fig. 13 exhibits strong 3D flow behaviour, the side-wall interferences remain negligible again. Moreover, the flow structure obtained in the 3D computations – assuming that such a "comparison" is eligible - is completely different. To conclude at this point: the measured side wall boundary layers in the 13.3° case do not contaminate the two-dimensionality of the flow.

Thus, more and refined work on numerical (and why not on experimental?) grounds might be needed to come to a more sound conclusion. But obviously, the computations and here in particular the turbulence modelling in areas where two different types of boundary layers merge have to be investigated more deeply to understand the flow behaviour in corners where two boundary layers merge.

19 NACA0012 – DNS Approach

S. Bourdet, Y. Hoarau, M. Braza, A. Bouhadji, A. Barthet, IMFT, and D. Ruiz, ENSEEIHT

Abstract

The objective of the present study is to analyse the main compressibility terms involved in the modelling of the turbulence kinetic energy equation for the upper transonic flow regime that is of priority interest for the aeronautical industry. It is well known that mostly the modelling of this equation comes from incompressible flow assumptions and for a long time it was admitted that compressibility effects become significant only beyond Mach numbers higher than 4 (Morkovin hypothesis). However, this stands only for steady flows. In case of unsteady boundary-layer-shock wave interaction and of appearance of recirculation regions near the wall, the van Driest law and Morkovin hypothesis for the turbulent fluctuations are no longer valid. For these reasons, it is also well known that the attempts to predict transonic flows in the upper transonic regime by simply 'transposing' the turbulence modelling hypotheses from the incompressible flows give very poor prediction of the shock position and of the drag coefficient. There are few developments for modelling of compressibility effects (Sarkar, (1992), (1995), Zeman, (1990)) among other. However, these developments had given poor improvements in the prediction of transonic flows around bodies. The aim of the present study is to address the terms to be modelled in the energy equation as well as to evaluate these terms in their exact form and according to the different suggested modelled assumptions in order to investigate the efficiency of the modelling assumptions. This can be done essentially by means of a DNS database, because the experimental measurement of compressibility effects is very difficult, although there remains the restriction for DNS concerning the low Reynolds number range of their realisation. IMFT has provided in the FLOMANIA programme a detailed DNS data base for the unsteady transonic flow around a NACA0012 wing at zero angle of incidence. Furthermore, the study of onset of the turbulent motion in its early stages has been carried out for the incompressible flow around a NACA0012 wing at 20°. In previous studies of the same research group (Bouhadji and Braza (2003a,b)) it had been shown that the increase of the Mach number in the upper transonic regime causes a natural amplification of a von Kàrmàn instability and of a strong shock-boundary layer and shock-vortex interaction near the trailing edge. Therefore, this test case is an ideal one for the investigation of compressibility terms under the effect of unsteadiness, for aerodynamic flows around bodies. Thanks to a fully parallelised version of the EMT2/IMFT code ICARE and to the use of a consequent grid size and small time step, it has been possible to perform DNS at Reynolds numbers 5,000 to 10,000 for the flow around the NACA0012 wing. The outcomes of this study are described in the following.

19.1 Test-case description and numerical method; configuration

The DNS study has been carried out for a NACA0012 wing of constant test-section along the span, submitted to infinite flow conditions in the two spanwise edges. The Reynolds number is 10,000 and the angle of incidence is 20° for the incompressible case and 0° concerning the compressible case. The main DNS (3D) study has been carried out for an incompressible flow and for Mach number 0.85. The code ICARE of the group EMT2/IMFT has been employed. Typical grids are of order (413 x 70x 90) to (1000x 400x90).

The physical investigation of the secondary instability under the effect of small-scale turbulence is carried out, at low Reynolds number, Re=1200 by DNS, based on the study by (Hoarau *et al* (2003)). The evaluation of the preferential wavelengths inherently appearing in the dynamic system is achieved. The modification of the organised spanwise vortex pattern due to the non-linear interaction with the small-scale turbulence is analysed and compared with the low-Reynolds number unsteadiness. The different classes of coherent vortices along the span are identified and tracked in space-phase by means of the Proper Orthogonal decomposition. In figures 3 and 4 the bi-orthogonal decomposition (Aubry, (1991), Berkooz *et al,* (1993)) has been applied in the DNS time-space 3D data in the near-wake, according to the DNS study by (Hoarau *et al* (2003)). The identification of the predominant organised modes is therefore performed, as well as the P.O.D. reconstruction of the flow pattern, in the sense of the low dimensioning of the system.

Concerning the Mach number effects, a previous study of this research group (Bouhadji and Braza (2003a)), (Bouhadji and Braza (2003b)) depicted the crucial Mach number regime (0.7-0.85) in which important transonic phenomena arise: The increase of the von Kàrmàn instability, the onset of buffeting, the drastic increase of drag and the attenuation of these modes as the Mach number increases. Especially, the progressive transformation of the global instability to a convective one under the sweeping effect of the shock wave that moves progressively downstream with the increase of the Mach number has been clearly investigated. Furthermore, the three-dimensional modification of the trailing-edge shock-vortex interaction and the amplification of the chock-vortex instability have been investigated, (Bourdet, Bouhadji, Braza and Thiele (2003)) by the DNS in the Reynolds number range 5000. This data base has been furthermore used in the FLOMANIA research program to extract the terms of the kinetic energy equation in their exact form and according to different modelling approaches.

19.2 Theoretical aspects in the context of compressible flow modelling

In the literature there are modifications in the turbulence modelling of the kinetic energy transport equations to take into account of the compressibility effects. These are mainly derived by means of the rapid distortion theory. Among the different modelling schemes, we refer to the works of (Sarkar, (1992)), of (Zeman (1990)) and more recently, (Sarkar (1995)). During the European research program ETMA, Efficient Turbulence Models in Aeronautics, (Dervieux, Braza and Dussauge, (1998)) the predictive capabilities of these modelling approaches have been investigated, especially for the shock-boundary layer interaction, for supersonic mixing layers and for the shock reflection. The main conclusions were that there were not achieved significant predictive improvements by the use of these models. The present study investigates these closures by comparing the exact terms to the modelled ones, by means of the present DNS. Although the general problem of DNS is that the Reynolds number is low comparing to the application of this kind of modelling, the DNS gives a first theoretical insight of the physical nature of the terms and of the modelling assumptions under the compressibility effects in the transonic regime. These are the main objectives of the present study whose results are presented below.

The exact form of the kinetic energy transport equation in homogeneous shear flow is

$$dK/dt = P - \varepsilon_s - \varepsilon_c + p'd'/\rho$$

where $K=0.5\,\tilde{u}_i'\tilde{u}_i'$ the turbulence kinetic energy, $P = -S\tilde{u}_1'\tilde{u}_2'$ the turbulence production, $\varepsilon_s = \nu\,\overline{\omega_i'\omega_i'}$ the solenoïdal dissipation rate, ω_i' the fluctuating vorticity, $\varepsilon_c = (4/3)\,\overline{\nu d'^2}$ the compressible (dilatational) dissipation rate, d' the fluctuation dilatation $(\nabla . u)$, $\overline{p'd'}$ the pressure- dilatation correlation.

The symbols "~" and "–" designate the Favre averaging and the Reynolds averaging respectively.

The compressibility corrections mentioned in the above references had been derived in *free* shear flows by DNS but have bee applied up to now even in wall flows. These corrections are briefly recalled in the following.

The dilatation-dissipation models

$\varepsilon_c = (4/3)\,\nu d'^2$ is the exact term representing the compressible (dilatational) dissipation rate. According to Sarkar, the modeled term is:

$$\varepsilon_c = \alpha_1 \bullet Mt^2 \bullet \varepsilon_s$$

where Mt is the turbulent Mach number, $Mt=(2K)^{0.5}/(\gamma RT)^{0.5}$ and K the turbulent kinetic energy.

α_1 is a constant that was calibrated by DNS results for homogeneous shear flows, to the value of 0.5.

$\varepsilon_c = (4/3)\,\nu d'^2$ is the exact term, for the compressible (dilatational) dissipation rate. According to Zeman an alternative for modeling of this term is suggested: ε_c is caused by 'turbulent shocklets'. The model is based on a threshold concerning the shocklets' appearance:

$$\varepsilon_c = Cd \bullet F(Mt) \bullet \varepsilon_s$$

$$F(Mt) = 0 \quad Mt<Mt_0$$

$$F(Mt)=1-e^{-[(Mt-Mt_0)/\sigma_0]^{0.5}}$$

$Cd = 0.75,\quad Mt_0 = 0.1,\quad \sigma_0 = 0.6$ ('first' Zeman's model)

$Cd = 0.75,\quad Mt_0 = 0.25,\quad \sigma_0 = 0.6$ ('second' Zeman's model for shear layers)

The aforementioned dilatation-dissipation models are applied to the k-ε closures by modifying *the turbulence time-scale:*

$$Tt = k/\varepsilon \quad \text{where now: } \varepsilon = \varepsilon_s + \varepsilon_c$$

The eddy-viscosity is defined by:

$$\mu_t = C\mu\,[\rho K^2/(\varepsilon_s + \varepsilon_c)]$$

19.3 Results

The onset of turbulent motion and the filling up of the energy spectrum are first analysed for the incompressible flow around the NACA0012 wing at Re=1200. Figure 2 shows the successive steps of the turbulent motion development. The energy spectrum is filled-up by a number of wavelengths with a predominant one of value λz=0.62. A comparison with the early stages of turbulent motion for the cylinder flow can be done by an analogy with the wing, where the vertical characteristic length is $c * \sin(\alpha°)$. In this way, it can be seen that the wavelength formation corresponds to the mode B development and is in good agreement with the cylinder DNS studies, (Persillon and Braza (1998)).

The POD analysis of the flow system in the sense of a "*Low Order Dimensioning*", LOD system has been performed by means of the DNS data, stored with a considerable detail *in space and time.*

From a data set $U(X)$, the Proper Orthogonal Decomposition analysis consists in searching the function $\Phi(x)$ that is most similar to the members of $U(X)$ on average (Berkooz & al. (1993)). This is done by solving an eigenvalue problem where the kernel is the two point correlation tensor $R(x,x') = \overline{u(x)u(x')}$.

The BOD (bi-orthogonal decomposition) has been applied in the present complete data set, according to (Aubry et al (1991)), because of the detailed representation of the results basis offered by the DNS, instead of the snapshots method usually applied in experimental data (Barthet, (2004)), (Faghani (1996)). Figures 1 and 3 show the BOD reconstruction of the flow system and the energy drop of the POD modes. It is shown that a 20-mode reconstruction is sufficient to describe the coherent structures and the secondary instability of the present system.

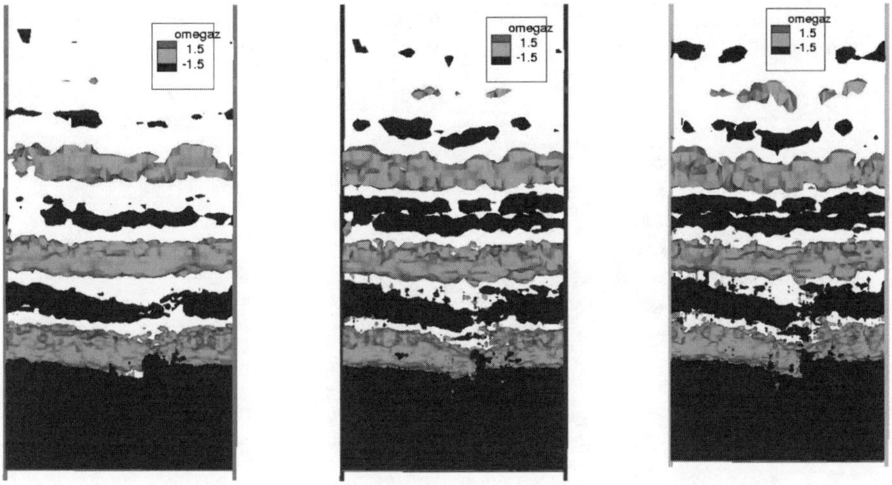

Figure 1 Same as in figure 3: 10, 15 and 20 modes, POD reconstruction, *(x,z)* plan

Concerning the Mach number effects in turbulence modelling, in the following, the 'S1' model means the Sarkar modelling as a function of Mt and 'S2' modelling designates the use of the gradient Mach number, Mg, instead of the turbulent Mach number, (Sarkar, (1995)):
$Mg= Sl/c$ with $S = dU/dy$ and l the turbulence length scale, $l \sim k^{1.5}/\varepsilon$.
Figures 4,5 show the DNS iso-Mach lines around the NACA0012 airfoil at zero angle of incidence at an instant corresponding to the formation of alternating, slightly supersonic pockets around the leading edge. This is the onset of buffeting effect at a predominant frequency ten times lower than the von Kàrmàn one (see the asymmetric pressure variations at two symmetrically located points on the suction and pressure sides), fig.4. The von Kàrmàn mode corresponds to the vortex shedding downstream and to the separated bondary-layer oscillation. There is a weak interaction between the buffeting and von Kàrmàn modes because of the highly different ranges of these two frequencies. However, the amplitudes of the oscillations drastically increase with the presence of the buffeting mode and the drag coefficient also displays a very significant increase in the Mach number range 0.7-0.85. As the Mach number increases to M=0.9, fig.5, the shock wave moves downstream in the wake and 'sweeps' the von Kàrmàn mode, that no more affects the near-wall region. This behaviour corresponds to the progressive transformation of the absolute character of the instability to a convective one.

The three-dimensional vortex pattern at M=0.85 and Re=5000 firstly displays the organisation of the *w* velocity component along the span, according to counter-rotating cells, (fig.6). This step is followed by the amplification of the secondary instability characterised by the appearance of a well-formed undulation of the von Kàrmàn vortex rows. This undulation is further modified under the non-linear interaction among two adjacent vortex rows and leads to the appearance of a breaking pattern in the continuity of the 'spinal column' of the undulated von Kàrmàn vortex, that displays a vortex dislocation. These successive steps occur downstream of the trailing edge shock and are similar to the ones developed in incompressible flows (Braza, Faghani and Persillon, (2001)).

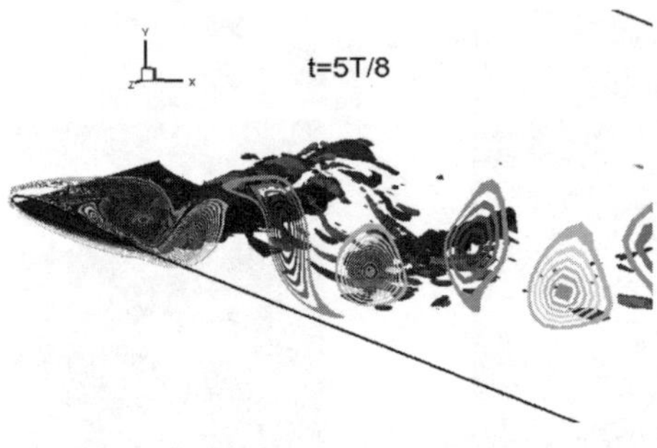

DNS, iso-vorticity ωz and ωx

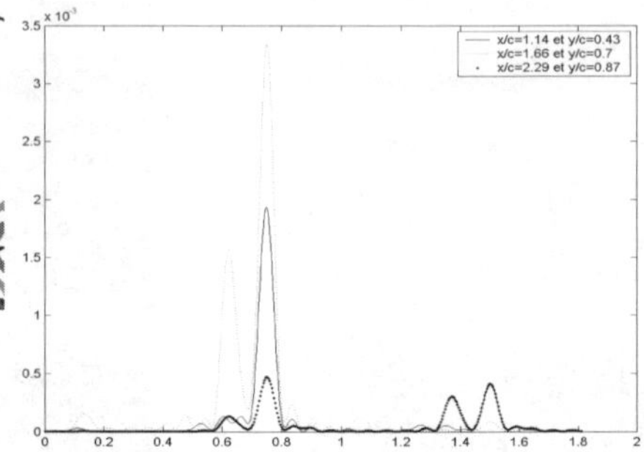

FFT, space variation of ωz; wavelength λz

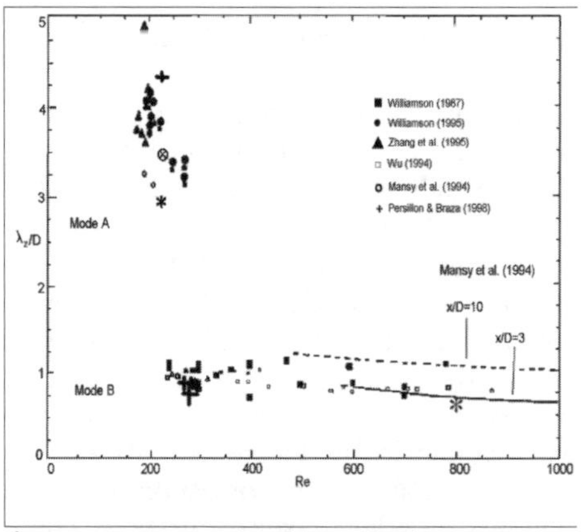

λz comparison with the cylinder case

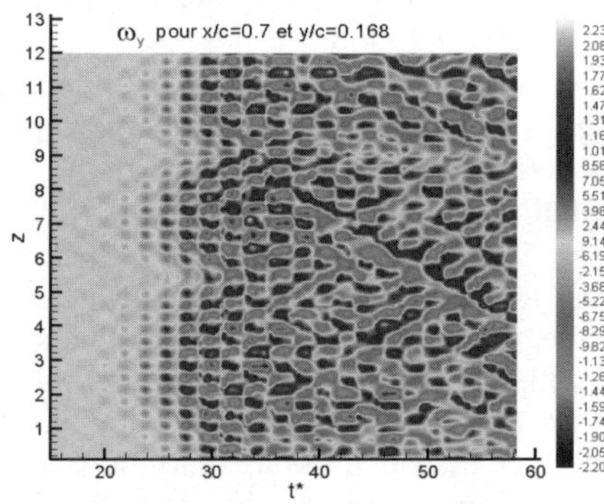

Instantaneous *(x,z)* variation of ωy

Figure 2 DNS, NACA0012, 20°, RE=1200; onset of turbulent motion, filling-up of the enrgy spectrum by well-distinct wavelengths; analogy with the cylinder flow

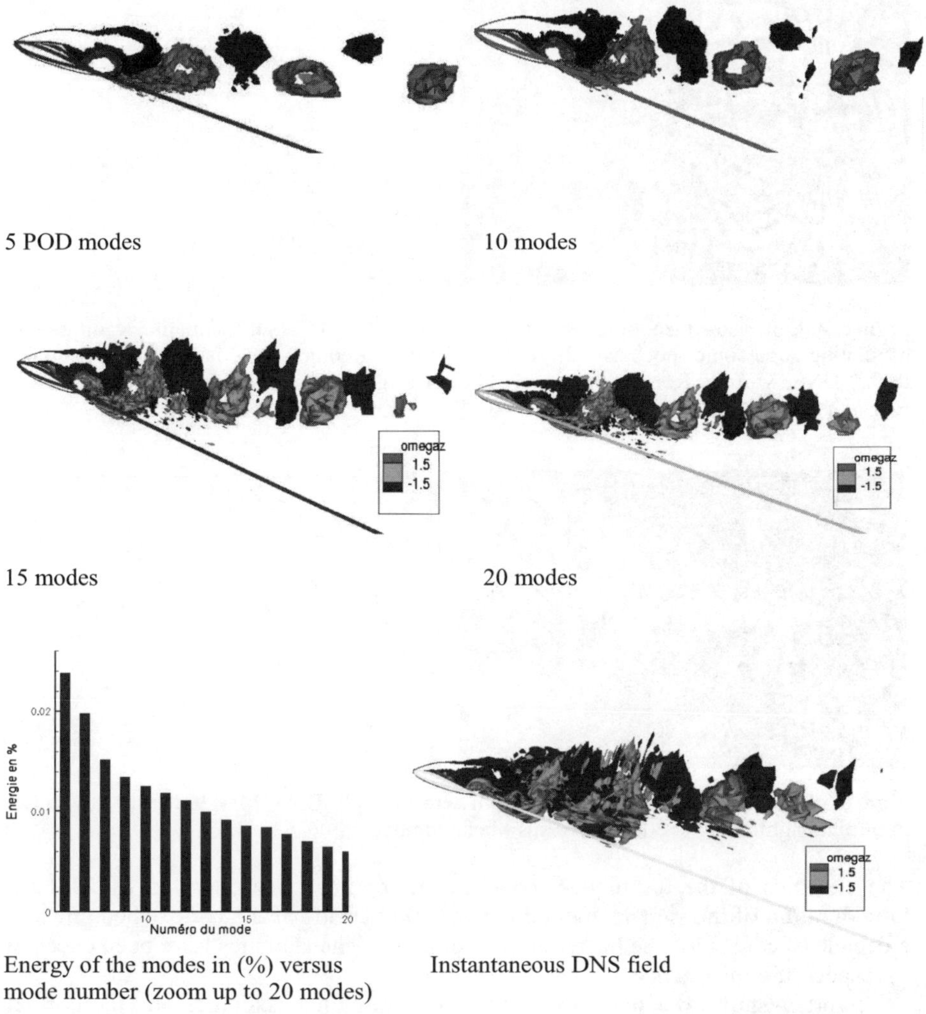

Figure 3 POD reconstructions of the 3D secondary instability pattern; iso-vorticity lines; upper row: 5 and 10 POD modes; middle: 15 and 20 modes; lower row: instantaneous DNS field and energy decay of the principal modes

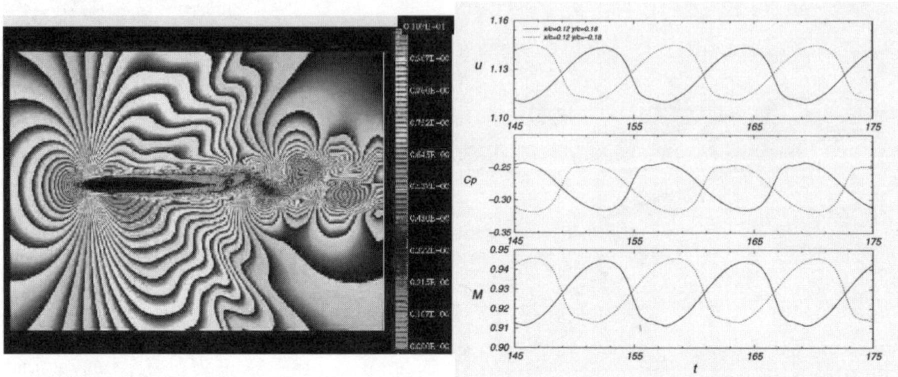

Figure 4 Left: iso-Mach lines, varying from zero to 1.07 (values in the leading-edge oscillating supersonic 'pockets'). Inside the alternating eddies, iso-Mach varying from 0.4 to 0.5. DNS, M=0.8, Re=10,000; onset of buffet. Right: Pressure coefficient versus time at two symmetric positions, *x/c*=0.12, *y/c*=0.18, *x/c*=0.12, *y/c*=-0.18

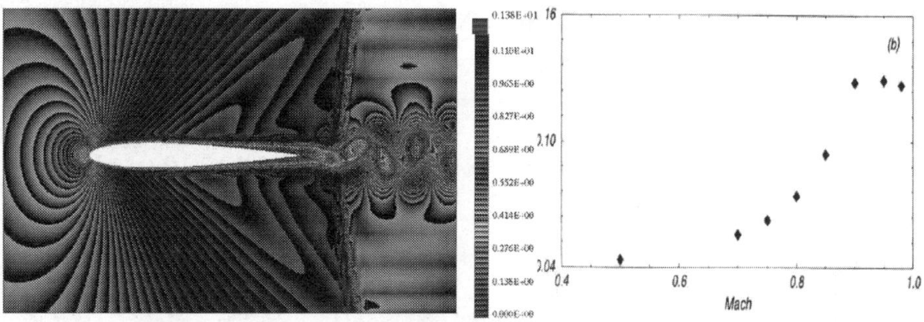

Figure 5 Left: iso-Mach lines, varying from zero to 1.38. DNS, M=0.9, Re=10,000; onset of buffet. Right: Drag coefficient versus Mach number

The statistics of the turbulence modelling terms have been derived by taking a long duration of the signals, including over 100 periods of the vortex shedding and a decade of cycles for the buffeting phenomenon. The statistics have been checked in respect of convergence.

Figure 7 shows the statistical averaged longitudinal velocities and the normal stresses. They clearly represent the trace of the alternating vortices downstream of the trailing-edge shock-vortex interaction. Moreover, the normal stress is organised in counter-rotating cells according to a preferential wavelength, l~0.5c. This averaged pattern illustrates the impact of the secondary instability in the same way as in the instantaneous iso-*w* field (fig. 6) by DNS.

Figure 8 shows the turbulent kinetic energy, turbulence production, the cross-term of the turbulence anisotropy tensor and the pressure-dilatation correlation. The turbulence kinetic energy presents the characteristic two-lobe structure issued from the averaged fluctuating pattern in respect of the alternating eddy formation. The turbulence anisotropy tensor presents a strongly three-dimensional character, especially in the shock-boundary-layer and shock-vortex interaction, indicating the need of more advanced turbulence modelling closures involving the compressibility effects. The pressure-fluctuation correlation shows an undulated three-dimensional pattern along the span and indicates a direct influence of this term by the secondary instability.

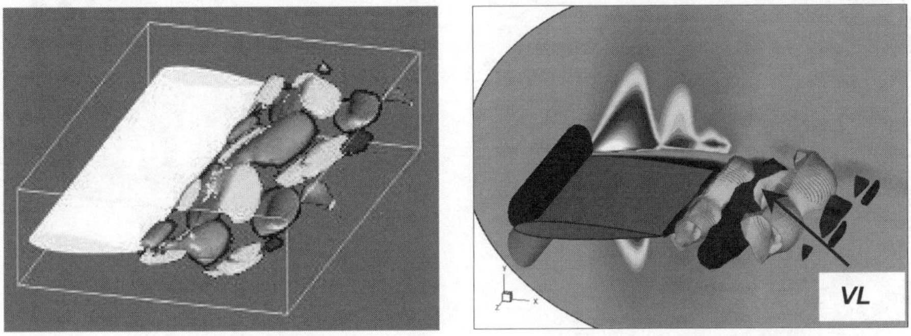

Figure 6 Left : iso-*w* velocity lines, DNS, M=0.85, Re=5000; right, iso-vorticity surfaces showing the secondary instability ; undulation of the von Kàrmàn vortex rows and vortex dislocation (*VL*)

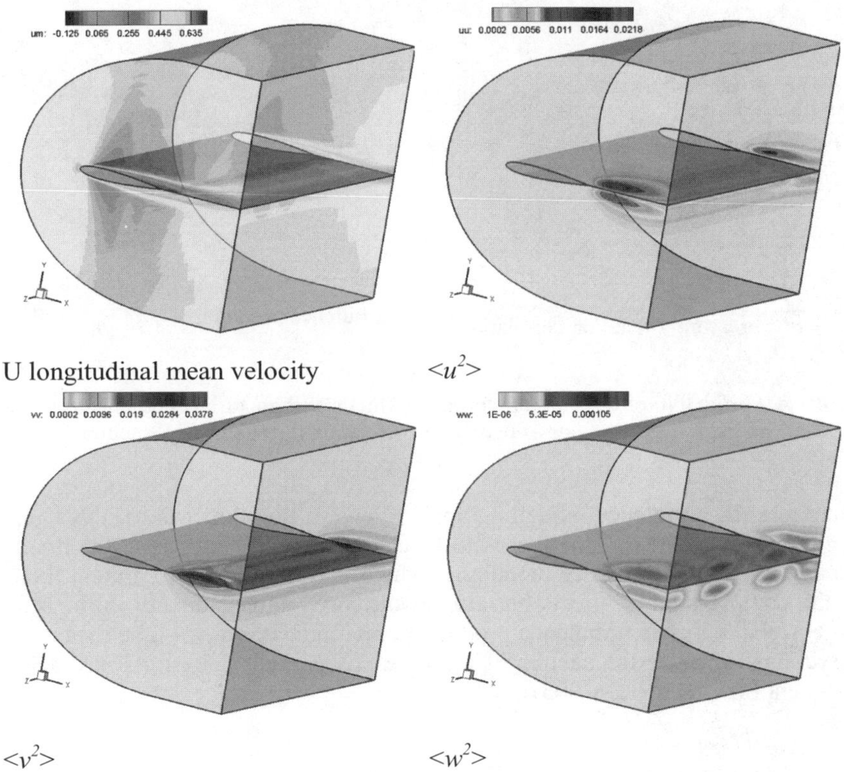

Figure 7 Statistical mean *U* velocity and normal stresses

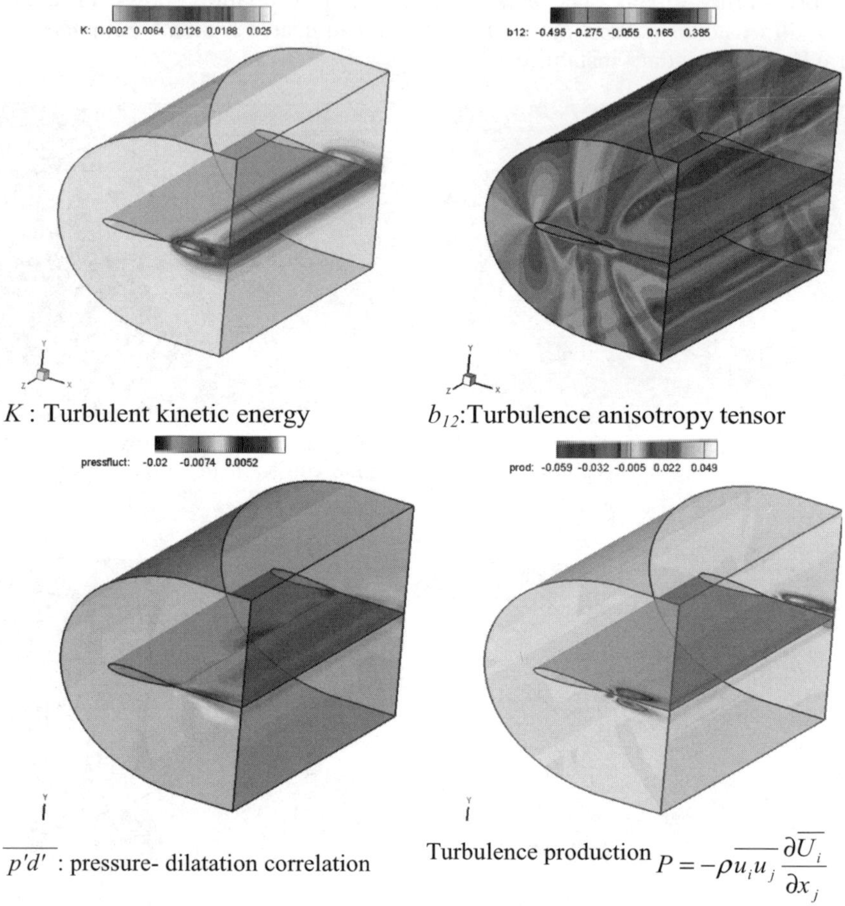

Figure 8 From left to right: on top, turbulent kinetic energy, turbulence anisotropy tensor, <b12>= - uv/K; bottom: pressure-fluctuation correlation, turbulence production.

Therefore, in turbulence modelling applications in two-dimensions, this character is completely omitted. This fact is to take into account concerning the predictive capability of the turbulence modelling, which is not only due to a good inclusion of the compressibility effects but also of the three-dimensionality in the sense of the correlations. The turbulence production rate indicates a two-lobe structure with maximum values in the centres of the lobes, as typically observed in other wake turbulent flows (Perrin, (2005)).

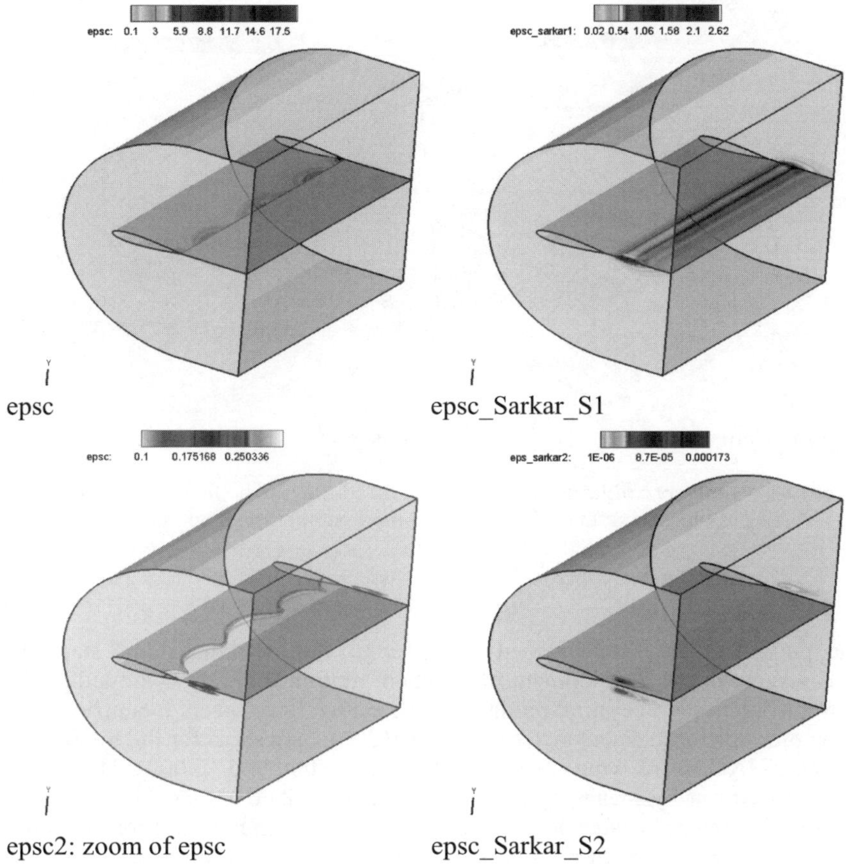

Figure 9 From left to right: on top, the exact form of the epsc term (dilatational dissipation), the modelled form according to S1; on bottom, zoom of the exact term variations and modelling as a function of the *Mg*.

Figure 9 presents the exact dilatational dissipation term and its modelled forms according to S1 et S2. It is remarkable that the modelling considerably attenuates the three-dimensional nature of the compressible dissipation and that the order of magnitude of this term is not correctly captured by the models. In fig.10, the Zeman's first model is shown for this term. The order of magnitude is better captured by this modelling, but the three-dimensionality is attenuated, less than in S1, S2 modelling. The exact solenoïdal (incompressible part) of the dissipation is shown in figure 10. This term shows also a two-lobe structure as in case of the turbulence production rate. It displays a quasi-two-dimensional behaviour. Its order of magnitude is comparable to the compressible dissipation term. This points out the importance of the compressible dissipation modelling in the upper transonic regime.

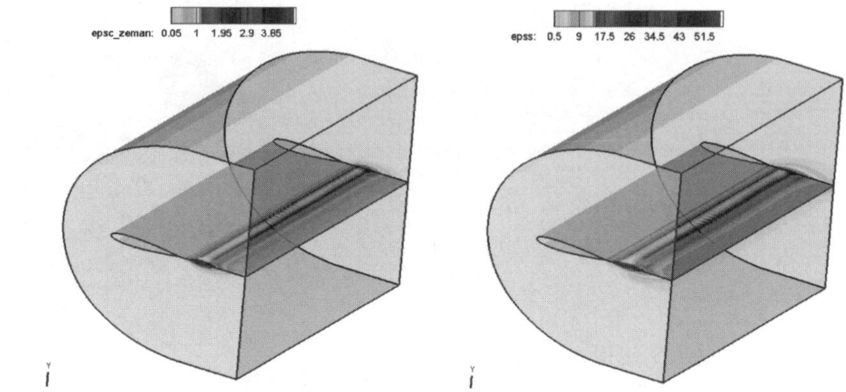

epsc_Zeman eps_s

Figure 10 Left: epsc (dilatational dissipation) according to the first Zeman model. Right. eps_s: the solenoidal (incompressible) dissipation rate

19.4 Conclusions

The present study has identified the successive three-dimensional steps of the early stages for the development of turbulent motion in the flow around a wing of constant section. The preferential wave-lengths have been quantified for the incompressible case (incidence of 20°, Re=1200), as well as for the transonic flow around a NACA0012 wing (incidence of 0°, Re=5000 and 10,000). This study has been carried out by means of DNS. The Mach number 0.8 has been selected for the transonic case, because it corresponds to the buffeting interaction with the von Kàrmàn mode, interesting especially the domain of aeronautical applications concerning this specific transonic regime. The shock-vortex interaction, the von Kàrmàn instability and the buffeting onset have been studied in detail. The fluctuating flow statistics at Re=5000 have been provided, according to the Reynolds and Favre averaging, on the purpose to study the exact terms involved in the turbulent kinetic energy equation. These terms have been evaluated under the compressibility effects of the upper transonic regime and have been compared with suggested turbulence modelling terms for compressible flows. The comparison has given differences in the orders of magnitude between the exact and the modelled terms, that had been derived for free shear-layer flows. A need of a more adequate compressibility modelling is depicted for the transonic flows, taking into account the near-wall and near-wake flow unsteadiness and three-dimensionality, in case of *non-homogeneous, non-equilibrium* turbulence. Although the DNS deals with the early stages of the turbulent motion, this study can be used as a first step, towards the improvement of existing modelling approaches concerning the compressibility effects. Furthermore, the impact of three-dimensionality has been shown important and this has to be taken into account in the prediction of transonic flows around bodies.

20 DLR F6

D. Schwamborn, DLR

Abstract

This chapter deals with the validation of a number of turbulence models for industrial relevant configurations, i.e. the DLR F6 wing-body model with and without pylon and through-flow nacelles. In the comparison, which employs structured and unstructured grids of different resolution, some light is shed on the performance of the models, the influence of transition and of the grids used.

20.1 Test Case Description

This test case of the so-called DLR F6 model employs actually two models which have been tested in the S2 tunnel of ONERA at a variety of conditions. While one model is a relatively simple Airbus-like wing body model referred to as the WB model in the following, the second more complex one is additionally equipped with through-flow nacelles (see Figure 1) and is thus abbreviated as WBPN, for wing-body-pylon-nacelle.

Both configurations have also been employed in the "Second Drag Prediction Workshop" (DPW, 2003) which was set up to evaluate the predictive capabilities for a variety of RANS-based numerical methods in drag computations. More than 20 participants have contributed to the workshop test cases using mandatory structured or unstructured meshes at various grid resolutions.

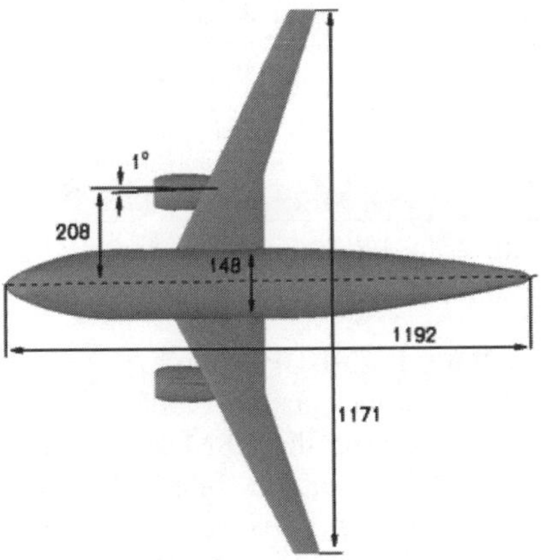

Figure 1 Major dimensions of F6 configuration in mm (left) and view of the model in the ONERA S2 wind tunnel (right)

Since several partners of FLOMANIA intended to participate in the DPW, it was suggested to adopt some of the workshop test cases also as FLOMANIA test cases. It was decided to employ both configurations, i.e. WB and WBPN, at a

Mach number Ma = 0.75 and a Reynolds number of Re = 3 million (based on c=141.2 mm) and to compute one or both of the following cases:

- Single point grid convergence study for a lift coefficient c_L = .5 ±.001 assuming a fully turbulent solution
- Drag Polar for angles of attack of α = -3°, -2°, -1.5°, -1°, 0°, 1°, 1.5° on the medium grid. If possible this should be run assuming boundary layer transition at 25% chord on the lower surface and at a given border on the upper surface. This border is a line from 5 per cent chord at the wing root to 15 at the kink, then to 15 at 84.4 per cent half span and finally to 5 per cent at the tip. Alternatively, 10 per cent could be used if one cannot vary the trip location OR fully turbulent calculation if one cannot specify a trip location.

20.2 Turbulence models and grids used

Table 1 shows a list of the solver-type and turbulence models used by the four partners that calculated this test case indicating also which cases were treated, i.e. whether WB and/or WBPN (indicated by an X) and whether the polar was computed (indicated by a P).

Table 1 List of grid types and models used by the partners and test cases computed

Partner		solver-type	turb. model	WB	WBPN
CFX	CFX-5	hybrid	SST	X,P	X,P
DLR	TAU	hybrid	SAE	X,P	X,P
	FLOWer	struct.	SSG/LRR-ω	-	X
FOI		hybrid	SA, Wi88, SST, Peng, WJ-st-ω	X,P	X,P
NUMECA		unstruct. hexa	SA	X,P	-
X: Test case has been treated; P: complete Polar has been computed					

Three partners used structured grids generated with ICEM-CFD HEXA for the drag prediction work shop, namely CFX (now ANSYS Inc., using its hybrid code CFX-5), DLR (for its structured solver FLOWer employing the RSM model) and NUMECA, who used the structured grid as an unstructured one. The grid dimensions of these grids are given in Table 2. FOI used the hybrid unstructured grids generated for the workshop by DLR using CENTAUR from CentaurSoft as the grid generator for the coarse grid. Finer hybrid grids were then obtained by the adaptation (refinement) feature of the TAU code, indicated as "unstructured" in Table 2. The grid refinement was performed based on solutions obtained with TAU for the constant lift case (c_L =.5). In this case a 30 per cent increase in number of points was allowed after each solution to obtain the adapted grid for the next calculation. In addition to refinement an adaptation of the prismatic grid part to $y^+ = 1$ in the first layer was also done in this process.

DLR employed a set of three refined grids where 50 per cent increase in points (compare Table 2) was allowed in each of the three refinements to arrive at grid sizes similar to those used in the structured grids. The size of all grids with respect to the number of grid points involved is given in Table 2. All grids used have 25 prismatic layers with approximately 13-16 layers in the boundary layer. The average y+ of the first cell of the wall is about two for the coarsest grids and one for the medium and finer grids.

Table 2 List of grids and grid sizes used in the computations (No. of points in millions)

	structured		unstructured		DLR (unstruct.)	
	WB	WBPN	WB	WBPN	WB	WBPN
Coarse	3.45	4.89	2.39	3.68	3.8	5.5
Medium	5.82	8.43	3.16	4.79	5.8	8.2
Fine	10.13	13.68	4.05	6.38	8.8	12.3

Note, however, that despite similar sizes of some of the grids, e.g. the coarse structured WBPN grid and the medium unstructured WBPN used by FOI or the medium structured grids and medium unstructured grids of DLR, there is quite an essential difference in the resolution obtained with these grids. While the structured grids waste a considerable amount of points in areas relatively far away from the surface compared to unstructured grids they are very efficient along the leading and trailing edges of the wing and along the fuselage due to the use of high aspect ratio cells (chord to span). Due to limitations in the aspect ratio of triangles along the surface (stemming from accuracy and stability considerations) unstructured grids are currently not very efficient in those areas at least when the flow gradients are in mainly one direction of the structured grid, i.e. a certain amount of points is wasted in these areas . However, except maybe for the coarsest grid, the unstructured grids are adapted to the flow and its features at $c_L = 0.5$ which is certainly a significant advantage, at least for this angle of attack, in addition to their higher flexibility in grid generation in the case with the nacelle.

20.3 Discussion of results

In the following we present first results for the integral quantities lift, drag and pitching moment in comparison with the available experimental data for the complete polars, before we come to the discussion of the pressure distribution for the case of constant lift. Finally, we will also try to shed some light on the influence of the turbulence modelling based on a FOI study [Peng, 2004; Peng & Eliasson, 2004] for the WBPN case at a constant angle of attack resulting in a lift close to the target lift.

20.3.1 Lift and drag polars

Figure 2 shows both lift and drag coefficient from all calculations of the WB case as a function of angle of attack and lift, respectively in comparison with the experimental data. It is obvious that the results obtained by CFX with the SST model on structured grids are closest to the experimental data, slightly over predicting the lift for angles of attack below 0.5° and under predicting it for larger ones. In other words, the slope of the lift ($dc_L/d\alpha$) is under predicted compared to the experiment. The fact that their three results for the single point grid study are quite close together, even for the moment coefficient, seems to indicate that there is not a large dependence of the solution on the grid size, although this conclusion is somewhat weak since the average increase in point number from one grid to the next finer is only about 70 per cent in total or about 20 per cent per coordinate

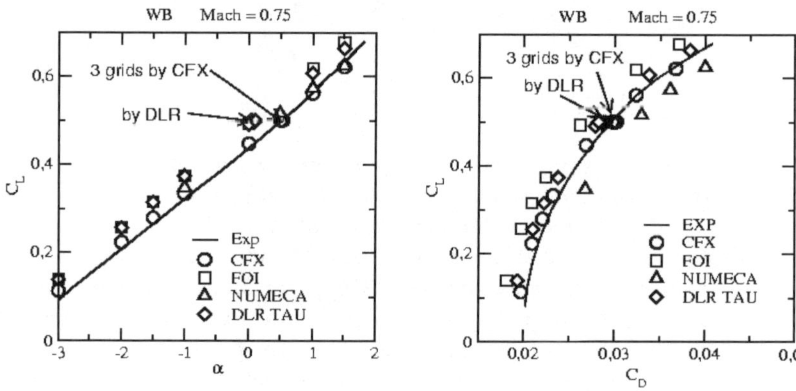

Figure 2 Integral coefficients for WB configuration; lift vs. angle of attack (left), lift vs. drag (right)

direction. Moreover, transition does not seem to have a large influence, since these three fully turbulent results coincide almost with the data along the polar obtained on the medium grid with the prescribed transition location. While the data along the polar obtained by NUMECA on the coarse structured grid seem to support the above findings on grid sensitivity with respect to lift over angle of attack, it is not completely clear whether the too high drag in their results is only due to the different turbulence model (SA) used or whether there is an additional influence resulting from differences in the code, like e.g. different numerical dissipation. The higher drag is, however, in agreement with the tendency of SA to produce a higher drag (mostly friction drag) than SST as we will see from the comparison of different turbulence models by FOI, where the computations have been performed with the same mesh using the FOI EDGE code.

Although the data of FOI and DLR were obtained on their corresponding medium grids which have quite different resolution the results for the lift over angle of attack coincide quite nicely with each other at least for a lift coefficient lower than 0.5. However, both sets of data overestimate the experimental lift by an almost constant value. As can be seen from the lift over drag polar in Figure 2 the drag obtained by FOI with the WJ-st-ω model is considerably lower than for the SAE results of DLR, which coincide closely with the experimental data. One might be tempted to attribute this to the different resolution of the grids or to the fact that FOI performed fully turbulent calculations while the experiments and the results of DLR were obtained with tripped transition as indicated above. However, this can hardly be the case since the results of the grid sensitivity study of DLR indicate that at least for the fine set of grids the influence of the resolution seems to be quite limited even for the pitching moment which is usually very sensitive to

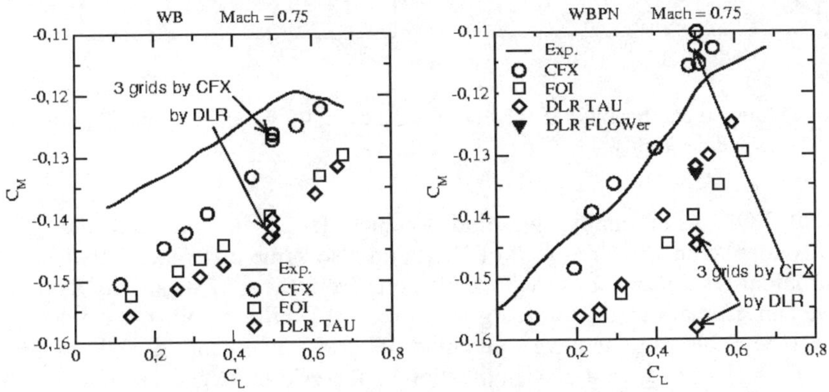

Figure 3 Moment coefficients versus lift coefficient; WB (left) and WBPN (right) configuration

differences in the computation. Noting that this sensitivity study was also made without boundary layer tripping and taking the relatively small difference between the fully turbulent and the transitional results of DLR into account, it seems likely that most of the difference in drag stems from the turbulence model.

As we shall see later this is in good agreement with the findings of FOI for the difference between the models for the same angle of attack and grid (Table 3). Taking a last look at the moment coefficients for the WB configuration given in Figure 3, we note that all computational results under estimate the moment quite severely, with the result of CFX being much closer to the experimental data but missing the slope, while the results of DLR and FOI (although being farther away) exhibit the correct slope at least for lift coefficients below 0.55.

Turning to the WBPN configuration (Figure 3) the situation for the moment coefficient is similar as for WB except that now the numerical results of CFX are much closer to the experiment, while the slope is still too steep. The single result obtained with the DLR FLOWer code on the structured grid employing the SSG/LRR-ω RSM implemented during the project shows here some over prediction of the lift with respect to angle of attack but performs very well with regard to thc drag over lift polar. The results obtained with the DLR TAU code are also closer to the experiment now again indicating the right slope just as the

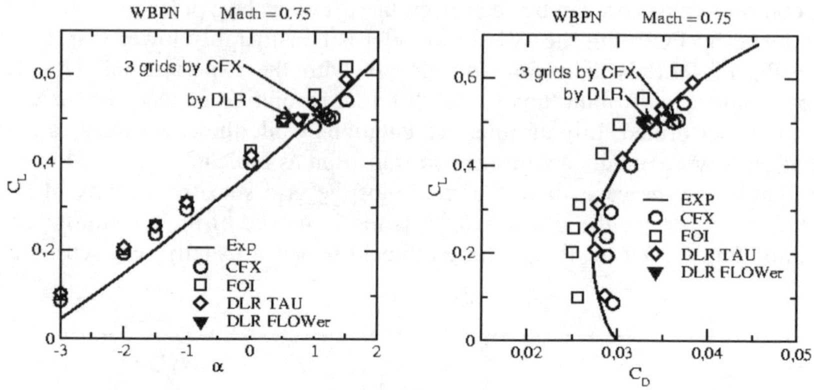

Figure 4 Integral coefficients for WBPN configuration; lift vs. angle of attack (left), lift vs. drag (right)

results of FOI. Additionally, we find a much larger influence of the grid, especially for the unstructured grid of DLR and also some difference between the fully turbulent results and those with transition. From the grid sensitivity study of DLR we can estimate that a refined grid of about 12 million points is needed for this case (see Table 2) to achieve some kind of "grid convergence", while the grid study of CFX doesn't give a clear indication of the grid influence for the moment coefficient. The results of DLR show in addition an indication that fully turbulent flow leads to a more negative moment. With respect to lift and drag, however, resolution seems not to be a very critical issue just as in the WB case grid (see Figure 4). The conclusions here are similar to those drawn on WB above. All results are similarly close to each other and to the experiment in both diagrams with a somewhat larger spread between FOI, DLR and CFX for the drag. From the drag over lift plot where the results of the DLR TAU code as well as the singular results of FLOWer with the RSM are closest to the experiment, it is obvious that the main difference between the computed polars is an almost constant delta in drag which might be mostly due to changes in friction drag resulting from different turbulence models. This would be in line with the findings from the study of turbulence models by FOI already mentioned above and further discussed below.

20.3.2 *Pressure distribution at selected wing sections*

In order to obtain a somewhat deeper insight into the different numerical results we turn to the pressure distributions for some selected span-wise wing sections, at 15, 37.7, 51.4 and 84.7 per cent of the half-wing span, where experimental data are available. It has to be kept in mind in this comparison, that the angles of attack from the experiment for the case c_L=0.5 (about 0.5° and about 1.0° for the WB and WBNP case, respectively) are - as seen above - not the ones found in constant lift computations most of the partners performed. For the WB case we find at the innermost station (see Figure 5) which is close to the fuselage that the results of both CFX (c_L= .50 at α=0.50°) and the NUMECA (c_L= .516 at α=0.49°) show a larger discrepancy from the experimental data for the rear 60 per cent of the upper wing surface indicating a separation area of 20 per cent near the trailing edge. The results of FOI (c_L= .494 at α=0.0°) and DLR TAU (c_L= .4999 at α=0.101°) exhibit also an indication of separation limited to the last 10 per cent of the chord, but the general agreement with the experimental data is good.

At the other three stations the situation changes in that now the results obtained on the structured grid by CFX and NUMECA are closer to the experiments except for the too forward position of the shock which is slightly upstream of the experiment and also more smeared. It is not clear whether the relatively small

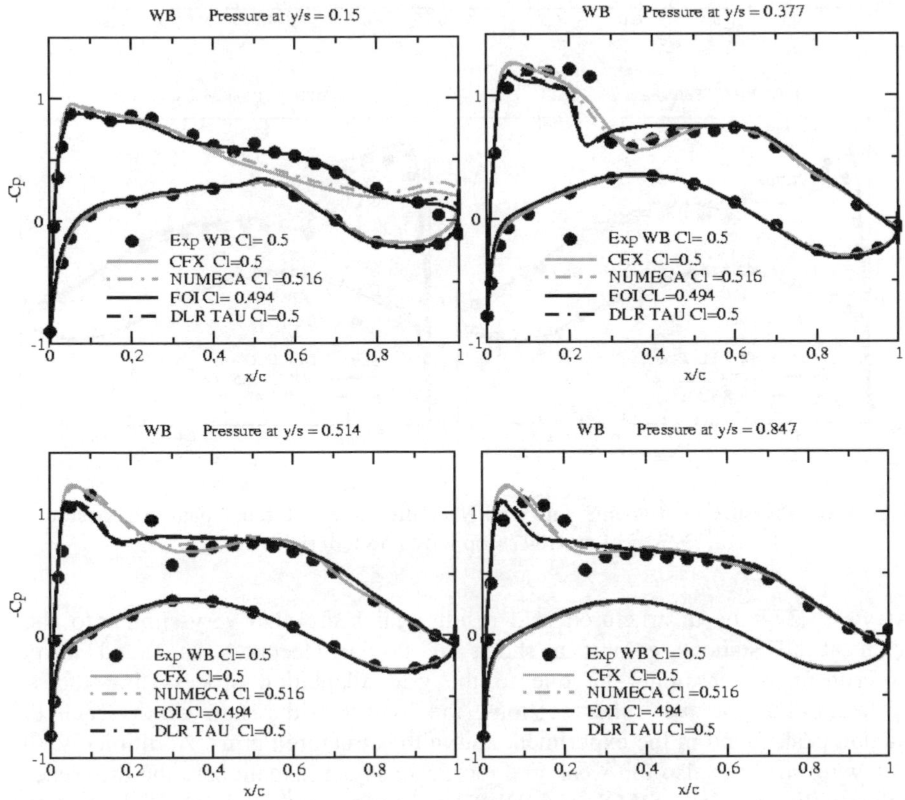

Figure 5 Pressure coefficients for WB configuration at four span-wise positions; y/s = .15, .377, .514 and .847 (top left to bottom right)

local differences found between the CFX and NUMECA results have to be attributed to the different grid sizes, the difference in turbulence model or a combination of both.

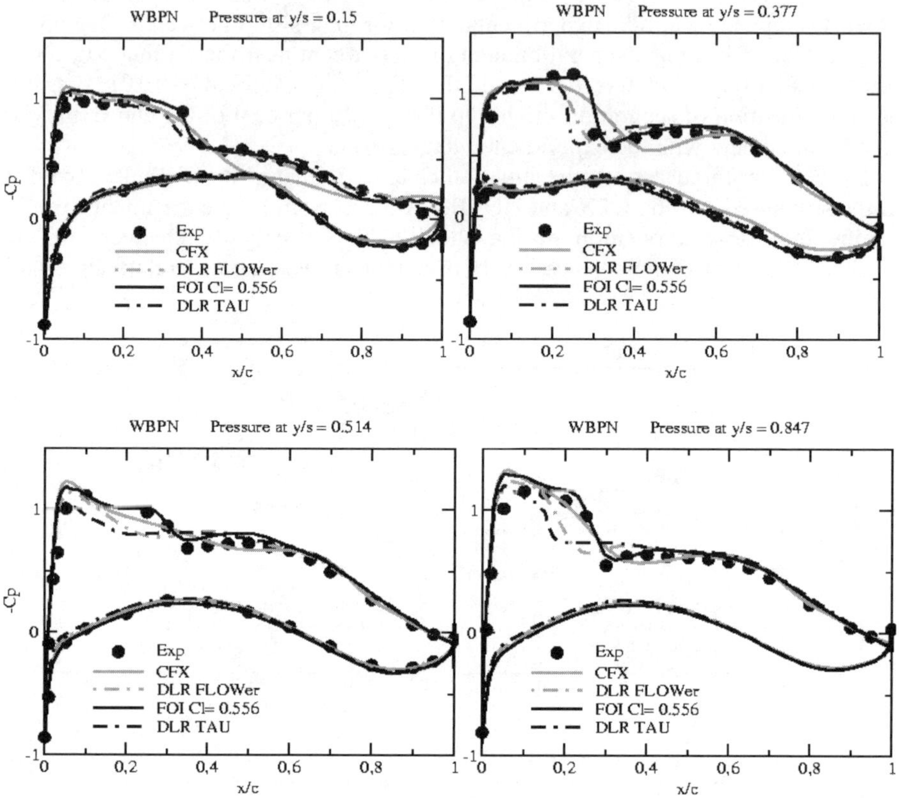

Figure 6 Pressure coefficients for WBPN configuration at four span-wise positions; y/s = .15, .377, .514 and .847 (top left to bottom right)

At y/s = .377 the unstructured grid results which are also very similar to each other at all stations exhibit a shock of similar steepness as found in the experiment, which might be due to the grid adaptation process, but too far upstream. For the two other sections, the hybrid grid results don't reach the suction peak found in the experiment and in the structured grids. Additionally, the following shock is also too weak and too far upstream. Again here the differences between the results of FOI and DLR are quite small and local, despite the differences in turbulence model and grid. While the latter is in line with the small grid influence discussed above one might have expected at least some larger local differences. The better agreement of the unstructured grid results with the experiment at the inner station as well as the better ones on the structured more outboard seems to be mainly an effect of the principle difference in angles of attack, which near 0° in the unstructured case and near 0.5° for the structured one.

Turning the results obtained on structured grid for the WBPN case we are faced with very similar findings as for the WB case as far as the CFX results are concerned: Again a larger discrepancy to the experiment for the section near the root indicating a larger separation and shocks that are more smeared than in the experiment for the other sections. However, the results predicted with the RSM model and the DLR FLOWer code on the structured grid are much more similar to those obtained with the DLR TAU code and the SAE model on the unstructured grid than to those of CFX despite the same structured grid was used.

Near the root section we find a good agreement with the experiments for both DLR results, while at 37.7 per cent of the half span the shocks are predicted too much upstream with some difference in the steepness of the shock, where the shock in the TAU calculation is probably steeper due to the adaptation effect. In the middle section at y/c =.514 the differences between FLOWer and TAU become larger, i.e. while FLOWer predicts about the correct experimental suction peak, the TAU's calculation shows too low a suction which is in line with the WB results. This tendency is also seen at the outboard position where TAU predicts the most upstream shock position. From these results it can be stated that both CFX and DLR TAU show a consistent behaviour for both the WB and WBPN case, despite the deficiencies seen either in the inboard or in the other sections.

Contrary to the WB case the data of FOI show here a much better agreement with the experiment, which could almost be called perfect in all sections shown (for the sections not shown here compare to [Peng, 2004; Peng & Eliasson, 2004]). However, it has to be noted that the FOI results were **not** obtained for the requested lift of c_L=0.5, but with an angel of attack of α= 1° which is about the angle used in the experiment to obtain a lift coefficient of .5. For the FOI computation this angle results in an 11 per cent too high lift of .556. Taking into account that the results for the pressure distributions of FOI and of the DLR TAU code are quite similar for the WB case one can conclude that most probably neither the different grid resolution nor the different turbulence model is responsible for the differences seen. And due to the very good pressure distribution over a large part of the wing it can not be precluded that the solution on the fuselage has quite some influence on the deficiencies in the results.

20.3.3 Influence of different turbulence models

In order to shed some light on the influence of the turbulence model alone FOI performed a study on their medium grid (see Table 2) for the five models mentioned in Table 1. In order to rule out the influences of other possibly factors,

Table 3 Variation of lift and drag with the turbulence model used (Peng, 2003)

Model	SA	SST	Wi88	Peng	WJ-st-ω	Exp.
c_L	0.5267	0.5480	0.5698	0.5713	0.5563	0.5005
c_D	0.0352	0.0316	0.0349	0.0365	0.0328	0.0338
$c_{D,\ pressure}$	0.0195	0.0197	0.0210	0.0207	0.0199	-
$c_{D,\ viscous}$	0.0157	0.0119	0.0139	0.0158	0.0129	-

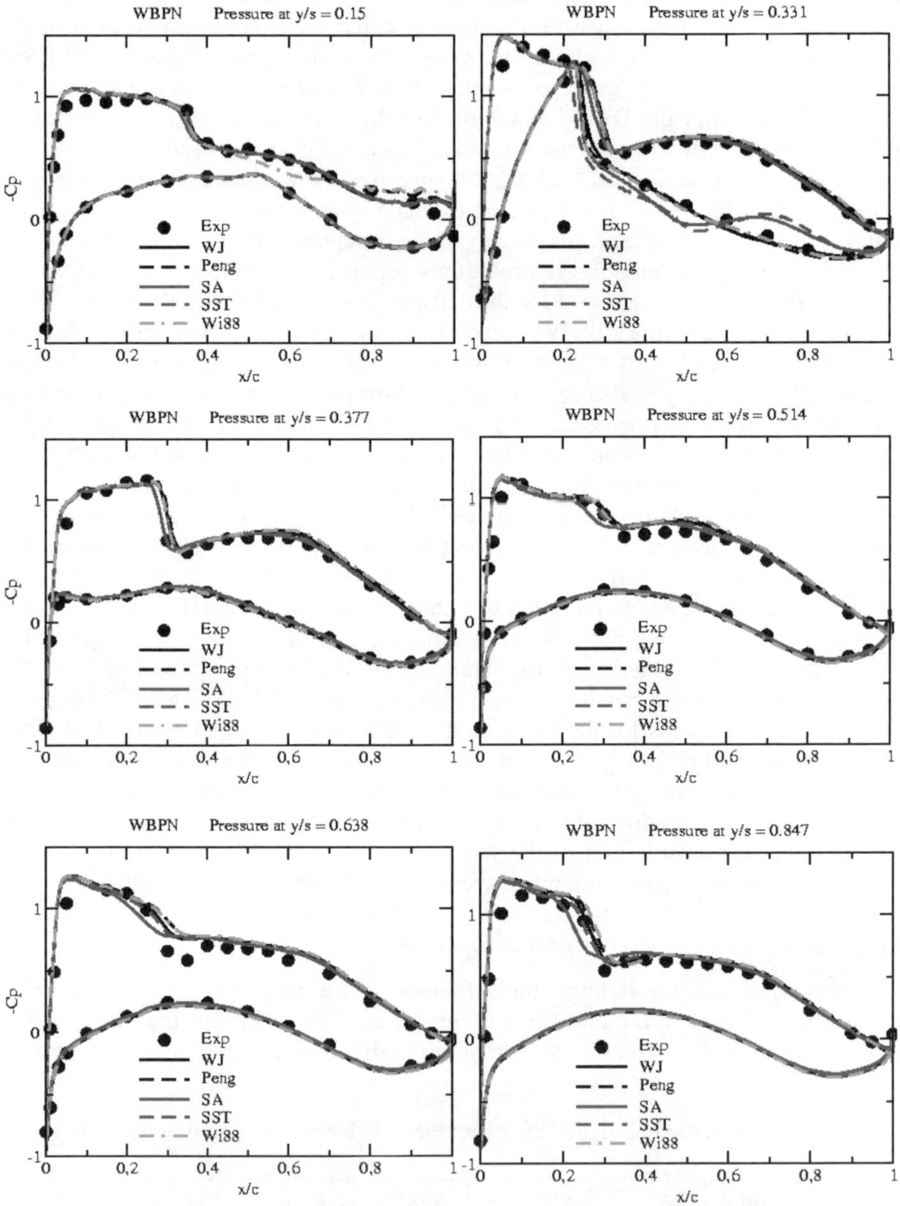

Figure 7 Comparison of pressure coefficients provided by FOI for different turbulence models at six span-wise positions of the WBPN configuration; y/s = .15, 0.331, .377, .514, .638 and .847 (top left to bottom right)

all the calculations were carried out using the same numerical setup assuming fully turbulent flow and employing a fixed angle of attack of 1°. In the following we will give a brief summary of the findings of FOI. For an in depth discussion of the results the reader is referred to [Peng, 2004].

Table 3 shows a comparison of the drag and lift resulting from the different models. As can be seen all the models over predict lift by 5 to 14 per cent as compared with the measured value. The Peng model produces the largest c_L value

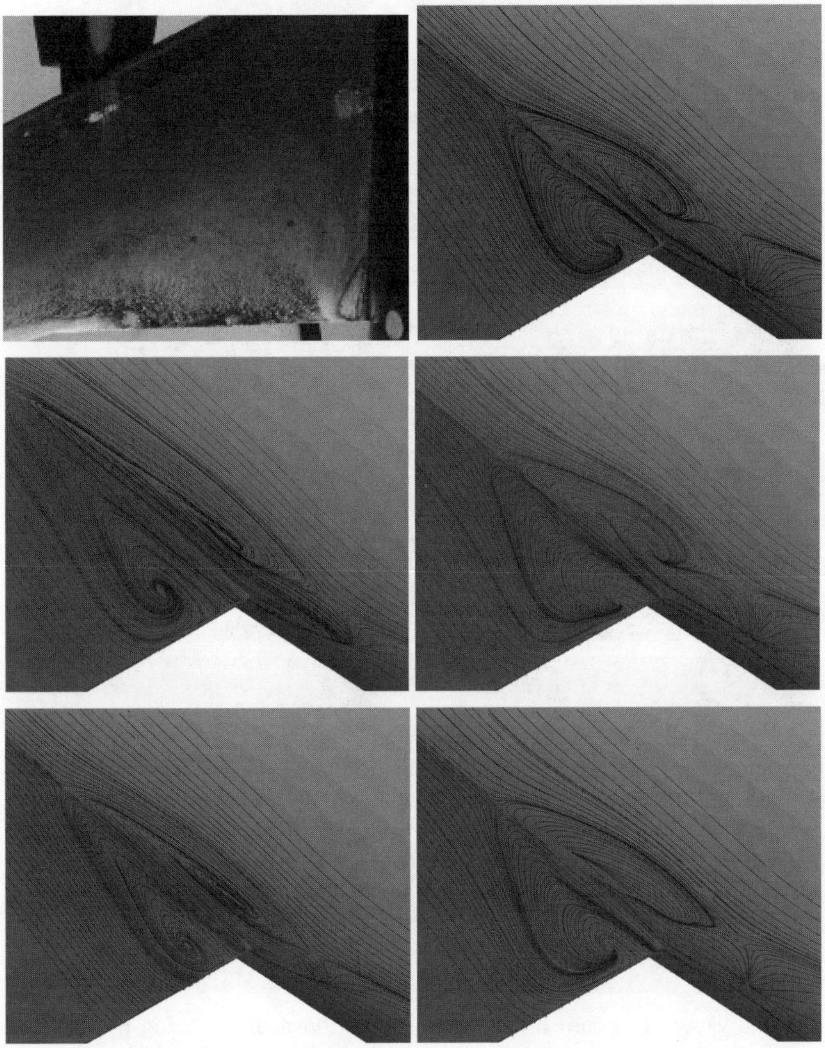

Figure 8 View of stream lines in the separation in the rear part of the wing body junction: Experiment, SA,Wi88, SST, Peng, WJ-st-ω (from upper left to lower right)

and the SA model the smallest one, although these two models both deliver the largest drag values at almost the same viscous drag (coefficient $c_{D,\ viscous}$ in Table 3). The viscous drag is obviously the main reason for the large variation in total drag, since it varies by 28 per cent (based on the average of $c_{D,visous}$) between the models. The pressure drag varies only about 7 percent and it is interesting to note, that the ratio of $c_{D,\ pressure}/c_L$ is practically constant (or $c_L/c_{D,\ pressure} \sim 27.5$).

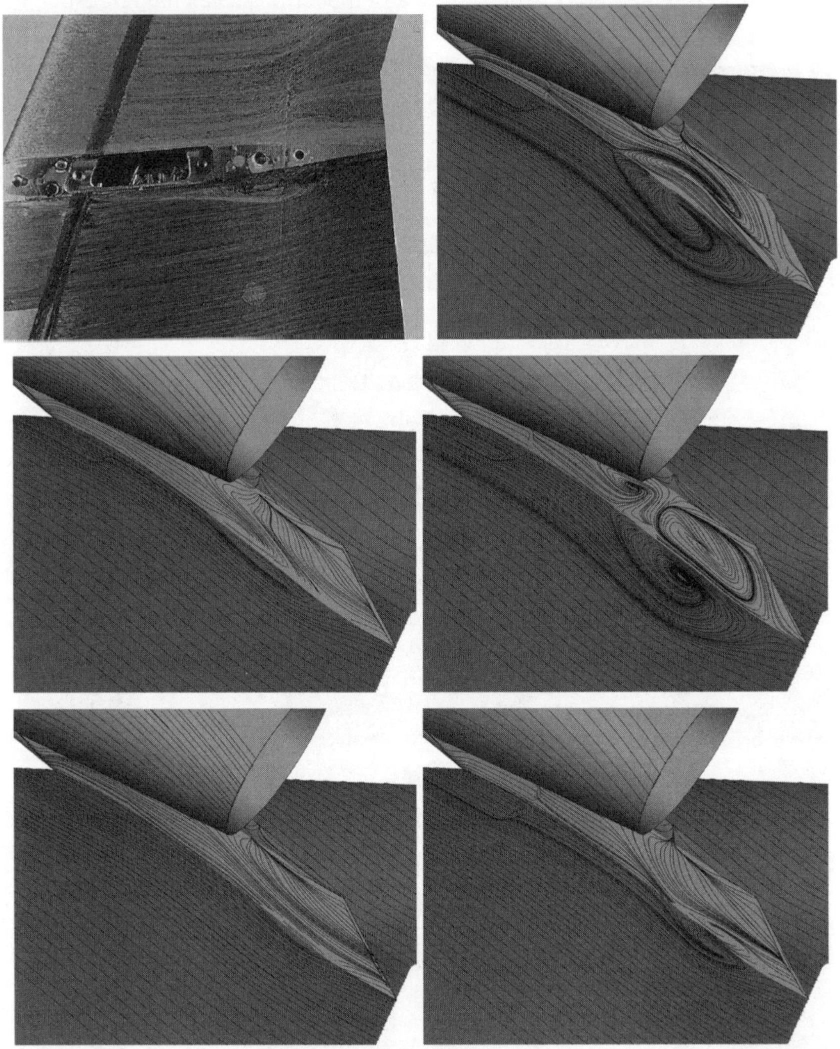

Figure 9 View of stream lines in the separation near the wing pylon junction junction: Experiment, SA,Wi88, SST, Peng, WJ-st-ω (from upper left to lower right)

In addition an analysis of c_L and c_D for each body element of the aircraft model was carried out in [Peng, 2004]. It turned out that all turbulence models estimate an about 87.5 per cent contribution of the wing to the total integrated lift. Thus we can't attribute too high total lift to effects on, e.g. the wing or fuselage alone. The wing is also a main source of drag generation, contributing about 45 – 50 per cent of the total drag depending on the model.

Figure 7 presents a comparison of the pressure distributions computed with the different models in six representative sections. The SA model returns a shock position which is generally more upstream than that of the other models, which are close to the experimental shock position for the inner sections up to y/c = 0.411, but a little too downstream from y/c = 0.514 toward the wing tip. The most significant variation in the results occurs near the wing root (at y/c = 0.15) and for the pylon inboard section at y/c = 0.331, respectively.

Near the root all models over predict the leading edge suction peak and indicate some flow separation through a pressure plateau near the trailing edge. The experimental data, however, show no sign of separation in the pressure distribution. Viewing the integrated skin friction distribution and a flow visualisation from the experiment in Figure 8 it becomes obvious that there is indeed a separation in the rear wing-body junction close to the position of pressure tapes. The reason why the separation is not apparent in the nearby pressure measurement might be found in the fact that its extent in the experiment is about 30 per cent smaller than in any of the simulations. Its length is about 18 per cent of the root chord, while the simulations show a variation between 25 and almost 50 per cent.

Comparing the different models at this station the Wi88 model which shows the largest discrepancy from the experiment produces the largest separation. The Peng model with the second largest extent of this separation exhibits also the second largest deviation from the experimental pressure. All the other models show an again smaller separation of similar size and an almost identical pressure distribution. Despite the differences in the separation all models exhibit an almost identical trailing-edge pressure which is considerably lower than the experimental one.

For the pylon inboard section at y/c = 0.331, the situation is different from that at the root: now the SA and the SST model exhibit the largest discrepancies compared to the experimental pressure distributions (see fig. 7) on the lower surface of the wing, whereas the other models have indicate reasonable agreement. This is also reflected in skin friction /oil flow visualisation near and at the pylon shown in Figure 9. While the other models indicate a separation about the size of the experimental visualisation it is now the SST and SA models which produce an excessive separation. And it seems that this effect exists already at a lower angle of attack, since almost exactly the same pressure distribution can be found in this section in the TAU results for the SA model at the lift coefficient of 0.5 and the correspondingly lower angle of attack (not shown here).

20.4 Conclusions

The comparison of the different results obtained by the partners has lead to a number of conclusions:

- From the comparison of some results derived with transition and fully turbulent flow, respectively, one can conclude that transition seems only to have some limited influence, such that there is no need to distinguish between the results with and without transition.
- From the grid sensitivity studies it seems that at least the medium grids used by CFX and DLR (Table 1) are sufficient to obtain reasonable solutions independent of the numerical method used. And even reducing the grid size further, at least if it is an adapted grid (medium one of FOI) seems not to result in large changes of the solution (as does the turbulence model variation). This is an indication that the solvers are more sensitive to the models than to grid influence, but does not mean that grid convergence could be reached even for the finest meshes as can be seen e.g. from the changes in the moment coefficient.
- With respect to global force coefficients there seems to be a general tendency to over predict the lift for a specified angle of attack as compared to the experiment, with the SA giving the closed match at least for the unstructured codes. For the structured code at least the SST model performs quite well, maybe even better than the SA model would do. Both models seem to perform also best with respect to the lift to drag ratio (about 15 for the experiment and both models following Table 3.), i.e. a reasonable drag polar is computed. Unfortunately, the other three models tested by FOI, namely Wi88, Peng and WJ-st-ω seem to deliver too low skin friction drag. The Reynolds stress model used by DLR in the FLOWer calculation seems to give quite good an answer; however this is difficult to judge with just one result available here.
- Since no experimental data are available on the fuselage, it is hardly possible to estimate the influence of the body on the force coefficients, although it seems likely that part of the deficiencies stems also from that part of the configuration.
- With respect to the pressure distributions it seems that none of the models really perform well everywhere. While SA gives better results near the root for the case with a lift coefficient of 0.5 it shows under prediction of suction peaks and too upstream shocks elsewhere. If one accepts that there is a shift in lift with respect to angle of attack, the SA result can be considered to perform quite well as seen from FOIs results, with the exception of the critical near pylon section. Considering the results of FOI further both SA and SST have problems near the pylon, while the Peng and Wi88 model have problems in the root section. The remaining WJ-st-ω model performs reasonable inboard but produces too upstream shocks in the more outboard sections.
- Taking all this deficiencies into account it seems that the SA and SST model give an overall better performance than the other models maybe with the exception of the SSG/LRR-ω model used by DLR where the amount of data is not sufficient to obtain enough insight.

V Summary of experience

U. Bunge, C. Mockett, B. Aupoix, W. Haase, F. Menter, D. Schwamborn, and K. Weinman

Abstract

The final chapter aims to summarise the experience gained with respect to the performance of turbulence models over the variety of applications considered in the FLOMANIA project. To achieve a level of summary whilst adhering to the case-specific nature of the problem, the discussion is based on groups of flows. This may also assist the potential users seeking to apply the knowledge gained in FLOMANIA to their own applications. As a significant amount of effort has been dedicated to hybrid methods, in particular DES, background models used by these methods are addressed in addition to pure RANS investigations. Furthermore, any derivable best practice guidelines are presented, and the inherent limitations of the applied evaluation procedure are outlined.

1 Introduction

Acknowledging that we are far from obtaining a general turbulence model, and that numerous parameters exist that significantly influence numerical results, it is difficult to draw representative conclusions or to provide any kind of rating for the models used.

Nevertheless, an attempt must be made to summarise the experienced gained throughout this sizeable project. It has been decided that the best approach to achieve this is to base the discussion on families of test cases and model types. This is intended to facilitate the extrapolation of findings from FLOMANIA onto the user's own related applications, whereby the expert user will no doubt be aware of the uncertainties inherent in this.

The authors are well aware that definite conclusions on model performance can only be drawn following the application of rigorous best practice procedures (grid convergence studies, elimination of differences between CFD codes etc.). Due to the large number of test cases and CFD methods, the strict application of such procedures was however deemed unfeasible. In particular, some uncertainties remain in the cross-comparison of different codes and model implementations. Different numerics, different model implementations, different wall treatments, and – as a result – different model parameter settings are still a formidable barrier to the absolute comparison of codes. Despite this, in several of the test cases a series of meshes have been employed with grid convergence obtained. In other cases mandatory meshes have been used based on existing experience, thereby enabling a mature level of comparison.

Therefore it is hoped that the overview of model behaviour across the various applications will enable the reader to evaluate the data available and to judge the reliability of the evaluations carried out.

In the sub-chapters for the application groups, if it is deemed acceptable to issue a judgement on the models applied, a systematic categorisation as "recommended", "applicable" or "not recommended" has been adopted. This information has mostly been obtained from the partners coordinating each test case presented in chapter IV, and is intended to serve as a first-guess for the reader.

In summary, the aim of this chapter is to provide some insight to (not necessarily novice) users on how turbulence models behave for certain classes of applications. The issuing of absolute statements is in general avoided because this is thought to be impossible. Rather, an open and honest discussion of the factors limiting such analysis is offered (see e.g. section 5.3), to equip the reader with the information necessary to determine the extent to which the conclusions drawn are transferable.

2 Statistics

Over the course of the FLOMANIA project 17 partners applied a grand total of 47 different turbulence models or model variations to 20 test cases of varying complexity. Furthermore, a number of models have also been used as background models for hybrid methods, such as DES, and to reflect this, information is also provided for these methods. As the only hybrid method used by more than one partner is the DES method, any interpretation of results is confined to DES in the following sections. Nonetheless, an overview of the experience gained for other methods is also provided.

A summary is presented in Table 1, where the figures in the matrix denote the number of partners who applied each model to the corresponding test case (abbreviated in the following as "TC"). If only one figure is given, it denotes the number of RANS applications per model and test case. If the model was also used as a DES model, two figures are given, the first denoting the number of RANS, and the second the number of DES applications. Each column therefore represents a test case and each row represents one turbulence model using the acronyms defined in chapter III. According to chapter III, the table is subdivided into zero-equation, one-equation and two-equation linear EVMs, non-linear models and EARSMs, and DRSMs. Moreover, the two-equation models are split up into k-ε, v2-f, k-ω, k-g and k-L based formulations, as seen in chapter III.1.

It can readily be seen that the SA-based one-equation models and the k-ω models dominate with respect to the number of test cases they were applied to. In particular, the SA and SST models have been applied to eleven test cases followed by the Wi88 model with nine cases. For SA and SST, the number of applications is higher than the number of test cases to which they have been applied, reflecting the fact that these models have been used by more than one partner for at least one test case. This indicates also that these models can be considered as a kind of standard – for industry in particular. Additionally, the SA and SST models are also widely used as background models for DES.

Table 1 Models vs. test cases with numbers indicating how many partners used the model for the corresponding case (one number: RANS only, two numbers: RANS/DES version). Numbers in brackets behind the acronym give the total number of test cases and applications, respectively, for the corresponding model.

Test case	**1** Rounded Tip NACA	**2** Rotor 37	**3** Asymmetric Diff.	**4** ONERA M6 wing	**5** OAT15A airfoil	**6** Stalled NACA0012	**7** MRTT	**8** Generic Train	**9** Car Mirror	**10** Forward swept wing	**11** TUM Delta wing	**12** 3D circular cylinder	**13** AS28 wing-body	**14** Generic air-intake	**15** Ahmed car body	**16** 3D hill	**17** 2D hill	**18** A-airfoil	**19** NACA0012 DNS	**20** DLR F6
0-equation model																				
BL (1,1)														1						
1-equation models																				
SA (11,21)		1		1		1/4		1/1	1/1			0/1	1	1	3/1			2		1
SAE (3,3)											1									1
SAED (,)																				
SARC (3,5)														1	2/1					
SAERC (2,2)											1									
SALSA (4,6)	1					1/1	1		1/1											
2-equation models: k-ε																				
JL (3,4)								1/1	1										1	
LS (4,6)	1										1				2		2			
CP (1,1)				1																
JB (3,3)						1						0/1							1	
YS (3,3)		1		1											1					
2-equation models: v2-f																				
V2F_Du (1,1)																	1			
V2F_Li (3,3)		1	1														1			
V2F_La (2,2)			1														1			
2-equation models: k-ω																				
Wi88 (9,9)				1		0/1				1			1	1		1	1	1		1
Peng (1,1)																				1
SST (10,15)				1		0/1				1	1	1			4/1	1	1	1		2
SST-R (1,1)											1									
TNT(2,2)													1	1						
LLR (3,5)	1					1/1			1/1											
2-equation models: k-g																				
k-g(2,2)	1						1													

Test case	1 Rounded Tip NACA	2 Rotor 37	3 Asymmetric Diff.	4 ONERA M6 wing	5 OAT15A airfoil	6 Stalled NACA0012	7 MRTT	8 Generic Train	9 Car Mirror	10 Forward swept wing	11 TUM Delta wing	12 3D circular cylinder	13 AS28 wing-body	14 Generic air-intake	15 Ahmed car body	16 3D hill	17 2D hill	18 A-airfoil	19 NACA0012 DNS	20 DLR F6
2-equation models: k-L																				
Sm95(3,3)					1								1	1						
Sm97(1,1)					1															
EARSM & NLEVM																				
AL (,)																				
AJL-ε (2,2)															1		1			
AJL-ω (1,1)															1					
WJ (3,3)				1							1						1			
WJ-ω (1,1)			1																	
WJ-st-ω (3,3)			1													1				1
WJ-Rt (3,3)	1										1									
WJ-C (,)																				
WJ-H (1,1)					1															
WJ-G (1,1)	1																			
LEA (2,2)				1														1		
CEASM (2,3)	1					1/1														
DRSM																				
SSG (3,7)			5												1		1			
SSG-Ch (3,4)			2												1		1			
SSG-ω (2,3)			2													1				
IP (,)																				
HJ (,)																				
LRR (2,2)						1													1	
Sh (1,1)																1				
Wi-ω (2,2)				1														1		
Wi-BSL (,)																				
SSG/LRR-ω (5,5)			1	1							1							1		1
TCL (1,1)	1																			

3 Limitations of the approach

As is inevitable in any attempt to summarise the findings of a project of this size, not to mention the inherent complexity of the subject matter, the extraction and compilation of the important details is no simple task and not without drawbacks. Most obviously, the necessity of reducing the total volume of information brings with it the risk that a misunderstanding of the findings may emerge. It is therefore the role of this section to address this issue by highlighting the principle limitations inherent in the method of analysis, providing a clear guide as to how the information presented is to be interpreted, and thereby enabling better evaluation procedures for future projects.

As outlined earlier, rating information in the form of three performance indicators was collected by the coordinating partner of each test case, which form the basis of the test-case specific findings. It is naturally to be expected that the reasoning behind each rating is based on the relative performance of the models employed in the specific case, and not on some universal set of principles. Therefore, any attempt to compare the performance of each model between test cases can potentially lead to the drawing of erroneous conclusions. It was with the intention of reducing the scope for over-interpretation of fine differences in such situations, and thereby providing the maximum level of comparability, that only three levels of performance rating ("recommended", "applicable", "not recommended") were allowed. This of course carries with it the negative effect of a severe data reduction. However, this summary chapter is not intended as a stand-alone information source, rather it is considered essential that it be studied in conjunction with reference to the individual test case chapters.

Specific aspects of the information lost in such a procedure include:

- Particular mathematical, physical or theoretical considerations in the model formulation
- Individual peculiarities specific to the test case
- Practical aspects, such as the relative numerical expense of different models, ease of implementation and model robustness

In the first case, it can be debated that this represents a strength of the approach – any level of "expectation" for a particularly advanced model should not result in an over-generous rating, should more simple models happen to perform better. However, it can equally be the case that some peculiarity of the test case (such as problems with or a lack of experimental data, or a "cancellation of errors" e.g. TC17) might cause an "unexpected" result. Therefore, although forming the basis of the information summary, case-specific details cannot be ignored. To address this, some level of repetition of such details is made and merged where relevant in the discussion of the following sections. These considerations underline the fact that any kind of averaging of ratings for each model across the test cases would be invalid and dangerous.

Inherent to the very nature of the theoretical simplifications made in turbulence modelling is the fact that different models exhibit different levels of suitability for different classes of flows: there is no "common model". Therefore, a potential user of the turbulence models summarised here must compare their anticipated application with the FLOMANIA applications to determine the extent to which this rating can assist in the choice of an appropriate model. Provided the underlying flow physics is similar, this evaluation can be a guideline, but an extrapolation to applications with different flow physics is impossible.

4 Test-case specific discussion of findings

An attempt shall now be made to provide an application-specific review of turbulence model performance. Although this is intended to be a guideline for the reader's own investigations corresponding to the six different applications to follow, the authors would like again to emphasise that any application – even if seemingly fitting within the classifications – might involve important differences resulting in different turbulence model behaviour. Therefore, the potential user must compare their own applications closely with the FLOMANIA application to determine the extent to which this information can help in choosing an appropriate model. Provided the underlying flow physics is similar, this model performance evaluation can be a guideline, however generally applicable conclusions and/or recommendations are out of reach.

The six main application categories are specified as (with the test case numbers in brackets):

1. Airfoils (TC5, TC18, TC19)
2. Wings/delta wings (TC1, TC4, TC11)
3. Wing-body cases (TC7, TC10, TC13, TC20)
4. Bluff bodies, counting the circular cylinder and the generic intake in cross-wind as bluff bodies (TC6, TC8, TC9, TC12, TC14, TC15)
5. Transonic compressor rotors (TC2)
6. Internal (separating & re-attaching) flows (TC3, TC16, TC17)

The main criterion for the comparison of models and their behaviour is agreement to experimental data, for which surface pressure distributions and integral parameters as well as in some cases the flow topology are used for comparison, often on a qualitative basis. In addition, and where applicable, additional factors such as computational cost, pressure loss, and/or reference computations with a different model or method have also been used as performance indicators.

4.1 Airfoil applications

Among the 20 test cases of FLOMANIA there are four airfoils, two of which involve the NACA0012 airfoil. However, as the NACA0012 airfoil in test case TC6 is at an angle of attack of 60° and accordingly exhibits fully separated flow, it is counted as a bluff body flow and discussed in 5.4.4. The other test case (TC19) employing a NACA0012 airfoil concerns DNS simulation at very low Re numbers. As such, it suffices to state that RANS models devised for high Reynolds numbers are not suitable for such cases and cannot be expected to give reasonable results.

This leaves us with test cases TC5 and TC18, the OAT15A and the A-airfoil respectively. In both cases some uncertainties are apparent, e.g. the influence of the outlet boundary condition for TC5 and the two sets of experimental data for the A-airfoil. Earlier suggestions about 3D effects in the measurements (as are erroneously apparent in RANS computations) have not been verified by the experimental findings, as discussed in chapter IV.18.

Unfortunately there is no overlap of any of the models used in these two cases, which would, of course, have been interesting with respect to more general conclusions as one test case exhibits transonic flow with a shock, while the other is a subsonic high angle of attack flow. Most models applied to TC18 are not even from the same group of models as those used in TC5, with the exception of the

WJ-H and the LEA EARSM models, which were both regarded as "applicable" for their respective case. This was however the case for most of the models applied (SA, Wi88, Wi-ω Sm97). Only the Sm95 was rated as "not recommended" (for TC5), while both the SST and the SSG/LRR-ω were classed as "recommended" based on TC18.

4.2 Wing/delta wing applications

Turning to three-dimensional cases, we find three wing test cases, namely the well-known ONERA M6 wing (TC4), the rounded wing tip (TC1) and the TUM delta wing (TC11). While TC4 is again a transonic flow with shocks, the two others deal with vortical flows.

As is often the case for three-dimensional calculations, we are faced with the fact that some of the meshes used in these calculations are fairly coarse. Indeed, some grid influence is found in the results which makes conclusions on the model performance very difficult, especially for TC1 and partly for TC11. In total, almost 20 models were applied to the three cases (see Table 1), however there is unfortunately no model that was used in all three cases and only 6 models were employed for two cases.

In the group of SA-based models, the SA model itself was found to be "applicable" for both wing cases, with a further note that the rotation correction appears not to deliver an improvement over the SA results for the delta wing. For the wing tip the SALSA model was rated as "not recommended" based on the performance with respect to the vortex core. As no rotation correction was used in this case, it can only be speculated whether this might have improved the situation.

While the only tested k-ε model (LS) was judged "not recommended" for both vortical flow cases, the k-ω-type models applied were rated as "applicable", although only the SST model receives this rating for two cases, i.e. TC5 and TC11, thus giving an indication of its often postulated superiority.

Among the EARSM, the WJ-Rt was rated as "applicable" for both vortical flows while the WJ was rated "applicable" for the vortical flow over the delta wing and even "recommended" for the transonic M6 wing. Also LEA (for TC4) and CEASM (for TC1) were seen as "applicable", thus giving the impression that the group of EARSM perform quite reasonably.

An even better performance was found for the full differential Reynolds stress model: here the SSG/LRR-ω model was rated as "recommended" for both TC4 and TC11, but unfortunately not applied to TC1.

4.3 Wing-body applications

Turning to the most complex wing-type flows treated in FLOMANIA, i.e. the combination of a wing with an aircraft fuselage, four test cases come under consideration, two of which resemble transport aircraft configurations. These are the AS28 wing-body configuration (TC13) and the DLR F6 wing-body configuration (TC20) which is used in two variants; one with a clean wing and one with a pylon and nacelle attached to the wing. The two other configurations are a wing-body configuration with a forward swept wing (TC10) and a wing with a pylon-mounted pod and a nearby fairing (TC7).

Unfortunately, only single grids with relatively low resolution are used in both TC7 and TC13 giving no indication of the grid influence to be expected. The two models applied to TC7 were found to be "not recommended", however due to uncertainties with this test case and as these models are not used in any other wing-type flow, no concrete conclusions regarding these models or this test case would be valid.

All models applied to TC13 (SA, Wi88, TNT, Sm95) were used both with and without wall functions noting that wall functions significantly reduced the discrepancies between the models. Interestingly, it was also found that the k-ω-type models gave results where the lift was nearly independent of the wall treatment. The drag for the same models was found, however, to be very sensitive to the boundary condition. The behaviour for the Sm95 was the other way around and overall the SA model gave globally the best results, leading to a "recommended" for SA, with all others rated "applicable" with caution.

The two other test cases – for which some fairly fine grids of up to 12 million points were used – give an indication that the results from the fine or even medium grids can be more safely used to evaluate the model performance than other results from the 3D wing-type flows. Unfortunately, SA-type models are only used in TC20, while Wi88 and SST are used in both cases making conclusions about the latter somewhat safer than those for SA or the other models used only in TC20 (Peng, WJ-st-ω). The results are relatively clear that Wi88 receives a "not recommended" while SST is at least "applicable" or even "recommended" based on TC10. The SA and SSG/LRR-ω model are only "applicable", as it is at this point unknown how they would perform for TC10. The SSG/LRR-ω model seems very promising in TC20 but due to too few results available a recommendation is not justified.

As is the case for the Wi88 model, both Peng and the EARSM receive a "not recommended" due to under-predicted friction drag and problems with the separation in the wing root area of TC20.

To summarise this discussion of wing-type flows it is once again emphasised that our recommendations are based only on the available data, and as such cannot be interpreted as absolute truths. Bearing this in mind it seems safe to say that both the SA and SST models are the first choice for wing-type flows, perhaps with some slight preference towards the SST model. Additionally, there is an indication that the SSG/LRR-ω may be somewhat superior to the other models of the EVM or EARSM group. However, the latter group, taking the WJ model as an example, can not be ruled out completely as we have seen that these models were rated as "applicable" in several cases.

From the point of view of the practitioner, using SST or SA (or maybe a DRSM such as the SSG/LRR-ω) should normally lead to reasonable results. Nevertheless, these should be critically scrutinised, especially in the case of new applications.

4.4 Bluff body applications

The test cases considered are TC6 (NACA0012 beyond stall), TC8 (prediction of cross-wind stability for a generic train), TC9 (analysis of sound sources for a generic car mirror), TC12 (3D circular cylinder), TC14 (generic air-intake) and TC15 (Ahmed car body). Of these, TC6, TC9, TC12 and TC15 are clearly bluff bodies, all involving massively separated flow behind a blunt downstream face (although this is preceded by flow separation and reattachment along the slant in the case of TC15). More debatable are TC8 and TC14, where the flow physics involved deviates from this somewhat. For TC8, although the flow is separated on the leeward side, strongly stable streamwise vortices are a dominating feature. In the case of TC14, the region of separated flow in the duct is limited due to the high axial velocities through the intake.

A dominating theme in the simulation of these test cases (with exception of TC14) is the evaluation of DES techniques in comparison to RANS, URANS and in one case LES methods. It will therefore be attempted to generalise as far as is possible the conclusions obtainable across the appropriate cases. Statistical models are only tested in TC12, TC14 and TC15, and the conclusions of these investigations will be handled separately.

It can clearly be stated, that for flows featuring unambiguous, massive separation (TC6, TC9, TC12), DES has been demonstrated to offer considerable increases in predictive accuracy compared to URANS. The clearest example of this is TC6, where all of the DES force coefficients (five partners and background models) lay within a 10% error margin compared to experimental values. All URANS results on the other hand demonstrated errors exceeding 10%, and a similar improvement was seen with the Strouhal number prediction. The same story is seen for TC9, with sound pressure levels in good agreement with the experiment returned by DES, and consistent underestimates of around 20dB returned by URANS on the same grids. In both test cases, the spectral overlap of modelled and resolved scales causes well-known problems for URANS approaches, whereas the LES-type treatment of the turbulent flow in the separated regions opens up the entire spectrum of resolved scales down to the order of magnitude of the grid spacing in DES. For the cylinder flow of TC12, the DES computation also delivered impressive agreement with the experimental data.

DES implementations based on a wide range of background models (one-equation, two-equation and NLEVM) have been tested. However on the basis of these cases, no strong evidence has emerged to suggest the clear superiority or otherwise of any one of these. Indeed, evidence to the contrary was collected in TC6, where probability density functions of the resolved wake showed a high degree of similarity. This is consistent with the idea that the influence of the SGS model on larger scales should not be significant for a properly resolved LES. Furthermore, in all of the cases discussed so far, the role played by the LES-mode region of massively separated flow is dominant.

These considerations naturally raise the question of grid resolution requirements for DES. In the separated, LES-mode region, there is no reason to expect that a coarser grid can be used for DES than for LES (Spalart, 2001), and it is reminded that the lower numerical expense of DES is purely due to the RANS handling of attached boundary layers. Unfortunately, reliable LES resolution criteria in such regions are elusive, and the dominant best practice recommendation is the conduction of a systematic grid refinement study. Although three grids were computed for TC6, this does not count as such, as only weak refinement was conducted in individual grid directions.

Finally, a mention should be made concerning the role of the underlying numerical scheme in DES computations. Existing experience from RANS and LES reveals contradictory demands on the numerical convection scheme, posing a problem for hybrid RANS-LES methods. An appropriate local blending between upwind and central-based schemes is therefore recommended, such as the hybrid scheme of (Travin et al., 2002), and the benefit of this was demonstrated in TC9.

Attention shall now be turned to the remaining cases for which DES has been applied; the generic train (TC8) and the Ahmed car body (TC15). As discussed, for each of these cases a deviation from the standard flow physics of bluff body vortex shedding is present. Perhaps as a result of this, such a clear-cut statement about the performance of DES as made about the earlier cases cannot be given. For TC8, for which unsteady predictive capability is central to the underlying engineering task, the URANS simulation failed to deliver an unsteady solution, the force values converging to a steady-state. This was attributed to a significant overlap of the modelled and resolved turbulent frequencies. The DES however returned force coefficients with strongly unsteady character, and the time averaged values showed a better agreement to the experiment than those of the other methods. However, upon examination of the vortex structures predicted by DES, it was seen that large regions showed little or no unsteady character, particularly near the front portion of the first car, and it was noted that the extent of LES-mode operation was correspondingly minimal in these regions. Further aft, the vortex structures show higher degrees of unsteadiness coinciding with the growth of significant zones of LES-mode operation. On the basis of the high Reynolds number (of the order of 10^7), the evidence is strong that a much finer computational grid would be required to obtain a solution compliant with the underlying methodology of DES. Furthermore, the large time step, chosen to resolve only the relatively low frequencies of engineering interest would also need to be reduced to enable a valid LES. The picture then, is of an ambitious test case where the DES method is applied with questionable validity. Nonetheless, it is a case where the prediction of unsteady behaviour is essential, to which end DES proved unquestionably advantageous to the goals of the engineer.

The final bluff body case involving DES is the Ahmed car body (TC15). The region of massively separated wake flow is preceded by a highly sensitive flow separation and reattachment along the 25° slant. Additionally, strong and stable streamwise vortices generated at the side of the slant interact with this, giving rise to a complex three-dimensional flow topology. The fully-resolved LES, although at a lower Reynolds number, showed near perfect agreement with the experimental data, although requiring some 16.5 million grid points. The DES computations, based on three different background models (SA, SARC and SST) were conducted on two considerably coarser grids. In contrast to the experience from TC6, strong variations between all DES implementations were observed, with the SA-DES failing to predict flow separation on the slant, and SARC-DES and SST-DES over-predicting the separation (although to a lesser extent for SST-DES). Interestingly, this trend is the same as seen in the RANS results for each model. For this test case then, in contrast to the pure bluff body flows discussed first, the critical flow feature is handled by the RANS mode of the DES, which explains the higher degree of model dependency.

For the DES solutions in which flow separation along the slant is predicted, flow reattachment does not occur. Part of the reason for this is an insufficient grid resolution in the separated flow region (the grids being essentially designed for URANS), which is substantiated by the much lower turbulent stresses than those of the experiment and LES. A more subtle issue which certainly plays a role, is the transfer from modelled to resolved turbulence in the separated shear layer, for which no explicit mechanism is defined in DES.

Having discussed the findings of the DES applications for the bluff body test cases, the RANS investigations shall now be addressed. In TC14, a range of RANS models have been compared using a common code and grid, and a very strong variation in the results was observed. This can be explained in terms of the sensitivity of the flow field to the prediction of flow separation and reattachment and the well-established model sensitivity of RANS closures in such situations. Unfortunately, no experimental data is available for this case, rendering the extraction of deeper conclusions beyond the acknowledgement of strong model variation impossible.

TC12 has been used to test the novel Organised-Eddy Simulation (OES) method, the development of which is motivated by the poor performance of URANS methods when spectral overlap occurs between clearly-defined coherent and non-coherent motion. Two-dimensional simulations show much improved agreement with the experimental drag coefficient by OES compared to the SST model. This promising performance certainly warrants further investigation of this technique in the future, and in general, the detailed experimental data offers considerable potential for detailed model validation.

In TC15, some general trends have been observed between different model types, backed up by the evidence of numerous codes and grids. All k-ε models were found to falsely predict attached flow along the entire slant. This resulted in a relatively good representation of the true flow topology; due to the lack of separation, the prediction of flow reattachment is unnecessary. All results for four separate implementations of the SST model gave separated flow along the slant, however none predicting the flow reattachment. This coincides with the general experience of models with an optimised capability for the prediction of separation onset. The majority of the SA model results failed to predict separation, with one implementation predicting separation without reattachment.

The two NLEVM tested, the AJL model based on the ω and ε equations, both gave fully-separated flow on the slant, similar to the SST results. Two different implementations of the SSG DRSM model were also used, one employing wall functions, and one being solved down to the wall. The version with wall functions did not separate on the slant, whereas the directly solved version predicted flow separation and, uniquely, reattachment (albeit late, right at the end of the slant). The SSG results again underline the dominant role played by wall functions in DRSM implementations and the general sensitivity of this class of model to implementation issues.

4.5 Transonic compressor rotor applications

The only test case falling under this category is the NASA rotor 37 (described in detail in chapter IV.2). A multi-block low-Reynolds mesh was used for this test case represented by a H-O-H topology with about 615,000 grid points allowing for four grid levels, but keeping y^+ lower than 1 in the finest mesh.

Three types of turbulence models have been tested:

- the Spalart-Allmaras model
- the low Reynolds Yang-Shih version of the k-ε model
- the code friendly version of the v^2-f model.

In order to investigate grid resolution effects, computations have also been carried out on the coarser mesh level (number of grid points divided by two in all directions, i.e., about 77,000 grid points) for both the v^2-f and the Spalart-Allmaras model.

For both coarse mesh computations, the choke mass flow was under-predicted in comparison to experimental data, however the overall performance was improved on the finer mesh for both v^2-f and Spalart-Allmaras models:

- The choke mass flow became closer to the experimental value.
- The peak efficiency was increased by about 0.5% for the Spalart-Allmaras and 1% for the v^2-f model.
- The overall efficiency was significantly increased.
- The prediction of the stability limit close to stall was improved.

Despite this, the pressure ratio profile was too flat at low mass flow in comparison to experimental data, whereas the slope was well predicted on the coarser mesh.

As a summary of the turbulence models used, the following comments might guide the potential investigator of this test case:

It appears that the Yang-Shih k-ε model has difficulties in predicting the choke mass flow, whereas the maximum mass flow is similar for the two other turbulence models. The k-ε model also under-predicts the pressure ratio, whereas both Spalart-Allmaras and v^2-f models produce a similar evolution of the pressure ratio with the mass flow. One of the major improvements of the v^2-f model in comparison to the Spalart-Allmaras model resides in its improved prediction of the isentropic efficiency profile. Thus the conclusion can be drawn that both the Spalart-Allmaras (SA) and the v^2f (V2F_LI) models receive ratings as "recommended" turbulence models for this test case (or class of flows), while the YS k-ε model, perhaps unsurprisingly, is not recommendable.

4.6 Internal flow applications

Three test cases are dealt with in this section, TC3, TC16 and TC17. In all cases, the flow is confined by an upper and a lower wall throughout the complete computational domain. Before an attempt is made to evaluate the findings in general, each case is summarised separately.

TC3, the asymmetric plane diffuser is characterised by a statistically two-dimensional flow exhibiting separation, reattachment and recirculation, and was declared mandatory for all partners applying DRSM. It has been computed by eight partners using four different models.

The extremely high sensitivity of this case to modelling details was noted in the test case chapter (IV.3), and an accurate prediction of separation, reattachment and recovery is required in the presence of a geometry-driven adverse pressure gradient.

The DRSM models do demonstrate improved representations of the flow physics in comparison to linear viscosity models, however second-moment closures cannot be considered generally superior. In particular, the solutions are sensitive to wall effects and it has been observed that wall function implementations often return better results in comparison to some low-Reynolds number solutions. In particular, the SSG-ω based models generally give a poorer representation of the stream-wise normal stress, suggesting weaknesses in the damping of stresses in the wall region. Examination of the wall pressure, skin friction and velocity fields show that the all implementations follow the general trend, but that some solutions require careful review and analysis.

Thus, the data gathered from TC3 suggests that DRSM application is justified but that a great deal of care is required in the construction and implementation of the method, especially within a generic CFD code. As noted earlier in the section on NLEVM, the task of fully understanding the sources and magnitudes of error contributions from numerics, model and implementation has not been addressed in this work in as much detail as would be necessary to give more specific guidelines.

As every partner employed a slightly different model variant (see chapter IV.3 for details), no model-specific recommendations or conclusions can be provided. It can only be said that such details are critical in the implementation of DRSM and that it has been shown how laborious and difficult it is to obtain an accurate and consistent implementation of second-moment closures with equal results, even for such a geometrically simple case on a common numerical grid.

TC16, the three-dimensional axisymmetric hill, involves a complex, three-dimensional, separated flow topology downstream of the hill. LES, RANS and hybrid LES-RANS have been applied to this case by six partners.

The dominant conclusion to be drawn is that all RANS simulations, even with DRSM, fail to capture the correct flow topology: the recirculation zone is too large (with some minor differences between models), and the secondary vortex rotation appears to be in the opposite direction compared to experimental data. In fact, this vortex (viewed at a plane normal to the flow direction) is not rotating in the wrong direction. Two vortex pair structures are found in the flow; one emanating from the upstream side of the hill and corresponding to a "tip or trailing" vortex, and the other vortex emanating from the lee-side of the hill ("tornado-like" vortex). In the experiments, the tip vortex (or trailing vortex) pair dominates, whereas in the RANS simulations, the tornado-like vortex is dominant (as reflected in the exaggerated separation region). As these two vortex systems have an opposite turning direction, the impression is given that the vortex direction is incorrectly predicted. At present, no satisfactory explanation can be given for this behaviour, and a discussion of the distinction between the RANS models is deemed superfluous in the light of this universal failure. Furthermore, the aspect of grid resolution effects were not investigated. In contrast to the RANS simulations, LES and hybrid RANS-LES yield results in good agreement to the experiment. DES has not been applied to this case.

When it comes to making specific recommendations concerning this test case, it can only be concluded that all RANS models applied are "not recommended", and only hybrid LES-RANS or LES are "recommended".

TC17, the separation behind two-dimensional hills, has been computed by two partners with eight models with an LES result serving as a reference computation. Rapidly changing velocity profiles pose a severe challenge to any model and anisotropy resolving models are required to capture them.

Second-moment closure with full wall resolution yields acceptable results, despite some inaccuracies at the hill-top and reattachment point. The NLEVMs tend to give a better representation of the turbulent stresses, but only the AJL model gives a close representation of the shape and length of the recirculation region. The two-equation models are unable to produce the correct levels of shear stress in the separating shear layer and tend to give either a late separation, or an excessively long recirculation zone. The k-ω model gives a good overall representation of the flow, with a better prediction of the size of the separation zone than the more advanced SST model. It is however acknowledged that models that are tuned for the prediction of separation onset, such as SA or SST, typically produce exaggerated separation zones. This is a result of their inability to properly predict the high turbulent stress levels in the separating shear layer. Most RANS models that predict a more realistic separation length do so due to a delayed separation onset prediction, and not because of an improved representation of the critically high stress levels in the separating shear layer. Therefore, this seemingly favourable result is caused by cancellation of model-inherent errors, a phenomenon which must always be carefully considered. The one exception in the model comparison is the AJL model, which gives much improved shear stresses in the separating shear layer.

In conclusion, for TC17 only the NLEVM AJL is recommended, having shown results superior even to the DRSMs applied. However, a satisfactory explanation for this superiority is elusive, and as a result of this it is suggested that the AJL model warrants further investigation.

In general, it cannot be concluded that DRSM are superior to linear or non-linear EVM. Taking into account the effort required to implement and apply DRSM, its application as a general approach cannot be recommended. However, if DRSM are to be applied, the use of wall functions should be considered. Furthermore, on the basis of TC3 and TC17, the low-Re formulation of the SSG model should in general be avoided, at least when separation is to be simulated.

In has also emerged that when the computational effort is permissible and unsteady boundary conditions pose no problem, hybrid LES-RANS or pure LES simulations deliver superior physical accuracy, and are preferential when separation is present.

NLEVMs give a better representation of the stresses for TC17, with the AJL giving the closest agreement to the LES data. These models present a viable alternative to full DRSM or expensive turbulence-resolving methods such as LES, DES or other hybrid LES-RANS. Nonetheless, being very often derived from DRSM, NLEVMs also have to be carefully implemented and validated. Any explicit algebraic stress model suffers to some extent from its conceptual simplification by neglecting substantial derivatives when derived from Reynolds-stress transport equations. These terms only vanish in a stream-line fitted frame of reference, so the model cannot account for stream-line curvature effects unless some curvature or rotation correction is applied. More comparison of DRSMs and NLEVMs with and without these corrections is required to estimate the potential of NLEVMs.

5 Conclusion

A large number of different test cases spanning a broad range of flow physical phenomena have been investigated by multiple partners applying the full spectrum of RANS model types as well as LES and hybrid RANS-LES methods. To provide an optimal level of summary, these individual combinations of models and test cases have been grouped into classifications based on the similarities or contrasts between them. As such, whilst acknowledging the limitations inherent in the approach, it has been possible to summarise the considerable amount of experience gained in a manner amenable to the transfer of this experience to a wider audience of potential users. Within the FLOMANIA consortium, the large number of partners has allowed the comparison of turbulence model implementations in different codes, resulting in a higher generalisation of the findings. Furthermore, the mixture of partners from universities, research establishments and industry has facilitated an important flow of technology and expertise.

In general, it has been shown that the suitability of different models and approaches strongly depend on the application at hand. Whereas some well-tuned linear EVM (such as the SST or SA models) have proven satisfactory for predominantly attached, high Reynolds number wing-type flows, these consistently fail for separating-reattaching internal flows. For the simpler, two dimensional examples of the latter, a better reproduction of the flow physics has been obtained using NLEVM and DRSM, although far from universally. When the topology of the flow separation and reattachment becomes complex and three-dimensional (as seen in the 3D hill of TC16), all RANS models, including DRSM, fail entirely, and acceptable results have only been produced by LES or hybrid LES-RANS approaches. This suggests that resolution of the unsteady dynamics may be essential in such situations.

Although demonstrating, in some cases, a superior predictive capability, the dominant conclusion resulting from the experience gained with DRSM must be the great importance of fine implementation details. It has proven very difficult to obtain reliable and consistent results from different implementations even on simple geometries described by common grids. Furthermore, a strong sensitivity to the choice of wall treatment has been observed, with wall functions tending to reduce the model sensitivity. As such, it cannot be said that DRSM are universally better than linear or non-linear EVM, and indeed promising results have been obtained using NLEVM without the problems arising from the inherent complexity of DRSM. These findings, made possible by the broad consortium, have important implications for the industrial applicability of full second moment closures, and underline the necessity of further, more detailed investigations in this area.

Another key outcome of the FLOMANIA project has been the first steps taken towards the investigation of hybrid RANS-LES approaches, in particular DES. This approach was found to deliver clear improvements in predictive accuracy for flows dominated by massive flow separation (such as bluff bodies) compared to unsteady RANS. Furthermore, in these cases a significantly reduced dependency on the background RANS model was observed. In cases where the boundary layer physics play an important role (such as the Ahmed car body), the opposite has been found to be true, reflecting the inherent hybrid nature of DES. However, much work remains to be done before reliable best practice guidelines can be established, including the investigation of grid resolution effects, an indication of the computational expense and the range of applicability of DES. To this end, FLOMANIA has laid the groundwork for deeper analysis of these issues in the future.

VI References

Abe, K., Kondoh, T. and Nagano, Y. (1977): On Reynolds-stress expressions and near-wall scaling parameters for predicting wall and homogeneous turbulent flows. International Journal of Heat and Fluid Flows. Vol. 18. pp. 266-282

Abe, K., Yang, J., Leschziner, M.A. (2003): An investigation of wall-anisotropy expressions and length-scale equations for non-linear eddy-viscosity models. International Journal of Heat and Fluid Flows, Volume 24, pp. 181-198.

Abernathy, F.H. (1961): Flow over an inclined plate, ASME J. Basic Eng., V. 61, pp 380-388.

Ahmed S.R. and Ramm G. (1984): Some salient features of the time-averaged ground vehicle wake. SAE technical paper 840300.

Almeida G.P., Durao D.F.G., and Heitor M.V... Wake flows behind two dimensional model hills. Exp. Thermal and Fluid Science, page 87, 1992

Apsley, D.D., Leschziner, M.A. (1998):A new low-Reynolds-number non-linear two-equation turbulence model for complex flows. International Journal of Heat and Fluid Flows, Volume 19, pp. 209-222.

Apsley, D.D and Leschziner, M.A. (1999), "Advanced turbulence modelling of separated flow in a diffuser", Flow Turbulence and Combustion, 63, 81-112.

Archambeau F., Mechitoua N., Sakiz M. (2004), A Finite Volume Code for the Computation of Turbulent Incompressible Flows – Industrial applications, I.J. Finite Volumes, http://averoes.math.univ-paris13.fr/IJFV.

Arina, R., Ceresola, N., Pianta, P.G. (1996): Application of a two-equation turbulence model to the numerical prediction of the transonic buffet of an airfoil. Proceedings 20th ICAS Congress.

Aubry, N., Guyonnet, R. & Lima, R. (1991) Spation-temporal analysis of complex signals: theory and applications. Journal of Statistical Physics 64 (3/4), 683-739.

Baldwin, B.S., Lomax, H. (1978): Thin layer approximation and algebraic model for separated turbulent flows. AIAA Paper 78-257, 16th Aerospace meeting, Huntsville, Alabama, January 16-18.

Barche, J. (Ed.) (1979): Experimental Data Base for Computer Program Assessment. Report of the Fluid Dynamics Panel Working Group 04. AGARD-AR-138

Bardina, J., Ferziger, J.H., Rogallo, R.S. (1985): Effect of rotation on isotropic turbulence: computation and modelling. Journal of Fluid Mechanics, Volume 154, pp. 321-336.

Barthet, A. (2003) :Analyse structurelle des propriétés de la turbulence 3D instationnaire autour d'une aile par simulation numérique directe, DEA INPT, 25 June 2003.

Basara B. and Jakirlic S. (2003): A new hybrid turbulence modelling strategy for industrial cfd. Int. J. Numer. Meth. Fluids, 42:89–116.

Batten, P., Goldberg, U., Chakravarthy, S. (2004), "Interfacing Statistical Turbulence Closures with Large-Eddy Simulation", AIAA J., Vol. 42, No. 3, pp. 485-492.

Benhamadouche S. and Laurence D. (2003): Les, course les, and transient rans comparisons on the flow across a tube bundle. International Journal of Heat and Fluid Flow, 2003.

Berkooz G., Holmes P. & Lumley J. L. (1993): The proper orthogonal decomposition in the analysis of turbulent flows, *Annual Rev. Flui Mech*. **25**, 539-75.

Berton, E., C. Allain, D. Favier, and C. Maresca, 'Experimental methods for subsonic flow measurements'. J. Notes on Num. Fluid Mech. and Multidisciplinary Design, dedicated volume "Progress in Computational Flow-Structure Interaction" **81**, 251–260. Sci. Eds. W. Haase, V. Selmin, B. Winzell, Publisher Springer.

Betts P.L. and Bokhari H. (1995). New experiments on natural convection of air in a tall cavity.*4th UK National conference on heat transfer*, 213.217. IMechE.

Bouhadji, A., Braza, M. (2003): Physical analysis by numerical simulation of organised modes and shock-vortex interaction in transonic flows around an aerofoil. Part I: Mach number effect", pp. 1-30, J. Computers & Fluids, 32/9, pp. 1233-1260, 2003.

Bouhadji, A., Braza, M. (2003): Physical analysis by numerical simulation of organised modes and shock-vortex interaction in transonic flows around an aerofoil. Part II: Reynolds number effect", pp. 1-23, J. Computers & Fluids, 32/9, pp. 1261-1281, 2003.

Bouhadji, A., Bourdet, S., Braza, M., Hoarau, Y., Rodes, P. and zabiras., M (2002): Tuirbulence modelling of unstaedy flows with a pronounced periodic character. Notes on Numerical Fluid Mechanics, Vol. 81, pp. 87-96, W; Haase, V. Selmin and B. Winzell scientific editors, Springer.

Bourdet, S., Braza, M.,Bouhadji, A., Thiele, F. (2003): Direct Numerical Simulation of the three-dimensional transition to turbulence in the transonic flow around a wing, J. Flow Turbulence and Combustion, 71, 203-220.

Braza, M., Faghani, D., Persillon, H. (2001): The role of natural vortex dislocations in three-dimensional wake transition, J. Fluid Mechanics, 439, pp. 1-41.

Bredberg J., (2000): On the Wall Boundary Condition for Turbulence Models. Internal report 00/4 Chalmers University of Technology.

Breitsamer C. (1997): Turbulente Stroemungsstrukturen an Flugzeugkonfigurationen mit Vorderkantenwirbeln (Turbulent flow structures on aircraft configurations exhibiting leading edge vortices), PhD thesis, Munich

Breitsamer C., Laschka B. (1993): Velocity measurements with hot-wires in a vortex-dominated flowfield. Wall interferences, support interferences and flow field measurements, AGARD-CP-535, pages 11-1-11-13, Brussels, Belgium.

Breitsamter, C. and Laschka, B. (1998): Vortical Flow Field Structures at Forward Swept Wing Configurations. ICAS Proceedings, 21st Congress, Melbourne, Australia.

Breitsamter, C. and Laschka, B. (2001): Vortical Flow Field Structure at Forward Swept-Wing Configurations. Journal of Aircraft, Vol. 38, No.2, pp. 193-207.

Buice, C.U. and Eaton, J.K. (1997): Experimental investigation of flow through an asymmetric plane diffuser. Report TSD-107, Department of Mechanical Engineering, Stanford University.

Bunge, U., Mockett, C., Thiele, F. (2003): Calibration of Different Models in the Context of Detached-Eddy Simulation. AG STAB Mitteilungen, DGLR, Göttingen.

Byun, G., Simpson, R., 2005. Structure of three-dimensional separated flow on an axi-symmetric bump. AIAA paper 2005-0113, Reno, N.V.

Byun, G., Simpson, R., Long, C.H., 2004. A study of vortical separation from three-dimensional symmetric bumps. AIAA Journal 42(4), 754--765.

Cantwell B. & Coles D. (1983), An experimental study of entrainment and transport in the turbulent near wake of a circular cylinder, *J. Fluid Mech.*, **136**, 321-374.

Catris, S. and Aupoix, B. (2000): Towards a calibration of the length-scale equation. International Journal of Heat and Fluid Flows. Vol. 21, part 5, pp 606-613

Cazalbou, J.B. and Chassaing, P. (2002): The structure of the solution obtained with Reynolds-stress-transport models at the free-stream edge of turbulent flows. Physics of Fluids, vol. 14, part 2, pp. 597-611

Cazalbou, J.B., Spalart, P.R. and Bradshaw, P. (2002): On the behaviour of two-equation models at the edge of a turbulent region. Physics of Fluids A, Vol. 6, part 5, pp. 1797-1804.

Cebeci, T. and Smith, A.M.O. (1974): Analysis of Turbulent Boundary Layers. Applied Mathematics and Mechanics, Vol. 15, Academic Press.

CFX-5 (2004): User Manual, ANSYS Inc.

Chalot F., Mallet M., Ravachol M. (1994), A comprehensive finite element Navier-Stokes solver for low and high-speed aircraft design, AIAA 94-0814, Reno, Nevada.

Chalot F., Dinh QV, Mallet M., Naïm A., Ravachol M. (1997): A multi-platform shared- or distributed-memory Navier-Stokes code, Parallel CFD '97, Manchester, UK.

Chen, H.C., Yang, Y.J., Han, J.C. (2000): Computation of heat transfer in rotating two-pass square channels by a second moment closure model. International Journal of Heat and Mass Transfer, Volume 43, pp. 1603-1616.

Chen, V.C., Patel, H.C. (1988): Near-wall turbulence models for complex flows including separation. AIAA Journal, Volume 26, part 6, pp. 641-648.

Chien, K.Y. (1982): Prediction of channel and boundary layer flows with a low-Reynolds-number turbulence model. AIAA Journal, vol. 20, part 1, pp 33-38.

Chow, J., Zilliac, G. and Bradshaw, P. (1997, a): Turbulence Measurements in the Near Field of a Wingtip Vortex. NASA Technical Memorandum 110418.

Chow, J., Zilliac, G. and Bradshaw, P. (1997, b): Mean and Turbulence Measurements in the Near Field of a Wingtip Vortex.. AIAA Journal, Vol. 35, No. 10, pp. 1561-1567.

Cid E., Cazin S. & Drouin V. (2002), Validation de PIV stéréoscopique et application a un écoulement aérodynamique de sillage 3D. *8e Congres Francophone de Vélocimétrie Laser.*

Comte-Bellot, G., S. Corrsin, S. (1971): Simple Eulerian time correlation of full- and narrow-band velocity signals in grid-generated, 'isotropic' turbulence. Journal of Fluid Mechanics, Vol. 48, pp. 273-337.

Constantinescu, G., Squires, K. D.(2000): LES and DES investigations of turbulent flow over a sphere. AIAA Paper 2000-0540.

Cooper, K. R. (1985): The effect of front-edge rounding and rear-edge shaping on the aerodynamic drag of bluff-body vehicles in ground proximity. SAE-Paper 850-288.

Craft, T.J., Gerasimov, A.V., Iacovides, H., Launder, B.E. (2003): Application of a new wall-function strategy to turbulent mixed and natural convection in vertical flows. 8th UK National Heat Transfer Conference (UKHT'03), Oxford.

Craft, T.J., Deevy, M., Jang, Y-J, Leschziner, M.A. and Wang, C (2004): Turbulence modelling of near-wall flows subjected to skewing strain and separation. RAeS Aerospace Aerodynamics Research Conference, London.

Craft, T.J., Ince, N.Z. and Launder, B.E. (1996): Recent developments in second-moment closures for buoyancy-affected flows. Dynamics of Atmosphere and Oceans, vol. 23, pp. 99-114.

Craft, T.J., Launder, B.E. and Suga, K. (1996): Development and application of a cubic eddy-viscosity model of turbulence, Int. J. Num. Meth. In Fluids,17, 108-115.

Dacles-Mariani, J., Zilliac, G.G., Crow, J.S., Bradshaw, P. (1995): Numerical/experimental study of a wingtip vortex in the near field. AIAA Journal, Volume 3, No. 9, pp. 1561-1568.

Dahlström, S., (2003): Large eddy simulation of the flow around a high-lift airfoil. Ph.D. thesis, Dept. of Thermo and Fluid Dynamics, Chalmers University of Technology, Göteborg, Sweden.

Dahlström, S., Davidson, L., (2003): Hybrid RANS-LES with additional conditions at the matching region. In: Hanjalic, K., Nagano, Y., Tummers, M. (Eds.), Turbulence Heat and Mass Transfer 4. begell house, inc., New York, Wallingford (UK), pp. 689--696.

Daly, J.B. and Harlow, F.H. (1970): Transport equations in turbulence. The Physics of Fluids. Vol. 13, No 11, pp. 2634-2649.

Davidson, L., Billson, M., (2004): Hybrid LES/RANS using synthesized turbulence for forcing at the interface. In: Neittaanmäki, P., Rossi, T., Korotov, S., Onate, E., Periaux, J., Knörzer, D. (Eds.), ECCOMAS 2004. July 24-28, Finland.

Davidson, L., Billson, M., (2005): Hybrid LES/RANS Using Synthesized Turbulence for Forcing at the Interface (submitted), Int. J. Heat Fluid Flow.

Davidson, L, Dahlström, S (2005): Hybrid LES-RANS: An approach to make LES applicable at highr Reynolds number. International Journal of Computational Fluid Dynamics (to appear).

Dervieux, A., Braza, M., Dussauge, J.P. Scientific Editors (1998): Computation and Comparison of Efficient Turbulence Models for Aeronautics, 65, Notes on Numerical Fluid Mechanics, Vieweg.

Diedrichs, B., Ekequist, M., Stichel, S. and Tengstrand, H. (2003): Quasi-static modeling of wheel rail reactions due to crosswind effects for various types of high-speed rolling stock. Sub. to J. Rail and Rapid Transit.

Dol, H.S., Kok, J.C., Oskam, B.: Turbulence modelling for leading-edge vortex flows. AIAA-Paper 2002-0843, Reno Conference, 14-17 January 2002.

DPW, "2nd AIAA CFD Drag Prediction Workshop," http://aaac.larc.nasa.gov/tsab/cfdlarc/aiaa-dpw/ , Orlando, FL., June 21-22, 2003.

Dunham, J. (1995): CFD validation for propulsion system components, AGARD-AR-355.

Durand, L., Kuntz, M. and Menter, F.R. (2002), Validation of CFX for the Ahmed Car Body, CFX Validation Report.

Durbin, P.A. (1991): Near-wall turbulence closure modelling without "damping functions". International Journal of Theoretical and Computational Fluid Dynamics. Volume 3, pp. 1-13.

Durbin, P.A. (1993): Application of a near-wall turbulence model to boundary layers and heat transfer. International Journal of Heat and Fluid Flows, vol. 14, pp. 316-323.

Durbin, P.A. (1995): Separated flow computations with the k-ε-v2 turbulence model. AIAA Journal, Volume 33, pp. 659-664.

Durbin, P.A., Pettersson Reif, B.A. (2001): Statistical Theory and Modeling for Turbulent Flows, John Wiley & Sons, LTD.

Edwards, J.R., and Chandra, S. (1996): Comparison of eddy-viscosity turbulence models for three-dimensional, shock-separated flowfields. AIAA Journal, vol. 34, No. 4, pp 756-763.

Eliasson P., Wallin S. (2000): A Positive Multigrid Scheme for Computations with Two-Equation Turbulence Models. ECCOMAS 2000.

Eliasson P., Wallin S. (2003): First report on Best Practice Guidelines for turbulence model implementation in Flomania. FLOMANIA deliverable 5.4-12-A.

Esch, T., Menter, F.R.,Vieser, W. (2003): Heat Transfer Predictions based on Two-Equation Turbulence Models. The 6th ASME-JSME Thermal Engineering Joint Conference, March 16-20, 2003.

Faghani,D. (1993) : Etude des structures tourbillonaires de la zone proche d'un jet plan: approche non stationnaire multidimensionnelle. *PhD Thesis INPT-Toulouse,* 1996.

Fahy, F. (2001): Foundations of Engineering Acoustics. Academic Press.

Ferrey, P. and Aupoix, B. (2004): A Reynolds stress-intermittency turbulence model. Advances in turbulence X, p. 818

Ferrey, P. and Aupoix, B. (2005): Behaviour of turbulence models near a turbulent/non-turbulent interface revisited. Ercoftac International Symposium on Engineering Turbulence Modelling and Measurements ETMM6, Sardinia

Ferry, P. (2004): Modèles aux tensions de Reynolds avec prise en compte de l'intermittence de frontière. Thèse de l'Université de Poitiers.

Forsythe, J., Hoffmann, K., and Dieteker, J.-F.(2000): Detached-eddy simulation of a supersonic axisymmetric base flow with an unstructured flow solver. AIAA Paper 2000-2410.

Gatski, T.B. and Speziale, C.G. (1993): On explicit algebraic stress models for complex turbulent flows, Journal of Fluid Mech., 254, 59-78.

Gerhold, T., Evans, J. (1999): "Efficient Computation of 3D-Flows for Complex Configurations with the DLR TAU-Code Using Automatic Adaptation", in Notes on Numerical Fluid Mechanics, Vol. **72**, Vieweg, (1999).

Gerolymos, G.A. and Vallet, I. (2001): Wall-Normal-Free Reynolds-Stress Model for compressible rotating flows applied to turbomachinery, AIAA (in print).

Gilliéron P. and Chometon F.(1999): Modeling of stationary three-dimensional separated air flows around an Ahmed reference model. ESAIM Symp.

Gleyzes, C., Capbern, P. (2003): Experimental study of two AIRBUS/ONERA airfoils in near stall conditions. Part I: Boundary layers, Aerospace Science and Technology, vol. 7, pp. 439–449.

Grotjans, H., Menter, F.R. (1998): Wall functions for general application CFD codes. ECCOMAS 98, Papailou et al. editors, pp. 1112-1117, John Wiley & Sons.

Grundestam, O., Wallin, S., Johansson, A.V. (2003): A generalised EARSM based on a non-linear pressure strain model. Proceedings of the Third International Symposium on Turbulence and Shear Flow Phenomena, Sendai, Japan.

Guimet, V. and Laurence, D. In A linearised turbulent production in the k-ε model for engineering applications, 2002. ETMM5, Mallorca, Spain.

Haase, W., Wagner, B., Jameson, A. (1983): Development of a Navier-Stokes method based on a finite volume technique for the unsteady Euler equations. In: Notes on Numerical Fluid Mechanics, Vieweg, Vol. 7, pp 99-107.

Hadzic I., Hanjalic K. and Laurence D. Modeling the response of turbulence subjected to cyclic irrotational strain. Physics of fluids, 13(6):1740–1747, 2001.

Hakimi, N., Hirsch, C., Pierret, S. (2000): Presentation and application of a new extended k-ε model with wall functions. ECCOMAS 2000, Barcelona.

Hanjalic, K., Hadziabdic, M., Temmerman, L., Leschziner, M.A. (2004): Merging LES and RANS strategies: zonal or seamless coupling? Direct and Large-Eddy Simulation V, editors: Friedrich, R., Guerts, B.O, Metais, O. Kluwer Academic Publishers, pp. 451--464.

Hanjalic, K., Jakirlic, S. (1998) Contribution towards the second-moment closure modelling of separated turbulent flows. Computers & Fluids, Volume 27, No. 2, pp. 137-156.

Hanjalic K, D. R. Laurence D, M. Popovac M and J.C. Uribe J.C. (2005), V2/k-f turbulence model and its application to forced and natural convection. Submitted to ERCOFTAC ETMM-6 conference, Sardinia, Mai 2005.

Hellsten, A. (2004): New advanced k-ω turbulence model for high-lift aerodynamics. AIAA Paper 2004-1120, 42nd Aerospace Sciences Meeting and Exhibit, Reno, Nevada.

Hoarau,Y. (2002) : Analyse Physique par Simulation Numérique et Modélisation des écoulements Décollés Instationnaires autour de Surfaces Portantes". PhD-IMFT/INPT, June 2002.

Hoarau, Y., Braza, M. (2002): Note on the Organised Eddy Simulation – equilibrium and non-equilibrium turbulence regions, Flomania progress report, September 2002.

Hoarau, Y., Braza, M., Ventikos, Y., Faghani, O., Tzabiras, G. (2003): Organised modes and the 3D transition to turbulence in the incompressible flow around a NACA0012 wing, J. Fluid Mechanics, 496, pp.63-72.

Hoarau, Y., Faghani, D., Braza, M., Perrin, R., Anne-Archard, D. (2003): Direct Numerical Simulation of the three-dimensional transition to turbulence in the incompressible flow around a wing, J. Flow Turbulence and Combustion, **71**, 119-132.

Hoarau, Y., Braza, M., Revell, A.J., Laurence, D., Barthet A. (2005a): Physical analysis and modelling of turbulent unsteady flows around a wing, Proceedings International Symposium "Bluff Body Wakes and Vortex Induced Vibrations-4", June 2005 Marie Curie and IMFT presentation.

Hoarau, Y., Braza M., Rodes P., Tzabiras G., Allain C., Favier D., Berton D. and Maresca M.. (2002): Turbulence modelling of unsteady flows with a pronounced periodic character around an airfoil. In Proceedings, IUTAM symposium "Unsteady Separated Flows", Toulouse, France.

Hoarau, Y., Braza, M., Ventikos, Y., Faghani, D., Tzabiras, G. (2003): Organised modes and the 3D transition to turbulence in the incompressible flow around a NACA0012 wing, J. Fluid Mechanics, 496, pp.63-72.

Höld, R., Brenneis, A., Eberle, A., Schwarz, V., and Siegert, R. (1999): Numerical simulation of aeroacoustic sound generated by generic bodies placed on a plate: Part I – Prediction of aeroacoustic sources. AIAA Paper 99- 1896.

Hoerner, S.F. (1965): *Fluid-Dynamic Lift* and *Fluid-Dynamic Drag*, Hoerner Fluid Dynamics, Bakersfield, CA.

Hussain A. K. M. F. & Reynolds W. C. (1975) Measurements in fully developed turbulence channel flow. J. Fluid Eng. Vol. **97**, p. 568.

Jakirlic, S. and Hanjalic, K. (1995): A second-moment closure for non-equilibrium and separating high- and low-Re-number flows. Proc. 10th Symp. Turbulent Shear Flows. Pennsylvania State University, August 14-16.

Jakirlic S., Jester-Zurker R., and Tropera C., editors. (2001): *9th ERCOFTAC/IAHR/COST Workshop on refined turbulence modelling*.

Jameson, A., Schmidt, W., Turkel, E. (1981): Numerical solutions of the Euler euqations by finite volume methods using Runge-Kutta time-stepping schemes. AIAA Paper 81-1259.

Jang, Y.J., Leschziner, M.A., Abe, K. and Temmerman, L. (2002), Investigation of anisotropy-resolving turbulence models by reference to highly-resolved LES data for separated flow, Flow, Turbulence and Combustion, Vol 69 pp. 161-203.

Jang, Y-J. and Leschziner, M.A. (2004), An Investigation of higher-order closures in the computation of the flow around a generic car body, *4th European Congress on Computational Methods in Applied Sciences and Engineering, ECCOMAS 2004*, Jyväskylä, Finland.

Jang, Y.J., Leschziner, M.A., Abe, K. and Temmerman, L. (2002), Investigation of anisotropy-resolving turbulence models by reference to highly-resolved LES data for separated flow, Flow, Turbulence and Combustion, Vol 69 pp. 161-203.

Jang, Y-J., Temmermann, L., and Leschziner, M.A. (2001): Investigation of anisotropy-resolving turbulence models by reference to highly resolved LES data for separated flow, *ECCOMAS Computational Fluid Dynamics Conference*, Swansea, September 4-7.

Jarrin N., S. Benhamadouche S., Addad Y. and Laurence D. (2003), Synthetic turbulent inflow conditions for Large-Eddy Simulation, in Turbulence, Heat and Mass Transfer 4, Hanjalic, Nagano, Tummers Edts. Begell House inc., 467-474.

Jeong, J. and Hussain, F (1995): On the identification of a vortex. J. Fluid Mech. Vol. 285, pp. 64–94.

Jin, G., Braza, M. (1994): A two-equation turbulence model for unsteady separated flows. AIAA Journal, Volume 32, No. 11, pp. 2316-2320.

Jones, W.P., Launder, B.E. (1972): The prediction of relaminarization with a two-equation model of turbulence. International Journal of Heat and Mass Transfer, Volume 15, pp 301-314.

Kalitzin, G. (1999): Application of the v2f Model to Aerospace Configurations. *Center for Turbulence Research. Annual Research Briefs.*

Kalitzin, G., Gould, A.R.B., Benton, J.J. (1996): Application of two-equation turbulence models in aircraft design. AIAA Paper 96-0327, 34th Aerospace Sciences Meeting and Exhibit, Reno, Nevada.

Kaltenbach, H.J, Fatica, M, Mittal, R., Lund, T.S. and Moin, P. (1999) Atudy of flow in a planar asymmetric diffuser using large-eddy simulation, J. Fluid Mech., 390, 151-185.

Karki, K.C., Patankar, S.V. (1989): Pressure based calculation procedure for viscous flows at all speeds. AIAA Journal, Vol. 27, pp. 1167-1174.

Kok, J.C. (2000): Resolving the dependence on freestream values of the k-ω turbulence model. AIAA Journal, Volume 38, pp. 1292-1295, 2000

Kok, J.C., Brandsma, F.J. (2000): Turbulence model based vertical flow computations for a sharp edged delta wing in transonic flow using the full Navier-Stokes equations. NLR Report, NLR-CR-2000-342.

Krajonovic, S. and Davidson, L. (2004), "Large eddy simulation of the flow around a simplified car model", 2004 SAE World Congress, SAE paper No. 2004-01-0227, Detroit/USA.

Krajnović S., Davidson L. (2005a): Flow Around a Simplified Car, Part I, Large Eddy Simulation, September 2005, *ASME: Journal of Fluids Engineering*, 127, no.5.

Krajnović S., Davidson L. (2005b): Flow Around a Simplified Car, Part II, Understanding the Flow, September 2005, *ASME: Journal of Fluids Engineering*, 127, no. 5.

Krajnović S. and Davidson L. (2005c), "Influence of Floor Motions in Wind Tunnels on the Aerodynamics of Road Vehicles" (in press), Journal of Wind Engineering and Industrial Aerodynamics

Langtry, R. (2004): "Drag Prediction of Engine-Airframe Interference Effects with CFX-5", AIAA 2004-0391.

Launder, B.E., Reece, G.J., Rodi, W. (1975): Progress in the development of a Reynolds-Stress turbulence closure. Journal of Fluid Mechanics, Volume 68, No 2, pp. 537-566.

Launder, B.E., Sharma, B.I. (1974): Application of energy-dissipation model of turbulence to the calculation of flow near a spinning disc. Letters in Heat and Mass Transfer, Volume 1, pp 131-138.

Laurence, D.R., Uribe, J.C., Utyuzhnikov, S.V. (2004): A robust formulation of the v^2f model. Journal of Flow, Turbulence and Combustion to appear. (prepublication, www.kluweronline.com, date: 14/04/2004).

Lecordier B. & Trinite M. (2003), Advanced PIV Algorithms with Image Distortion - Validation and Comparison from Synthetic Images of Turbulent Flow, *PIV03 Symposium Busan, Korea.*

Leder A. (1991), Dynamics of fluid mixing in separated flows, *Physics of Fluids* **A 3**, 1741-1748.

Lien, F.S., Durbin, P.A., Parneix, S. (1997): Non-linear k-ε-v^2 modelling with applications to aerodynamic flows. Proceedings of the 11th Turbulent Shear Flow Symposium, Grenoble.

Lien, F.S., Kalitzin, G. (2001): Computations of transonic flows with the v^2f turbulence model. International Journal of Heat and Fluid Flow, pp. 53-61.

Lien, F.S., Leschziner, M.A. (1993): Computational modelling of 3D turbulent flow in S--diffuser and transition ducts. Engineering Turbulence Modelling and Experiments, Vol. 2, pp. 217-228.

Lien, F.S. and Leschziner, M.A (1994a): A General non-orthogonal collocated finite volume algorithm for turbulent flow at all speeds incorporating second-moment turbulence-transport closure, Part I: Computational implementation. *Comput. Methods Appl. Mech. Engrg* 114, 123-148.

Lien, F.S. and Leschziner, M.A. (1994b), Upstream monotonic interpolation for scalar transport with application to complex turbulent flows, Int. J. Num. Meths in Fluids, 19, 527-548.

Lienhart H., Stoots C. and Becker S. (2000): Flow and turbulence structures in the wake of a simplified car model (Ahmed model). DGLR Fach Symp . der AG Stab, Stutgart University.

Lienhart, H. and Becker, S., (2003): Flow and turbulence structure in the wake of a simplified car model. SAE Technical Paper, 2003-01-0656.

Loyau, H., Batten P. and Leschziner, M.A. (1998): Modeling Shock/Boundary-Layer Interaction with Nonlinear Eddy-Viscosity Closures", Flow, Turbulence and Combustion, 60, 257-282.

Lübcke, H., Rung, T., Thiele, F. (2002): Prediction of the spreading mechanism of 3D turbulent wall jets with explicit Reynolds-Stress closures. Engineering Turbulence Modelling and Experiments 5, pp. 127-145.

Manceau, R.2002): "10th joint ERCFOTAC (SIG-15)/IAHR/QNET-CFD Workshop on Refined Turbulence Modelling", Poitiers, October 10-11.

Manceau R. and Bonet J.P., editors (2002). *10th ERCOFTAC/IAHR/COST Workshop on refined turbulence modelling.*

Manceau, R., Wang, M., Laurence, D. (2001) Inhomogeneity and anisotropy effects on the redistribution term in Reynolds-averaged Navier Stokes modeling. J. Fluid Mech., 438, 307-338, July 2001. CUP.

McCroskey, W:J:; "A Crtical Assesment of Wind Tunnel results for the NACA0012 Airfoil", AGARD_CP429, 1987.

Mellen, C., Froehlich, J., Rodi, W. (2000): Large Eddy Simulation of the flow over periodic hills, *16th IMACS World Congress,* Lausanne.

Menter, F.R. (1992): Influence of free-stream values on k-ω turbulence model predictions. AIAA Journal, vol. 30, No. 6, pp. 1657-1659.

Menter, F.R. (1994): Two-equation eddy-viscosity turbulence models for engineering applications. AIAA Journal, Volume 32, No 8, pp 1598-1605.

Menter, F., Esch, T. and Kubacki, S. (2002): Transition modeling based on local variables, 5th Int. Symp. on Enginnering Turbulence Modelling and Measurements (ETMM5), pp. 1-10, Elsevier Press.

Menter, F., Kuntz, M. (2002) Implementation Specification for the SSG Turbulence Model for Flomania, *FLOMANIA, Deliverable D4.1-12*, October

Menter, F.R., Kuntz, M., Bender, R. (2003 a): A Scale-Adaptive Simulation Model for Turbulent Flow Predictions. AIAA Paper 2003--0767.

Menter. F. R., Kuntz, M. and Durand. L. (2002): Adaptation of Eddy Viscosity Turbulence Models to Unsteady Separated Flow Behind Vehicles. In Lecture Notes in Applied and Computational Mechanics / The Aerodynamics of Heavy Vehicles: Trucks, Buses, and Trains, R. McCallen, F. Browand, J. Ross (Eds.), Springer Verlag, ISBN 3-540-22088-7.

Menter, F.R., Kuntz, M. and Langtry R. (2003): Ten Years of Industrial Experience with the SST Turbulence Model. Turbulence, Heat and Mass Transfer 4, K. Hanjalic, Y. Nagano and M. Tummers (Editors).

Moin, P. (1998): Numerical and physical issues in large eddy simulation of turbulent flows. JSME Int. J., Series B, Vol. 41, No.2, pp. 454-463.

Naot,D, Shavit, A., Wolfshtein, M. (1973): Two-point correlation and the redistribution of turbulence. The Physics of Fluids, vol. 16, No. 6, pp. 738-743.

Norberg C. (1998), LDV-measurements in the near wake circular cylinder. In: Conference on Bluff Body Wakes and Vortex Induced Vibrations Presented at ASME Fluids Engineering Division (Annual Summer Meeting), Washington DC (FEDSM98-5202).

Obi, S., Aoki, K. and Masuda, S. (1993): Experimental and computational study of turbulent separating flow in an asymmertric plane diffuser. Proceedings 9th Symposium on Turbulent Shear Flows, Kyoto, Japan, August 16-18.

Obi, S., Perić, M., Scheurer, M. (1991): Second moment calculation procedure for turbulent flows with collocated variable arrangement. AIAA Journal, Vol. 29, pp. 585-590.

Papoulis, A., (1984): "Probability, Random Variables, and Stochastic Processes", McGraw-Hill Book Co.

Peng, S-H., Davidsson, L. Holmberg, S. (1997): A modified low-Reynolds number k-ω model for recirculating flows. ASME Journal of Fluids Engineering. Volume 119, pp. 867-875.

Peng, S., Eliasson, P. (2004) "A Comparison of Turbulence Models in Prediction of Flow Around the DLR-F6 Aircraft Configuration", AIAA Paper 2004-4718, 22nd AIAA Applied Aerodynamics Conference, Providence, Rhode Island, 16-19 August 2004.

Perrin, R. (2005) : Analyse physique et modélisation d'écoulements incompressibles instationnaires turbulents autour d'un cylindre circulaire à grand nombre de Reynolds, Thèse de Doctorat INPT, 5Juillet 2005

Persillon, H., Braza, M. (1998): Physical analysis of the transition to turbulence in the wake of a circular cylinder by three-dimensional Navier-Stokes simulation, J. Fluid Mechanics, 365, pp.23-89.

Piomelli, U., Balaras, E., Pasinato, H., Squire, K.D. Spalart, P.R. (2003), "The Inner-Outer Layer Interface in Large-Eddy Simulations With Wall-Layer Models", Int. J. Heat Fluid Flow ", Vol. 24, pp. 538-550.

Pope, S.B. (1975): A more general effective-viscosity hypothesis, Journal of Fluid Mech., 72, 331-340.

Pope, S.B., (2001): "Turbulent Flows", Cambridge University Press.

Revell A. J. Benhamadouche S., Craft T., Laurence D. and Yaqobi, K. (2005), A stress-strain lag eddy viscosity model for unsteady mean flow. ETMM6, Sardinia.

A.J. Revell, A.J., Braza, M., Perrin, R., Hoarau, Y. (2004): EMT2/IMFT group seminar "Turbulence modelling for unsteady aerodynamic flows in non-equilibrium", 18 december 2004.

Reynolds W. C. & Hussain, A. K. M. F (1972), The mechanics of an organized wave in turbulent shear flow. Part 3. Theoretical models and comparisons with experiments, *J. Fluid Mech*, **54**, 263-288.

Rodde, A.M., Archambaud, J.P. (1994) OAT15A Airfoil Data, AGARD FDP AR 303.

Rodi W., Mansour N.N. (1993): Low Reynolds number k-ε modelling with the aid of direct simulation data. J. Fluid Mechanics **250**, p. 509.

Roe, P.L. (1981): Approximate Rieman solvers, parameter vectors and difference schemes, Journal of Computational Physics, vol. 1, No. 46, pp. 357-378.

Rogers, S.E., Kwak, D. (1988): An upwind differencing scheme for the time-accurate incompressible Navier-Stokes equations, AIAA Paper, AIAA 88-2583-CP.

Roshko A. (1961), Experiment on the flow past a circular cylinder at very high Reynolds number, *J. Fluid Mech*, **10**, 345-356.

Rubinstein, R:, Rumsey, C.L., Salas, M.D., Thomas, J.L. (2001): Turbulence Modelling Workshop. NASA-CR-2001-210841.

Rung, T., Thiele, F. (1996): Computational modelling of complex boundary-layer flows. Proceedings of the 9th International Symposium on Transport Phenomena in Thermal-Fluid Engineering, Singapore, pp. 321-326.

Rung, T., Lübcke, H., Franke, M., Xue, L., Thiele, F., Fu, S. (1999): Assessment of explicit algebraic stress models in transonic flows. Engineering Turbulence Modelling and Experiments 4, pp. 659-668.

Rung, T., Lübcke, H., Thiele, F. (2000): Universal wall-boundary conditions for turbulence-transport models, ZAMM-Z, Math. Mech 81 (2001).

Rung, T., Bunge, U., Schatz, U., Thiele, F. (2003): Restatement of the Spalart-Allmaras eddy-viscosity model in a strain-adaptive formulation. AIAA Journal, Volume 51, No 7, pp. 1396-1399.

Sarkar, S. (1992): The pressure-dilatation correlation in compressible flows. Phys. Fluids A 4 2674-2682.

Sarkar, S. (1995): The stabilising effect of compressibility in turbulent shear flow. J. Fluid Mech., 282, 163-186.

Schewe, G. (1983): "On the force oscillations on a circular cylinder in cross flow from subcritical up to transitional Reynolds numbers". *J. Fluid Mech.*, 1983, v.133, pp.265-285.

Schmitt, V., Charpin, F. (1979): Pressure distributions on the ONERA-M6-Wing at transonic Mach numbers. In: Barche, J. (Ed.): Experimental Data Base for Computer Program Assessment. Report of the Fluid Dynamics Panel Working Group 04. AGARD-AR-138, pp. B1-1 – B1-44.

Schwamborn, D., Gerhold, T., Hannemann, V. (1999): "On the validation of the DLR TAU-Code´´, in Notes on Numerical Fluid Mechanics, Vol. **72**., Vieweg, (1999).

Shabbir, A., Celestina, M.L., Adamczyk, J.J., Strazisar, A.J. (1997) The effect of hub leakage flow on two high speed axial flow compressor rotor, ASME Paper 97-GT-346.

Shima, N. (1998): Low-Reynolds-number second-moment closure without wall-reflection redistribution terms. International Journal of Heat and Fluid Flows. Volume 19, pp. 549-555.

Shur, M., Spalart, P.R., Strelets, M., Travin, A. (1999): Detached-Eddy simulation of an airfoil at high angle of attack. Engineering Turbulence Modelling and Measurements, Vol. 4, pp. 669-678.

Shur, M.L., Strelets, M., Travin, A. and Spalart, P.R. (1997): On the sensitization of turbulence models to rotation and curvature. Aerospace Science and Technology, No. 5, pp. 297-302

Shur, M., Strelets, M., Travin, A., Spalart, P.R. (2000): Turbulence modelling in rotating and curved channels ; Assessing the Spalart-Shur correction term. AIAA Journal, Volume 38, No 5, pp 784-782.

Siegert, R., Schwarz, V., and Reichenberger, J. (1999): Numerical simulation of aeroacoustic sound generated by generic bodies placed on a plate: Part II – Prediction of radiated sound pressure. AIAA Paper 99-1895.

Simpson, R.L., Long, C.H. and Byun, G., (2002):. Study of vortical separation from an axismmetric hill. *Int. J. Heat and Fluid Flow* 23, 582-591.

Smagorinsky, J., 1963. General circulation experiments with the primitive equations. Monthly Weather Review 91, 99--165.

Smith, B.R. (1990): The k - kL turbulence model and wall layer model for compressible flows. AIAA Paper 90-1483 21st Fluid and Plasma Dynamics Conference, Seattle, Washington.

Smith, B.R. (1994): A near wall model for the k-l two equation turbulence model. AIAA Paper 94-2386 25th Fluid Dynamics Conference, Colorado Springs, Colorado.

Smith, B.R. (1995): Prediction of hypersonic shock wave turbulent boundary layer interactions with the k-l two equation turbulence model. AIAA Paper 95-0232, 33rd Aerospace Sciences Meeting & Exhibit, Reno, Nevada.

Smith, B.R. (1997): A non-equilibrium turbulent viscosity function for the k-l two equation turbulence model. AIAA Paper 97-1959 28th AIAA Fluid Dynamics Conference, Snowmass Village, Colorado.

Sohankar, A., Norberg, C., and Davidson, L. (1996): "A numerical study of unsteady two-dimensioanl flow around rectangular cylinders at incidence". *Dept. Thrmo and Fluids.*, Chlamers University of Technology, Göteborg, Sweden, Internal Report 96/25, May 1996.

Spalart, P.R. (2001): "Young person's guide to Detached-Eddy Simulation grids", http://techreports.larc.nasa.gov/ltrs/PDF/2001/cr/NASA-2001-cr211032.PDF

Spalart, P.R., Allmaras, S.R. (1992): A one-equation turbulence model for aerodynamics flows. AIAA Paper 92-0349, 30th Aerospace Sciences Meeting and Exhibit, Reno, Nevada.

Spalart, P.R., Allmaras, S.R. (1994): A one-equation turbulence model for aerodynamics flows. La Recherche Aérospatiale, Volume 1, pp 5-21.

Spalart, P.R., Deck, S., Shur, M.L., Squires, K.D., Strelets, M.Kh., Travin, A. (2005): A New Version of Detached-Eddy Simulation, Resistant to Ambigious Grid Densities, to be published in: Theoretical and Computational Fluid Dynamics.

Spalart, P.R., Jou, W.-H., Strelets, M. and Allmaras, S.R. (1997): Comments on the feasibility of LES for wings and on a hybrid, RANS/LES approach. In Liu, C. and Liu, Z. (eds) Advances in DNS/LES, Proceedings of 1st AFOSR International Conference on DNS/LES, Ruston, LA, August, 4-8, Greyden Press, Columbus, OH, pp. 137-147.

Spalart, P.R. and Shur, M.L (1997): On the sensitization of turbulence models to rotation and curvature, Aerospace Science and Technology, vol. 1, No. 5, 1997, pp. 297-302.

Speziale, C.G., Sarkar, S., Gatski, T.G. (1991): Modelling the pressure-strain correlation of turbulence: An invariant dynamical systems approach. Journal of Fluid Mechanics, Volume 227, pp. 245-272.

Speziale, C.G. and Xu, X.H. (1996): Towards the development of second-order closure models for nonequilibrium turbulent flows, Int. J. Heat and Fluid Flow, 17, 238-244.

Spohn A. and Gillieron P. (2002), Flow separations generated by a simplified geometry of an automobile vehicle. IUTAM Symposium: Unsteady Separated Flows, April 8-12, Toulouse, France.

Stein E., De Borst R., Hughes T. (2004): Encyclopedia of Computational Mechanics, Wiley.

Strelets, M. (2001): Detached Eddy Simulation of Massively Separated Flows. AIAA paper 2001-0879, 39th Aerospace Sciences Meeting and Exhibit, Reno, NV.

Swalwell, K., (2004): Private communication with TUB.

Swalwell, K.E., Sheridan, J. and Melbourne W.H. (2003):Frequency Analysis of Surface Pressure on an Airfoil after Stall. Presented at the 21st AIAA Applied Aerodynamics Conference, AIAA Paper 2003-3416.

Taulbee, D.B. (1992): An improved algebraic Reynolds stress model and corresponding non-linear stress model. The Physics of Fluids A, Vol. 4, No 11, pp. 2555-2561.

Temmerman, L., Chen, W. and Leschziner, M.A. (2004), A comparative study of separation from a three-dimensional hill using LES and second-moment-closure RANS modeling, *4th European Congress on Computational Methods in Applied Sciences and Engineering, ECCOMAS 2004*, Jyväskylä, Finland.

Temmerman, L., Hadziabdic, M., Leschziner, M.A., and Hanjalic, K., (2004), "A hybrid two-layer URANS-LES approach for large eddy simulation at high Reynolds numbers", Int. J. Heat and Fluid Flow (in press).

Temmerman, L., Leschziner, M.A. (2001): Large Eddy Simulation of separated flow in a streamwise periodic channel construction, *Int. Symp. on Turbulence and Shear Flow Phenomena*, Stockholm, June 27-29.

Temmerman, L., Leschziner, M., Mellen, C. and Froehlich J. (2003) Investigation of subgrid-scale models and wall-function approximations in Large Eddy Simulation of separated flow in a channel with streamwise periodic constrictions, Int. J. Heat Fluid Flow, Vol 24, pp. 157-180.

Travin, A., Shur, M., Strelets, M., and Spalart, P.R. (2000): Detached-eddy simulations past a circular cylinder. Int. J. Flow, Turbulence and Combustion, Vol. 63, pp. 293-313.

Travin, A., Shur, M., Strelets, M., and Spalart P.R. (2002): Physical and numerical upgrades in the Detached-Eddy Simulation of complex turbulent flows. In: Fluid Mechanics and its Applications, Vol. 65, pp. 239-254. Advances in LES of Complex Flows, R.Friederich and W.Rodi (editors). Proc. of EUROMECH Colloquium 412, Kluwer Academic Publishers, Dordrecht / Boston/ London.

Utyuzhnikov, S.V. (2005): Some new approaches to building and implementation of wall-functions for modeling of near-wall turbulent flows, *International Journal Computers & Fluids* (to be published). (Prepublication in www.ScienceDirect.com, data 30/11/2004).

Utyuzhnikov, S.V. (2005): Generalized wall-functions and their application for simulation of turbulent flows", *International Journal Numerical Methods in Fluid* (to be published). (Prepublication in www3.interscience.wiley.com, data 07/01/2005).

Viegas, J.R and Rubesin, M.W. (1983): Wall function boundary conditions in the solution of the Navier-Stokes equations for complex compressible flows. AIAA Paper 83-1694. AIAA 16th Fluid and Plasma Dynamics Conference, Danvers, Massachusetts.

Viegas, J.R., Rubesin, M.W., Hortsman, C.C. (1985): On the use of wall functions as boundary conditions for two-dimensional separated compressible flows. AIAA Paper 85-0180, 23rd Aerospace Sciences Meeting, Reno.

Vieser, W., Esch, T., Menter, F. (2002): Heat transfer predictions using advanced two-equation turbulence models. Technical Memorandum CFX-VAL10/0602.

Wallin, S. (2000): Engineering turbulence modelling for CFD with a focus on explicit algebraic Reynolds stress models. KTH Doctoral thesis.

Wallin, S., Johansson, A.V. (2000): An explicit algebraic Reynolds stress model for incompressible and compressible turbulent flows. Journal of Fluid Mechanics, Volume 403, pp. 89-132.

Wallin, S., Johansson, A.V. (2002): Modelling streamline curvature effects in explicit algebraic Reynolds stress turbulence models. International Journal of Heat and Fluid Flows, Volume 23, No 5, pp. 721-730.

Wang, C., Jang, Y.J. and Leschziner, M.A. (2004), "Modelling two- and three-dimensional separation from curved surfaces with anisotropy-resolving turbulence closures, Int. J Heat and Fluid Flow, 25, 499-512.

Weiss, J. M., Smith, W.A. (1995): Preconditioning applied to variable and constant density flows, AIAA Journal, vol. 33, No.11, pp.2050-2057.

Wilcox, D.C. (1988): Reassessment of the scale-determining equation for advanced turbulence models. AIAA Journal Volume 26, No 11, pp 1299-1310.

Wilcox, D.C. (1993): Turbulence Modelling for CFD. DCW Industries, La Canada, California.

Wilcox, D.C. (1994): Simulation of transition with a two-equation turbulence model, AIAA Journal, 32, 247-255.

Willert C. (1997), Stereoscopic digital particle image velocimetry for application in wind tunnel flows, *Meas. Sci. Tech.* **8**, 1465-149.

Wlezien R. W. & Way J. L. (1979), Tecniques for the experimental investigation of the near wake of a circular cylinder, *AIAA Journal*, **17:6**, 563-570.

Wolfshtein, M. (1969): The velocity and temperature distribution in one-dimensional flow with turbulence augmentation and pressure gradient. International Journal of Heat and Mass Transfer, vol. 12

Xue, L. (1998): Entwicklung eines effizienten parallelen Lösungsalgorithmus zur dreidimensionalen Simulation komplexer turbulenter Strömugen. PhD thesis, TU-Berlin.

Yang, Z., Shih, T.H. (1992): A k-□ calculation of transitional boundary layers. NASA TM ICOMP-92-08, CMOTT-92-05.

Yap, C.R. (1987): Turbulent heat and momentum transfer in recirculating and impinging flows, Ph.D. thesis, UMIST.

Zeman, O. (1990): Dilatation dissipation: The concept and application in modelling compressible mixing layers. Phys. Fluids A 2 (2), 178-188.

Notes on Numerical and Fluid Mechanics and Multidisciplinary Design

Available Volumes

Volume 94: W. Haase, B. Aupoix, U. Bunge, D. Schwamborn (eds.): FLOMANIA - A European Initiative on Flow Physics Modelling - Results of the European-Union funded project 2002 - 2004. ISBN 3-540-28786-8

Volume 93: Y. Shokin, M. Resch, N. Danaev, M. Orunkhanov, N. Shokina (eds.): Advances in High Performance Computing and Computational Sciences - The 1th Kazakh-German Advanced Research Workshop, Almaty, Kazakhstan, September 25 to October 1, 2005. ISBN 3-540-33864-0

Volume 92: H.-J. Rath, C. Holze, H.-J. Heinemann, R. Henke, H. Hönlinger (eds.): New Results in Numerical and Experimental Fluid Mechanics V - Contributions to the 14th STAB/DGLR Symposium Bremen, Germany 2004. ISBN 3-540-33286-3

Volume 91: E. Krause, Y. Shokin, M. Resch, N. Shokina (eds.): Computational Science and High Performance Computing II - The 2nd Russian-German Advanced Research Workshop, Stuttgart, Germany, March 14 to 16, 2005. ISBN 3-540-31767-8

Volume 90: K. Fujii, K. Nakahashi, S. Obayashi, S. Komurasaki (eds.): New Developments in Computational Fluid Dynamics - Proceedings of the Sixth International Nobeyama Workshop on the New Century of Computational Fluid Dynamics, Nobeyama, Japan, April 21 to 24, 2003. ISBN 3-540-27407-3

Volume 89: N. Kroll, J.K. Fassbender (eds.): MEGAFLOW - Numerical Flow Simulation for Aircraft Design - Results of the second phase of the German CFD initiative MEGAFLOW, presented during its closing symposium at DLR, Braunschweig, Germany, December 10 and 11, 2002. ISBN 3-540-24383-6

Volume 88: E. Krause, Y.I. Shokin, M. Resch, N. Shokina (eds.): Computational Science and High Performance Computing - Russian-German Advanced Research Workshop, Novosibirsk, Russia, September 30 to October 2, 2003. ISBN 3-540-24120-5

Volume 86: S. Wagner, M. Kloker, U. Rist (eds.): Recent Results in Laminar-Turbulent Transition - Selected numerical and experimental contributions from the DFG priority programme 'Transition' in Germany. ISBN 3-540-40490-2

Volume 82: E.H. Hirschel (ed.): Numerical Flow Simulation III - CNRS-DFG Collaborative Research Programme, Results 2000-2002. ISBN 3-540-44130-1

Volume 81: W. Haase, V. Selmin, B. Winzell (eds.): Progress in Computational Flow-Structure Interaction - Results of the Project UNSI, supported by the European Union 1998-2000. ISBN 3-540-43902-1

Volume 80: E. Stanewsky, J. Délery, J. Fulker, P. de Matteis (eds.): Drag Reduction by Shock and Boundary Layer Control - Results of the Project EUROSHOCK II, supported by the European Union 1996-1999. ISBN 3-540-43317-1

Volume 79: B. Schulte-Werning, R. Grégoire, A. Malfatti, G. Matschke (eds.): TRANSAERO - A European Initiative on Transient Aerodynamics for Railway System Optimisation. ISBN 3-540-43316-3

Volume 78: M. Hafez, K. Morinishi, J. Periaux (eds.): Computational Fluid Dynamics for the 21st Century. Proceedings of a Symposium Honoring Prof. Satofuka on the Occasion of his 60th Birthday, Kyoto, Japan, 15-17 July 2000. ISBN 3-540-42053-3

Volume 77: S. Wagner, U. Rist, H.-J. Heinemann, R. Hilbig (eds.): New Results in Numerical and Experimental Fluid Mechanics III. Contributions to the 12th STAB/DGLR Symposium Stuttgart, Germany 2000. ISBN 3-540-42696-5

Volume 76: P. Thiede (ed.): Aerodynamic Drag Reduction Technologies. Proceedings of the CEAS/DragNet European Drag Reduction Conference, 19-21 June 2000, Potsdam, Germany. ISBN 3-540-41911-X

Volume 75: E.H. Hirschel (ed.): Numerical Flow Simulation II. CNRS-DFG Collaborative Research Programme, Results 1998-2000. ISBN 3-540-41608-0

Volume 66: E.H. Hirschel (ed.): Numerical Flow Simulation I. CNRS-DFG Collaborative Research Programme. Results 1996-1998. ISBN 3-540-41540-8

Printing: Krips bv, Meppel
Binding: Stürtz, Würzburg